Lecture Notes in Computer Science   5760

Commenced Publication in 1973
Founding and Former Series Editors:
Gerhard Goos, Juris Hartmanis, and Jan van Leeuwen

Susanne Albers   Helmut Alt
Stefan Näher (Eds.)

Efficient Algorithms

Essays Dedicated to Kurt Mehlhorn
on the Occasion of His 60th Birthday

 Springer

Volume Editors

Susanne Albers
Universität Freiburg, Institut für Informatik
Georges-Köhler-Allee 79, 79110, Freiburg, Germany
E-mail: salbers@informatik.uni-freiburg.de

Helmut Alt
Freie Universität Berlin, Institut für Informatik
Takustr. 9, 14195 Berlin, Germany
E-mail: alt@mi.fu-berlin.de

Stefan Näher
Universität Trier, Fachbereich IV - Informatik
54286 Trier, Germany
E-mail: naeher@uni-trier.de

The illustration appearing on the cover of this book is the work of Daniel Rozenberg
(DADARA)

Library of Congress Control Number: Applied for

CR Subject Classification (1998): F.2, G.1, G.2, I.1.2, G.4, E.5

LNCS Sublibrary: SL 1 – Theoretical Computer Science and General Issues

ISSN   0302-9743
ISBN-10   3-642-03455-1 Springer Berlin Heidelberg New York
ISBN-13   978-3-642-03455-8 Springer Berlin Heidelberg New York

springer.com

© Springer-Verlag Berlin Heidelberg 2009
Printed in Germany

Typesetting: Camera-ready by author, data conversion by Scientific Publishing Services, Chennai, India
Printed on acid-free paper   SPIN: 12731985   06/3180   5 4 3 2 1 0

Kurt Mehlhorn - Born 1949

Preface

"Effiziente Algorithmen" was the title of the first book by Kurt Mehlhorn in 1977. It was meant as a text for graduate students and published in German by Teubner-Verlag.

We decided to adopt this title 32 years later for this Festschrift in honor of Kurt on the occasion of his 60th birthday. It contains contributions by his former PhD students, many of whom are now university teachers themselves, and colleagues with whom he cooperated closely within his career. It is our pleasure that even Kurt's former PhD advisor, Bob Constable from Cornell University, kindly agreed to contribute. Many of the contributions were presented at a colloquium held in Kurt's honor on August 27 and 28, 2009 in Saarbrücken, Germany.

This Festschrift shows clearly how the field of algorithmics has developed and matured in the decades since Kurt wrote his book with the same title.

The classic approach based on discrete mathematics and computability and complexity theory continues to be the foundation of the field with ever new and important challenges as the first chapters of this Festschrift show. Kurt has contributed significantly to classical algorithmics and gained worldwide reputation. Starting from research in computability theory in his PhD thesis he made major contributions to complexity theory, graph algorithms, data structures, and was one of the first to recognize the significance of computational geometry contributing one of the early textbooks on the subject.

In spite of the success of classical algorithmics, by the 1990s more and more researchers recognized that in order to have their results acknowledged in the scientific community as a whole and applied commercially it was necessary that they took care of the implementation of algorithms themselves. It turned out that this aspect of algorithmics created challenging new theoretic problems, in particular, concerning software engineering, a closer investigation of heuristic methods, the numerical robustness of algorithms, and as a possible solution to this problem, exact computation. Peter Sanders' contribution to the Festschrift, for example, describes this new field of algorithm engineering and Chee Yap's contribution is a convincing plea why numerical computing is of great importance to the field of algorithmics.

Kurt Mehlhorn was one of the initiators of this new development of algorithmics and became one of its leaders and driving forces worldwide. In particular, he and his group created the software library LEDA, a uniquely comprehensive collection containing implementations of all the classical algorithms and being used extensively in academia and industry. In addition, considerable research was done by them on the theoretical aspects of implementing algorithms especially concerning robustness and exactness of computation.

Kurt has been a leader not only in scientific research but also in scientific organizations, in particular during his years as a vice president of the Max Planck society. The administrative work never prevented him from being a productive researcher which he continues to be up to this day.

So let us honor this eminent scientist, whom to our privilege we have had as a teacher and a friend.

Happy Birthday, Kurt!

August 2009

Susanne Albers
Helmut Alt
Stefan Näher

Acknowledgements

We would like to thank everybody who contributed to this Festschrift: the authors for their interesting articles, the colleagues and PhD students who helped proofread the contributions, Marc Scherfenberg for collecting, editing, and compiling the files of all the authors, and Wolfgang J. Paul and his team for organizing the colloquium in Saarbrücken.

The Editors

Table of Contents

I Models of Computation and Complexity

Building Mathematics-Based Software Systems to Advance Science and
Create Knowledge 3
   Robert L. Constable

On Negations in Boolean Networks 18
   Norbert Blum

The Lovász Local Lemma and Satisfiability 30
   Heidi Gebauer, Robin A. Moser, Dominik Scheder, and Emo Welzl

Kolmogorov-Complexity Based on Infinite Computations 55
   Günter Hotz

Pervasive Theory of Memory 74
   Ulan Degenbaev, Wolfgang J. Paul, and Norbert Schirmer

Introducing Quasirandomness to Computer Science.................. 99
   Benjamin Doerr

II Sorting and Searching

Reflections on Optimal and Nearly Optimal Binary Search Trees 115
   J. Ian Munro

Some Results for Elementary Operations 121
   Athanasios K. Tsakalidis

Maintaining Ideally Distributed Random Search Trees without Extra
Space 134
   Raimund Seidel

A Pictorial Description of Cole's Parallel Merge Sort 143
   Torben Hagerup

Self-matched Patterns, Golomb Rulers, and Sequence Reconstruction ... 158
   Franco P. Preparata

III   Combinatorial Optimization with Applications

Algorithms for Energy Saving 173
  Susanne Albers

Minimizing Average Flow-Time 187
  Naveen Garg

Integer Linear Programming in Computational Biology 199
  Ernst Althaus, Gunnar W. Klau, Oliver Kohlbacher,
  Hans-Peter Lenhof, and Knut Reinert

Via Detours to I/O-Efficient Shortest Paths 219
  Ulrich Meyer

IV   Computational Geometry and Geometric Graphs

The Computational Geometry of Comparing Shapes................... 235
  Helmut Alt

Finding Nearest Larger Neighbors: A Case Study in Algorithm Design
and Analysis 249
  Tetsuo Asano, Sergey Bereg, and David Kirkpatrick

Multi-core Implementations of Geometric Algorithms 261
  Stefan Näher and Daniel Schmitt

The Weak Gap Property in Metric Spaces of Bounded Doubling
Dimension 275
  Michiel Smid

On Map Labeling with Leaders 290
  Michael Kaufmann

The Crossing Number of Graphs: Theory and Computation 305
  Petra Mutzel

V   Algorithm Engineering, Exactness, and Robustness

Algorithm Engineering – An Attempt at a Definition 321
  Peter Sanders

Of What Use Is Floating-Point Arithmetic in Computational
Geometry? 341
  Stefan Funke

Car or Public Transport—Two Worlds 355
    Hannah Bast

Is the World Linear? 368
    Rudolf Fleischer

In Praise of Numerical Computation 380
    Chee K. Yap

Much Ado about Zero 408
    Stefan Schirra

Polynomial Precise Interval Analysis Revisited 422
    Thomas Gawlitza, Jérôme Leroux, Jan Reineke, Helmut Seidl,
    Grégoire Sutre, and Reinhard Wilhelm

Author Index 439

Part I

Models of Computation and Complexity

Building Mathematics-Based Software Systems to Advance Science and Create Knowledge

Robert L. Constable

Cornell University

Abstract. Kurt Mehlhorn's foundational results in computational geometry provide not only a basis for practical geometry systems such as Leda and CGAL, they also, in the spirit of Euclid, provide a sound basis for geometric truth. This article shows how Mehlhorn's ideas from computational geometry have influenced work on the logical basis for constructive geometry. In particular there is a sketch of new decidability results for constructive Euclidean geometry as formulated in computational type theory, CTT. Theorem proving systems for type theory are important in establishing knowledge to the highest standards of certainty, and in due course they will play a significant role in geometry systems.

1 Introduction

It is an honor to be associated with a scientist of the stature of Kurt Mehlhorn, and I knew his potential before most other computer scientist because I supervised his excellent PhD thesis [1] finished at Cornell in 1974, a copy of which still sits on my bookshelf. I've enjoyed several occasions to open this thesis since 1974 – once to aid my explanation on the telephone to a famous logician who in the 1990's had tried unsuccessfully to prove one of Kurt's theorems and wanted to know about the thesis proof.

In this short article I focus on a common career long interest which Kurt and I share. We have both spent considerable effort on the design, implementation, and deployment of software systems that *automate important intellectual tasks* used in applications of computer science to mathematics and other sciences. Moreover, our systems share the feature that they are based on computational mathematics and required many new results in theoretical and experimental computer science to build and extend them. They are in themselves contributions to the science of computing. I will focus on the Leda [2] system and related work [3, 4, 5, 6, 7] associated with Kurt and his colleagues and on the Nuprl [8, 9] system associated with me and my research group.

The task of assembling a team to design, build, and deploy large software systems requires management and marketing skills as well as technical expertise, and it may be noteworthy that Kurt and I have both spent part of our careers in high level science administration, perhaps because we learned from assembling our research groups how to manage research and market ideas. We also learned to advocate effectively for computer science, in part by recognizing that while

S. Albers, H. Alt, and S. Näher (Eds.): Festschrift Mehlhorn, LNCS 5760, pp. 3–17, 2009.

it is one of the youngest sciences, it has become essential to progress in all the others. Computing became essential through its technologies and through the new scientific results required to ensure that systems designed to aid scientific thought led to *reliable methods and certain knowledge.*

At this point in the development of computer science, the field is much admired for its technologies and artifacts. *I see this as a good time to stress that while computer science has created stunning new technologies used in science, engineering, government, business, and everyday life, it has also created deep fundamental knowledge and even more significantly, entirely new ways of knowing.* These aspects of computer science will be stressed in this article. I especially enjoy the contributions computing has made to some of the oldest investigations in science and mathematics, those concerned with geometry – its core truths and their applications.

2   Mathematics-Based Software Systems

Background. Here is what I mean by a *mathematics-based* software system. First, the objects manipulated by the system can be defined mathematically and go beyond numbers to include functions, polynomials, equations, matrices, polygons, Voronoi diagrams, types, sets, algebraic systems, proofs, and so forth – even formalized theories. These objects have a clear mathematical meaning that forms the basis of a precise semantics which the system must respect. For example, there might be a data type of infinite precision computable real numbers that should precisely relate to the conventional mathematical definition of real numbers; the notion of a polynomial should relate to the algebraic one, and the notion of a polygon should be recognizable, etc.

Second, the system provides algorithms for manipulating the data, and these algorithms satisfy properties that can be precisely formulated as *design specifications.* The algorithms should be efficient and clear. For example, an algorithm to find integer square roots should behave in the standard way, e.g. the root r of n should be such that $r^2 \leq n$ and also $n < (r + 1)^2$.

Third, there is a notion of what it means for an algorithm to be *correct* according to the standard semantics for its specification, and correctness can be grounded in a firm notion of *mathematical truth.*

Fourth, the system should collect and organize its information in a way that contributes to a library of scientific knowledge.

Fifth, the system should present information using notations that conform to standard mathematical practice, e.g. can be presented in Tex or MathML, etc. In general the understanding necessary to use such systems should be grounded in mathematics education, and it should be possible to use them in teaching mathematics at the college level.

There are other characteristics of these systems that are important but not defining. For example, there will be a clear relevance to mathematical problems that arise in science, applied mathematics, and engineering. There will be a long history of the concepts that can be respected in the cultural aspects of the system. The system can support many levels of abstraction and translate between them, say relating infinite precision reals to multiple precision reals or relating planar regions of $\mathbb{R}^2$ to discrete graphs.

As these systems have been built, used, and studied we have seen that *they have become models of larger issues in computer science* such as the problem of providing a precise machine-useable semantics to concepts from science. They have raised new questions in computing theory, such as how to define the computational complexity of higher-order functions and operators and how to define the notion of a mathematical type. In general these systems have made ideas from *computational mathematics* very precise and practical.

Outline. Geometry is one of the oldest branches of mathematics, and its axiomatic presentation in Euclid's *Elements* [10] circa 300 BCE also marks a critical time in the development of logic, applying Aristotle's ideas from the *Organon* circa 350 BCE to mathematics. I will use computational geometry and logic to illustrate the transformative power of computer science on the sciences and on the very means for constructing objective knowledge. For both of these topics there are mathematics-based systems that support computational work. Leda [2, 11] is one of the most influential systems for *computational geometry*, and Nuprl [8, 9] has been an influential system for *computational logic*. Using ideas from Leda, in the last section (Section 6), I sketch a new result for constructive Euclidean geometery, namely that equality and congruence of line segments is decidable, as is equality of points.

3   Computational Geometry, Leda and CGAL

Computational Euclidean geometry is grounded in analytic geometry where points are represented by tuples of real numbers; thus even the simplest geometric facts depend on real arithmetic, and this poses deep theoretical and practical problems. Consider the following very simple example from the Leda book. Suppose we have two lines l_1 and l_2 which pass through the origin and have different positive slopes; then we know that they intersect in exactly one point and that in the positive quadrant, one line remains above the other as we see in Figure 1.

If we examine points on these lines at increments of 0.001 on the x-axis starting at 0, we would expect to see that at 0 they intersect, but at all other points, those on one line are above the other, say l_1 points are above those on l_2. However, as the Leda account shows, if we test this belief using C^{++} *floating point arithmetic*, we find something else entirely – the lines intersect multiple times and which line is on top changes as well. This bizarre behavior results from the character of floating point arithmetic. Floating point representation of lines leads to what is called *braided lines* as in Figure 2. When we look in detail at the floating point representation of a straight line, we see that it is a step function because there are only finitely many floating point values. In the Leda book, Kurt and Stephan Näher use this diagram in Figure 2 to explain the problem.

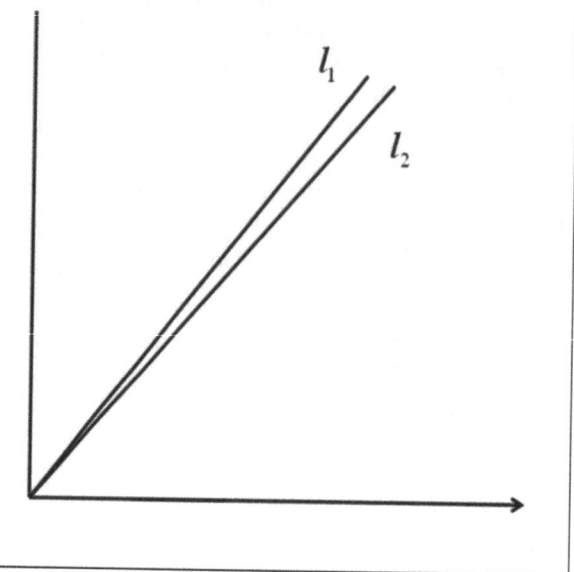

Fig. 1.

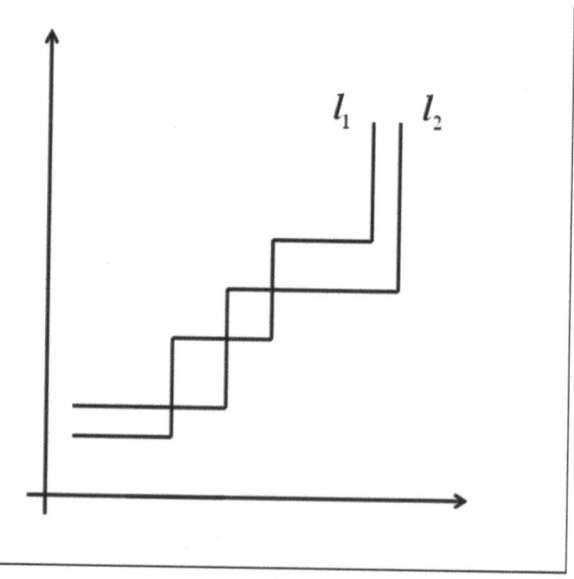

Fig. 2.

A considerable amount of theoretical work was devoted to providing definitions of real number sufficient to support a computational geometry that is logically sound and yet efficient. Leda has a robust concept of infinite precision real numbers [12] called *bigfloats*. These generalize floating point arithmetic, but equality and order are not decidable relations on these numbers, and computations with them are slow. This is a problem with all attempts that I know about to use infinite precision reals in serious computations. However, Leda can use bigfloats as a tool to implement another notion of real number.

The Leda system also incorporated *algebraic numbers* defined by the type *real*. These correspond to the numbers used in constructive analysis [13, 14, 15] and in the work of Edwards [16] on algebra, i.e. those real numbers that are roots of polynomials with rational coefficients. These numbers are formed using addition, subtraction, multiplication, division, and $k - th$ roots of rational numbers for all natural numbers k. The rationals are formed with infinite precision integers.

It is possible to compute the sign of reals, but not the sign of bigfloats. Leda uses a theorem that determines a bound on the number of bigfloat digits whose mantissa can be calculated from the symbolic representation of the real. The idea is that if r is an expression for a real, then we can find an expression $sep(r)$ called the *separation bound* such that $val(r) \neq 0$ implies that the absolute value of $val(r)$ is greater than or equal to $sep(r)$, and we can easily calculate $sep(r)$. This result is important for the *logic of computational geometry*.

4　Computational Logic and Nuprl

We have seen that there is a need for logically sound computable real numbers in geometry systems like Leda and CGAL. The need also arises in computer algebra systems in order to prove that various algebraic algorithms over the reals are correct [17, 18]. This has led to developing formal logical theories of computable real numbers. There is a long history of this subject in mathematics going back to the 18th as part of the *arithmetization of analysis*.

For theorem provers such as Coq [19] and Nuprl which are based on constructive logic, it is important to formulate theorems about the reals that are constructively valid. For the Cornell group this work is based on the book *Constructive Analysis* [13, 20] and other writings of Errett Bishop. Even simple results such as the *Intermediate Value Theorem* (IVT) do not hold in their classical form. Figure 3 suggests a simple counterexample to IVT based on the idea that if we could find the root of the simple piecewise linear function in the diagram, then we could decide whether the real number a was positive or negative, but this cannot be decided for constructive real numbers. Just below we provide a constructively valid version of IVT [21] proved in Nuprl.

Some definitions and theorems in the Nuprl theory of reals. Mark Bickford has recently defined constructive reals that include the rational numbers and integers as subtypes, $\mathbb{N} \sqsubseteq \mathbb{Z} \sqsubseteq \mathbb{Q} \sqsubseteq \mathbb{R}$. This is equivalent to Bishop's definition which was formalized exactly by Forester [21] and is repeated below by copying the definition directly from the Nuprl library.

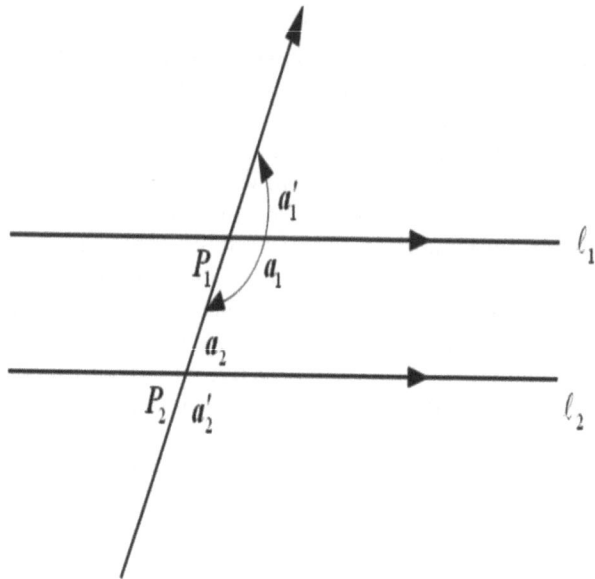

Fig. 3.

Bishop Reals

$$\mathbb{R} == \{x : Seq((\mathbb{Q}) \mid (\forall i : \mathbb{N}^+ . |x(i) - x(j)| \leq 1/i + 1/j)\}.$$

Equality

$$(r = s) == \forall n : \mathbb{N}^+ . |r(n) - s(n)| \leq 2/n$$

Definition as an extension of Rationals. A real number is either a rational or a pair of functions $r(n)$, $\delta(r;n)$ that gives the nth approximation to the real and its "error bar".

```
ℝ ==
  ℚ ∪ {r:ℕ → ℚ × Infinitesimal|
       let x,err = r in ∀i,j:ℕ.  (|(x i) - (x j)| ≤
                                   (err(i) + err(j)))} + Void.
```

The + **Void** makes reals be marked with *inl*, so that we can later use things marked *inr* for complex numbers. An infinitesimal is just a decreasing sequence of rationals with limit zero.

```
Infinitesimal ==
  {p:ℕ → ℚ × (ℕ+ → ℕ)|
    let f,g = p in
       (∀i:ℕ. (0 ≤ (f i))) &
       (∀i,j:ℕ.  ((i ≤ j) ⇒ ((f j) ≤ (f i)))) &
       (∀n:ℕ+. f (g n) < (1/n))}.
```

Bickford proved Cantor's theorem in Nuprl with his definition.

Cantor's theorem

```
∀z:N → R. ∀x,y:R. ((x < y) ⇒ (∃u:R. ((x < u) & (u < y) &
   (∀n:N. u ≠ z n)))).rpositive(r) ==  ∃n:N. δ(r;n) < r(n).x<y
   ==  rpositive(y - x).
x ≠ y ==  (x < y) ∨ (y < x).
```

Continuous functions

To state the Intermediate Value Theorem, we need the notion of a *continuous* function on an interval $[a, b]$.

$\exists w : \mathbb{R} \to \mathbb{R}$. $\forall \epsilon : \mathbb{R}^+$. $w(\epsilon) > 0$ & $(\forall x : [a, b]. \forall y : [a, b]. |x - y| \leq w(\epsilon) \Rightarrow |f(x) - f(y)| \leq \epsilon)$.

Theorem IVT

$\forall a : \mathbb{R}$. $\forall b : \mathbb{R}$. $a < b \to (\forall f : \mathbb{R} \to \mathbb{R}$. f cont on $[a, b] \Rightarrow$
$f(b) > 0 \Rightarrow -(f(a)) > 0 \Rightarrow (\forall \epsilon : \mathbb{R}^+. \exists c : [a, b]. |f(c)| < \epsilon))$.

The formal proof provides a procedure which picks the midpoint c of $[a, b]$ and tests whether $|f(c)| < \epsilon$. If so, the result is established, otherwise $|f(c)| > 0$ and it follows that either $f(c) > 0$ or $-f(c) > 0$. Now we repeat the procedure on the intervals $[c, b]$ or $[a, c]$. The procedure terminates because the diameter of the intervals goes to 0 and by continuity, the diameter of the range of f goes to 0 as well.

The proof is inductive with the following induction hypothesis.

Bisecting

bisects (a; b; f; ϵ; n) == $(\exists c : [a, b]$. $|f(c)| < \epsilon) \vee$
$(\exists \alpha : [a, b]. \exists \beta : [a, b]. \beta - \alpha = 1/2^n * b - a$ &
$f(\beta) > 0$ & $- (f(\alpha)) > 0)$.

Complex Numbers. In *Constructive Analysis* [13, 20] Bishop and Bridges develop a substantial amount of complex analysis over the complex numbers as ordered pairs of reals. They give a constructive proof of the Fundamental Theorem of Algebra (FTA) and the Riemann Mapping Theorem. The FTA has been formalized in the provers Coq and HOL [22], and articles about the efforts are forthcoming. We will discuss below the relationship of Bishop's work to geometry and to the type of algebraic numbers used in Leda.

5   Constructive Euclidean Geometry

Ever since Descartes introduced analytic geometry and translated geometric questions into analysis, mathematicians have realized that there are other rigorous ways to study geometry beyond Euclid's. Nevertheless Euclidean geometry has been the subject of sustained axiomatic and logical analysis right up to the present. New proofs in the style of Euclid are being published even now such as the *Steiner-Lehmus* theorem that if two angle bisectors of a triangle are equal in length, then the triangles must be isosceles [23]. Two of the most cited works are Hilbert's *Foundations of Geometry* [24],Tarski's decision procedure for geometry based on the first-order theory of *real closed fields* [25], and von Plato's axiomatization in type theory [26]. I am also thinking along these lines using computational type theory (CTT), and I sketch below some new ideas on decidability that were influenced by Kurt's work and which we will implement via Nuprl in due course. I appreciate that constructive Euclidean geometry captures the spirit of Euclid's work in a 21st century style, continuing the 2,300 year tradition of teaching and investigating this body of work. Interestingly the logical approach connects well to Descartes deep study of the nature of *certain knowledge*, and here I have connected the Leda and Nuprl work to the topic of certain knowledge. I mean this account of geometry to be intuitive and strongly tied to Euclid's approach because that is such a widely known starting place for geometry – and thus a vehicle to explain computational type theory.

5.1   Primitive Abstractions, Definitions, and Displays

We use these types *Point, Ray, Line, Angle, Figure*, and these primitive terms, $point\{p : atom\}$, $between(a; p; b)$, $ray(a; b)$, $origin(r)$, $dest(r)$, $line(r)$, $on(p; l)$, $angle(a; b; c)$, $right(p; l)$, $left(p; l)$, $ipoint(r; p; q)$, $nonparallel(l_1; l_2)$, $circle(p; r)$, $ctr(c)$, $rad(c)$, $leftcpt(c_1; c_2; r)$, $rightcpt(c_1; c_2; r)$.

Point. Points are the basic data type; we will see later that we can reduce every concept to points in some sense. The elements of the type *Point* include *given points* which are canonical or irreducible elements of the type. We use the terms $point\{a : atom\}$ for these where a is just a name. There are other terms for points that arise from axioms that tell under what conditions we have a point. For example, if we know that two lines will intersect, they can be extended to intersect in a *constructed point*. One of these "straightedge" constructions is given by $nonparallel(l_1; l_2)$ which is created (or found) by extending straight line segments. Another constructed point is found by connecting given points p and q forming a line known to intersect the ray r, the constructed point is denoted by $ipoint(r; p; q)$. Axiom 6, a continuity axiom, brings this term into the theory as a constructed point in the type *Point*.

  As with all types, *Point* comes with an equality relation, $p = q$ in *Point*. Given points are equal only if they have the same names. If a term reduces to a point, as we see in the next definition, it is equal to that point.

Rays. *Ray* is a primitive type whose elements are *directed straight line segments* called rays which have a direction from an origin point a to a destination point b. We can imagine a ray as a one dimensional object consisting of points constructed using a idealized *straightedge*. The canonical term denoting this construction is $ray(a; b)$, and that concrete term defines the *logical object*, quite distinct from the imagined ideal object.

We say that a is the *origin* of the ray $ray(a; b)$ and b is the *destination*; we write these for a ray r as $origin(r)$ and $dest(r)$. Two rays are equal, $r_1 = r_2$ *in Ray*, iff they have the same origin and the destination. They are *congruent* if they have the same length, see below. The term $origin(ray(a; b))$ reduces by a computation rule of the logic to the term a which will be a point if $ray(a; b)$ is known to be in the type *Ray*, likewise $dest(ray(a; b))$ reduces to b.

We say that the origin and destination points are *on the ray*, $on(origin(r); r)$, $on(dest(r); r)$, and also we have the betweenness relation for points between the origin and destination that are on the ray, say p so that $between(origin(r); p; dest(r))$.

In addition to equality on rays, we introduce another primitive equivalence relation called *congruence*(Euclid's terminology). We say that one ray is congruent to another, $r_1 \simeq r_2$, meaning that they have the same length; but we do not treat length as a numerical quantity. If we did, it would be a positive real number since the origin and destination of a ray are distinct points.

Segments. Intuitively segments are *finite straight lines*, what Euclid calls a line and many authors call a segment. We define the type *Segment* from the type *Ray* using a *quotient operation.*

$Segment == Ray//x = y \; iff \; (origin(x) = origin(y) \; \& \; dest(x) = dest(y) \; \vee \; origin(x) = dest(y) \; \& \; dest(x) = origin(y)).$

The type used to define segments is the *quotient type*. In general if T is a type and eq is an equivalence relation on T, then $T//eq$ is the type whose elements are from T but whose equality is given by eq instead of the equality of T. This is an elegant way to define such notions as the integers modulo a number. In this case, the quotient operation hides the direction of the ray. In classical mathematics this quotient construction is accomplished using equivalence class, but that is not a good definition for computation.

The identity function maps from *Ray* to *Segment*, and we say that *Ray* is a *subtype* of *Segment*, written $Ray \sqsubseteq Segment$. Given any construction with rays, it can also be done by reversing all the rays, and the results map out to segments as well. Lines have no direction, but we can orient them by generating them by rays.

Line. A *straight line* is another primitive concept, an element of the atomic type *Line*. Straight lines are unbounded objects consisting of points "that lie continuously on a line." We can imagine them as infinite in extent, but they are created by extending a ray. Given a ray r, the line specified is given by the primitive $line(r)$. If we do not want to use the orientation of the line, we can treat r as a segment. Associated with lines and points is the atomic relation $on(p; l)$ where p is a point and l is a line. By generating lines from rays, we settle an issue about lines, that they have at least two points on them, namely $on(origin(r), line(r))$ and $on(dest(r), line(r))$.

Figures. Intuitively a *circle* is the set of all points in a plane at an equal positive distance, called the *radius*, from a point called the *center*. To say this we use congruence. Given a point p for the center, and a ray r for the radius, we will have an axiom that allows us to construct a circle, imagine using a *compass* to do it. The canonical circle is $circle(p; r)$. There are operators to pick out the center of a circle, $ctr(circle(p; r))$ reduces to p, and $rad(circle(p; r))$ reduces to r. We will also know informally that:

$$circle(p; r) \ == \ \{x : Point| \ \exists r' : Ray. \ origin(r') = p \ \& \ dest(r') = x \ \& \ ray(p; x) \simeq r\}.$$

We can define the *interior* of a circle as the center and those points which are at a distance less than the radius. The points *on the circle* are the circle, and the points *outside* are those whose distance from the center are greater than the radius. We say that a circle belongs to the type of *Figure*.

There are the usual rectilineal figures in the type *Figure*, and we will define one of them below, triangles.

Angles. An *angle* at p is formed by two non-collinear rays r_1 and r_2 emanating from p. The destination points of the rays, $dest(r_1)$ and $dest(r_2)$ are used with p to name the angle as in $angle(dest(r_1), p, dest(r_2))$ or $angle(dest(r_2), p, dest(r_1))$, these are equal names for the same object in the type *Angle*.

If one of the rays r forming an angle at p is extended in the opposite direction from p on the line formed by the ray, forming a new ray r', then another angle is created at p, and this is called the *adjacent* angle to the first one (also the *complement* to the first).

Interior angles are defined in terms of two lines l_1 and l_2 (generated by rays r_1 and r_2) respectively and a ray r intersecting both lines, say l_1 at p_1, and l_2 at p_2, say l_1 above l_2 in the direction of r. Consider the adjacent angles on the right side of r at p_1, say a_1' and a_1 and the adjacent angles on the right side of r at p_2, say a_2 and a_2'. The angles between the lines l_1 and l_2 are called the *interior angles*. Let a_1 be the interior angle at p_1 and a_2 the interior angle at p_2. See Figure 4 below where we arranged that the interior angles are a_1 and a_2.

Extracts. In computational type theory [9] and related logics, e.g. [27, 19], axioms and theorems come with terms that express their *computational content*. This is a feature of the *propositions-as-types* semantics for type theory as systematized by deBruijn [28] and Martin-Löf [29]. The axioms for Euclidean geometry come with these extracts. Here is a simple example. If we claim that there is a point on a given line l, we can write $\exists p : Point. \ (p \ on \ l)$ and to witness this claim there should be a term in the logic for the point p and a term witnessing that $(p \ on \ l)$. The witness for the point could be $point\{a : atom\}$, and if the witness for its being on l is the term pf, then we would write the following to show that a certain ordered pair is a witness to the claim:

$$\exists p : Point. \ (p \ on \ l) \ \textbf{extract} \ pair(point\{a : atom\}; pf).$$

In much of what follows, the witness for relations such as $(p \ on \ l)$ and equality relations such as $p = q$ will simply be the term *axiom* which tells us that the witness is primitive and carries no computational information, i.e. is "axiomatic".

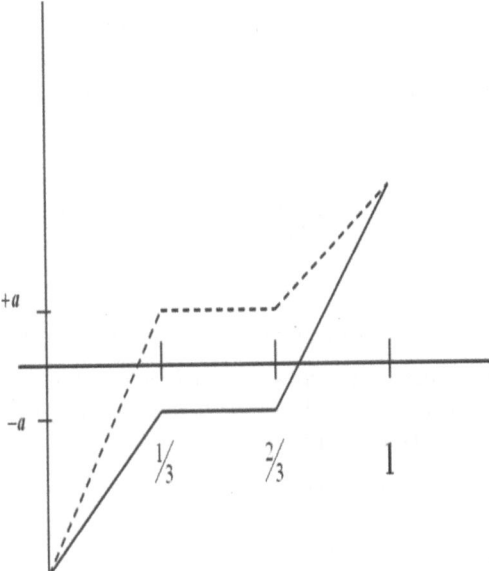

Fig. 4.

Many of the extracts in the axioms are functions, and they are given using lambda notation, so for example, $\lambda(x.\ x)$ is the notation for the identity function. This notation is especially simple for the first axiom where the extract is simply a function that takes two points p_1 and p_2 as inputs and then takes some proof neq that the points are not equal, and returns a ray connecting the points and two axiomatic equalities about how to obtain the origin and destination of the ray.

5.2   Euclid-Like Axioms

Axiom 1. To draw a straight line segment from any point to any other point.

$\forall p_1, p_2 : Point.\ p_1 \neq p_2\ \Rightarrow \exists r : Ray.\ origin(r) = p_1\ \&\ dest(r) = p_2.$
extract $\lambda(p_1, p_2.\lambda neq.\ pair(ray(p_1; p_2); pair(axiom; axiom))).$

Axiom 2. To produce a finite line continuously in a straight line.

$\forall r : Ray.\ \exists l : Line.\ \forall p : Point.\ (p\ on\ r \Rightarrow p\ on\ l).$ **extract**
$\lambda(r.\ pair(line(r)); \lambda(p.\ \lambda(hyp.\ axiom))).$

The line l is oriented by the ray r whose origin and destination are two points on l. We define the relations $p\ right\ l$ and $p\ left\ l$ with reference to the direction of r.

We can prove Euclid's axiom by using the segment given by the ray, namely

$\forall s : Seg.\ \exists l : Line.\ \forall p : Point.\ (p\ on\ s \Rightarrow p\ on\ l).$ **extract**
$\lambda(s.\ pair(line(r)); \lambda(p.\ \lambda(hyp.\ axiom))).$

Axiom 3. To describe a circle with any center and distance.

$\forall p : Point.\ \forall r : Ray.\ \exists c : Figure.\ c = circle(p; r).$ **extract**
$\lambda(p.\lambda(r.\ pair(circle(p; r); axiom))).$

Axiom 4. That all right angles are equal to one another.

$\forall p : Point.\ \forall r_1, r_2 : \{x : Ray | origin(x) = p\}.\ (dest(r_1) \neq dest(r_2)\ \&\ not$
$(colinear(p, dest(r_1), dest(r_2))) \Rightarrow \exists a : Angle.\ \forall a' : Angle.\ (right(a)\ \&\ right(a')$
$\Rightarrow a = a' in\ Angle)).$
extract
$\lambda(p.\lambda(r_1, r_2.\lambda(andhyp.\ pair(angle(dest(r_2), p, dest(r_1)); \lambda(a'.\ \lambda(h.axiom)))))).$

Axiom 5. If ray r intersects segments l_1 and l_2 such that the interior angles on the same side of r are less than right angles, then l_1 and l_2 intersect on the side where the angles are less than right.

$\forall r : Ray.\ \forall l_1, l_2 : Seg.\ (r\ intersects\ l_1\ at\ p_1)\ \&\ (r\ intersects\ l_2\ at\ p_2)\ \&$
$(interior\ angles\ a_1, a_2\ are\ acute) \Rightarrow \exists p : Point.\ intersect(l_1; l_2; p)\ \&\ p\ right\ r.$
extract $\lambda(r.\ \lambda(l_1, l_2.\ \lambda(hyp.\ pair(nonparallel(l_1; l_2); axiom)))).$

Axiom 6 Continuity. Euclid did not have this axiom nor anything equivalent to any of the cases of the axiom. Only two cases are given and only informally because these versions of continuity may not be the most general approach, though they match the important work of W. Killing [30].

Let l be a line determined by a ray r and let p be a point on the left of l and q a point on the right, then the segment connecting p and q intersects l in a point which has the primitive name $ipoint(r; p; q)$.

Let C_1 and C_2 be two circles such that a point of C_2 is inside C_1 and the center is outside, and let r be a ray connecting the center of C_1 to the center of C_2, then C_1 and C_2 intersect in exactly two points, one on the left of r and one on the right of r. These points are called $leftcpt(c_1; c_2; r)$ and $rightcpt(c_1; c_2; r)$.

5.3   Theorems

The first proposition in Euclid's *Elements* is the construction of an equilateral triangle. The construction can be used to bisect a side of the triangle.

A triangle is a figure with three sides, a three sided *rectilinear figure*. We say that a list of three rays a, b, c forms a triangle iff the $dest(a)$ is $origin(b)$, the $dest(b)$ is $origin(c)$, and the $dest(c)$ is $origin(a)$. The triangle is *equilateral* iff $a \simeq b \simeq c$. Euclid says "On a given finite straight line to construct an equilateral triangle."

Proposition 1 $\forall r : Ray.\ \exists T : Figure.\ (EquilateralTriangle(T)\ \&\ \forall s : \{x : Ray | side(T, x)\}.\ s \simeq r).$

Proof

Let a be $origin(r)$ and b be $dest(r)$, thus $r = ray(a; b)$. Using **Axiom 3** construct circles $circle(a; r), circle(b; r)$, call them A and B. Notice that a is inside A and on circle B, thus the condition for invoking **Axiom 6** on circles is satisfied.

By continuity, **Axiom 6**, there are two points p and q at which the circles meet, let p be $leftcpt(circle(a;r); circle(b;r); r)$; now form the rays by **Axiom 1** $ray(b;p)$ and $ray(p;a)$. These two rays along with $ray(a;b)$ form a triangle T. If we write r as $ray(a;b)$ then T is $[ray(a;b), ray(b;p), ray(p;a)]$ and the triangle property is easily checked by inspecting the endpoints of the rays. Notice that each of $ray(b;p)$ and $ray(p;a)$ is a radius of one of the circles, thus they are each congruent to $ray(a;b)$, thus T is equilateral.   **Qed**

Notice that the *extract* of the theorem is a triangle. The first ray in the list is r, the second ray, called $ray(b;p)$ is constructed from $dest(r)$ and $leftcpt(circle(a;r); circle(b;r); r)$. The term for the second ray is $ray(dest(r); leftcpt(circle(origin(r);r))); circle(dest(r);r); r)$ for the second ray. The third ray is similar. We can see the structure of the construction from the terms.

We can find the midpoint of a ray by bisecting it, using a construction like the one we just described.

$\forall r \; : \; Ray. \; \exists p \; : \; Point. \; between(origin(r); p; dest(r)) \; \& \; ray(origin(r);p) \; \simeq$ $ray(p; dest(r))$. By repeating this construction we can create an unbounded sequence of points on any ray, segment, or line.

We can provide proofs like this for many propositions in Euclid's books I through IV. Using the ideas of Section 6, we can give constructive proofs for all theorems of these books.

5.4   Reduction to Points and to Numbers

We have presented the concepts in a style similar to Euclid's. A more modern approach would be to build everything from the type *Point* in the manner of Tarski [25]. We would define a ray to be an ordered pair of distinct points. We could avoid segments altogether, and take a line to be a pair of points with a tag, distinguishing it from a ray. A circle is a tagged pair of point and ray. Rectilineal figures are lists of rays. Angles would be defined as two pairs of points thought of as rays with a common origin.

Once we have settled on a single primitive type of points, it is easy to reduce the theory further to numbers by defining a point to be a *constructive real number* *a la* Bishop. In this analytical geometry, point, line and angle equality would not be decidable. We could not decide whether a point was on a line or whether two lines met at a point.

However, just as with Kurt's work on Leda, we can consider the option to use constructive algebraic numbers for points. In this case the whole theory changes dramatically because it is known that equality on these numbers is decidable. We discuss this point in the next section.

6   Logical Issues in Constructive Euclid-Like Geometry

Euclid was very careful in his choice of propositions and proofs. There is no proposition saying "Given two line segments AB and BC, to decide whether they are congruent." Nor does he try to prove that one can decide whether a given angle is a right angle, nor does he try to compare angles. He does not try to decide whether points are equal or whether a point is on a line or whether given two circles with the same center one is inside another. If any one of these questions could be decided, they all could be.

It appears as if he confines himself to statements that can be proved constructively and that he is centrally concerned with construction. However, not all of his proofs are constructive because he uses *reductio absurdum* to prove positive statements, and he argues by cases when it is not possible to decide among the cases.

The definitions and axioms above are all interpreted in *Computational Type Theory* (CTT) [9], the type theory implemented by Nuprl. So it is a fully constructive theory, thus not all of the theorems from the Elements books I to IV can be proved in this theory, but many can be. It is a genuinely constructive core of the *Elements*.

In reading Kurt's work in Leda, I came to realize that we could extend this theory by changing the definition of the elements of Points to name them with the constructive algebraic numbers, $point\{a : real\}$, precisely the numbers of the Leda type *real*. By doing this we have a theory *Constructive Euclid*.

Result. In Constructive Euclid, all plane geometry theorems of the *Elements* can be constructively proved because equality of points, lines, angles, and circles are all decidable. These relations are decidable because of the results of Bishop's development of the constructive complex numbers and the 1978 result of Julian, Mines, and Richman [14] that *equality of algebraic numbers is decidable*.

This Constructive Euclid-like Geometry would lead to many new theorems and to shorter proofs of existing theorems in the *Elements,* and all results would be fully constructive. I am now examining this theory with an eye to formalizing and implementing parts, just for the joy of it. Such a theory would follow the spirit and axiomatization of Euclid but use at its base a constructive logic and a richer set of constructions beyond straightedge and compass that would involve effective measuring and comparing.

References

1. Mehlhorn, K.: Polynomial and abstract subrecursive classes. Journal of Computer and System Sciences, 148–176 (1976)
2. Mehlhorn, K., Näher, S.: LEDA — A platform for combinatorial and geometric computing. Cambridge University Press, Cambridge (1999)
3. Mehlhorn, K.: Data Structures and Algorithms 3: Multi-Dimensional Searching and Computational Geometry. Springer, Heidelberg (1984)
4. Burnikel, C., Fleischer, R., Mehlhorn, K., Schirra, S.: A Strong and Easily Computable Separation Bound for Arithmetic Expressions Involving Square Roots. In: Proceedings of the 8th ACM-SIAM Symposium on Discrete Algorithms (SODA 1997) (1997)
5. Burnikel, F., Mehlhorn, S.: Efficient exact geometric computation made easy. In: Annual ACM Symposium on Computational Geometry, pp. 341–350 (1999)
6. Mehlhorn, K., Osbild, R., Sagraloff, M.: Reliable and efficient computational geometry via controlled perturbation. In: Bugliesi, M., Preneel, B., Sassone, V., Wegener, I. (eds.) ICALP 2006. LNCS, vol. 4051, pp. 299–310. Springer, Heidelberg (2006)

7. Fortune, S.J., Van Wyk, C.: Static analysis yields efficient exact integer arithmetic for computational geometry. Transactions on Graphics, 223–248 (1996)
8. Constable, R.L., Allen, S.F., Bromley, H.M., Cleaveland, W.R., Cremer, J.F., Harper, R.W., Howe, D.J., Knoblock, T.B., Mendler, N.P., Panangaden, P., Sasaki, J.T., Smith, S.F.: Implementing Mathematics with the Nuprl Proof Development System. Prentice-Hall, NJ (1986)
9. Allen, S.F., Bickford, M., Constable, R., Eaton, R., Kreitz, C., Lorigo, L., Moran, E.: Innovations in computational type theory using Nuprl. Journal of Applied Logic 4, 428–469 (2006)
10. Heath, S.T.L.: The thirteen books of Euclid's Elements. Dover, New York (1956)
11. Kettner, L., Näher, S.: Two computational geometry libraries: LEDA and CGAL. In: Handbook of Discrete and Computational Geometry, 2nd edn., pp. 1435–1464. CRC Press LLC, Boca Raton (2004)
12. Boehm, H., Cartwright, R.: Exact real arithmetic, formulating real numbers as functions. Research Topics in Functional Programming, 43–64 (1990)
13. Bishop, E.: Foundations of Constructive Analysis. McGraw Hill, New York (1967)
14. Julian, W., Mines, R., Richman, F.: Polynomial and abstract subrecursive classes. Pacific Journal of Mathematics 1, 92–102 (1978)
15. Mines, R., Richman, F., Ruitenburg, W.: A Course in Constructive Algebra. Springer, New York (1988)
16. Edwards, H.M.: Essays in Constructive Mathematics. Springer, New York (2005)
17. Buchberger, B.: Theory exploration with Theorema. Analele Universitatii din Timisoara, Ser Matematica-Informatica, XXXVIII 2, 9–32 (2000)
18. Jackson, P.B.: Enhancing the Nuprl Proof Development System and Applying it to Computational Abstract Algebra. PhD thesis, Cornell University, Ithaca, NY (1995)
19. Bertot, Y., Castéran, P.: Interactive Theorem Proving and Program Development; Coq'Art: The Calculus of Inductive Constructions. Texts in Theoretical Computer Science. Springer, Heidelberg (2004)
20. Bishop, E., Bridges, D.: Constructive Analysis. Springer, New York (1985)
21. Forester, M.B.: Formalizing constructive real analysis. Technical Report TR93-1382, Computer Science Department, Cornell University, Ithaca, NY (1993)
22. Harrison, J.: Theorem Proving with the Real Numbers. Springer, Heidelberg (1998)
23. Gilbert, A., MacDonnell, D.: The Steiner-Lehmus theorem. American Math. Monthly 70, 79–80 (1963)
24. Hilbert, D.: The Foundations of Geometry. Open Court Publishing, London (1921)
25. Tarski, A., Givant, S.: Tarski's system of geometry. Bulletin of Symbolic Logic 5(2), 175–214 (1999)
26. von Plato, J.: The axioms of constructive geometry. Annals of Pure and Applied Logic 76, 169–200 (1995)
27. Schwichtenberg, H.: Constructive analysis with witnesses. In: Proof Technology and Computation. Natio Science Series, pp. 323–354. IOS Press, Berlin (2006)
28. de Bruijn, N.G.: The mathematical vernacular, a language for mathematics with typed sets. In: Nederpelt, R.P., Geuvers, J.H., Vrijer, R.C.D. (eds.) Selected Papers on Automath. Studies in Logic and the Foundations of Mathematics, vol. 133, pp. 865–935. Elsevier, Amsterdam (1994)
29. Martin-Löf, P.: Constructive mathematics and computer programming. In: Proceedings of the Sixth International Congress for Logic, Methodology, and Philosophy of Science, pp. 153–175. North-Holland, Amsterdam (1982)
30. Killing, W.: Einführung in die Grundlagen der Geometrie. Druck un Verlag von Ferdinand Schöningh, Mainz (1893)

On Negations in Boolean Networks

Norbert Blum

Informatik V, Universität Bonn,
Römerstr. 164, D-53117 Bonn, Germany
blum@cs.uni-bonn.de

Abstract. Although it is well known by a counting argument that relative to the full basis most Boolean functions need exponentially many operations, for explicit Boolean functions only linear lower bounds with small constant factors are known. For monotone networks (i.e., networks without negations) exponential lower bounds for explicit monotone Boolean functions have been proved. We describe the state of the art and give some arguments why techniques developed for the proof of lower bounds for monotone networks cannot easily be extended to Boolean networks with negations.

Keywords: Boolean function, monotone network, non-monotone network, network complexity.

1 Introduction

First, we will give some basic definitions. $\mathcal{B}_{n,m} := \{f \mid f : \{0,1\}^n \to \{0,1\}^m\}$ is the set of all n-ary Boolean functions with m outputs. Instead of $\mathcal{B}_{n,1}$ we write $\mathcal{B}_n$. The ith $variable$, $1 \leq i \leq n$, is denoted by $x_i : \{0,1\}^n \to \{0,1\}$. Let $V_n := \{x_i \mid 1 \leq i \leq n\}$ and $V'_n := V_n \cup \{\neg x_i \mid 1 \leq i \leq n\}$. Variables and negated variables are called $literals$. A function $m : \{0,1\}^n \to \{0,1\}$ which is the product of some literals is called a $monomial$. The empty product is the constant function 1. For $f, g \in \mathcal{B}_n$ we define: $f \leq g :\Leftrightarrow f \wedge g = f$, and then we call f a $subfunction$ of g. $IM(f) = \{t \mid t$ monomial, $t \leq f\}$ is the set of $implicants$ of the function f. An implicant $t \in IM(f)$ is a $prime$ $implicant$ of f if $\forall t' \in IM(f) : [t \leq t' \leq f \Rightarrow t = t']$. $PIM(f) \subseteq IM(f)$ is the set of all prime implicants of f. Let $a := (a_1, a_2, \ldots, a_n)$, $b := (b_1, b_2, \ldots, b_n) \in \{0,1\}^n$. We write $a \leq b$ iff $a_i \leq b_i$ for $1 \leq i \leq n$. A function $f = (f_1, f_2, \ldots, f_m) \in \mathcal{B}_{n,m}$ is $monotone$ iff for all $a, b \in \{0,1\}^n$, $a \leq b$ implies $f_i(a) \leq f_i(b)$, $1 \leq i \leq m$. Let $\mathcal{M}_{n,m}$ denote the set of monotone Boolean functions in $\mathcal{B}_{n,m}$. We also write $\mathcal{M}_n$ for $\mathcal{M}_{n,1}$. $\mathcal{B}_2$ is the set of $basic$ $operations$. Let $\Omega \subseteq \mathcal{B}_2$. An Ω-network β is a directed, acyclic graph such that each node has indegree ≤ 2. The nodes u with indegree 0 are input nodes and are labelled with $op(u) \in V_n$. Each non-input u is labelled by an $op(u) \in \Omega$. A node with outdegree 0 is an $output node$. For a node u in β let $suc(u) := \{v \mid u \to v$ is an edge in $\beta\}$ and $pred(u) := \{v \mid v \to u$ is an edge in $\beta\}$ be the sets of direct successors and direct predecessors. With

S. Albers, H. Alt, and S. Näher (Eds.): Festschrift Mehlhorn, LNCS 5760, pp. 18–29, 2009.

each node u we associate a function $res_\beta(u) : \{0,1\}^n \to \{0,1\}$ (n is the number of input nodes of β):

$$
res_\beta(u) := \begin{cases} op(u) & \text{if } u \text{ is an input,} \\ \neg res_\beta(v) & \text{if } op(u) = \neg, \text{ where} \\ & v \text{ is the direct predecessor of } u, \\ res_\beta(v)\, op(u)\, res_\beta(w) & \text{otherwise, where} \\ & v, w \text{ are the direct predecessors of } u. \end{cases}
$$

For $u \in \beta$, $res_\beta(u)$ is the function computed at node u in β. Let $G \subset \mathcal{B}_n$. The minimal number of gates in an Ω-network which computes G where negations are not counted is the Ω-complexity $C_\Omega(G)$ of G. Let $\Omega_0 := \{\wedge, \vee, \neg\}$ and $\Omega_m := \{\wedge, \vee\}$. An Ω_m-network is also called a *monotone network*. Note that exactly the monotone functions can be computed by a monotone Boolean network.

Although sixty years ago, Shannon [37] has proved by a counting argument that at least a fraction $(1 - 2^{2^n n^{-1} \log \log n})$ of the functions in $\mathcal{B}_n$ has $\mathcal{B}_2$-complexity strictly larger than $2^n/n$, for explicit Boolean functions only linear lower bounds with small constant factors for their $\mathcal{B}_2$-complexity or their Ω_0-complexity have been proved. In 1974, Schnorr [36] has given the first $2n$-lower bound for the $\mathcal{B}_2$-complexity of a function in $\mathcal{B}_n$. Next, Paul [27] proved a $2.5n$-lower bound for the $\mathcal{B}_2$-complexity of an $(n + \log n)$-ary Boolean function. Then Stockmeyer [39] proved that the lower bound of Paul holds for a larger class of functions. In 1982, Blum [8] has given a $3n$-lower bound for the $\mathcal{B}_2$-complexity of an $(n + 3 \log n + 1)$-ary function. This lower bound is still the largest lower bound for the $\mathcal{B}_2$-complexity of an explicit Boolean function. With respect to the base Ω_0 better linear lower bounds are known. In 1974, Schnorr has also given the first $3n$-lower bound for the Ω_0-complexity of an n-ary Boolean function. In 1988, Zwick [50] has proved a $4n$-lower bound for the Ω_0-complexity of a function in $\mathcal{B}_n$. Iwama, Lachish, Morizumi and Raz [17] claim to have a $5n - o(n)$-lower bound for the Ω_0-complexity of an n-ary Boolean function. All these linear lower bounds use the so-called "gate-elimination method". The gate-elimination method uses induction. By an assignment of some variables with values from $\{0,1\}$, a specific number of gates are eliminated in each step and the resulting function is of the same type as the function before the assignment. Over the years, the case analyses used in the proofs have become more and more complicated and have culminated to a case study of large depth in [17]. Since each operation in $\mathcal{B}_2$ can be realized within the base Ω_0 using at most three operations from $\{\wedge, \vee\}$ and some negations, a lower bound of the Ω_0-complexity larger than $6n$ would imply a lower bound for the $\mathcal{B}_2$-complexity larger than $3n$. Note that for functions in $\mathcal{B}_{n,m}$, $m > 1$ no better lower bounds are known. I am convinced that the gate-elimination method alone will not lead to the proof of a nonlinear lower bound for the Ω_0-complexity of any explicit Boolean function.

The inability to prove lower bounds for the Ω_0-complexity of explicit Boolean functions has led to the consideration of resticed models of Boolean networks like monotone or bounded-depth Boolean networks. For both restricted models, exponential lower bounds for the complexity of an explicit Boolean function are

known. We will give an overview of the progress obtained with respect to monotone Boolean networks in the next section. Section 3 will discuss the attempts to
extend techniques developed for monotone networks to Ω_0-networks. Textbooks
on the complexity of Boolean functions are [47, 13, 35, 12].

2   Monotone Boolean Networks

By an estimation of the number of n-ary monotone Boolean functions and an
application of Shannons counting argument one can also prove that nearly all
monotone Boolean functions have exponential (monotone) network complexity.
In contrast to the non-monotone complexity of Boolean functions nonlinear lower
bounds for the monotone network complexity of explicit monotone Boolean functions have been proved. All these lower bound proofs use the following property
of each monotone network β which computes a function $f = (f_1, f_2, \ldots, f_m) \in \mathcal{M}_{n,m}$:

For all nodes u in β, the function $res_\beta(u)$ which is computed at the node u
can be written as a polynomial; i.e., $res_\beta(u) = \bigvee_{j=1}^{t} m_j$ where each m_j is a
monomial. Starting at the input nodes of the network, we can compute these
polynomials in the obvious way by applying the properties of the Boolean operations. We call this representation of $res_\beta(u)$ the *polynomial expansion* of $res_\beta(u)$.
Let u_i, $1 \le i \le m$ be the output node of the network β which computes the
function f_i. For the polynomial $res_\beta(u_i) = \bigvee_{j=1}^{t_i} m_j$ which is computed at u_i,
the following hold:

1. For $1 \le j \le t_i$, the monomial m_j is a an implicant of the function f_i.
2. For all prime implicants p of f_i there is a $j \in \{1, 2, \ldots, t_i\}$ with $m_j = p$.

If the first property is not fulfilled then there is an input $(a_1, a_2, \ldots, a_n) \in \{0, 1\}^n$
such that $f_i(a_1, a_2, \ldots, a_n) = 0$ but $res_\beta(u_i)(a_1, a_2, \ldots, a_n) = 1$. If the second
property is not fulfilled then there is an input $(a_1, a_2, \ldots, a_n) \in \{0, 1\}^n$ such that
$f_i(a_1, a_2, \ldots, a_n) = 1$ but $res_\beta(u_i)(a_1, a_2, \ldots, a_n) = 0$.

Most functions considered for proving lower bounds are homogeneous. A
Boolean function $f \in \mathcal{M}_{n,m}$ is called *k-homogeneous* if all prime implicants of f
are of length k. Examples for homogeneous functions are the Boolean matrix multiplication, the Boolean convolution and the clique function. The first nonlinear
lower bound for the monotone network complexity of an explicit Boolean function
has been proved by Neciporuk [25] in 1969 for a function in $\mathcal{M}_{n,n}$. He has considered 1-homogeneous functions in $\mathcal{M}_{n,n}$, the so-called Boolean sums. Neciporuk
considers Boolean sums which have "nothing in common" such that nothing can
be gained by using conjunctions or overlap. "Nothing in common" means that
two distinct Boolean sums have at most one prime implicant in common. We
say then that the Boolean sum is $(1, 1)$-*disjoint*. A well known construction of
Kővári, Sós and Turán [20] leads to such a Boolean sum $f = (f_1, f_2, \ldots, f_n)$ with
$\Omega(n^{1.5})$ prime implicants such that an $\Omega(n^{1.5})$ lower bound for the monotone
complexity of this Boolean sum has been proved. Some years later, Pippenger [28]
and Mehlhorn [24] have generalized the approach of Neciporuk to Boolean sums

which have "little in common" such that only little can be gained by using conjunctions or overlap. Mehlhorn has used an approach introduced by Wegener [44] where some functions besides the variables are given for free. "Little in common" means that three different Boolean sums have at most two prime implicants in common. We say then that the Boolean sum is $(2,2)$-*disjoint*. Using a construction of Brown [10] such a Boolean sum with $\Omega(n^{5/3})$ prime implicants has been constructed such that an $\Omega(n^{5/3})$ lower bound for the monotone complexity of this Boolean sum has been proved.

In 1974, Pratt [30] has shown that each monotone network computing the product of two $n \times n$ Boolean matrices contains at least $\frac{1}{2}n^3$ $\wedge$-gates. Mehlhorn and Galil [23] and Paterson [26] have refined the method of Pratt and have proved that the school-method for the Boolean matrix multiplication is the unique optimal monotone network for Boolean matrix multiplication. Furthermore, the paper of Mehlhorn and Galil contains several general theorems about local transformations in monotone networks. These local transformations are essentially replacement rules, where in a monotone network β for the computation of a function f, we can replace the function $res_\beta(u)$ which is computed at a gate u by the function which we obtain after the deletion of some monomials from the polynomial expansion of $res_\beta(u)$ which are not part of any prime implicant of f. Let Y be the Boolean matrix product of the matrix X_1 with the transposed matrix X_2. Then we have $y_{ij} = 1$ iff the ith row of X_1 and the jth row of X_2 have a common one. In 1979, Wegener [44] has generalized this to the "direct product" of m $M \times N$-matrices $X_1, X_2, \ldots, X_m$. For each choice of one row of every matrix the corresponding output is one iff the chosen rows have a common one. He proved a $\frac{2}{m}NM^m$ lower bound using the elimination method and the pigeon hole principle. Choosing appropriate values for m, M and N, this leads to an $\Omega((\frac{n}{\log n})^2)$ lower bound for a function in $\mathcal{M}_{n,n}$. One year later, Wegener [46] has improved that bound to $\frac{1}{2}NM^m$ introducing a new method for proving lower bounds for Boolean networks. This improves the bound above to $\Omega(\frac{n^2}{\log n})$. He has defined a suitable value function to estimate the contribution of each gate for the computation of the outputs. At each gate he distributes at most the value 1 among the prime implicants. Then he has proved the necessity to give to each prime implicant at least the value $\frac{1}{2}$. Hence, at least $\frac{1}{2}NM^m$ gates are needed.

All these sets of functions have some disjointness properties. The Boolean sums are $(1,1)$- or $(2,2)$-disjoint. A monotone function $f : X \cup Y \to \{0,1\}^m$ is *bilinear* if each prime implicant of f consists of one variable from X and one variable from Y. The Boolean matrix multiplication is a set of disjoint bilinear forms. The generalization of the Boolean matrix product of Wegener is a set of disjoint multilinear forms [46]. If we consider monotone functions which do not have such disjointness properties, the situation becomes more difficult. The so-called semi-disjoint bilinear forms are such Boolean functions. A bilinear form f is *semi-disjoint* if each variable is contained in at most one prime implicant of f_i, $1 \le i \le m$, and $PIM(f_i) \cap PIM(f_j) = \emptyset$ for $1 \le i < j \le m$. The Boolean convolution is a semi-disjoint bilinear form. The first approaches for proving lower bounds for the monotone network complexity of semi-disjoint bilinear forms use graph-theoretical properties of monotone networks realizing such functions. Pippenger and Valiant [29] have studied shifting graphs and have proved that each monotone network for some monotone functions like the Boolean convolution or sorting has to be a shifting graph obtaining an $\Omega(n \log n)$ lower bound for the monotone network complexity of these functions. Independently, Lamagna [21] has obtained a general $\Omega(n \log n)$ lower bound for some semi-disjoint bilinear forms by a combination of the graph-theoretical approach with the gate-elimination method. The hope to prove nonlinear lower bounds for the non-monotone complexity of Boolean functions using graph theoretical arguments only has been destroyed by the construction of superconcentrators of linear size [42]. In 1981, Blum [9] has introduced the technique of normal form transformation in Boolean complexity for proving an $\Omega(n^{4/3})$ lower bound for the number of $\wedge$-gates in any monotone network computing the nth degree convolution. Starting with any monotone network β_0 computing the nth degree convolution, the network is transformed into a normal form network β_1 which computes a number of subfunctions of the convolution. The normal form transformation enlarges the number of $\wedge$-gates at most by a constant factor. During the transformation, some $\wedge$-gates of β_0 are counted. If after the normal form transformation the amount of counted $\wedge$-gates is not large enough, an application of the gate-elimination method proves the desired lower bound. In 1982, using the gate-elimination method and some information flow arguments Weiß [48] has proved an $\Omega(n^{3/2})$ lower bound for the number of $\vee$-gates in any monotone network for the computation of the Boolean convolution. Using the technique in [9] in connection with a little bit more sophisticated application of the gate-elimination method one can also prove an $\Omega(n^{3/2})$ lower bound for the number of $\wedge$-gates in any monotone network for the Boolean convolution. Although it is widely believed that the optimal monotone network for the computation of the nth degree convolution contains n^2 $\wedge$-gates and $n^2 - n$ $\vee$-gates, no lower bound better than $\Omega(n^{3/2})$ for the monotone network complexty has been proved so far. I conjecture that a combination of the methods in [9], [48] and [46] might lead to a proof of an $\Omega((\frac{n}{\log n})^2)$ lower bound.

Although since 1969 nonlinear lower bounds for the monotone network complexity for explicit functions in $\mathcal{M}_{n,m}$ where $m = \theta(n)$ has been proved, the best lower bound for the monotone network complexity of an explicit function in $\mathcal{M}_n$ before 1985 was of size $4n$ [41]. All nonlinear lower bound proofs for the monotone network complexity of functions in $\mathcal{M}_{n,m}$ strongly depend on the fact that a set of functions has to be computed. With respect to single output monotone Boolean functions, no technique for counting a nonlinear number of gates has been developed before 1985. In 1985, Razborov [31, 32] and Andreev [4] have succeeded to get the breakthrough. Razborov has developed the so-called "method of approximation" and given an $n^{\Omega(\log n)}$ lower bound for the monotone complexity of the clique function and a lower bound of the same size for the perfect matching function. Nearly at the same time, Andreev used different but similar methods for proving an exponential lower bound for another monotone function in NP. Some months later, Alon and Boppana [1] have strengthened the combinatorial arguments of Razborov and proved an exponential lower bound for the clique function. We will sketch the approximation method as developed by Razborov.

$P(\{0,1\}^n)$ denotes the power set of $\{0,1\}^n$. Note that $P(\{0,1\}^n)$ with the operations $\cup$ and $\cap$ is a lattice. For a function $f \in \mathcal{M}_n$ let $A(f) := \{a \in \{0,1\}^n \mid f(a) = 1\}$. Note that $A(0) = \emptyset$ and $A(1) = \{0,1\}^n$. Furthermore, for $f, g \in \mathcal{M}_n$, $A(f \vee g) = A(f) \cup A(g)$ and $A(f \wedge g) = A(f) \cap A(g)$. Given any monotone Boolean network β for a function $f \in \mathcal{M}_n$, we obtain a network β' which computes $A(f)$ if we replace each input x_i, $1 \le i \le n$ by $A(x_i)$, each $\wedge$-gate by an $\cap$-operation and each $\vee$-gate by an $\cup$-operation. Razborov's idea was to replace in β' the operations $\cap$ and $\cup$ by two operations $\sqcap$ and $\sqcup$ which have the property that $M \sqcap N \subseteq M \cap N$ and $M \cup N \subseteq M \sqcup N$. After doing this, the network does not compute $A(f)$ but an approximation of $A(f)$. Given the two operations $\sqcup$ and $\sqcap$, we define the *legitimate model* $\mathcal{S}$ to be the smallest subset of $P(\{0,1\}^n)$ such that

1. $A(0), A(1), A(x_1), A(x_2), \ldots, A(x_n) \in \mathcal{S}$ and
2. $\mathcal{S}$ is closed under the operations $\sqcup$ and $\sqcap$.

For $M, N \in \mathcal{S}$ let

$$\delta_{\sqcup}(M, N) := (M \sqcup N) \setminus (M \cup N) \quad \text{and} \quad \delta_{\sqcap}(M, N) := (M \cap N) \setminus (M \sqcap N).$$

For $f \in \mathcal{M}_n$ and the legitimate model $\mathcal{S}$, we define the *distance* $\rho(f, \mathcal{S})$ from f to $\mathcal{S}$ to be the minimal t such that there are $M, M_1, N_1, M_2, N_2, \ldots, M_t, N_t \in \mathcal{S}$ such that

$$A(f) \subseteq M \cup \bigcup_{i=1}^{t} \delta_{\sqcap}(M_i, N_i) \quad \text{and} \quad M \subseteq A(f) \cup \bigcup_{i=1}^{t} \delta_{\sqcup}(M_i, N_i).$$

The distance $\rho(f, \mathcal{S})$ from f to $\mathcal{S}$ is a lower bound for the monotone network complexity of f. To see this, we consider any monotone Boolean network β computing f. Let $g_1, g_2, \ldots, g_t$ be the gates in β numbered in a topological order. Consider the network β' which we obtain from β by replacing each $\vee$ by $\sqcup$, each $\wedge$ by $\sqcap$, each 0 by $A(0)$ and each 1 by $A(1)$. The network β' computes elements of $\mathcal{S}$. Let M_i, N_i, $1 \le i \le t$, be the elements of $\mathcal{S}$ computed at the inputs of the gate g_i in β', and let M be the element of $\mathcal{S}$ computed at the output gate of β'. Then it is easy to prove by induction that

$$A(f) \subseteq M \cup \bigcup_{i=1}^{t} \delta_{\sqcap}(M_i, N_i) \quad \text{and} \quad M \subseteq A(f) \cup \bigcup_{i=1}^{t} \delta_{\sqcup}(M_i, N_i).$$

Hence, the size of β is a upper bound for the distance $\rho(f, \mathcal{S})$ from f to $\mathcal{S}$. Note that distance measure $\rho(f, \mathcal{S})$ is very strong. All elements of $\mathcal{S}$ are given for free such that it is not required that the approximating sets M, M_i and N_i, $1 \le i \le t$, can be computed by a $\{\sqcup, \sqcap\}$-network β'. The proof sketched above also works with respect to the following weaker distance measure:

For $f \in \mathcal{M}_n$ and the legitimate model $\mathcal{S}$, we define the *weak distance* $\rho'(f, \mathcal{S})$ from f to $\mathcal{S}$ to be the minimal t such that there is a $\{\sqcup, \sqcap\}$-network β' with gates $g_1, g_2, \ldots, g_t$ numbered in a topological order where $M_i, N_i, 1 \leq i \leq t$, are the elements of $\mathcal{S}$ computed at the inputs of gate g_i and M is the element of $\mathcal{S}$ computed at the output gate g_t such that

$$A(f) \subseteq M \cup \bigcup_{i=1}^{t} \delta_{\sqcap}(M_i, N_i) \quad \text{and} \quad M \subseteq A(f) \cup \bigcup_{i=1}^{t} \delta_{\sqcup}(M_i, N_i).$$

Note that $\rho(f, \mathcal{S})$ is a lower bound of $\rho'(f, \mathcal{S})$ but not vice versa. Obviously, the weak distance $\rho'(f, \mathcal{S})$ from f to $\mathcal{S}$ is also a lower bound for the monotone complexity of f. The idea now is to choose appropriate operations $\sqcup$ and $\sqcap$ such that $\rho'(f, \mathcal{S})$ is large with respect to the considered monotone function f.

In the following years, generalizations of the approximation method [19, 49] and seemingly other methods [15, 18, 2, 6, 16] have been developed. Particularly the so-called "bottleneck counting method" introduced by Haken [15] in 1995 has become popular. Using the bottleneck counting method, simplified proofs of known lower bounds [6, 2] and better lower bounds [16] have been obtained. In 1997, Simon and Tsai [38] have proved the equivalence of the bottleneck counting method and the approximation method with respect to the weak distance measure.

3   From Monotone to Non-monotone Complexity

As mentioned above, in [32] Razborov has proved an $n^{\Omega(\log n)}$ lower bound for the monotone complexity of the perfect matching function. Since a maximum matching of a graph can be computed in polynomial time, an Ω_0-network of polynomial size for the perfect matching function exists. Hence, the gap between monotone and non-monotone network complexity is at least $n^{\Omega(\log n)}$. In 1986, Tardos [40] has shown that this gap is indeed exponential. Hence, it is not always possible to obtain large lower bound for the Ω_0-complexity from a large lower bound of the monotone complexity of a monotone Boolean function. But this does not exclude the possibility that techniques developed for the proof of lower bounds for the monotone complexity can also be useful for the proof of lower bounds for the Ω_0-complexity of Boolean functions.

Given any Ω_0-network β, we can convert β to an equivalent Ω_0-network β' where all negations occur only at the input nodes. Moreover, the size of β is at most doubled. For doing this, we start at the output nodes and apply deMorgan rules for bringing the negations to the inputs. Since gates can be simultaneously negated and not negated, some gates have to be doubled. The resulting network is a so-called *standard network* where only variables are negated. The *standard complexity* of a function $f \in \mathcal{B}_{n,m}$ is the size of a smallest standard network which computes f. Note that the standard and the Ω_0-complexity of a function f differs at most by the factor two. Hence, for proving nonlinear lower bounds for the Ω_0-complexity of Boolean functions we can restrict us to the consideration of standard networks.

Do there exist explicit monotone Boolean functions where negations are almost powerless? Already in 1982, Berkowitz [7] has given an affirmative answer to this question. The idea is to replace in a standard network for a function $f \in \mathcal{M}_{n,m}$ each negated variable $\neg x_i$ by a function $h_i \in \mathcal{M}_n$ without changing the function computed by the network. Then h_i is called a *pseudo-complement* for x_i with respect to f. Since $\neg x_i$ is not monotone, it is not clear that functions in $\mathcal{M}_{n,m}$ with pseudo-complements exist. A function $f \in \mathcal{B}_n$ is a *k-slice* if $f(x) = 0$ for all $x \in \{0,1\}^n$ with less than k ones and $f(x) = 1$ for all inputs $x \in \{0,1\}^n$ with more than k ones. $f = (f_1, f_2, \ldots, f_m) \in \mathcal{B}_{n,m}$ is a *k-slice* if f_i, $1 \leq i \leq m$, are k-slices. Obviously, slice functions are monotone. Only for inputs with exactly k ones, a k-slice might be nontrivial. Berkowitz has proved that pseudo-complements for x_i with respect to a k-slice f can be obtained using threshold functions. Threshold functions are monotone and all threshold functions needed for the replacements of the negated variables in a standard network for f can be realized by a monotone network of size $O(n^2 \log n)$. Valiant [43] has improved the obvious upper bound to $O(n \log^2 n)$. The proof of a larger lower bound than $\Omega(n \log^2 n)$ for a slice function in $\mathcal{M}_{n,m}$ would imply a nonlinear lower bound for the $\mathcal{B}_2$-complexity of the same function. The trick with respect to the pseudo-complement for x_i for k-slices is that for inputs with exactly k ones, the value of the pseudo-complement is equal to $\neg a_i$ where a_i is the value assigned to the variable x_i. If the number of ones in the input is not equal to k then the considered function is trivial. On the kth slice of $\{0,1\}^n$, everything what we can do with $\neg x_i$ we can also do with the pseudo-complement for x_i. Hence, I believe that the consideration of slice functions does not help for proving a nonlinear lower bound for the Ω_0-complexity of a Boolean function. It is better to consider standard networks directly.

In a standard network we can replace the negated inputs $\neg x_1, \neg x_2, \ldots, \neg x_n$ by a network with inputs $x_1, x_2, \ldots, x_n$ and outputs $\neg x_1, \neg x_2, \ldots, \neg x_n$. We call such a network which negates n given variables an *inverter* I_n. Already 1958, Markov [22] has shown that inverters I_n using $\lceil \log(n+1) \rceil$ negations exit and that this number of negations is also necessary. He has not considered the complexity of the constructed inverter. In 1974, Fischer has constructed an inverter I_n of size $O(n^2 \log^2 n)$ and depth $O(\log^2 n)$ using $\lceil \log(n+1) \rceil$ negations. This has been improved to size $O(n \log n)$ and depth $O(\log n)$ by Beals, Nishino and Tanaka [5]. This suggests the consideration of negation-limited network complexity. Indeed, for the clique function Amano and Maruoka [3] have proved a nonpolynomial lower bound for Ω_0-networks using only $\Omega(\log \log n)$ negations. Nevertheless, for the same reasons why I believe that the consideration of slice functions does not help, I prefer to consider standard networks directly.

Next we will extend the approximation method as done by Razborov in [33]. For doing this we restrict us to standard networks. Given the two operations $\sqcup$ and $\sqcap$, we define the *legitimate model* S to be the smallest subset of $P(\{0,1\}^n)$ such that

1. $A(0), A(1), A(x_1), A(x_2), \ldots, A(x_n), A(\neg x_1), A(\neg x_2), \ldots, A(\neg x_n) \in S$ and
2. S is closed under the operations $\sqcup$ and $\sqcap$.

For $M, N \in \mathcal{S}$ we do not need that $M \sqcap N \subseteq M \cap N$ or $M \cup N \subseteq M \sqcup N$. Hence, we define more generally

$$\delta_{\sqcup}^{+}(M, N) := (M \sqcup N) \setminus (M \cup N) \quad \text{and} \quad \delta_{\sqcup}^{-}(M, N) := (M \cup N) \setminus (M \sqcup N)$$

and

$$\delta_{\sqcap}^{+}(M, N) := (M \sqcap N) \setminus (M \cap N) \quad \text{and} \quad \delta_{\sqcap}^{-}(M, N) := (M \cap N) \setminus (M \sqcap N).$$

For $f \in \mathcal{B}_n$ and the legitimate model $\mathcal{S}$, we define the *distance* $\rho(f, \mathcal{S})$ from f to $\mathcal{S}$ to be the minimal t such that there are $M \in \mathcal{S}$ and triples $\langle op_1, M_1, N_1 \rangle$, $\langle op_2, M_2, N_2 \rangle, \ldots, \langle op_t, M_t, N_t \rangle$ with $op_i \in \{\sqcup, \sqcap\}$, $M_i, N_i \in \mathcal{S}$ such that

$$A(f) \subseteq M \cup \bigcup_{i=1}^{t} \delta_{op_i}^{-}(M_i, N_i) \quad \text{and} \quad M \subseteq A(f) \cup \bigcup_{i=1}^{t} \delta_{op_i}^{+}(M_i, N_i).$$

Exactly as in the monotone case we can prove that $\rho(f, \mathcal{S})$ is a lower bound for the standard complexity of f. Note that with respect to non-monotone approximations, the distance measure $\rho(f, \mathcal{S})$ is also very strong. All elements of $\mathcal{S}$ are given for free such that it is not required that the approximating sets M, M_i and N_i, $1 \leq i \leq t$, can be computed by a $\{\sqcup, \sqcap\}$-standard network β'. Analogously to the monotone case, we can define the following weaker distance measure:

For $f \in \mathcal{B}_n$ and the legitimate model $\mathcal{S}$, we define the *weak distance* $\rho'(f, \mathcal{S})$ from f to $\mathcal{S}$ to be the minimal t such that there is a $\{\sqcup, \sqcap\}$-standard network β' with gates $g_1, g_2, \ldots, g_t$ numbered in a topological order where $M_i, N_i, 1 \leq i \leq t$, are the elements of $\mathcal{S}$ computed at the inputs of gate g_i, op_i is the operation of gate g_i, and M is the element of $\mathcal{S}$ computed at the output gate g_t such that

$$A(f) \subseteq M \cup \bigcup_{i=1}^{t} \delta_{op_i}^{-}(M_i, N_i) \quad \text{and} \quad M \subseteq A(f) \cup \bigcup_{i=1}^{t} \delta_{op_i}^{+}(M_i, N_i).$$

Note that $\rho(f, \mathcal{S})$ is a lower bound of $\rho'(f, \mathcal{S})$ but not vice versa. Obviously, the weak distance $\rho'(f, \mathcal{S})$ from f to $\mathcal{S}$ is a lower bound for the standard complexity of f. The idea now is to choose appropriate operations $\sqcup$ and $\sqcap$ such that $\rho'(f, \mathcal{S})$ is large with respect to the considered function f.

In 1989, Razborov [33] has shown that the largest lower bound which can be obtained with the approximation method using the distance measure ρ for a Boolean function in $\mathcal{B}_n$ is of size $\Omega(n^2)$. The proof given in [33] uses the fact that all elements of the model $\mathcal{S}$ are given for free. Hence, the proof cannot be applied to the approximation method which uses the weak distance measure ρ'. Therefore, it might be possible to prove a superpolynomial lower bound for a Boolean function using the approximation method with the weak distance measure.

In 1994, Razborov and Rudich [34] have introduced the notion of "natural proof". They have shown that natural proofs cannot be used for separating P from NP unless hard pseudorandom number generators do not exist. It seems to me that this famous result has discouraged a lot of researchers to work hardly on proving lower bounds for the network complexity of Boolean functions. Next, we will sketch the result of Razborov and Rudich.

A *combinatorial property* is a subset $\{C_n \subset \mathcal{B}_n \mid n \in \mathbb{N}\}$ of Boolean functions. C_n is called *natural* if there is a stronger property $C_n^* \subseteq C_n$ which satiesfies:

1. For all $f \in \mathcal{B}_n$ it can be decided in $2^{O(n)}$ time if $f \in C_n^*$. (*constructiveness*)
2. $|C_n^*| \geq 2^{-O(n)}|\mathcal{B}_n|$. (*largeness*)

The first property means that the characteristic function of C_n^* can be computed in polynomial time in the size of the truth table of the input function $f \in \mathcal{B}_n$. The second property says that a function randomly chosen from $\mathcal{B}_n$ is contained in C_n^* with non-negligible probability. $P/poly$ is the set of languages which are recognizable by a family of Boolean networks of polynomial size. Note that $P \subseteq P/poly$. A combinatorial property is *useful against $P/poly$* if the network complexity of any sequence $f_1, f_2, \ldots, f_n, \ldots$ where $f_n \in C_n$ is superpolynomial; i.e., for all $k \in \mathbb{N}$ there is $n_k \in \mathbb{N}$ such that the network complexity of f_n is greater than n^k for all $n > n_k$. A proof that a Boolean function does not have polynomial network complexity is *natural against $P/poly$* if the proof uses a natural combinatorial property C_n which is useful against $P/poly$. The fundamental result of Razborov and Rudich is the proof that any large and constructive C_n which separates P from NP would imply that 2^{n^ε}-hard pseudorandom number generators do not exist. It is a widely believed conjecture that such pseudorandom number generators exist. They show that almost all non-relativizing, non-monotone and superlinear lower bounds proved up to that time are natural. They also mention that strong lower bound proofs for the monotone network complexity are not natural. Recently, Chow [11] has defined "almost-natural proofs" by weakening the largeness condition slightly and proved that almost-natural and useful properties exist.

The result of Razborov and Rudich has no discouraging influence to me for the following reasons: Bounding the depth of the network to be constant seems to be a much harder restriction than allowing only monotone networks. Most natural proofs are for constant-depth Boolean networks. All natural proofs are with respect to Boolean functions of small complexity. For separating P and NP, the techniques developed for proving exponential lower bounds for the monotone network complexity of the characteristic function of an NP-complete problem seems to be more suitable than techniques developed for proving exponential lower bounds for the constant-depth network complexity of a Boolean function of small complexity. The result of Razborov [33] does not exclude the approximation method with weak distance measure or the bottleneck counting method. These methods are not natural.

Understanding the power of negations is one of the most challenging problems in complexity theory. Even when someone would prove a superpolynomial lower bound for the Ω_0-complexity of a Boolean function in NP solving the famous P versus NP-problem, much would remain open with respect to the understanding of the power of negations. Can we multiply two integers in linear time or can we prove an $\Omega(n \log n)$ lower bound for the Ω_0-complexity of the multiplication of two n-bits numbers? What is the Ω_0-complexity of the nth degree convolution or of the Boolean matrix multiplication? Note that these are Boolean functions with many outputs. For working on such problems, techniques developed for proving small lower bounds for the monotone complexity remain to be of interest.

References

1. Alon, N., Boppana, R.B.: The monotone circuit complexity of Boolean functions. Combinatorica 7, 1–22 (1987)
2. Amano, K., Maruoka, A.: The potential of the approximation method. SIAM J. Comput. 33, 433–447 (2004)
3. Amano, K., Maruoka, A.: A superpolynomial lower bound for a circuit computing the clique function with at most $(1/6)\log\log N$ negation gates. SIAM J. Comput. 35, 201–216 (2005)
4. Andreev, A.E.: On a method for obtaining lower bounds for the complexity of individual monotone functions. Soviet Math. Dokl. 31, 530–534 (1985)
5. Beals, R., Nishino, T., Tanaka, K.: On the complexity of negation-limited Boolean networks. SIAM J. Comput. 27, 1334–1347 (1998)
6. Berg, C., Ulfberg, S.: Symmetric approximation arguments for monotone lower bounds without sunflowers. Comput. Complex. 8, 1–20 (1999)
7. Berkowitz, S.J.: On some relationships between monotone and non-monotone circuit complexity, Tech. Report, Comput. Sci. Dept., Univ. of Toronto (1982)
8. Blum, N.: A Boolean function requiring $3n$ network size. TCS 28, 337–345 (1984)
9. Blum, N.: An $\Omega(n^{4/3})$ lower bound on the monotone network complexity of the nth degree convolution. TCS 36, 59–69 (1985)
10. Brown, W.G.: On graphs that do not contain a Thompson graph. Canad. Math. Bull. 9, 281–285 (1966)
11. Chow, T.Y.: Almost-natural proofs. In: Proc. 49th FOCS, pp. 72–77 (2008)
12. Clote, P., Kranakis, E.: Boolean Functions and Computation Models. Springer, Heidelberg (2002)
13. Dunne, P.E.: The Complexity of Boolean Networks. Academic Press, London (1988)
14. Fischer, M.J.: The complexity of negation-limited networks - a brief survey. In: Brakhage, H. (ed.) GI-Fachtagung 1975. LNCS, vol. 33, pp. 71–82. Springer, Heidelberg (1975)
15. Haken, A.: Counting bottlenecks to show monotone $P \neq NP$. In: Proc. 36th FOCS, pp. 36–40 (1995)
16. Harnik, D., Raz, R.: Higher lower bounds on monotone size. In: Proc. 32nd STOC, pp. 191–201 (2000)
17. Iwama, K., Lachish, O., Morizumi, H., Raz, R.: An explicit lower bound of $5n - o(n)$ for Boolean circuits (manuscript, 2005)
18. Jukna, S.: Combinatorics of monotone computations. Combinatorica 19, 65–85 (1999)
19. Karchmer, M.: On proving lower bounds for circuit size. In: Proc. 8th Structure in Complexity Theory, pp. 112–118 (1993)
20. Kővári, T., Sós, V.T., Turán, P.: On a problem of K. Zarankiewicz. Colloq. Math. 3, 50–57 (1954)
21. Lamagna, E.A.: The complexity of monotone networks for certain bilinear forms, routing problems, sorting and merging. IEEE Trans. Comput. 28, 773–782 (1979)
22. Markov, A.A.: On the inversion complexity of a system of functions. J. ACM 5, 331–334 (1958)
23. Mehlhorn, K., Galil, Z.: Monotone switching circuits and Boolean matrix product. Computing 16, 99–111 (1976)
24. Mehlhorn, K.: Some remarks on Boolean sums. Acta Inform. 12, 371–375 (1979)
25. Neciporuk, E.I.: On a Boolean matrix. Systems Theory Res. 21, 236–239 (1971)

26. Paterson, M.S.: Complexity of monotone networks for Boolean matrix product. TCS 1, 13–20 (1975)

27. Paul, W.J.: A $2.5n$ lower bound on the combinational complexity of Boolean functions. SIAM J. Comput. 6, 427–443

28. Pippenger, N.: On another Boolean matrix. TCS 11, 49–56 (1980)

29. Pippenger, N., Valiant, L.G.: Shifting graphs and their applications. J. ACM 23, 423–432

30. Pratt, V.R.: The power of negative thinking in multiplying Boolean matrices. SIAM J. Comput. 4, 326–330 (1974)

31. Razborov, A.A.: Lower bounds on the monotone complexity of some Boolean functions. Soviet Math. Dokl. 31, 354–357 (1985)

32. Razborov, A.A.: A lower bound on the monotone network complexity of the logical permanent. Math. Notes Acad. Sci. USSR 37, 485–493 (1985)

33. Razborov, A.A.: On the method of approximation. In: Proc. 21st STOC, pp. 167–176 (1989)

34. Razborov, A.A., Rudich, S.: Natural proofs. JCSS 55, 24–35 (1997)

35. Savage, J.E.: Models of Computation: Exploring the Power of Computing. Addison-Wesley, Reading (1998)

36. Schnorr, C.P.: Zwei lineare untere Schranken für die Komplexität Boolescher Funktionen. Computing 13, 155–171 (1974)

37. Shannon, C.E.: The synthesis of two-terminal switching circuits. Bell Syst. Techn. J. 28, 59–98 (1949)

38. Simon, J., Tsai, S.-C.: On the bottleneck counting argument. TCS 237, 429–437 (2000)

39. Stockmeyer, L.: On the combinational complexity of certain symmetric Boolean functions. Math. Systems Theory 10, 323–336 (1977)

40. Tardos, É.: The gap between monotone and non-monotone circuit complexity is exponential. Combinatorica 8, 141–142 (1988)

41. Tiekenheinrich, J.: A $4n$ lower bound on the monotone Boolean complexity of a one output Boolean function. IPL 18, 201–202 (1984)

42. Valiant, L.G.: Graph-theoretic properties in computational complexity. JCSS 13, 278–285 (1976)

43. Valiant, L.G.: Negation is powerless for Boolean slice functions. SIAM J. Comput. 15, 531–535 (1986)

44. Wegener, I.: Switching functions whose monotone complexity is nearly quadratic. TCS 9, 83–97 (1979)

45. Wegener, I.: A new lower bound on the monotone network complexity of Boolean sums. Acta Informatica 13, 109–114 (1980)

46. Wegener, I.: Boolean functions whose monotone complexity is of size $n^2/\log n$. TCS 21, 213–224 (1982)

47. Wegener, I.: The Complexity of Boolean Functions. Wiley-Teubner, Chichester (1987)

48. Weiß, J.: An $n^{3/2}$ lower bound on the monotone network complexity of the Boolean convolution. Information and Control 59, 184–188 (1983)

49. Widgerson, A.: The fusion method for lower bounds in circuit complexity. In: Combinatorics, Paul Erdős is eighty. Elsevier, Amsterdam (1993)

50. Zwick, U.: A $4n$ lower bound on the combinational complexity of certain symmetric Boolean functions over the basis of unate dyadic Boolean functions. SIAM J. Comput. 20, 499–505 (1991)

The Lovász Local Lemma and Satisfiability[*]

Heidi Gebauer, Robin A. Moser, Dominik Scheder, and Emo Welzl

Institute of Theoretical Computer Science, ETH Zürich
CH-8092 Zürich, Switzerland

Abstract. We consider boolean formulas in conjunctive normal form (CNF). If all clauses are large, it needs many clauses to obtain an unsatisfiable formula; moreover, these clauses have to interleave. We review quantitative results for the amount of interleaving required, many of which rely on the Lovász Local Lemma, a probabilistic lemma with many applications in combinatorics.

In positive terms, we are interested in simple combinatorial conditions which guarantee for a CNF formula to be satisfiable. The criteria obtained are nontrivial in the sense that even though they are easy to check, it is by far not obvious how to compute a satisfying assignment efficiently in case the conditions are fulfilled; until recently, it was not known how to do so. It is also remarkable that while deciding satisfiability is trivial for formulas that satisfy the conditions, a slightest relaxation of the conditions leads us into the territory of NP-completeness.

Several open problems remain, some of which we mention in the concluding section.

1 Introduction

SAT, the problem of deciding whether a boolean formula in conjunctive normal form (CNF) is satisfiable by a truth assignment, is the classical NP-complete problem. Such a *CNF formula* is obtained as a conjunction of clauses, where a *clause* is the disjunction of literals, with a *literal* either a boolean variable or its negation; we require that variables in a clause do not repeat (neither with the same nor complementary signs). A CNF formula is *satisfiable* if there is a true-false assignment to the variables so that every clause has at least one literal that evaluates to true. Consider e.g.

$$(x_1 \vee x_2 \vee x_3) \wedge (x_1 \vee x_3 \vee \overline{x_4}) \wedge (\overline{x_1} \vee \overline{x_2} \vee x_4) \wedge (x_2 \vee \overline{x_3} \vee x_4) \wedge (\overline{x_2} \vee \overline{x_3} \vee \overline{x_4}),$$

a 3-CNF formula with 5 clauses over the variables $\{x_1, x_2, x_3, x_4\}$ (for k a nonnegative integer, a *k-CNF formula* is a CNF formula where every clause contains *exactly* k literals). This formula is satisfiable, e.g. by the assignment $(x_1, x_2, x_3, x_4) \mapsto (\text{true}, \text{true}, \text{false}, \text{true})$. But, actually, it can be recognised as satisfiable even without any close inspection, simply because 5, the number of clauses, is less than 2^3. This is because a simple probabilistic argument for instance demonstrates that

---

[*] Research is supported by SNF Grant 200021-118001/1.

S. Albers, H. Alt, and S. Näher (Eds.): Festschrift Mehlhorn, LNCS 5760, pp. 30–54, 2009.
© Springer-Verlag Berlin Heidelberg 2009

it needs at least 2^k clauses to construct an unsatisfiable k-CNF formula.

For, suppose that some k-CNF formula has fewer than 2^k clauses, then an assignment sampled uniformly at random violates each clause with probability 2^{-k} and, by linearity of expectation, the expected total number of violated clauses is then smaller than 1, implying that some of the assignments have to satisfy the whole formula. While this result may reveal some of the beauty of probabilistic reasoning (cf. [1]), it is not very striking in its own. But let us discover that it can be extended to yield something much more powerful.

The statement becomes miraculous as soon as we observe that the constraint on the formula size need not only be satisfied globally but even *locally*. What do we mean by global and local? Suppose you have a formula of arbitrary size. Now pick any of its clauses, say C. We will say that the *neighbourhood of C*, denoted by $\Gamma(C)$, is the set of clauses that share variables with C. These are, in a sense, those clauses that relate to C, since, if we have C violated by some given assignment and change some values of variables within it to remedy that problem, then the clauses in $\Gamma(C)$ are exactly the ones we might harm. Now our intuition suggests the following: if we can change values in a clause C without causing too much damage in its surroundings and if this local property holds everywhere, then most probably we can find a globally satisfying assignment by just moving around violation issues until they disappear. And this intuition proves to be absolutely correct. In order to construct an unsatisfiable k-CNF formula, not only do we need at least 2^k clauses in total, but those clauses need to be, at least somewhere in the formula, concentrated densely around some clause. For one can prove the following:

If every clause in a k-CNF formula, $k \geq 1$, has a neighbourhood of size at most $2^k/e - 1$, then the whole formula admits a satisfying assignment.

This statement is known as the *Lovász Local Lemma* from 1975 ([2], cf. [1]), formulated in terms of satisfiability. Before we present two proofs in the next section, let us discuss other variants of the theme

> *"In an unsatisfiable CNF formula clauses have to interleave –
> the larger the clauses, the more interleaving is required."*

First, it is clear that clauses sharing variables of the same sign will not get us in major trouble in a search for a satisfying assignment. To reflect this, we define the *conflict-neighbourhood of a clause C* in a CNF formula as the set of clauses which share variables with C, at least one with opposite sign. The so-called *lopsided Local Lemma* shows that the above mentioned condition for neighbourhoods holds actually for conflict-neigbourhoods. As an aside we cannot resist mentioning the fact that if each pair of clauses in a CNF formula either has no conflict or a conflict along at least two variables, then this formula is satisfiable, unless it contains the empty clause. For a reader familiar with resolution, the mystery can be resolved instantaneously: Try resolution!

Second, it is easily seen that a clause with a large neighbourhood requires some variable to occur often in a formula. To make this precise, we call the number of occurrences of a variable x (with either sign) in a CNF formula the *degree of x*. Then we have:

If every variable in a k-CNF formula, $k \geq 1$, has degree at most $2^k/(ek)$, then the formula is satisfiable.

Is $2^k/(ek)$ tight? While we do not believe this to be true, one can show that it cannot be increased by more than a constant factor. This holds also for the previously mentioned bounds for the (conflict-)neighbourhood size, but while this is certified by the simple example of a k-CNF formula that contains all possible 2^k clauses over a given set of k variables, the degree bound requires a more elaborate construction and therefore this had been open for some time.

Third, what can be said if we constrain the quality of interleaving rather than the quantity? For this we consider *linear*[1] *CNF formulas*, i.e. CNF formulas where any two clauses share at most one variable. Here is an example of a linear 2-CNF formula:

$$(\overline{y_1} \vee \overline{y_2}) \wedge (y_1 \vee x) \wedge (y_2 \vee x) \wedge (z_1 \vee \overline{x}) \wedge (z_2 \vee \overline{x}) \wedge (\overline{z_1} \vee \overline{z_2}).$$

Since the first half of the formula forces x to be true in a satisfying assignment and the second forces it to be false, the formula is not satisfiable; it is the smallest unsatisfiable linear 2-CNF formula. Unsatisfiable linear k-CNF formulas can be constructed for all k, although their size needs to grow faster than 2^k, again a fact whose proof falls back on the Local Lemma.

Any linear k-CNF formula with at most $4^k/(4e^2k^3)$ clauses is satisfiable.

We will see that the bound in the condition is tight up to a polynomial factor. Via a probabilistic argument $8k^3 4^k$ clauses can be shown to suffice for unsatisfiability; the best explicit construction we know, however, delivers formulas of tower-like size (2 to the 2 to the 2 ... k times).

Algorithms, finally: Whenever the easily checkable conditions formulated above are satisfied, then the algorithmic problem of deciding satisfiability becomes trivial. However, whenever the Local Lemma is invoked, it is by no means obvious how to actually *construct* a satisfying assignment. This tantalising fact was resolved only recently via a randomised local repair algorithm as indicated above. We will present and analyse this method in the next section.

We return to deciding satisfiability. For k a positive integer, let us define $f(k)$ as the largest integer, so that every k-CNF formula with no variable of degree exceeding $f(k)$ is satisfiable; we know that $f(k) = \Theta(2^k/k)$. Clearly, satisfiability of k-CNF formulas with maximum variable degree at most $f(k)$ is trivially decidable in polynomial time. We might hope that slight violation of the bound may still allow for an efficient decision procedure. However, one can show that, provided $k \geq 3$, even for k-CNF formulas with max-degree at most $f(k) + 1$ the satisfiability problem becomes already NP-complete. This *sudden*

---

[1] The term "linear" is borrowed from hypergraph theory, where this must have been inspired by the behaviour of lines.

jump behaviour in complexity at $f(k)$ can be shown, although $f(k)$ is not known for k exceeding 4 (it is not even known whether the function f is computable). A similar immediate transition from trivial to NP-complete can be observed for the related problem for the conflict-neighbourhood size.

The remainder of this paper will treat the topic outlined above in more detail, mostly with proofs. We will also supply references and more of the historical background of the developments to today's state of knowledge.

Notation. We will assume (and have assumed) some familiarity with basic notions for boolean formulas in propositional logic and in discrete mathematics. Still, for the remaining more technical treatment, we want to clarify some notation and terminology. We like to regard clauses as sets of literals, formulas as sets of clauses. Let us actually go through a succinct recapitulation of our set-up: Given a set V of boolean variables, we set $\overline{V} := \{\overline{x} \mid x \in V\}$ and call the elements of $V \cup \overline{V}$ *literals over* V with V the *positive literals* and $\overline{V}$ the *negative literals*. A *clause* C *over* V is a set of literals over V with no pair x and $\overline{x}$ appearing simultaneously. A *CNF formula* F *over* V is a set of clauses; if all clauses in F have the same cardinality k, we call F a *k-CNF formula*. Although we regard formulas and clauses as sets, we sometimes return to the logic notation, writing $F \wedge C$ (instead of $F \cup \{C\}$) or even $F \wedge \neg C$ or similar.

An *assignment* α over variable set V is a mapping $\alpha : V \to \{0, 1\}$ that extends to $\overline{V}$ via $\alpha(\overline{x}) := 1 - \alpha(x)$ for $x \in V$ (1 for "true," 0 for "false"). α *satisfies* a clause if at least one of its literals evaluates to 1 under α. And α *satisfies* a CNF formula if it satisfies all of its clauses. A CNF formula is *satisfiable* if a satisfying assignment exists.

We denote the set of variables that occur in a clause C by $\mathrm{vbl}(C)$; for a CNF formula F, $\mathrm{vbl}(F) := \bigcup_{C \in F} \mathrm{vbl}(C)$. For a clause $C = \{u_1, u_2, \ldots, u_k\}$, we write $\overline{C} := \{\overline{u_1}, \overline{u_2}, \ldots, \overline{u_k}\}$ (note $\overline{C} \neq \neg C$, unless $k = 1$). The *neighbourhood of a clause* C in a CNF formula F is defined by $\Gamma(C) = \Gamma_F(C) := \{D \in F \mid \mathrm{vbl}(D) \cap \mathrm{vbl}(C) \neq \emptyset\}$. Analogously, the *conflict-neighbourhood of* C is $\Gamma'(C) = \Gamma'_F(C) := \{D \in F \mid C \cap \overline{D} \neq \emptyset\}$. The *degree of a variable* x in a *CNF formula* F is set to $\deg(x) = \deg_F(x) := |\{C \in F \mid x \in \mathrm{vbl}(C)\}|$.

We have already encountered $f(k)$, which we defined to be the largest integer d such that every k-CNF formula with maximum variable degree at most d is satisfiable. Similarly, let $l(k)$ be the largest integer d such that every k-CNF formula F for which $|\Gamma_F(C)| \leq d$, for all $C \in F$, is satisfiable. Let $lc(k)$ be defined analogously, but with $|\Gamma'_F(C)| \leq d$, for all $C \in F$, instead.

A *hypergraph* H is a pair (V, E) with V a finite set and $E \subseteq 2^V$; it is *k-uniform* if $|e| = k$ for all $e \in E$. H is called *2-colourable* (or has *property B*) if there is a colouring of the vertices in V by red and blue so that no hyperedge in E is monochromatic. Extremal problems for 2-colourable hypergraphs have been considered since Erdős' papers [3, 4] in 1963. They relate to satisfiability of CNF formulas in that $H = (V, E)$ is 2-colourable iff the CNF formula $E \cup \{\overline{e} \mid e \in E\}$, with V now considered as set of boolean variables, is satisfiable. And, therefore, they will make their appearance during this presentation.

2   Local Lemma in Terms of SAT – Proof and Algorithm

Theorem 1. *Let $k \in \mathbb{N}$ and let F be a k-CNF formula. If $|\Gamma(C)| \leq 2^k/e - 1$ for all $C \in F$, then F is satisfiable.*

The statement was first formulated in the famous paper [2] by Erdős and Lovász, in terms of its application to the hypergraph 2-colouring problem. Its wide applicability to combinatorial questions soon became apparent. Nowadays it is usually formulated in general probabilistic terms in the following fashion.

Theorem 2 (Lovász Local Lemma, symmetric form)
Let $\mathcal{A} = \{A_1, A_2, \ldots, A_m\}$ be any collection of events in a probability space, each one having probability at most p and such that each event is mutually independent of all but at most d of the other events. If $ep(d + 1) \leq 1$, then with positive probability, none of the events in $\mathcal{A}$ occur.

The SAT formulation, Theorem 1, follows as an immediate corollary. Considering the random experiment of sampling truth assignments to the CNF formula F at random and defining A_i to be the event that clause number i becomes violated, each event has probability 2^{-k} and the desired bound follows. This way, it is a natural extension of the simple probabilistic argument bounding from below the total number of clauses in an unsatisfiable formula.

Theorem 1 is asymptotically tight. This is most simple to see as the CNF formula consisting of all 2^k clauses of size k over k variables is clearly unsatisfiable and has neighbourhoods of size $2^k - 1$ at each clause. In Section 3, we indicate how an unsatisfiable k-CNF formula having neighbourhoods of size 2^{k-1} each can be constructed, tightening even further the constant gap between the known lower and upper bounds. Note that in the general probability space setting as in Theorem 2 the constant e is known to be tight [5].

In the sequel, we give two proofs for Theorem 1. The first "existential" proof (from [2]) is beautifully short and astounding, but suffers from the mentioned shortcoming that it is non-constructive and so does not reveal how a satisfying assignment should be efficiently found. Whether this is in any way possible used to be a long-standing open problem until in 1991, Beck achieved a breakthrough by proving in [6] that a polynomial-time algorithm exists which finds a satisfying assignment to every k-CNF formula in which each clause has a neighbourhood of at most $2^{k/48}$ other clauses. His approach was deterministic and used the non-constructive version of the Local Lemma as a key ingredient, basically proving that even after truncating clauses to a 48th of their size (a step used to simplify the formula and make it fall apart into small components), a solution remains guaranteed that can then be looked for by exhaustive enumeration. Alon simplified Beck's algorithm and analysis by introducing randomness and presented an algorithm that works up to neighbourhoods of $2^{k/8}$ in size [7]. Czumaj and Scheideler later demonstrated that a variant of the method can be made to work for the non-uniform case where clause sizes vary [8]. In 2008, Srinivasan improved the bound of what was polynomial-time feasible to essentially $2^{k/4}$ by a more accurate analysis [9]. Later that year, Moser published a polynomial-time algorithm that can cope with neighbourhood sizes up to $\mathcal{O}(2^{k/2})$ [10], and some-what later an improved variant that allows for 2^{k-5} neighbours [11], which is asymptotically optimal with a constant gap.

The second proof we present here, finally, is a fully constructive version published by Moser and Tardos [12] which does not suffer from any gap to the existential version anymore. While it is general enough so as to apply to many applications of the Local Lemma, we will formulate it in terms of satisfiability here. The proof formalises the intuitive idea mentioned in the introduction: that the simplest possible method starting at a random point and then applying some corrections, thereby moving around violated clauses in the formula, always converges to a solution.

2.1 First Proof of Local Lemma – Existence

Let F be our k-CNF formula and let us require that each clause has a neighbourhood of at most $d := 2^k/e - 1$ other clauses. Suppose we select an assignment α of truth values to the variables uniformly at random. What is the probability that α satisfies F? If we can prove that probability to be positive, then F has to be satisfiable. Let us try to do so.

Let $F' \subset F$ be any subformula that arises from F by removing at least one clause. Let $C \in F \backslash F'$ be one of the clauses removed. α has a certain probability $Pr(F')$ of satisfying F'. We are interested to compute the drop in probability if we add back C as an additional constraint and we claim that this drop be bounded by a factor of $(1 - e2^{-k})$, that is $Pr(F' \wedge C) \geq (1 - e2^{-k})Pr(F')$ (or, equivalently, $Pr(F' \wedge \neg C) \leq e2^{-k}Pr(F')$). No matter what the factor exactly is, as long as it is positive, this readily gives what we have claimed, since the empty formula is satisfied with probability 1 and then successively adding back all of F's clauses diminishes that probability by a positive factor each step, leaving a positive probability in the very end.

So let us prove the auxiliary claim. We proceed inductively. Suppose the auxiliary claim has been proved for all subformulas F' up to a given size and now we would like to establish it for larger subformulas. First of all, there is a trivial special case. If the constraint C that we join back to F' is independent, that is, does not have any variables in common with F', then the events that F' or C are satisfied, respectively, are independent from one another and the probability decreases by a factor of exactly $(1 - 2^{-k})$. We have to understand now why lowering that factor to $(1 - e2^{-k})$ is sufficient to account for the (restricted) amount of possible dependencies that we might encounter. So, given the more problematic case that C shares some variables with F', let us get rid of those dependencies by removing, additionally, all clauses from F' that neighbour C; let $F'' := F' \backslash \Gamma(C)$. Now F'' and C are independent. Clearly, in this case

$$Pr(F'' \wedge \neg C) = 2^{-k}Pr(F'').$$

Note that we have removed from F' at most d clauses (due to the global hypothesis). By induction, we can add back all of these clauses one by one to F'' to get F' and thereby obtain

$$Pr(F') \geq (1 - e2^{-k})^d Pr(F'') \geq e^{-1} Pr(F'').$$

On the other hand, since every assignment satisfying F' satisfies F'', we have

$$Pr(F' \wedge \neg C) \leq Pr(F'' \wedge \neg C) = 2^{-k} Pr(F'').$$

The two results yield $Pr(F' \wedge \neg C)/Pr(F') \leq 2^{-k}/e^{-1}$, as claimed. $\square$

2.2   Second Proof of Local Lemma – Algorithm

Recall our intuitive understanding of the problem setting. If we start with a randomly chosen assignment, then a 2^k-th of the clauses are, on average, violated. Now suppose that we continue in the most naive fashion: repeatedly select any of the violated clauses and just select new uniformly random values for each of the variables occurring in that clause until a satisfying assignment is reached. Such a strategy of successive local corrections might fail if correcting a violated clause causes lots of new clauses to be violated. But since the influence of a clause correction is restricted to the neighbourhood of that clause, then if such neighbourhoods are always sufficiently small, the strategy sounds intuitively promising. We will demonstrate that under the hypothesis of the Local Lemma, it converges to a satisfying assignment in an expected polynomial number of steps. The existence of such an assignment then follows with the correctness of the procedure.

Let us execute the algorithm and observe what it does, recording a *log* of what corrections are being applied, that is a mapping $L : \mathbb{N}_0 \to F$ with the meaning that in step t, the algorithm selects clause $L(t)$ for correction. We hope for the algorithm to terminate quickly, in particular after a finite number of steps, but in order to be rigorous, we have, for the moment, to allow for an infinite log and then prove that we will not ever encounter one. Moreover, let $N : F \to \mathbb{N}_0 \sqcup \{\infty\}$ be random variables that count the number of times a given clause occurs in the log, that is for $C \in F$, we define $N(C) := |\{t \in \mathbb{N}_0 | L(t) = C\}|$. Again, we a-priori have to allow for such a counter to take infinity as a value, but we will show that it never does. In fact, what we prove now is that for each clause $C \in F$, the expected value $E[N(C)]$ is upper bounded by a constant. Note that this implies everything we claim: Since in the expected case, each clause is corrected at most a constant number of times, the total number of clauses corrected is, in the expected case, bounded by $\mathcal{O}(|F|)$. So not only does the algorithm always terminate after a finite number of steps (implying the existence of a solution), it even returns after a polynomial number of operations.

Bounding the expected value $E[N(C)]$ is strikingly simple once we introduce a concept that goes back to Beck and Alon [6, 7]. The concept we are talking about is the one of witness trees. A *witness tree* is an unordered, rooted tree T along with a labelling $\sigma : V(T) \to F$ of its vertices $V(T)$ by clauses from F. Given a specific run of the algorithm and thus a log L, a witness tree can serve as a justification for the necessity of any of the executed correction steps. What do we mean? Let t be any time index such that $L(t)$ is defined. Now let us build a witness tree in the following sense. Start with a root vertex r and label that vertex $\sigma(r) := L(t)$, that is by the clause corrected in step t. Now traverse the log backwards and for each time step $s = t - 1, t - 2, ..., 0$, check if the clause $L(s)$ has any variables that it shares with any of the labels in the tree built so far. If $L(s)$ is independent from all clauses currently serving as labels, discard it. If there *are* nodes in the tree that have variables in common with $L(s)$, then select any deepest of those nodes and create a new child node of it, labelling that new child $L(s)$. Once arriving at $s = 0$ we have built a witness tree $T(t)$ that justifies correction step t. In the following sense.

If we look at a witness tree $T(t)$, thereby forgetting everything else we have seen while the algorithm was running, we can reconstruct a significant portion of the execution history. Traversing the tree $T(t)$ in a bottom-up and level-by-level fashion (as in a reverse breadth-first-search that starts at the root), we obtain a sequence of clauses that is essentially a subsegment of the execution log. Each node we encounter during such a traversal represents some correction step in L with the label of the node being the clause corrected in that step. And the way we defined the witness $T(t)$ immediately assures us of two things: firstly, the ordering in which the corrections have taken place is similar to the ordering in which we traverse the nodes. It isn't identical, but it preserves what we will be interested in: Whenever two nodes v_1 and v_2 are labelled with clauses that depend on each other, i.e. that have common variables, then v_1 occurs before v_2 in the traversal if and only if v_1 represents a correction step occurring before v_2. Secondly, when we traverse some node v representing correction step t, then all correction steps $t' < t$ that relate to step t in the sense that $L(t)$ and $L(t')$ share common variables do occur in the tree and have therefore been traversed before.

What these two properties imply is the following: If we traverse our tree in the described way and we count the number of times some variable x has occurred so far in labelling clauses, then that number corresponds to the number of times x has been reassigned new values before the corresponding correction step. So if we have seen variable x already 10 times before we traverse a node v labelled $\sigma(v) = C$, then this means that at the time the correction v represents took place, x had its 10th new random value and was then assigned its 11th one. This in turn means that we can reconstruct, by just looking at the tree, all the 10 values x had been assigned before. This is because node v represents a time step where clause C was selected for correction, that is a time step when C was violated and thus the 10th value of x has to have been the one that dissatisfies the corresponding literal we find in C. The same holds for all other variables in the clause and for all other nodes we traverse.

Now suppose you are given a fixed witness tree T. What is the probability that exactly this tree can occur as witness for some correction step? As we have seen before, if we traverse T bottom-up we can reconstruct for each node the values the k variables in the corresponding clause were assigned before the correction step represented, that is we can reconstruct k of the random bits the algorithm has used. If the tree has n vertices, we can reconstruct nk bits in total, just looking at the tree. Since those bits are uniformly random, the probability that all of them sample such that T can be constructed (we will say that T is *valid* if they evaluate as needed) is exactly 2^{-nk}. On the other hand, let us count how many witness trees there exist in total. Let us fix some clause $C \in F$ and some number n and let us count the number of witness trees of order n which have C as the label of their root vertex. What restricts that number is the way in which witness trees were defined, which requires that if u is a child node of v, then the label $\sigma(u)$ must be a neighbour of the clause $\sigma(v)$. This allows us to embed each witness tree rooted at label C into an infinite tree that just enumerates neighbouring nodes: Consider an infinite tree with its root labelled C and such that each node v labelled $\sigma(v)$ has $|\Gamma(\sigma(v))|$ children labelled $\Gamma(\sigma(v))$. Such a tree is at most d-ary and each witness tree is clearly a subtree of it. A two-line counting exercise shows that an infinite rooted ($\leq d$)-ary tree has at most $(ed)^n$ subtrees of size n. Therefore there are at most $(ed)^n$ witness trees of order n that have C as their root label. Since each of them may occur with a probability of at most 2^{-nk}, the expected number of witness trees of size n that can occur is bounded by $(ed2^{-k})^n$. Plugging in d and summing over all possible sizes $n \geq 1$, this becomes a geometric series that converges to a constant. Hence, there is at most a constant expected number of valid witness trees rooted at C.

What does this mean? Clause C occurs $N(C)$ times $t_1, t_2, \ldots, t_{N(C)}$ in the execution log. For each of those times we can ask for a witness tree $T(t_1)$, $T(t_2)$, ..., $T(t_{N(C)})$ to justify that correction step. All of those trees have to be valid, and they are distinct since $T(t_{i+1})$ needs to have basically the same vertices as $T(t_i)$ (though maybe arranged differently) and at least one more (to represent step t_{i+1}). So $N(C)$ is at most as large as the number of valid witness trees rooted at C. Since the latter number is bounded by a constant in expectation, the former is so, too. And this concludes the argument.   □

2.3   A Stronger Variant – Conflicts

There is a slightly stronger version of the Lovász Local Lemma, referred to as the lopsided Local Lemma ([13, 1, 14]), which does not only distinguish between dependent and independent events but also discriminates between positive and negative correlations. In terms of satisfiability, this means that the bound on the maximum neighbourhood size is replaced by a bound on conflict neighbourhoods.

Theorem 3. *Let $k \in \mathbb{N}$ and let F be a k-CNF formula. If $|\Gamma'(C)| \leq 2^k/e - 1$ for all $C \in F$, then F is satisfiable.*

Both the purely existential and the constructive proof we detailed above can be adapted so as to demonstrate this statement. For the latter, the same algorithm will work and for the analysis it suffices to observe that witness trees built by attaching only lopsided neighbours during backward traversal of the log equally allow to reconstruct k bits of the randomness used per vertex, irrespective of the fact that a smaller amount of information might be encoded by the tree.

The lopsided Local Lemma has successfully been used to establish better bounds for the number of dependencies or the number of occurrences per variable that we can allow, still being guaranteed that all formulas within that class are satisfiable. Berman, Karpinski and Scott, e.g., have demonstrated in [15], using the lopsided Local Lemma, that every 6-, 7-, 8- or 9-CNF formula in which every variable occurs at most 7, 13, 23 or 41 times, respectively, is satisfiable. Their argument can be used to obtain bounds for other values of k as well and it can be made constructive by the methods we presented in the previous section. Let us now have a closer look at bounded occurrence instances of satisfiability.

3   Bounded Variable Degree

Let us call a k-CNF formula in which no variable occurs in more than d clauses a (k,d)-CNF formula. Recall $f(k)$ which we can now equivalently define as the unique integer so that all $(k, f(k))$-CNF formulas are satisfiable and an unsatisfiable $(k, f(k) + 1)$-CNF formula exists. For $k \geq 1$, a $(k, 0)$-CNF formula has to be empty and thus satisfiable. The CNF formula of all 2^k k-clauses over a given set of k variables constitutes an unsatisfiable $(k, 2^k)$-CNF formula, so $f(k)$ exists with $0 \leq f(k) \leq 2^k$.

The first to consider $f(k)$ was Tovey [16] in 1984 (with Christos Papadimitriou raising the question). He showed $f(k) \geq k$ (by an argument based on Hall's Theorem which we will provide later in Lemma 3); he suspected his bound to be weak and actually conjectured that all $(k, 2^{k-1} - 1)$-CNF formulas are satisfiable [16, Conjecture 2.5].

A clearer picture of $f(k)$ has evolved since then. For, if every variable occurs at most d times in a k-CNF formula, no clause can collect more than $k(d-1)$ neighbours. Thus, with the Local Lemma (as in Theorem 1), the inequality $k(d-1) \leq 2^k/e - 1$ implies that every (k, d)-CNF formula is satisfiable. This connection and the implied bound of $f(k) \geq \lfloor 2^k/(ek) \rfloor$ was first established by Kratochvíl, Savický, and Tuza [17] – still the best lower bound known for k large.

They supplied also an upper bound of $2^{k-1} - 2^{k-4} - 1$. Significant progress on the upper end was made when Savický and Sgall [18] showed $f(k) = O(k^{-0.26}2^k)$. This was further improved to $f(k) = O((2^k \log k)/k)$ by Hoory and Szeider [19], now only a log-factor shy of the lower bound. Recently, the gap has been closed up to a constant factor by Gebauer [20] and $f(k) = \Theta(2^k/k)$ is settled. (A brief discussion of the situation for small k is postponed to the end of this section.)

Theorem 4. *For k a large enough integer,*

$$\left\lfloor \frac{2^k}{ek} \right\rfloor \leq f(k) < \frac{2^{k+1}}{k}.$$

If k is a sufficiently large power of 2 we have $f(k) < 2^k/k$.

SAT connects to many (sometimes seemingly unrelated) problems. The proof of the upper bound in Theorem 4 is another example for this fact: The actual construction was originally developed for refuting a conjecture of Beck on Combinatorial games [21]. In such a game Maker and Breaker take turns in choosing vertices from a given hypergraph. Maker wants to completely occupy a hyper-edge and Breaker tries to prevent this. The problem is to find the minimum $d = d(k)$ such that there is a k-uniform hypergraph of maximum vertex degree d where Maker has a winning strategy.

One possible strategy Maker can use is to partition all but at most one of the vertices into pairs and whenever Breaker claims one vertex of a pair, Maker takes the other one. If Maker uses such a *pairing strategy*, this game on hypergraphs is in some sense equivalent to unsatisfiability. Indeed, given a hypergraph H together with a pairing P we can interpret this as a CNF formula F where the hyperedges of H are understood as clauses and the two vertices of a pair of P are considered as complementary literals. It is easily seen that Maker wins the game on H using the pairing strategy according to P if and only if F is unsatisfiable.

If there is a k-uniform hypergraph of maximum vertex degree d with a winning pairing strategy for Maker, then there is an unsatisfiable $(k, 2d)$-CNF formula.

This clearly shows the relation between the two problems. For the proof a third player enters the picture: binary trees. In this presentation we will proceed directly from binary trees to CNF formulas.

3.1   Trees with All Leaves Deep, But Few Leaves Close Below Any Node

We consider binary trees where every node has either two or no children. In such a binary tree we say that a leaf v is ℓ-*close* to a node w if w is an ancestor of v, at distance at most ℓ from v. For k and d positive integers, we call a binary tree T a (k, d)-*tree* if (i) every leaf has depth[2] at least $k - 1$ and (ii) for every node u of T there are at most d leaves $(k - 1)$-close to u; (from (i) it follows that every leaf is $(k - 1)$-close to exactly k nodes). A moment of reflection shows that every binary tree with all leaves at depth at least $k - 1$ is a $(k, 2^{k-1})$-tree. The following lemma motivates a search for (k, d)-trees with d smaller than 2^{k-1}.

Lemma 1. *Let T be a (k, d)-tree, k and d positive integers. Then there is a k-CNF formula $F = F(T)$ with the following properties.*

(a) *For m, the number of leaves of T, we have $|F| = 2m$.* (b) *Every literal occurs in at most d clauses of F.* (c) *If $\mathrm{vbl}(C) \cap \mathrm{vbl}(D) \neq \emptyset$ for two distinct clauses C and D in F, then these clauses are conflict-neighbours with a unique variable that appears in C and D with opposite signs.* (d) *F is unsatisfiable.* (e) *If, for every node u in T, there is at least one leaf that is $(k - 1)$-close to u, then $|F| = |\mathrm{vbl}(F)| + 1$ and F is minimal unsatisfiable.*

(1) *F is an unsatisfiable $(k, 2d)$-CNF formula; thus, $f(k) \leq 2d - 1$.*

(2) *F is an unsatisfiable k-CNF formula with $|\Gamma(C)| \leq kd$ for all clauses $C \in F$; hence, $l(k) \leq kd - 1$.*

---

[2] The root has depth 0, its children have depth 1, ...

Proof. We move to a binary tree $\widehat{T}$ by attaching the roots of two copies of T as the two children of a new root r. This yields a (k, d)-tree (actually, with all nodes of depth at least k). Obviously, it has $2m$ leaves and, as it goes with binary trees, $4m - 1$ nodes altogether. Two nodes are called *siblings*, if they share the same parent. Leaving aside the root, the remaining $4m - 2$ nodes of $\widehat{T}$ can be partitioned into $2m - 1$ sibling pairs.

For V some set of $2m - 1$ boolean variables, we label the nodes of $\widehat{T}$ other than the root by literals in $V \cup \overline{V}$ so that every literal appears exactly once and siblings get complementary literals. With every leaf v we associate a clause C_v by walking along a path of length $k - 1$ from v towards the root and collecting all labels encountered on this path (i.e. the labels of all nodes to which v is $(k - 1)$-close). The set of clauses C_v, over all leaves v of $\widehat{T}$, constitutes F.

The fact that every leaf of $\widehat{T}$ has depth at least k guarantees that paths of length $k - 1$ starting at leaves will never reach the root. So there are indeed always k literals to collect and F is a k-CNF formula. Also $|F| = 2m$ is obvious (therefore (a)). The defining property of (k, d)-trees guarantees that no label is collected more than d times (hence (b)). F is unsatisfiable (as claimed in (d)), for if an assignment α over V is given, it defines a path from the root to a leaf, say v, by always proceeding to the unique child whose label is mapped to 0 by α; C_v's fate of being violated by α is determined.

Let us settle the claimed neighbour property in (c). Suppose that $\mathrm{vbl}(C_u) \cap \mathrm{vbl}(C_v) \neq \emptyset$ for leaves u and v, $u \neq v$. If $x \in C_u$ and $\overline{x} \in C_v$, then the parent w of the siblings labelled by x and $\overline{x}$, respectively, is the lowest common ancestor of u and v (i.e. the node of maximum depth that appears on both paths from u and v, respectively, to the root); therefore x is unique. For the existence of a complementary pair in C_u and C_v, consider the lowest common ancestor w of u and v; w cannot be a leaf since $u \neq v$. The literals occurring in the two subtrees rooted at the children of w do not share any common variable other than the complementary pair placed at the children of w. Hence, the literals associated with these two children must appear in the clauses, one literal in C_u the other complementary literal in C_v, since otherwise their variable sets are disjoint. Assertion (c) is shown.

Next we prove (e). With the assumption given, all $2(2m - 1)$ literals appear in some clause, so $|\mathrm{vbl}(F)| + 1 = 2m = |F|$. It remains to show satisfiability of $F_v := F \setminus \{C_v\}$ for all leaves v. For v a leaf of $\widehat{T}$, consider the following procedure, in the course of which, besides defining an assignment, we make nodes responsible for clauses in F_v.

First, set the literals of all nodes on the path from v to root r (excluding r) to 0 and initialise S as the set of all siblings of these nodes. Now, while S is nonempty, (i) remove some $u \in S$ from S, (ii) set its literal to 1, (iii) choose a leaf v_u (maybe u itself) that is $(k - 1)$-close to u, (iv) set all literals of nodes on the path from v_u to u (excluding u) to 0, (v) add all siblings of these nodes to S, and, finally, (vi) make u responsible for C_{v_u}.

In this proceeding we see an invariant maintained: The subtrees rooted at the nodes in S are disjoint and they comprise exactly the nodes with their literals undefined. Thus the value of a literal is set at most once and, actually, at least once, since the node of minimal depth with its literal undefined would have to be in S (and therefore the procedure couldn't possibly have stopped already). Now it is easily seen that this procedure gives a valuation of the literals that is indeed an assignment of V. Whenever a node u is responsible for a clause, then this clause is satisfied due to u's literal. Also, a clause cannot have two nodes responsible for it, while every node with its literal set to 1 is responsible for some clause. It becomes obvious that the responsibility map is a bijection between the $2m - 1$ nodes with their literal set to 1 and the $2m - 1$ clauses in F_v. Hence, F_v is satisfied. And so are we.

Implication (1) follows from (b) and (d). (2) follows from (b-d): In particular, if we define $occ(u) := |\{C \in F \mid u \in C\}|$, then (c) allows us to write $|\Gamma(C)|$ as $\sum_{u \in C} occ(\bar{u})$ – hence, at most kd – for every clause $C \in F$. $\qquad \Box$

In a similar fashion, one can show that a (k, d)-tree yields a k-uniform hypergraph of maximum vertex degree d where Maker has a winning pairing strategy. We are left with the task of constructing (k, d)-trees with sufficiently small d.

Lemma 2. (i) A $(k, \lfloor 2^k/k \rfloor)$-tree exists for every sufficiently large k. (ii) If k is a sufficiently large power of 2 then a $(k, 2^{k-1}/k)$-tree exists.

Lemmas 1(1) and 2 imply the upper bounds in Theorem 4. Also Lemmas 1(2) and 2(ii) give us an upper bound of $l(k) < 2^{k-1}$ for large enough powers of 2. So let us now summarise also our findings for $l(k)$ and $lc(k)$.

Theorem 5. *We have*

$$\lfloor 2^k/e \rfloor - 1 \le lc(k) \le l(k) < 2^{k-1}$$

where the upper bound holds for k any sufficiently large power of 2 but can be replaced by $2^k - 1$ for all positive integers k.

Moreover, the relation to $l(k)$ tells us that we will not be able to find (k, d)-trees with $d < (l(k)+1)/k$ and, therefore, – along the lower bounds of $l(k)$ in Section 2 – the range $d < 2^k/(ek) - 1$ is inaccessible.

The proof of Lemma 2 is tedious and too long for this presentation. However, a weaker statement, still settling the asymptotics of $f(k)$, can be shown with less effort.

Proof of existence of $(k, 2^{k+1}/k)$-trees for k a power of 2. We have $2^{k+1}/k \ge 2^{k-1}$ for $k \le 4$, so we can assume that $k \ge 8$ in our proof. Let T' be a full binary tree of height $k - 1$ (i.e. a binary tree where all leaves have the same depth $k - 1$). We subdivide its leaves into intervals of length $k/2$. For $\{v_0, \ldots, v_{k/2-1}\}$ such an interval, we attach a full binary subtree of height i to v_i. Let T denote the resulting tree and let the *leaf-range* $r(v)$ of a node v denote the number of leaves $(k-1)$-close to v.

It suffices to show that $r(v) \le \frac{2^{k+1}}{k}$ for all nodes of T. We apply induction on the depth i of v. For $i = 0$ the claim holds. Indeed, note that out of each of the

$\frac{2^{k-1}}{k/2}$ intervals, exactly one vertex (v_0, respectively) is $(k-1)$-close to the root and, hence, the leaf-range of the root is $\frac{1}{2}\frac{2^{k+1}}{k}$. Now suppose that v has depth $i \in \{1,\dots,k/2-1\}$. Note that the set of descendants of v at depth $k-1$ can be subdivided into $\frac{2^{k-1-i}}{k/2}$ intervals, i.e. at least one interval for the values of k we consider. Let v' denote the parent of v. By construction the number of leaf descendants which have distance at most $k-2$ from v equals $r(v')/2$. Moreover, every interval $\{v_0,\dots,v_{k/2-1}\}$ gives rise to 2^i leaves on level $k-1+i$, implying that the number of leaf descendants of v which have distance exactly $k-1$ from v equals $\frac{2^{k-1-i}}{k/2} \cdot 2^i = \frac{1}{2}\frac{2^{k+1}}{k}$. So altogether $r(v) \le \frac{r(v')}{2} + \frac{1}{2}\frac{2^{k+1}}{k} \le \frac{2^{k+1}}{k}$. It remains to consider the case where v has depth at least $k/2$. By construction no leaf of T has depth larger than $k/2+k-2$, implying that the leaf-range of v is at most the leaf-range of its parent. $\qquad\square$

3.2   Small Values

Although the lower bounds on $f(k)$, $l(k)$ and $lc(k)$ we can derive via the Local Lemma grow exponentially, they are weak for small values of k. Here another classic from combinatorics enters the picture: Hall's marriage theorem.

Lemma 3. (1) $f(k) \ge k$ for $k \ge 1$ [16] and (2) $l(k) \ge lc(k) \ge k$ for $k \ge 2$.

Proof. (1) For $k \ge 1$, let F be a k-CNF formula over a variable set V with no variable occurring in more than k clauses of F. Consider the incidence graph between clauses and variables, i.e. the bipartite graph with vertex set $F \cup V$, where $\{C, x\}$ is an edge iff $x \in vbl(C)$. In this graph, clause-vertices have degree exactly k and by assumption variable-vertices have degree at most k. Therefore, Hall's condition for a matching covering all clause-vertices holds. An assignment is now defined by letting every variable x that is matched to a clause C map to the value so that it satisfies C. The matching property prevents conflicts in doing so. No matter how we complete the assignment for unmatched variables, it will satisfy all clauses.

(2) Let $k \ge 2$. We will actually prove a bound of $lc(k) \ge \lceil(f(k)+1)/2\rceil + k - 2$; with the bound on $f(k)$ from (1) this yields a lower bound of $\lceil(3k-3)/2\rceil \ge k$.

So we have to show that every unsatisfiable k-CNF formula F contains a clause C with $|\Gamma'_F(C)| \ge \lceil(f(k)+1)/2\rceil + k - 1$. First we pass to a minimal unsatisfiable k-CNF formula $G \subseteq F$. G has a variable x with $\deg_G(x) \ge f(k) + 1$; w.l.o.g. we assume that literal $\bar{x}$ occurs at least $\lceil(f(k)+1)/2\rceil$ times. Now choose a clause $C \in G$ with literal x. $\Gamma'_G(C)$ contains all clauses with $\bar{x}$. In addition, according to Lemma 4, for all variables $z \in C \setminus \{x\}$ there has to be a clause $D_z \in G$ with the property that z is the unique variable that appears in C and D_z with opposite signs. It follows that $|\Gamma'_G(C)| \ge \lceil(f(k)+1)/2\rceil + k - 1$. This concludes the argument, since $\Gamma'_F(C) \supseteq \Gamma'_G(C)$. $\qquad\square$

In fact, $f(k) = k$ is known for $k \le 4$ ([16] for 3 and [22] for 4). The best known bounds for $k = 5$ are $5 \le f(5) \le 7$, [23]. $k = 6$ is the first value for which the bound in Lemma 3(1) is known not to be tight, [15]: $7 \le f(6) \le 11$. See [23] for a further discussion of $f(k)$ for small k with currently best bounds known (and also Section 3.3 below). The bound for $lc(k)$ suffices our needs in the proof of Theorem 9 below, but perhaps even the lower bound of 3 for $lc(3)$ is not tight; we have not followed up this matter further.

We conclude with the missing lemma employed in the proof of Lemma 3(2) (in the spirit of the properties of so-called blocked clauses introduced by Kullmann [24, Definition 3.1]).

Lemma 4. *Let F be a minimal unsatisfiable CNF formula. Consider x and $C \in F$ with $x \in \mathrm{vbl}(C)$. Then there is a clause D with the property that x is the unique variable that appears in C and D with opposite signs.*

Proof. Since F is minimal, $F \setminus \{C\}$ has a satisfying assignment α. By assumption of unsatisfiability of F, α cannot satisfy C. Now switch the value of x in the assignment, thereby satisfying C and thus violating some other clause $D \in F$. Now, as easily seen, D serves the purpose.   □

3.3   A Special Class of Unsatisfiable CNF Formulas – MU(1)

MU(1) is defined as the set of all minimal unsatisfiable CNF formulas F with $|F| = |\mathrm{vbl}(F)| + 1$. This definition may appear somewhat arbitrary, so why care? First, when searching for extremal unsatisfiable CNF formulas with the properties we are interested in, we can confine ourselves to minimal unsatisfiable CNF formulas. Second, as observed by A. Tarsi (cf. [25]), $|F| - |\mathrm{vbl}(F)|$, called *deficiency*, is positive for every minimal unsatisfiable CNF formula F; hence, CNF formulas in MU(1) are minimal w.r.t. to deficiency. Formulas in MU(1) can be recognised efficiently [26, 27, 28] and they have been shown to be universal in the sense that every unsatisfiable CNF formula F has a $G \in \mathrm{MU}(1)$ with a homomorphism from G to F [29]. We do not define homomorphisms for CNF formulas here. They preserve unsatisfiability but alter other properties, e.g. due to "clause collapses" they can decrease neighbourhood sizes and variable degrees. Still, many extremal unsatisfiable CNF formulas can be drawn from MU(1).

With this in mind and, returning to our problem, given that we do not even know whether $f(k)$ is computable, Hoory and Szeider [23] headed for the more modest goal of investigating $f_1(k)$, the largest integer such that every $(k, f_1(k))$-CNF formula in MU(1) is satisfiable. They show that f_1 is computable, in fact, reasonably efficiently. With $f(k) \leq f_1(k)$, this allows them to derive the best known upper bounds for $f(k)$ for small k: $f(5) \leq 7, f(6) \leq 11, f(7) \leq 17, f(8) \leq 29, f(9) \leq 51$.

While the previous bound of $f(k) = O((2^k \log k)/k)$ in [19] was not established through formulas in MU(1), the constructions in [20] reside in MU(1); indeed, Lemma 1(e) provides this fact, since it is not too hard to see that if a (k, d)-tree exists, then we can modify it so that the assumption of Lemma 1(e) holds. Therefore, we can conclude that $f(k)$ and $f_1(k)$ are within a constant factor of each other. Whether $f(k) = f_1(k)$ for all k is an interesting open question.

4   Linear Formulas

We call a CNF formula F *linear* if $|\mathrm{vbl}(C) \cap \mathrm{vbl}(D)| \leq 1$ for any two distinct clauses C and D in F. For example, the formula $\{\{x,y\}, \{\bar{y}, z\}, \{\bar{x}, \bar{z}\}\}$ is linear, whereas the formula $\{\{x, y, z\}, \{\bar{x}, y, u\}\}$ is not. This class of formulas has been introduced in [30]. It is a natural analogue of the notion of *linear hypergraphs*: A hypergraph $H = (V, E)$ is linear if $|e \cap f| \leq 1$ for any two distinct edges $e, f \in E$. In this section we investigate the following questions: Are there unsatisfiable linear k-CNF formulas, for each k? If yes, how large do they have to be? There is the analogous question asking for k-uniform linear hypergraphs that are not 2-colourable (see last paragraph of introduction). Existence of those has been shown by Abbott [31], and Erdős and Lovász [2] give lower bounds on their size (in terms of number of vertices and hyperedges). Bounds for chromatic numbers exceeding 2 can be found in [32]. Given a k-uniform non-2-colourable hypergraph H with m hyperedges, we immediately obtain an unsatisfiable k-CNF formula $F(H)$ with $2m$ clauses (as described at the end of the introduction). However, for $k \geq 2$, even if H is linear, $F(H)$ is certainly not. Therefore, it is not clear (but true, as we will see) that bounds on the size of unsatisfiable linear k-CNF formulas are similar to those of non-2-colourable linear k-uniform hypergraphs.

Let $f_{\mathrm{lin}}(k)$ be the largest integer so that every linear $(k, f_{\mathrm{lin}}(k))$-CNF formula is satisfiable. Note $f_{\mathrm{lin}}(k) \geq f(k) \geq \lfloor 2^k/(ek) \rfloor$.

Theorem 6 ([33]). *Any unsatisfiable linear k-CNF formula has at least*

$$\tfrac{1}{k}\left(1 + f_{\mathrm{lin}}(k-1)\right)^2 > \tfrac{4^k}{4e^2 k^3}$$

clauses. There exists an unsatisfiable linear k-CNF formula with at most $8k^3 4^k$ clauses.

Remark. $\tfrac{1}{k}\left(1 + f_{\mathrm{lin}}(k-1)\right)^2 \leq 8k^3 4^k$ follows; thus $f_{\mathrm{lin}}(k-1) \in \mathcal{O}(k^2\, 2^k)$.

Proof. The proof of the lower bound is similar to the one for the size of non-2-colourable linear k-uniform hypergraphs in [2]. The following lemma is instrumental.

Lemma 5. *Let F be a linear k-CNF formula. If there are at most $f_{\mathrm{lin}}(k-1)$ variables of degree exceeding $f_{\mathrm{lin}}(k-1)$, then F is satisfiable.*

First we show how the lemma implies the lower bound. Let X be the set of variables x with $\deg_F(x) > f_{\mathrm{lin}}(k-1)$. If F is unsatisfiable, then by the lemma, $|X| > f_{\mathrm{lin}}(k-1)$. Therefore, the lower bound follows from

$$|F| = \tfrac{1}{k} \sum_{x \in \mathrm{vbl}(F)} \deg_F(x) \geq \tfrac{1}{k}(1 + f_{\mathrm{lin}}(k-1))|X| \geq \tfrac{1}{k}\left(1 + f_{\mathrm{lin}}(k-1)\right)^2$$

Proof (of the lemma). For a literal u, let $\deg_F(u)$ denote the degree of the variable underlying u in F. First we construct a linear $(k-1)$-CNF formula F' as follows: For every clause $C \in F$, let u_C be a literal of C that maximises $\deg_F(u_C)$ and write $C' := C \setminus \{u_C\}$; F' is obtained as $F' := \{C' \mid C \in F\}$. We claim that $\deg_{F'}(x) \leq f_{\mathrm{lin}}(k-1)$ for all variables x; thus, F' is satisfiable and therefore F is satisfiable.

Consider a variable x. Clearly, $\deg_{F'}(x) \leq \deg_F(x)$ and so if $\deg_F(x) \leq f_{\mathrm{lin}}(k-1)$, we are done. Otherwise, let $C'_1, \ldots, C'_t$, $t = \deg_{F'}(x)$, be the clauses in F' containing x or $\bar{x}$. There are clauses $C_1, \ldots, C_t$ in F such that $C'_i = C_i \setminus \{u_{C_i}\}$, $1 \leq i \leq t$. By choice of u_{C_i}, $\deg_F(u_{C_i}) \geq \deg_F(x) > f_{\mathrm{lin}}(k-1)$. Since F is linear, the u_{C_i}'s have to be distinct, thus by assumption, $t \leq f_{\mathrm{lin}}(k-1)$. □

We now prove the upper bound. Take a linear k-uniform hypergraph $H = (V, E)$ with n vertices and m edges, to be determined later. By interpreting the vertices of H as variables and the hyperedges as clauses, this is a (satisfiable) linear k-CNF formula. We now replace each literal in each clause by its complement with probability $\frac{1}{2}$, independently in each clause. Let F denote the resulting (random) formula. Any fixed assignment α has a $1 - 2^{-k}$ chance of satisfying a given clause of F, and thus

$$\Pr[\![\alpha \text{ satisfies } F] = (1 - 2^{-k})^m < e^{-m2^{-k}} .$$

There are 2^n distinct assignments, hence by the union bound

$$\Pr[\![\text{some } \alpha \text{ satisfies } F] < 2^n e^{-m2^{-k}} = e^{\ln(2)n - m2^{-k}} .$$

If $m/n \geq \ln(2)2^k$, the above expression is at most 1, and hence with positive probability, no assignment satisfies F. In other words, at least one F obtained in the above fashion is not satisfiable.

We construct a linear k-uniform hypergraph with few hyperedges, but with a large hyperedge-vertex ratio. Let $q \in \{k, \ldots, 2k\}$ be a prime power. Choose $d \in \mathbb{N}$ such that $q^2 \ln(2)2^k \leq q^d < q^3 \ln(2)2^k$ and set $n := q^d$. Consider the d-dimensional vector space $\mathbb{F}_q^d$ over the field $\mathbb{F}_q$. It has n elements, called *points*. In a vector space, there is a line through any pair of points, and a line has q elements. Hence there are exactly $\binom{n}{2}/\binom{q}{2} \geq \frac{n^2}{q^2}$ lines in $\mathbb{F}_q^d$. By choice of d, we have $n \ln(2)2^k \leq \frac{n^2}{q^2}$, hence we can choose $m := n \ln(2)2^k$ distinct lines in $\mathbb{F}_q^d$. From each such line arbitrarily select k points and form a hyperedge. Let E be the set of all m hyperedges formed this way. (The reader may easily check that two distinct lines cannot yield the same hyperedge.) Thus, $H = (\mathbb{F}_q^d, E)$ is a k-uniform hypergraph. It is a linear hypergraph, since any pair of distinct lines intersect in at most one point. By construction, $\frac{m}{n} = \ln(2)2^k$, and $m = n \ln(2)2^k \leq q^3 \ln(2)^2 4^k \leq \ln(2)^2 8k^3 4^k$, which proves the upper bound. □

This proof is simpler than the probabilistic construction of a non-2-colourable k-uniform linear hypergraph in [32]. This has a good reason: In our case, we inject randomness by choosing the signs of literals, whereas in the hypergraph case, there are no signs, and randomness comes only in the form of selecting hyperedges from some large set. This cannot be done independently for every hyperedge, since one has to guarantee linearity.

A question that might have formed in the reader's mind is what happens when one relaxes linearity to require that two clauses share at most one *literal*, i.e. $|C \cap D| \leq 1$, as opposed to the stricter requirement that $|\mathrm{vbl}(C) \cap \mathrm{vbl}(D)| \leq 1$. The answer is that little changes. We can prove almost the same lower bound as in Theorem 6, sacrificing only a constant factor. However, if we require that any two distinct clauses C, D have at most one conflict, i.e. $|C \cap \overline{D}| \leq 1$, things change dramatically, see Section 4.2 below.

What happens if we require that $|\mathrm{vbl}(C) \cap \mathrm{vbl}(D)| \leq \ell$, for some $\ell \geq 2$? Here our bounds also change significantly, and the exponential function in upper and lower bound will no longer be 4^k, but $2^{\frac{\ell+1}{\ell}k}$. If ℓ is a constant (i.e. does not grow with k), upper and lower bounds are still only a polynomial factor apart. The proof goes along similar lines as for Theorem 6 and the details can be found in [33]. The bounds comply nicely with those in [32] for 2-colourability of hypergraphs in which two hyperedges can intersect in at most ℓ vertices.

4.1   Why Are Small Explicit Constructions So Hard to Come Up with?

It is often surprisingly easy to obtain rather tight bounds through a probabilistic construction – and frustratingly difficult to come up with an *explicit* one. In the case of linear formulas, explicit constructions have been given in [34, 35]. However, the size of the constructed formulas is terrifying: For an unsatisfiable linear k-CNF formula, it takes $2^{2^{\cdot^{\cdot^2}}}$ clauses, where the size of the tower is k. Actually, we can provide some evidence for why small explicit constructions may be difficult. Loosely speaking, there are three ways to come up with unsatisfiable CNF formula: First, one can build formulas where unsatisfiability follows immediately from construction, by *local* considerations. This is the case for the "complete formula" mentioned in Section 2, and also for the formula constructed in the proof of Theorem 4. Second, unsatisfiability can follow from a probabilistic or counting argument, as in the proof of Theorem 6. Third, unsatisfiability can follow from some more *global* combinatorial principle (e.g. the pigeon hole principle or the fact that in a graph, the number of vertices having odd degree must be even). This is typically the case for formulas with provably large resolution complexity (see Ben-Sasson and Wigderson [36] for several beautiful proofs on resolution complexity). Normally, formulas obtained the first way have small resolution complexity, even short *treelike* resolution proofs. Therefore, it seems unlikely to obtain unsatisfiable linear k-CNF formulas of reasonable size going the first way, for one can prove the following:

Theorem 7 ([33]). *Any resolution tree of any unsatisfiable linear k-CNF formula has at least $2^{2^{(k-1)/2}-1}$ nodes.*

Let us recall the class MU(1) from Section 3.3. The extremal examples for most parameters considered so far are in MU(1): for f, l and lc via the tree derived formulas from Section 3; also an unsatisfiable k-CNF formula with 2^k clauses can be found in MU(1) (see argument for Proposition 1 below). For linear CNF formulas, the explicit constructions in [34, 35] can be adapted to result in formulas in MU(1), but they exhibit the tremendous size as mentioned. This is inherently so for linear CNF formulas in MU(1).

Theorem 8 ([33]). *For every $\epsilon > 0$ there exists some $c \in \mathbf{N}$ such that every linear k-CNF formula in MU(1) has at least $a^{a^{\cdot^{\cdot^{\cdot^{a}}}}}$ clauses, where $a = 2 - \epsilon$ and the size of the tower is $k - c$.*

4.2   1-Conflicts

Clauses C and D are called 1-*conflict neighbours*, if exactly one variable occurs in C and D with opposite signs,[3] i.e. $|C \cap \overline{D}| = 1$. We know already from the introduction (see also Lemma 4) that 1-conflicts have to occur in unsatisfiable CNF formulas (unless there is an empty clause); hence, they are crucial and deserve special attention. In the presence of linear formulas, it seems natural to consider CNF formulas where all conflicts are restricted to 1-conflicts, so let us call a CNF formula *conflict-linear* if each pair of clauses either has no conflict or has a 1-conflict. Here the typical questions we ask can be easily resolved.

Proposition 1. *For every nonnegative integer k, there is an unsatisfiable conflict-linear k-CNF formula with 2^k clauses.*

This is tight, since *all* k-CNF formulas with less than 2^k clauses are satisfiable.

Proof. Follow the construction of an unsatisfiable k-CNF formula $F(T)$ as in Lemma 1 starting from a full binary tree T of height $k - 1$ (with 2^{k-1} leaves). This readily delivers a CNF formula as required.   □

We define $lc_1(k)$ as the largest integer so that all k-CNF formulas with all 1-conflict neighbourhoods of size at most $lc_1(k)$ are satisfiable. Note that if we take all 2^k k-clauses over a given set of k variables, then in the resulting unsatisfiable k-CNF formula all 1-conflict neighbourhoods have size k. Therefore, $lc_1(k) \leq k - 1$. This bound is already tight: Given an unsatisfiable k-CNF formula F, consider a minimal unsatisfiable subset G of F. By Lemma 4, for all clauses in G the number of 1-conflict neighbours in G has to be at least k and 1-conflict neighbourhoods in F can only be larger. We have not only determined $lc_1(k) \geq k - 1$ in this way; in fact, we have shown that every unsatisfiable k-CNF formula has at least 2^k clauses with 1-conflict neighbourhoods of size at least k.

Proposition 2. $lc_1(k) = k - 1$.

5   A Sudden Jump in Complexity

Satisfiability of $(k, f(k))$-CNF formulas is trivially decidable in polynomial time. If the degree bound is relaxed, we agree that some inspection of instances is required, but we would hope that the problem does not immediately develop the

---

[3] Equivalently, C and D are 1-conflict neighbours iff their resolvent exists.

full computational complexity of SAT. Tovey [16], however, proved that for 3-CNF formulas with maximum variable degree $f(3)+1 = 4$ satisfiability is already NP-complete. Kratochvíl, Savický, and Tuza [17] generalised this sudden jump behaviour to general k: For every fixed $k \geq 3$, satisfiability of $(k, f(k) + 1)$-CNF formulas is NP-complete. It may be somewhat intriguing that one can prove such a result, given that we do not even know the values of $f(k)$ for $k \geq 5$; but we will see. Berman, Karpinski, and Scott [15] obtained similar results, showing that for $(k, f(k)+1)$-CNF formulas it is even hard to approximate the maximum number of clauses that can be simultaneously satisfied.

We will also approach the related problems for the size of neighbourhoods and conflict-neigbourhoods. While we can show that the latter performs a similar sudden jump, we have to leave a slack for the neighbourhood bound.

Theorem 9. *Let $k \geq 3$. Then,*

(1) *deciding satisfiability of k-CNF formulas with variable degrees at most $f(k)+$ 1 is NP-complete (cf. [17]),*
(2) *deciding satisfiability of k-CNF formulas with clause neighbourhoods of size at most[4] $\max\{k + 3, l(k) + 2\}$ is NP-complete, and*
(3) *deciding satisfiability of k-CNF formulas with clause conflict-neighbourhoods of size at most $lc(k) + 1$ is NP-complete.*

Before engaging in the proof, we describe a general construction that takes a k-CNF formula F and produces a CNF formula $\widehat{F}$ which is satisfiable iff F is satisfiable, so that $\widehat{F}$ is very sparsely interleaved – at the expense of the appearance of 2-clauses. We will later expand these 2-clauses to k-clauses in a fashion tailored to which of the three claims we wish to prove.

We first introduce a useful gadget. Given a set of $j \geq 2$ variables $U = \{x_0, x_1, \ldots, x_{j-1}\}$, the 2-CNF formula

$$\{\{x_0, \overline{x_1}\}, \{x_1, \overline{x_2}\}, \ldots, \{x_{j-2}, \overline{x_{j-1}}\}, \{x_{j-1}, \overline{x_0}\}\}$$

is called an *equaliser of* U; the equaliser of a singleton set U is the empty formula. As it is easily seen, such an equaliser is satisfied by an assignment to U iff all variables in U are mapped to the same value.

Now let F be a k-CNF formula, $k \geq 3$. For each variable $x \in \mathrm{vbl}(F)$, we replace every occurrence (as x or $\overline{x}$) by a new variable inheriting the sign of x in this occurrence. This yields a k-CNF formula F' with $|F|$ clauses over a set of $k|F|$ variables. Moreover, for each variable $x \in \mathrm{vbl}(F)$, we add an equaliser for the set of variables that have replaced occurrences of x. This gives an extra set F'' of at most[5] $k|F|$ 2-clauses. By the property of equalisers, $\widehat{F} := F' \cup F''$ is satisfiable iff F is satisfiable; $\widehat{F}$ can be readily obtained from F in polynomial time. Interleaving is sparse in that

---

[4] Note that $2^k/e - 1 \geq k + 1$ for $k \geq 5$. Therefore, $\max\{k + 3, l(k) + 2\} = l(k) + 2$ in that range and we actually suspect that this holds for all $k \geq 3$.

[5] F'' contains exactly $k|F|$ clauses unless there are variables in F occurring only once.

- every variable of vbl($\widehat{F}$) occurs at most 3 times in $\widehat{F}$,
- each k-clause in F' does not share variables with any other clause in F' and the number of its neighbouring 2-clauses in F'' is at most $2k$ – however, at most k of the 2-clauses are in the conflict-neighbourhood –, and
- each 2-clause in F'' neighbours two k-clauses in F' and at most two 2-clauses in F'' (all four clauses may be in the conflict-neigbourhood).

Proof of (1) (variable degrees). Let $k \geq 3$ and fix some minimal (w.r.t. set inclusion) unsatisfiable $(k, f(k) + 1)$-CNF formula G. Choose some clause C in G and replace one of its literals by $\overline{x}$ for a new variable x. This new formula, which we denote by $G(x)$, has the property that (i) it is satisfiable (otherwise G would not be minimal), (ii) every satisfying assignment has to set x to 0 (since otherwise G would be satisfiable), (iii) all variables have degree at most $f(k)+1$, and (iv) the newly introduced variable x has degree 1 in $G(x)$.

A reduction from satisfiability of general k-CNF formulas follows. Given such a k-CNF formula F we first generate $\widehat{F}$ as described above. Then we augment each 2-clause in $\widehat{F}$ by $k-2$ positive literals of new variables so that it becomes a k-clause. For each of the new variables x we add a copy of $G(x)$ to our formula; by renaming variables in G these copies are chosen so that their variable sets are pairwise disjoint. The new formula is k-CNF, it is satisfiable iff $\widehat{F}$ is satisfiable. Moreover, the maximum variable degree is $\max\{3, f(k) + 1\}$ which is $f(k) + 1$, since we assumed $k \geq 3$. This constitutes a polynomial reduction of satisfiability of general k-CNF formulas to satisfiability of k-CNF formulas with maximum variable degree $f(k) + 1$. Assertion (1) in Theorem 9 is established.   □

Proof of (2) (neighbourhoods). Again, let $k \geq 3$. Fix some minimal unsatisfiable k-CNF formula G where all neighbourhoods have size at most $l(k)+1$. We choose some clause C and replace one of its literals by $\overline{x}$ for a new variable x, resulting in a k-CNF formula $G(x)$ that forces x to 0 in every satisfying assignment.

Starting from a 3-CNF formula F (yes, we mean 3-CNF, not k-CNF) we proceed as before, first producing $\widehat{F}$ consisting of 3- and 2-clauses. Then we augment all clauses in $\widehat{F}$ to k-clauses along with disjoint copies of $G(x)$ for each new variable x. What happened to the neighbourhood sizes? A 3-clause in F' had 6 neighbours in $\widehat{F}$ and gained $k - 3$ new neighbours, so there are at most $k + 3$. A 2-clause, now extended to a k-clause, had 4 neighbours to begin with and gets an extra neighbour for each of the $k - 2$ new literals – which makes $k + 2$ neighbours. In a copy $G(x)$ all clauses stay with a neighbourhood of size at most $l(k) + 1$ except for the special clause C where we have planted the new literal $\overline{x}$. This clause may now have $l(k) + 2$ neighbours. Altogether a bound of $\max\{k + 3, l(k) + 2\}$ for neighbourhoods holds and the polynomial reduction from satisfiability of general 3-CNF formulas is completed.   □

Why did we miss a reduction to k-CNF formulas with neighbourhoods of size at most $l(k) + 1$? We could have succeeded, if we had a minimal unsatisfiable k-CNF formula G with neighbourhoods of size at most $l(k) + 1$ at our disposal, where at least one clause has a neighbourhood of size at most $l(k)$. Even if all neighbourhoods had size $l(k) + 1$, we would have been happy with one clause C in G which links to some other clause D via a single variable (we could then replace the literal of this variable in C, thereby leaving D as neighbour behind). Fortunately, this idea actually helps when we deal with conflict-neighbourhoods in the next proof.

We will also have to employ equalisers more carefully (wastefully, one might say). Given a variable set $U = \{x_0, x_1, \ldots, x_{j-1}\}$, $j \geq 2$, of concern, let $W = \{z_0, z_1, \ldots, z_{j-1}\}$ be a set of variables disjoint from U. The $(U \cup W)$-equaliser

$$\{\{x_0, \overline{z_0}\}, \{z_0, \overline{x_1}\}, \{x_1, \overline{z_1}\}, \{z_1, \overline{x_2}\}, \ldots, \{z_{j-2}, \overline{x_{j-1}}\}, \{x_{j-1}, \overline{z_{j-1}}\}, \{z_{j-1}, \overline{x_0}\}\}$$

is called a *stretched equaliser of U* – still serving the purpose of forcing all variables in U to the same value. If we use such stretched equalisers in the otherwise identical construction of $\widehat{F}$, we benefit in that

- the 2-clauses in stretched equalisers have a conflict with two other 2-clauses but to at most one of the k-clauses in F'.

Proof of (3) (conflict-neighbourhoods). For $k \geq 3$, fix some minimal unsatisfiable k-CNF formula G where conflict-neighbourhoods have size at most $lc(k) + 1$. As before, we want to replace a literal in a clause by a new literal, but now we want to be more careful about where we want to do this. Recall from Lemma 4 that G must have a pair of clauses, C and D, say, which share a unique variable y in a conflicting manner, i.e. $y \in C$ and $\overline{y} \in D$ (there may be other variables in $\mathrm{vbl}(C) \cap \mathrm{vbl}(D)$, but they have to appear with the same sign on either side). So here we do our little surgery: Choose a new variable x and replace y in C by $\overline{x}$. This gives the building block $G(x)$ forcing x to be 0. Note that the clause C' containing $\overline{x}$ (obtained from modifying C) has a conflict-neigbourhood of size at most $lc(k)$ since it lost the conflict-neighbour D.

A reduction from satisfiability of k-CNF formulas now follows in the manner customary. Given F, a k-CNF formula, we move on to $\widehat{F}$ – now with stretched equalisers – and then expand 2-clauses with the help of new variables that are forced to 0 by disjoint copies of $G(x)$. In the final product of this proceeding k-clauses in F' have at most k conflict-neighbours, k-clauses obtained from augmenting 2-clauses have at most $3 + (k - 2) = k + 1$ conflict-neighbours, and, finally, clauses in copies of $G(x)$ do not have conflict-neighbourhoods of size exceeding $lc(k) + 1$ due to our careful construction of $G(x)$. That is, the maximum size of a conflict neighbourhood is $\max\{k + 1, lc(k) + 1\}$ which equals $lc(k) + 1$ due to the lower bound of $lc(k) \geq k$ in Lemma 3.   □

6   Open Problems

We know $f(k)$, $l(k)$, and $lc(k)$ up to a constant, so one might hope to eventually determine them exactly. (The upper bounds in Theorems 4 and 5 can be improved by a factor of $\frac{63}{64}$ [20]). Progress on the lower bounds we would find very interesting.

Open Problem 1. *Is it possible to improve any of the known lower bounds on $f(k)$, $l(k)$, and $lc(k)$ by a constant factor?*

One possible approach would be to better understand how these functions depend on each other. For example, the current lower bound on $f(k)$ follows by a very simple argument from a lower bound on $l(k)$. Can this relation be improved?

Open Problem 2. *Is there a constant $c_0 > 1$ with $f(k) \geq c_0 l(k)/k$ for k large enough?*

So far, it seems hard to discriminate between $l(k)$ and $lc(k)$.

Open Problem 3. *Is there a constant $c_1 > 1$ such that $l(k) \geq c_1 lc(k)$ for k large enough?*

Note also that we have not even settled the computability of these functions.

Open Problem 4. *Are the functions $f(k)$, $l(k)$ and $lc(k)$ computable?*

While further progress as expressed in the "wish list" above is desirable, one should not forget about the broader picture. Our goal is to find interesting and at the same time simple combinatorial conditions on a CNF formula that entail satisfiability. The bounds on $l(k)$ (and $lc(k)$) express such conditions as degree bounds in the graph of dependencies (and graph of conflicts, resp.) of the clauses of a CNF formula. The consideration of linear CNF formulas imposes a restriction on the quality of dependencies – we have seen that this can make a significant difference in how many clauses are needed for an unsatisfiable k-CNF formula. A possible next step is to combine these two types of criteria.

We conclude with problems that arose in the course of the investigations. Firstly, the CNF formulas providing the best known upper bound on $f(k)$ for k large are obtained via the construction of suitable binary trees. Let $f_{\text{tree}}(k)$ be the largest integer d such that no (k, d)-tree exists. A detour to k-CNF formulas and employment of the Local Lemma shows $f_{\text{tree}}(k) \geq 2^k/(ek) - 2$. A more "direct approach" may allow improvement of this bound.

Open Problem 5. *Is there a constant $c_2 > 1$ such that $f_{\text{tree}}(k) \geq c_2 \frac{2^k}{ek}$ for infinitely many k?*

The known proof of small unsatisfiable linear k-CNF formulas is probabilistic. Known explicit constructions give huge formulas – not an unfamiliar situation.

Open Problem 6. *Give an explicit construction of an unsatisfiable linear k-CNF formula of singly exponential size.*

Acknowledgement. We thank Andreas Razen, Patrick Traxler, and Philipp Zumstein for many helpful discussions on the topic of this survey; Andreas Razen, in particular, for carefully reading the manuscript and many suggestions.

References

1. Alon, N., Spencer, J.H.: The Probabilistic Method, 3rd edn. John Wiley & Sons Inc., Chichester (2008)
2. Erdős, P., Lovász, L.: Problems and results on 3-chromatic hypergraphs and some related questions. In: Hajnal, A., Rado, R., Sós, V.T. (eds.) Infinite and Finite Sets (to Paul Erdős on his 60th birthday), vol. II, pp. 609–627. North-Holland, Amsterdam (1975)
3. Erdős, P.: On a combinatorial problem. Nordisk Mat. Tidskr. 11, 5–10, 40 (1963)
4. Erdős, P.: On a combinatorial problem. II. Acta Math. Acad. Sci. Hungar. 15, 445–447 (1964)
5. Shearer, J.B.: On a problem of Spencer. Combinatorica 5(3), 241–245 (1985)
6. Beck, J.: An algorithmic approach to the Lovász Local Lemma. I. Random Struct. Algorithms 2(4), 343–365 (1991)
7. Alon, N.: A parallel algorithmic version of the local lemma. Random Struct. Algorithms 2(4), 367–378 (1991)
8. Czumaj, A., Scheideler, C.: A new algorithm approach to the general Lovász Local Lemma with applications to scheduling and satisfiability problems. In: Proc. 32nd Ann. ACM Symp. on Theory of Computing, pp. 38–47 (2000)
9. Srinivasan, A.: Improved algorithmic versions of the Lovász Local Lemma. In: Proc. 19th Ann. ACM-SIAM Symp. on Discrete Algorithms, pp. 611–620 (2008)
10. Moser, R.A.: Derandomizing the Lovász Local Lemma more effectively. CoRR **abs/0807.2120** (2008)
11. Moser, R.A.: A constructive proof of the Lovász Local Lemma. CoRR **abs/0810.4812** (2008); Proc. 41st Ann. ACM Symp. on Theory of Computing (to appear)
12. Moser, R.A., Tardos, G.: A constructive proof of the general Lovász Local Lemma. CoRR **abs/0903.0544** (2009)
13. Erdős, P., Spencer, J.: Lopsided Lovász Local Lemma and Latin transversals. Discrete Appl. Math. 30(2-3), 151–154 (1991)
14. Lu, L., Székely, L.: Using Lovász Local Lemma in the space of random injections. Electron. J. Combin. 14(1), 13, Research Paper 63 (2007) (electronic)
15. Berman, P., Karpinski, M., Scott, A.D.: Approximation hardness and satisfiability of bounded occurrence instances of SAT. Electronic Colloquium on Computational Complexity (ECCC) 10(022) (2003)
16. Tovey, C.A.: A simplified NP-complete satisfiability problem. Discrete Appl. Math. 8(1), 85–89 (1984)
17. Kratochvíl, J., Savický, P., Tuza, Z.: One more occurrence of variables makes satisfiability jump from trivial to NP-complete. SIAM J. Comput. 22(1), 203–210 (1993)
18. Savický, P., Sgall, J.: DNF tautologies with a limited number of occurrences of every variable. Theoret. Comput. Sci. 238(1-2), 495–498 (2000)
19. Hoory, S., Szeider, S.: A note on unsatisfiable k-CNF formulas with few occurrences per variable. SIAM J. Discrete Math. 20(2), 523–528 (2006)
20. Gebauer, H.: Disproof of the neighborhood conjecture with implications to SAT. CoRR abs/0904.2541 (2009)
21. Beck, J.: Combinatorial Games: Tic Tac Toe Theory, 1st edn. Encyclopedia of Mathematics and Its Applications, vol. 114. Cambridge University Press, Cambridge (2008)

22. Stříbrná, J.: Between Combinatorics and Formal Logic, Master's Thesis. Charles University, Prague (1994)
23. Hoory, S., Szeider, S.: Computing unsatisfiable k-SAT instances with few occurences per variable. Theoret. Comput. Sci. 337(1-3), 347–359 (2005)
24. Kullmann, O.: New methods for 3-SAT decision and worst-case analysis. Theor. Comput. Sci. 223(1-2), 1–72 (1999)
25. Aharoni, R., Linial, N.: Minimal non-two-colorable hypergraphs and minimal unsatisfiable formulas. J. Comb. Theory, Ser. A 43(2), 196–204 (1986)
26. Davydov, G., Davydova, I., Kleine Büning, H.: An efficient algorithm for the minimal unsatisfiability problem for a subclass of CNF. Ann. Math. Artificial Intelligence 23(3-4), 229–245 (1998)
27. Kleine Büning, H.: An upper bound for minimal resolution refutations. In: Gottlob, G., Grandjean, E., Seyr, K. (eds.) CSL 1998. LNCS, vol. 1584, pp. 171–178. Springer, Heidelberg (1999)
28. Kleine Büning, H.: On subclasses of minimal unsatisfiable formulas. Discrete Appl. Math. 107(1-3), 83–98 (2000)
29. Szeider, S.: Homomorphisms of conjunctive normal forms. Discrete Appl. Math. 130(2), 351–365 (2003)
30. Porschen, S., Speckenmeyer, E., Randerath, B.: On linear CNF formulas. In: Biere, A., Gomes, C.P. (eds.) SAT 2006. LNCS, vol. 4121, pp. 212–225. Springer, Heidelberg (2006)
31. Abbott, H.: An application of Ramsey's Theorem to a problem of Erdős and Hajnal. Canad. Math. Bull. 8, 515–517 (1965)
32. Kostochka, A., Mubayi, D., Rödl, V., Tetali, P.: On the chromatic number of set systems. Random Struct. Algorithms 19(2), 87–98 (2001)
33. Scheder, D.: Unsatisfiable linear CNF formulas are large, and difficult to construct explicitely. CoRR abs/0905.1587 (2009)
34. Porschen, S., Speckenmeyer, E., Zhao, X.: Linear CNF formulas and satisfiability. Discrete Appl. Math. 157(5), 1046–1068 (2009)
35. Scheder, D.: Unsatisfiable linear k-CNFs exist, for every k. CoRR abs/0708.2336 (2007)
36. Ben-Sasson, E., Wigderson, A.: Short proofs are narrow—resolution made simple. J. ACM 48(2), 149–169 (2001)

Kolmogorov-Complexity Based on Infinite Computations

Günter Hotz

Fakultät für Mathematik und Informatik
Universität des Saarlandes

Abstract. We base the theory of Kolmogorov complexity on programs running on a special universal machine M, which computes infinite binary sequences $x \in \{0,1\}^\infty$. The programs are infinite sequences $p \in \{0,1\}^* \cdot 1 \cdot 0^\infty$. As length $|p|$ we define the length of the longest prefix of p ending with 1. We measure the distance $d(x,y) = 2^{-n}$ of $x, y \in \{0,1\}^\infty$ by the length n of the longest common prefix of x and y. $\Delta_M(x, 2^{-n})$ is the length of a minimal program p computing a sequence y with $d(x,y) \leq 2^{-n}$. It holds $\Delta_M(x, 2^{-n}) \leq \Delta_M(x, 2^{-(n+1)}) \leq n + 2$ for all n. We prove that the sets of sequences

$$K_{\Delta_M} := \bigcup_{c \in \mathcal{N}} \{x \in X^\infty : \Delta_M(x, 2^{-n}) > n - c \quad \text{for all n}\}$$

$$K_{\Delta_M}^{o(n)} := \{x \in X^\infty : n + 1 - \Delta_M(x, 2^{-n}) = o(n)\}$$

have the measure 1 for memoryless sources with equal probabilities for 0 and 1. The sequences in $K_{\Delta_M}^{o(n)}$ are Bernoulli sequences. The sequences in K_{Δ_M} define collectives in the sense of von Mises up to a set of measure 0 and the sequences in $K_{\Delta_M}^{o(n)}$ have this property in a certain resricted fom.

Keywords: monotone Kolmogorov complexity, infinite computations, collectives.

1 Introduction

Kolmogorov [1] based his concept to define binary random sequences on the prefix approximation of the infinite sequences and the size of shortest programs to compute the prefixes. An infinite sequence $x \in X^\infty$ is random in his sense if for all prefixes $x[n]$ of length n of x the difference $|n - |p_n||$ of n and the length $|p_n|$ of a shortest program p_n to compute the prefix on a universal Turing machine is bounded by a constant $c \in \mathcal{N}$ depending on x and the used universal machine. Martin-Löf [2] proved that such sequences x do not exist, but that the concept works, if one substitutes the $\forall n$ by $\forall^\infty n$. This set of random sequences has the measure 1. The reason for the discovered breakdowns of the prefix complexity is a relation between the length of the prefix and the content of the prefix.

This relation in our sense has nothing to do with the character of a sequence to be random but only with the interest on certain prefix lengths. So we change the

S. Albers, H. Alt, and S. Näher (Eds.): Festschrift Mehlhorn, LNCS 5760, pp. 55–73, 2009.

definition of Kolmogorov at another place as Martin-Löf did. We use the approximations of the random sequences by infinite computable sequences. This is close
to the concept of Schnorr, who used only programs p, which compute infinite sequences $M(p)$. But in this case too came in the connection with the length of the
prefix $x[i]$ of the sequence $x \in \{0,1\}^{\infty}$ by the condition $M(p)(i) = x[i]$. So if I
am interested in a shortest program p with $M(p)(i) = x[i]$ the computer has as
additional information i, i.e., $\log(i)$ bits, the reason for the non-monotone growing
length of the shortest programs to describe prefixes of the length i.

In our definition we do not expect that the computer on an argument i prints
out $x[i]$. We are looking for approximations of x by sequences $M(p)$ of a precision
$d(x, M(p)) < \epsilon$ and the size of shortest programs $p_{x,\epsilon}$ with this property. The
size of these programs fulfils the monotonicity relation $|p_{x,\epsilon}| \leq |p_{x,\epsilon^*}|$ for $\epsilon > \epsilon^*$.
We prove that the set of our random sequences has measure 1.

Additionally, we give a weaker definition for randomness and prove that each
sequence in this set is a Bernoulli sequence. This concept allows us, only based
on our complexity concept, to classify the sequences with convergent limit probabilities generated by information sources $(\{0,1\}, \mu)$ with $\mu : \{0,1\} \to \mathcal{R}$ a
probability distribution.

We additionally prove that a certain subset of these random sequences is
invariant under each blind selection of subsequences by computers, this means
that *Das Starke Gesetz der grossen Zahlen* for collectives of von Mises [8] holds.
The subset for which we are not able to prove this property has measure 0. This
we prove without any restriction for $x \in K_{\Delta_M}$ but in the case of $K_{\Delta_M}^{o(n)}$ under a
restriction, which has to do with the density of the components of x we select.

We do not need assumptions such as prefix free codes for the programs as
introduced by Chaitin [5] to get the original idea of Kolmogorov really working.
Essential for the simplicity of the theory is the existence of a print command, the
assumption of infinite computations and a special procedure technique. This is
not constructive in the classical sense. But the existing theory is not constructive
either. So why not use infinite computations if it simplifies the theory.

We use a special universal Turing machine M and because of proof technical
reasons two different nets of three such machines, which mathematically are
based on the scalar products of the machines and not on the machines simulated
by the universal machines. So the state sets are independent from the programs
running on the machines.

The books of Calude [7], Li and Vitányi [9], and Chaitin [6] give excellent
overviews about the existing theory. This paper is based on ideas presented
in [10] especially on the chapter $\Delta^* -$ Complexity. Our concept can easily be
extended to more general spaces.

2   Basic Definitions

2.1   The Universal Machine

Our machine is a special universal Turing-machine with **three** tapes: An input
tape, a computing tape and an output tape. Each tape uses the binary alphabet

$X := \{0,1\}$ and it is infinitely long to the right side. In the initial state each position on the computing tape and the output tape are equal to 0. X^* is the free monoid of finite words over the alphabet X and X^∞ the set of infinite sequences over X. We define $w \cdot v$ for $w \in X^*$ and $v \in X^* \cup X^\infty$ by concatenation. The inscription of the input tapes are elements of $X^* \cdot 0^\infty$. We define for $x \in X^* \cdot 0^\infty$ and $x \neq 0^\infty$

$$|x| := \max\{i \in \mathcal{N} : x(i) = 1\}$$

and

$$|x| := 1 \quad \text{for} \quad x = 0^\infty.$$

We consider only infinite converging computations. An infinite computation is converging iff the writing head visits each position of the output tape only once. We substitute the halting states of the traditional machine by the cycle state *move to the right*. This means that the terminating finite computations will be transformed into converging infinite computations.

We use different machine models. The machine M has three tapes: An input tape, which is the program tape. The machine has one computing tape and an output tape. The input tape is a read-only tape. The output tape is a write-only tape. Its writing head moves only to the right. It is special in the following sense: It allows a printing command, which we will define later precisely.

Our second machine $\widetilde{M}$ and the third machine $\overline{M}$ have three input tapes T_1, T_2, T_3, for three programs, which are read-only tapes two and three computing tapes, respectively, and one write-only output tape. The printing head of this tape moves only to the right side. The main program is on tape T_1. The tapes T_2 and T_3 are for procedures, which will be called by the main program. The program on tape T_3 of machine $\overline{M}$ is always a print instruction. The programs on tape T_2 may be print instructions or any other programs. The print instruction has the form $0p \in X^* \cdot 0^\infty$. The programs of the type $1p \in X^* \cdot 0^\infty$ will be interpreted as usual by the universal machine. The main program of $\overline{M}$ uses the computing tape C_1 and the program on tape T_2 uses the computing tape C_2.

The main program p_1 never writes on the output tape. It only calls and controls the activity of the programs p_2 on T_2 and p_3 on T_3. The programs p_2 and p_3 are never both active, but p_1 may be active together with each of the two other programs. If the activity of the programs p_2 and p_3 changes then the writing on the output tape will continue on the position it has been stopped. The machines $\widetilde{M}$ and $\overline{M}$ may be understood as a net of three machines of type M. One machine calls the others and controls them. But they have only one output tape. They are motivated by proof technical reasons. Based on these machines we get some lower bounds for the size of minimal programs of M. The constructions of these machines are mathematically based on standard cartesian products of the universal machines, not of the machines simulated on the universal machines. This is essential for the additivity of the complexities of the running programs.

All convergent computations of a program p of the machine M or of programs $p := p_1(p_2, p_3)$ of the machine $\widetilde{M}$ generate infinite sequences $x \in X^\infty$ on the output tape. We write in this case $x := M(p)$ or $x := \widetilde{M}(p)$, respectively. For the program $p := 0 \cdot w \cdot 0^\infty$, $w \in X^*$ on T_3 we write $print(w)$. It generates as output $w \cdot 0^\infty$. We look at our programs as infinite sequences with a finite number of 1's. The length of the program is defined as the length of the longest prefix ending with 1.

2.2   Infinite Sequences

We define $\bar{0} := 1$ and $\bar{1} := 0$ and for $x \in X^\infty$

$$\bar{x} := (\bar{x}(1), \bar{x}(2), \bar{x}(3), \ldots).$$

We write $x_i \oplus y_i$ for the addition modulo 2 for $x_i, y_i \in \{0, 1\}$ and for $x, y \in X^\infty$ we understand $x \oplus y$ as the componentwise application of the operation.

We define for $x \in X^\infty$

$$\|x\| := \sum_{i=1}^{\infty} \frac{x(i)}{2^i}$$

and as distance of $x, y \in X^\infty$

$$d(x, y) := \sum_{i=1}^{\infty} \frac{x(i) \oplus y(i)}{2^i}.$$

For $w \in X^*$ and $x \in X^\infty$ we write

$$w \prec x \quad \text{for} \quad w = x[n] := x(1) \cdot x(2) \cdot \ldots \cdot x(n)$$

where n is the length of w. For $x, y \in X^*$ it holds

$$d(x, y) < 2^{-n} \implies x[n] = y[n].$$

This means that x, y have a common prefix w of length n for $d(x, y) < 2^{-n}$. If x, y have w as common prefix, then it holds

$$d(x, y) \leq 2^{-n}.$$

We see that for $w \in X^n$ and $x \in X^\infty$ we get

$$d(w1x, w0x) = d(w1x, w1\bar{x}) = 2^{-n}.$$

Using the prefix distance we get for the expression on the right hand side as distance $2^{-[n+1]}$.

3   Kolmogorov - Complexity of Binary Sequences

3.1   Definitions and Simple Consequences

Given a sequence $x \in X^\infty$ we ask for computable sequences $y \in X^\infty$ such that $d(x, y) < \epsilon$ for a given $0 < \epsilon < 1$. We are interested in programs p of minimal length $|p|$, with $M(p) = y \in X^\infty$, which are approximations of a given precision of given $x \in X^\infty$. The program length of the program $p := p_1(p_2, p_3)$ with program p_i on tape T_i of $\widetilde{M}$ or $\overline{M}$ we define as

$$|p| := |p_1| + |p_2| + |p_3|.$$

Definition 1. *Let $x \in X^\infty$, $\epsilon > 0$ and M, $\widetilde{M}$, $\overline{M}$ our universal machines. We define*

$$\Delta_M(x, \epsilon) := \min\{|p| : d(x, M(p)) < \epsilon\},$$
$$\Delta_{\widetilde{M}}(x, \epsilon) := \min\{|p| : d(x, \widetilde{M}(p)) < \epsilon\},$$
$$\Delta_{\overline{M}}(x, \epsilon) := \min\{|p| : d(x, \overline{M}(p)) < \epsilon\}$$

Because each sequence $M(p)$, which is an approximation of precision $\epsilon_2 < \epsilon_1$ of x is also an approximation of precision ϵ_1, it follows

Lemma 1. *For $x \in X^\infty$ and $\epsilon_1 > \epsilon_2 > 0$ it follows*

$$\Delta_M(x, \epsilon_1) \le \Delta_M(x, \epsilon_2),$$
$$\Delta_{\widetilde{M}}(x, \epsilon_1) \le \Delta_{\widetilde{M}}(x, \epsilon_2) \quad and \quad \Delta_{\overline{M}}(x, \epsilon_1) \le \Delta_{\overline{M}}(x, \epsilon_2).$$

Remark: We may run on our machines $\widetilde{M}$ and $\overline{M}$ main programs p_1, which do nothing else as one call of the program p_2 or p_3. These programs may be programs for M which compute a best approximation for the given x and ϵ. If $c := |p_1|$, then we have

$$\Delta_M(x, \epsilon) < \Delta_{\widetilde{M}}(x, \epsilon) + c, \quad \Delta_{\overline{M}}(x, \epsilon) + c$$

holds for all $x \in X^\infty$ and $\epsilon > 0$.

We see that our variant of the original definition of Kolmogorov for $\epsilon := 2^{-n}$ leads to a monotone dependence of $\Delta_M(x, 2^{-n})$ from n.

Our *print*-operation guarantees a simple upper bound for the approximations, if we restrict to $\epsilon = 2^{-n}$.

Lemma 2. *Let be $x \in X^\infty$ and $\epsilon = 2^{-n}$ then it holds*

$$\Delta_M(x, \epsilon) < n + 1 \quad and \quad \Delta_{\widetilde{M}}(x, \epsilon) < n + 1 + c$$

with c independent from x and n.

3.2 (Δ_M, H)- Approximable Sequences

We define sets of sequences $x \in X^\infty$, which in a certain degree are approximable by computable sequences. Let H be the set of monotone, unbounded, and computable mappings $h : \mathcal{N} \to \mathcal{N}$.

Definition 2. *For $h \in H$ we define the sequence $x \in X^\infty$ approximable of degree h iff there exists $n_0 \in \mathcal{N}$ such that for all $n > n_0$*

$$n + 1 - \Delta(x, 2^{-n}) \ge h(n)$$

holds. We define

$$\Lambda_{\Delta_M}(h, n_0) := \{x \in X^\infty : \forall_{n > n_0}(n + 1 - \Delta_M(x, 2^{-n}) \ge h(n))\}$$

and

$$\Lambda_{\Delta_M}(H) := \bigcup_{h \in H, n_0 \in \mathcal{N}} \Lambda_{\Delta_M}(h, n_0)$$

$h : \mathcal{N} \to \mathcal{N}$. Λ_{Δ_M} *is the set of the (Δ_M, H)-approximable sequences.*

We first give a lower and upper bound for the average lengths of a minimal set of shortest programs $p \in X^n$, which compute sequences $M(p) \in X^\infty$ such that every word of X^n appears as prefix. An upper bound we get by $n+1$ the maximal length of *print*-commands, which compute the outputs $X^n \cdot 0^\infty$.

Counting the words $w1 \in X^k$ for $k = 1, 2, \ldots, n$ we get

$$\sum_{m=1}^{n-1} 2^m = 2^n - 1.$$

We did not count the *print*-command $print(0)$. So we have 2^n programs of length $\leq n$. By computing the average length of the 2^n shortest programs we get a lower bound for the average length

$$A_n := \sum_{w \in X^n} \frac{\Delta_M(w \cdot X^\infty, 2^{-n})}{2^n}$$

of the minimal programs we are interested in. There are 2^m of our programs of length $m+1$ available. So we get for the average length of the set of programs with length $\leq n-1$ not considering the trivial print command the expression $\frac{B_n}{2^n}$, where B_n is defined as follows:

$$B_n := \sum_{m=0}^{n-1} (m+1) \cdot 2^m = n \cdot 2^{n-1} + B_{n-1}.$$

We get as solution of this recursion

$$B_n = (n-1) \cdot 2^n + 1$$

Taking in account the trivial print command we get as a lower bound for A_n

$$A_n \geq \frac{(n-1) \cdot 2^n + 2}{2^n} = n + \frac{1}{2^{n-1}} - 1.$$

Using the existence of our *print*-commands we get $n+1$ as upper bound. So we proved the following lemma.

Lemma 3

$$n - 1 < \sum_{w \in X^n} \Delta_M(w \cdot X^\infty, 2^{-n}) \cdot 2^{-n} \leq n + 1$$

This is the base for the proof of the following lemma.

Lemma 4. *For the source* (X, μ) *with* $\mu(1) = \mu(0) = 2^{-1}$ *and each unbounded mapping* $h : \mathcal{N} \to \mathcal{N}$ *and each* $n_0 \in \mathcal{N}$ *it holds*

$$\mu(\Lambda_{\Delta_M}(h, n_0)) = 0$$

Proof: We define

$$\Lambda^n_{\Delta_M}(h) := \{x \in X^\infty : n + 1 - \Delta_M(x, 2^{-n}) \geq h(n)\}.$$

To decompose a following sum in two parts, we define

$$S_1 := \{w \in X^n : h(n) \leq n + 1 - \Delta_M(wX^\infty, 2^{-n})\}$$

and

$$S_2 := \{w \in X^n : h(n) > n + 1 - \Delta_M(wX^\infty, 2^{-n})\}$$

It follows from lemma 3

$$n - 1 < \sum_{S_1} \frac{\Delta_M(wX^\infty, 2^{-n})}{2^n} + \sum_{S_2} \frac{\Delta_M(wX^\infty, 2^{-n})}{2^n} \leq n + 1$$

and

$$n - 1 < (n + 1 - h(n)) \cdot \sum_{S_1} 2^{-n} + (n + 1) \cdot \sum_{S_2} 2^{-n}.$$

We use the abbreviation

$$\mu_n := \mu(\Lambda^n_{\Delta_M}(h))$$

and get

$$n - 1 < (n + 1 - h(n)) \cdot \mu_n + (n + 1) \cdot (1 - \mu_n)$$

and

$$\mu_n < \frac{2}{h(n)}$$

$h(n)$ is unbounded. Therefore there exists a sequence $n_0 < n_1 < n_2, \ldots$ with $n_i \in \mathcal{N}$ such that $h(n_i) < h(n_{i+1})$ for all $i \in \mathcal{N}$. It follows $\mu_{n_i} \to 0$. Obviously we have

$$\Lambda_{\Delta_M}(h, n_0) \subset \bigcap_{i=0}^\infty \Lambda^{n_i}_{\Delta_M}(h)$$

and therefore

$$\mu(\Lambda_{\Delta_M})(h, n_0) = 0$$

So it follows

Theorem 1

$$\mu(\Lambda_{\Delta_M}(H)) = 0.$$

We may substitute the set H of computable functions by a subset of H or we may extend H by non computable monotone mappings or by constant functions. The question is, which influence this has on a theory of random sequences. The following three lemmas give a first answer on this question.

Lemma 5. *To each countable set $\widetilde{H} := \{h_1, h_2, h_3, \ldots\}$ of unbounded monotone mappings there exists an unbounded monotone mapping $f : \mathcal{N} \to \mathcal{N}$, which is an asymptotic lower bound for all $h \in \widetilde{H}$.*

Proof: We define $f(n) := h_1(1)$ for

$$n < n_1 := \min\{k : h_1(k) > h_1(1) + 1, h_2(k) > h_1(1) + 1\}$$

and

$$f(n_1) := \min\{h_1(n_1), h_2(n_1)\} - 1$$

Let n_i and f be defined for $n \leq n_i$, then we define

$$n_{i+1} := \min\{k > n_i : h_1(k), h_2(k), \ldots h_{i+1}(k) > f(n_i) + 1\},$$

$$f(n) := h(n_i) \quad \text{for} \quad n_i < n < n_{i+1}$$

and

$$f(n_{i+1}) := \min\{h_1(n_{i+1}), \ldots h_{i+1}(n_{i+1})\} - 1$$

This defines f for all $n \in \mathcal{N}$ and it follows $f(n_i) < f(n_{i+1})$ and $f(n) < h_i(n)$ for $n \geq n_i$. h is asymptotically a lower bound for each $h \in \tilde{H}$ and it is not in $\tilde{H}$. This ends the proof of the lemma.

Obviously it holds

$$\Lambda_{\Delta_M}(h_i) \subset \Lambda_{\Delta_M}(f).$$

This result is true for each given counting of the set $\tilde{H}$. We define

$$\Lambda_{\Delta_M}(\tilde{H}) := \bigcup_{h \in \tilde{H}} \Lambda_{\Delta_M}(h)$$

and get

Lemma 6

$$\Lambda_{\Delta_M}(\tilde{H}) \subset \Lambda_{\Delta_M}(f)$$

The lemmas show that there is a gap between $\Lambda_{\Delta_M}(H)$ and $\Lambda_{\Delta_M}(\tilde{H})$, if $\tilde{H}$ includes not computable monotone mappings, which are lower bounds for all $h \in H$. We can close this gap by extending H by the constant functions as the following lemma shows.

Lemma 7. *To each monotone unbounded mapping $f : \mathcal{N} \to \mathcal{N}$ there exists an upward approximation by computable monotone and bounded functions*

$$a_{n_0}^f(n) := f(n) \quad \text{for} \quad n < n_0 \quad \text{and} \quad a_{n_0}^f(n) := f(n_0) \quad \text{for} \quad n \geq n_0$$

such that

$$\Lambda_{\Delta_M}(f) = \bigcap_{n_0 \in \mathcal{N}} \Lambda_{\Delta_M}(a_{n_0}^f)$$

The proof the lemma is obvious.

The lemma shows that we may restrict to monotone and computational mappings.

3.3   Closure Properties of Λ_{Δ_M}

We are interested in sets $\mathcal{H} \subset H$ such that $\Lambda_{\Delta_M}(\mathcal{H})$ has properties which are consistent with our intuition.

If $\Lambda_{\Delta_M}(\mathcal{H})$ is the set of algorithmically well approximable sequences then it should follow for $y, z \in \Lambda_{\Delta_M}(\mathcal{H})$ that $x \in \Lambda_{\Delta_M}(\mathcal{H})$ for the operations

$$x := y \vee z, \quad x := y \wedge z, \quad x := \overline{y}, \quad x := y \oplus z,$$

where $\oplus$ means the addition $(mod2)$. In other words $(\Lambda_{\Delta_M}(H), \vee, \wedge, \overline{\cdot})$ should be a boolean algebra.

It seems reasonable to look for a even stronger closure property: For $y \in \Lambda_{\Delta_M}(\mathcal{H})$ and each program p, which is able to use y as parameter in a reasonably restricted way it should follow $x := M(p; y) \in \Lambda_{\Delta_M}(\mathcal{H})$. This should hold not only for one but for each finite set of parameters. For shortness we discuss only the case of two parameters $y, z \in \Lambda_{\Delta_M}(\mathcal{H})$.

We extend our machine by two parameter tapes y, z, which are only readable and we allow for the reading heads only moves from left to right. We restrict these moves by the following condition. Let $r_x(t)$ the position of the printing head on the output tape of M at time t and $r_y(t), r_z(t)$ the position of the reading heads on y and z. We bound the possible moves of the reading heads on y and z as described by the relation

$$r_y(t), r_z(t) \leq r_x(t),$$

which does not allow the reading heads on the parameter tapes to move faster to the right as the writing head on the printing tape x.

If the extension of M by the parameter tapes y, z with program q under these conditions computes the output x we write

$$x := M(q; y, z).$$

The idea is to approximate x by use of programs p_y^n and p_z^n, which compute approximations of y and z, and the program q. We define

$$\mathcal{H} := \{h \in H : n + 1 - h(n) = o(n)\}.$$

$x \in \Lambda_M(\mathcal{H})$ is equivalent to the condition: There exists a computable function $g(n) = o(n)$ such that

$$\Delta_M(x, 2^{-n}) < g(n) \quad \text{for all} \quad n \in \mathcal{N}$$

For $y, z \in \Lambda_M(\mathcal{H})$ there exist $h_y, h_z \in \mathcal{H}$ such that for $g_y(n) := n + 1 - h_y(n)$ and $g_z(n) := n + 1 - h_z(n)$ there exist programs p_y^n, p_z^n with

$$|p_y^n| \geq g_y(n) = o(n), \quad d(M(p_y^n), y) < 2^{-n},$$

and

$$|p_z^n| \geq g_z(n) = o(n), \quad d(M(p_z^n), z) < 2^{-n}.$$

We use the programs q, p_y^n and p_z^n as procedures for a program p, which simulates the behavior of q to compute an approximation of x with the precision 2^{-n}. This can be done in the following way: p calls the program p_y^n, writes the results of the program on its computing tape, counts its write commands and stops it after n steps. Then it calls p_z and proceeds as before in the case of p_y. Having finished this process it calls q to use both arrays virtually extended by 0^∞ instead of the tapes y, z to compute a sequence $\tilde{x}$. We write for the result of this procedure

$$\tilde{x} := M(p(q, p_y^n, p_z^n)).$$

It approximates $x := M(q : y, z)$ with precision 2^{-n}.

We are able to realize this procedure technique with a program p^n with a length $o(n)$. It depends on n only that far as it has to count the mentioned n steps. So it is sufficient to show

$$|q| + |p_y^n| + |p_z^n| = o(n).$$

$|q|$ is constant and our assumption about y, z guarantees the existence of $h \in \mathcal{H}$ such that $g(n) := n + 1 - h(n)$ is an upper bound for $|p_y^n|$ and $|p_z^n|$. So it follows that there exists $h_1 \in \mathcal{H}$ such that $|p(q, p_y^n, p_z^n)| < n + 1 - h_1(n)$. This proves the following

Theorem 2

$$\Lambda_{\Delta_M}(\mathcal{H}) :=$$
$$\{x \in X^\infty : \exists_{h \in \mathcal{H}} \exists_{n_0 \in \mathcal{N}} (n + 1 - \Delta_M(x, 2^{-n}) \geq h(n) \quad \text{for all} \quad n > n_0)\}$$

is closed under the operation $x := M(p; y, z)$ *under the assumption of no preview on* y *and* z. *The closure under the boolean operations* $y \vee z, y \wedge z, y \oplus z, \bar{y}$ *are special cases of the first statement. The Operation* $\oplus$ *means the addition mod 2.*

It follows that $(\Lambda_{\Delta_M}(\mathcal{H}), \vee, \wedge, \bar{\cdot})$ is an infinite boolean algebra and $(\Lambda_{\Delta_M}(\mathcal{H}), \oplus)$ is an infinite abelian subgroup of the abelian group $(X^\infty, \oplus)$. We are interested in the quotient group $X^\infty / \Lambda_{\Delta_M}(\mathcal{H})$ because the elements $y, z \in X^\infty, y \equiv z (\mathrm{mod}\, \Lambda_{\Delta_M}(\mathcal{H}))$ are related under the aspect of the efficient computational approximation. If we consider y to be random then we may consider z as random, too.

Before we switch to the discussion of the random sequences we clarify the size of the set $\Lambda_{\Delta_M}(\mathcal{H})$ and that a restriction concerning the size of a preview is necessary.

There exists a simple algorithm to map X^∞ injectively to $\Lambda_{\Delta_M}(\mathcal{H})$.

Let $x \in X^\infty$ be a given sequence then we define $\delta : X^\infty \to \Lambda_{\Delta_M}(\mathcal{H})$ as follows: We first define for $i \in \mathcal{N}$ and $x_i \in \{0, 1\}$

$$\bar{\delta}_i(x_i) := x_i^{i+1}.$$

Inductively we define under the assumption that we have defined $y[n]$

$$y[n+1] := y[n] \cdot 1 \cdot \bar{\delta}(n+1, x_{n+1}) \quad \text{for} \quad x_{n+1} = x_n = 0,$$
$$y[n+1] := y[n] \cdot 0 \cdot \bar{\delta}(n+1, x_{n+1}) \quad \text{for} \quad x_{n+1} = x_n = 1,$$

$$y[n+1] := y[n] \cdot 10 \cdot \bar{\delta}(n+1, x_{n+1}) \quad \text{for} \quad x_n = 0, x_{n+1} = 1,$$
$$y[n+1] := y[n] \cdot 01 \cdot \bar{\delta}(n+1, x_{n+1}) \quad \text{for} \quad x_n = 1, x_{n+1} = 0.$$

This defines δ uniquely. It is obvious, that δ is injective.

We give an example: For $x[6] := 110100$ we get

$$11 \cdot 0 \cdot 111 \cdot 01 \cdot 0000 \cdot 10 \cdot 11111 \cdot 01 \cdot 000000 \cdot 1 \cdot 0000000$$

This sequence of length 35 is uniquely defined by the original sequence of length 6 and a short program p, which is the same for all n. In general we have a description of length $n + |p| = o(n^2)$ for a sequence of a length $\geq n^2$. It follows $\delta(x) \in \Lambda_{\Delta_M}(\mathcal{H})$. The cardinality of X^∞ and $\Lambda_{\Delta_M}(\mathcal{H})$ are equal. The not total inverse mapping δ^{-1} is computable, but it needs a preview on y of size $r(n) = O(n^2)$.

Observation: We are able by application of a very simple partial algorithm on $\Lambda_{\Delta_M}(\mathcal{H})$ by compressing sequences to generate the whole set X^∞.

3.4 Random Sequences

We define for $c \in \mathcal{N}$ motivated by Lemma 7 and following the idea of Kolmogorov

$$\Lambda_{\Delta_M}(c) := \{x \in X^\infty : \exists_{n_0 \in \mathcal{N}} (n + 1 - \Delta_M(x, 2^{-n}) \geq c \quad \text{for} \quad n > n_0)\}.$$

It follows in the notation used in the proof of Lemma 4 and on base of the same arguments

$$\mu(\Lambda_{\Delta_M})(c) < \frac{2}{c}.$$

We define

$$\Lambda_{\Delta_M}(\mathcal{N}) := \bigcap_{c \in \mathcal{N}} \Lambda_{\Delta_M}(c)$$

and get

$$\mu(\Lambda_{\Delta_M}(\mathcal{N})) = 0.$$

It follows for $K_{\Delta_M}(\mathcal{N}) := X^\infty - \Lambda_{\Delta_M}(\mathcal{N})$

$$\mu(K_{\Delta_M}(\mathcal{N})) = 1.$$

$K_{\Delta_M}(\mathcal{N})$ is the set of all sequences $x \in X^\infty$, for which exists a constant $c \in \mathcal{N}$ such that infinitely often $n + 1 - \Delta_M(x, 2^{-n}) < c$ holds.

The question is if we should consider the sequences $x \in K_{\Delta_M}(\mathcal{N})$ as random sequences. The following sections will give answers to this question. Our first question is how do the sequences behave globally over $\mathcal{N}$. This behavior is described by the function

$$f_x(n) := \max\{n + 1 - \Delta_M(x, 2^{-i}) : i \leq n\} \quad \text{for} \quad x \in K_{\Delta_M}(\mathcal{N}).$$

This function is not computable and we do not need to compute this function. An assumption about the behavior of this function has as consequence the existence of certain programs. In this sense we will use the function. We study our problem by applying our machines of type $\widetilde{M}$.

Let us assume that there exists $c_{n_0} \in \mathcal{N}$ such that

$$n + 1 - \Delta_M(n, 2^{-n}) < c_{n_0}$$

infinitely often and that $f_x(n)$ is unbounded. Then there exists for each $c \in \mathcal{N}$ a minimal $n_c \in \mathcal{N}$ such that $f_x(n_c) \geq c$. We define a program $\widetilde{p}^k := p_1(p_2, p_3^k)$ for $\widetilde{M}$ as follows: Let be $p_2 := p$, where $M(p)$ approximates x such that

$$n_c + 1 - \Delta_M(x, 2^{-n_c}) = f_x(n_c)$$

holds. The program $\widetilde{p}^k$ depends on n_c, too, but we construct for each n_c an infinite sequence of programs $\widetilde{p}^k$. The dependence on k comes in by the print program. We define the print programs by $p_3 := \mathbf{print}x[n_c : n_c + k]$ for $k = 2, 3, 4 \dots$. Remember that the print programm generates the infinite sequence $x[n_c : n_c + k] \cdot 0^\infty$. But if we use n_c as information for p_1, then our proof will not work. So we try to use $f_x(n_c)$ as information for p_1 to stop p_2 and to call p_3^k. The final definition will be given later.

The program p_1 first calls p_2 and counts the number n_t of moves of the printing head on the output tape depending on the number t of computing steps. It counts additionally the moves of the reading head of tape T_2 and computes the maximum $\max(t)$ of its positions after time t. If $\max(t) - n(t) = f_x(n_c)$, then p_1 stops the program p_2 and starts the program $p_3^k := \mathbf{print}x[\max(t) + 1 : \max(t) + k]$.

It may be that $\max(t) < n_c$ but this does not matter because for $k_0 := n_c - \max(t)$, $\widetilde{M}(\widetilde{p}^{k_0})$ approximates the sequence x with the same precision as $M(p)$ does. So instead of the program $p_2 := p$ we may choose the program $p_2 := p[1 : \max(t)]$ because the rest of the program p will not be used in $\widetilde{p}^k$. This means that we may assume $\max(t) = |p_2|$ and $m(t) = n_c$.

p_1 does not depend on k. So we have

$$|p_1| = \widetilde{c} + \log(f_x(n_c))$$
$$f_x(n_c) = n_c + 1 - |p_2|$$
$$n = n_c + k$$

where $\widetilde{c}$ is independent from k and c. It follows

$$n + 1 - \Delta_{\widetilde{M}}(x, 2^{-(n_c+k)}) \geq n + 1 - |\widetilde{p}^k|$$
$$= n + 1 - (|p_1^k| + |p_2| + |p_3^k|)$$
$$= n_c + k + 1 - (\widetilde{c} + \log(f_x(n_c)) + |p_2| + k + 1)$$
$$= n_c + k + 1 - \widetilde{c} - \log(f_x(n_c)) - |p_2| - k - 1$$
$$= n_c + 1 - |p_2| - 1 - \widetilde{c} - \log(f_x(n_c))$$
$$= f_x(n_c) - \log(f_x(n_c)) - (\widetilde{c} + 1)$$

For $c \to \infty$ it follows

$$n + 1 - \Delta_{\widetilde{M}}(x, 2^{-n}) \to \infty.$$

This contradicts our assumption: There exists a constant c_0 such that $n + 1 - \Delta_{\widetilde{M}}(x, 2^{-n}) < c$ infinitely often and it holds

$$\Delta_M(x, 2^{-n}) \geq \Delta_{\widetilde{M}}(x, 2^{-n}) - \bar{c}.$$

$\bar{c}$ is a constant defined by the length of a main program p_1 of $\widetilde{M}$ that calls a best approximation program p of M as $p_2 := p$. This program p_1 does not depend on n. It follows

Theorem 3. *For each $x \in K_{\Delta_M}$ there exists a constant c such that*

$$n + 1 - \Delta_M(x, 2^{-n}) < c.$$

for all $n \in \mathcal{N}$.

3.5   Other Classes of Random Sequences

We will discuss here a weaker condition for randomness. The class is of interest because we are able to prove that the sequences of this class are Bernoulli sequences. And the class is invariant under some restricted blind selections of subsequences by computers. These properties can be proved without any assumption concerning the existence of the limit behavior of mean values related to x.

We define

$$K_{\Delta_M}^{o(n)} := \{x \in X^\infty : n + 1 - \Delta_M(x, 2^{-n}) = o(n)\}$$

We prove first some elementary relations between the two sets of random sequences we are interested in.

Lemma 8

$$K_{\Delta_M}^{o(n)} = K_{\Delta_M}^{0(n)} \oplus \Lambda_{\Delta_M}(\mathcal{H})$$

Proof: From $0^\infty \in \Lambda_{\Delta_m}(\mathcal{H})$ it follows

$$K_{\Delta_M}^{o(n)} \subset K_{\Delta_M}^{0(n)} \oplus \Lambda_{\Delta_M}(\mathcal{H})$$

For the proof of the inclusion in the opposite direction we choose $x = y \oplus z$ with $x \in K_{\Delta_M}^{o(n)}$ and $z \in \Lambda_{\Delta_M}(\mathcal{H})$. It follows $x \oplus z = y$. Using a variant of the concept of the machine $\widetilde{M}$ to compute the approximation of y on base of minimal programs p_2 and p_3 to approximate x respective z with $c := |p_1|$ we get

$$n + 1 - \Delta_M(y, 2^{-n}) \geq n + 1 - \Delta_M(x, 2^{-n}) - \Delta_M(z, 2^{-n}) - c.$$

We divide the inequality by n and apply our assumptions about y and z. So it follows for $n \to \infty$

$$n + 1 - \Delta_M(x, 2^{-n}) = o(n)$$

this means $x \in K_{\Delta_M}^{o(n)}$ as claimed by the lemma.

It follows

$$K_{\Delta_M} \subset K_{\Delta_M}^{o(n)} = K_{\Delta_M}^{o(n)} \oplus \Lambda_{\Delta_M}(\mathcal{H})$$

and

Lemma 9
$$(K_{\Delta_M}(\mathcal{N}) \oplus \Lambda_{\Delta_M}(\mathcal{H})) \subset K_{\Delta_M}^{o(n)}$$

It is open if the lemma remains true if we substitute $\subset$ by $=$.

Bernoulli Sequences. In a first step we prove our results under some assumptions about the limit behavior about $x \in K_{\Delta_M}^o$. To formulate these conditions we define a characteristic function $\chi : A \times A \to \{0, 1\}$

$$\chi(v, w) = 1 \Leftrightarrow v = w.$$

and $x[l : m] := x_l \cdot \ldots \cdot x_m$ for $l < m$.

Theorem 4. *Under the assumption of the existence of the following limits we define*

$$p_w := \lim_{k \to \infty} \frac{\sum_{i=0}^{k-1} \chi(w, x[i \cdot n + 1 : (i+1) \cdot n])}{k} \quad for \quad w \in A^n.$$

We claim

$$x \in K_{\Delta_M}^{o(n)} \Rightarrow p_w = \frac{1}{2^n}$$

for all n and all $w \in A := X^n$

Proof: Let be $k, n \in \mathcal{N}$ and $A := X^n$. We consider the prefixes $x[n \cdot k] \in A^k$ of x and define for $w \in A$

$$n_w := \sum_{i=0}^{k-1} \chi(w, x[i \cdot n + 1 : (i+1) \cdot n])$$

and

$$p_{n,k}(w) := \frac{n_w}{k}$$

It follows

$$\sum_{w \in A} p_{n,k}(w) = 1 \quad \text{and} \quad \lim_{k \to \infty} p_{n,k}(w) = p_w.$$

We use the well known construction of minimal prefix free codes based on the Kraft inequality in the special case

$$c : A^* \to \{0, 1\}^*.$$

We define $i : A \to \mathcal{N}$ uniquely by

$$- \log p_{n,k}(w) \le i(w) < - \log p_{n,k}(w) + 1.$$

This is equivalent to

$$p_{n,k}(w) \ge 2^{-i(w)} > \frac{1}{2} \cdot p_{n,k}(w).$$

It follows

$$1 = \sum_{w \in A} p_{n,k}(w) \geq \sum_{w \in A} 2^{-i(w)} > \frac{1}{2}.$$

As we know from the coding theorem we can find a prefix free code c such that

$$- \log p_{n,k}(w) \leq |c(w)| = i(w) < - \log p_{n,k}(w) + 1 \quad \text{for} \quad w \in A. \tag{1}$$

Using the identy

$$c(x[n \cdot k]) = \prod_{l=0}^{k-1} c(x[l \cdot n + 1] : (l+1) \cdot n)$$

we get

$$|c(x[n \cdot k])| = \sum_{l=0}^{k-1} |c(x[l \cdot n + 1] : (l+1) \cdot n)| = \sum_{w \in A} |c(w)| \cdot n_w.$$

We define the entropy

$$H(p_{n,k}) := - \sum_{w \in A} p_{n,k}(w) \cdot \log p_{n,k}(w)$$

an get by summing up (1)

$$H(p_{n,k}) \leq \frac{1}{k} \cdot |c(x[n \cdot k])| = \frac{1}{k} \cdot \sum_{w \in A} n_w \cdot |c(w)| \tag{2}$$

$$< - \sum_{w \in A} \left(\frac{n_w}{k} \log \frac{n_w}{k} - \frac{n_w}{k} \right) = H(p_{n,k}) + 1 \tag{3}$$

We define $p_n := \lim_{k \to \infty} p_{n,k}$ and get

$$\lim_{k \to \infty} H(p_{n,k}) = H(p_n).$$

From (2) it follows

$$|c(x[n \cdot k])| \leq k \cdot H(p_{n,k})$$

and

$$\lim_{k \to \infty} \frac{|c(x[n \cdot k])|}{k} \leq H(p_n)$$

If $H(p_n) < n$, then it follows $n \cdot k - |c(x[n \cdot k])| \neq o(n \cdot k)$. c depends on k, but it can be computed and the application $c(w)$ is computable. So it follows that $n \cdot k - \Delta_M(x, 2 - n \cdot k) \neq o(n \cdot k)$ and x is not in $K^o_{\Delta_M}$.

This theorem can be generalized:

Theorem 5

$$x \in K^{o(n)}_{\Delta_M} \Rightarrow \{x \quad is \ a \ Bernoulli \ sequence\}$$

This means that we do not need the assumption about the existence of the limit.

Proof: If one of the limits we assumed to exist in the theorem above does not exist, then there exist at least two limit points. To each such limit point exists a subsequence $n_1 < n_2 < n_3 < \ldots \in \mathcal{N}$, for which a unique limit exists. Applying the construction our proof is based on, we get a compression of x by a factor $0 < \alpha \leq 1$. For at least one of the limit points we get a compression by a factor $\alpha < 1$. This contradicts the assumption $x \in K_{\Delta_M}^{o(n)}$ because the oscillations of $n + 1 - \Delta_M(x, 2^{-n})$ are bounded by $o(n)$. End of the proof.

This result can be generalized to

Theorem 6. *Let be* $\alpha, \epsilon \in [0,1]$, $\alpha = H(\epsilon, 1 - \epsilon)$ *the entropy and*

$$K_{\Delta_M}^{[\alpha]} := \{x \in X^\infty : \alpha \cdot n - \Delta_M(x, 2^{-n}) = o(n)\}.$$

For

$$x \in K_{\Delta_M}^{[\alpha]}$$

and each sequence

$$x_1 < x_2 < x_3 < \ldots \in \mathcal{N}$$

it holds

$$p_w := \lim_{l \to \infty} \frac{\sum_{i=0}^{k-1} \chi(w, x[i \cdot |w| + 1 : (i+1) \cdot |w|])}{k_l} = \epsilon^{m_1} \cdot (1 - \epsilon)^{|w| - m_1},$$

where m_1, k_l *are defined by*

$$m_1 := \sum_{i=1}^{|w|} \chi(1, w_i) \quad \text{and} \quad k_l := \lfloor \frac{n_l}{|w|} \rfloor$$

Proof: If there exist two different sequences $n_1 < n_2 < n_3 < \ldots$ with two different limit points α_1, α_2 then there exist oscillations of $\Delta_M(x, 2^{-n})$ of the size $|\alpha_1 - \alpha_2|$. This contradicts the restriction $\alpha \cdot n - \Delta_M(x, 2^{-n}) = o(n)$ for these oscillations.

Invariance Properties of K_{Δ_M} and $K_{\Delta_M}^{o(n)}$. A subsequence of a random sequence generated by an algorithm, which blindly selects the elements for the subsequence should be again a random sequence [8]. This should be even the case if the algorithm knows the prefix of the random sequence up to the position before the position i it has to decide "select or not select" x_i. We will prove here this invariance property only for the special case under additional assumptions. In the case of K_{Δ_M} we prove this not for each sequence, but only for sequences of the subset $K_{\Delta_{\overline{M}}} \subset K_{\Delta_M}$. In the case of $K_{\Delta_M}^{o(n)}$ we restrict the selection procedures by a density condition.

Let $i : \mathcal{N} \to \mathcal{N}$ and $j : \mathcal{N} \to \mathcal{N}$ be strictly monotone and computable mappings with the following properties:

$$i(\mathcal{N}) \cup j(\mathcal{N}) = \mathcal{N} \quad \text{and} \quad i(\mathcal{N}) \cap j(\mathcal{N}) = \emptyset,$$

We define

$$i(x) := (x_{i(1)}, x_{i(2)}, x_{i(3)}, \ldots) \quad \text{and} \quad j(x) := (x_{j(1)}, x_{j(2)}, x_{j(3)}, \ldots)$$

for $x \in X^\infty$ and

$$n_1 := \sharp\{k \in \mathcal{N} : i(k) \leq n\} \quad \text{and} \quad n_2 := \sharp\{k \in \mathcal{N} : j(k) \leq n\}.$$

If S is a finite set, then $\sharp S$ means the number of elements in S. It follows

$$n_1 + n_2 = n.$$

Theorem 7. *Under the assumption* $x \in K_{\Delta_{\bar{M}}}$, *where* $K_{\Delta_{\bar{M}}} \subset K_{\Delta_M}$ *for a net* $\bar{M}$ *of three machines it holds*

$$x \in K_{\Delta_M} \Rightarrow i(x), j(x) \in K_{\Delta_M}$$

and under the condition $n_1, n_2 = \Omega(n)$ *it holds*

$$x \in K_{\Delta_M}^{o(n)} \Rightarrow i(x), j(x) \in K_{\Delta_M}^{o(n)}$$

Proof: We get approximations of the precision 2^{-n} by computing approximations of $y := i(x)$ and $z := j(x)$ of precision 2^{-n_1} and 2^{-n_2}, respectively. Let p_y and p_z be programs, which compute infinite sequences to approximate y and z with precision 2^{-n_1} and 2^{-n_2}, respectively. p_1 and p_2 compute the mappings i and j, respectively. We describe a program p that uses variants of the programs p_1, p_2, p_y, p_z as procedures to compute an approximation $\tilde{x}$ of x with the precision 2^{-n}.

We substitute the programs p_y, p_z by programs $print(p_y), print(p_z)$, which we will define later. the program p we define as follows.

repeat infinitely often{ $i := 1$; $j := 1$; $k := 1$
    if $p_1(i) = k$ *then* $print(p_y)$; $i := i + 1$; $k := k + 1$
        else $print(p_z)$ $j := j + 1$; $k := k + 1$}

We modify p_y as follows.

The call $print(p_y)$ starts the program in its last state when it has been stopped.

The program state of p_y after the printing on the output tape will substituted by a stop state.

The procedure $print(p_z)$ we define analogously.

We see that the sizes of the programs p, p_1, p_2 are independent from n. The size of the programs p_y, p_z has not been changed. So we get for the size $|P_n|$ of $P_n := p(p_1, p_y, p_2, p_z)$ relative to the universal machine $\bar{M}$ consisting of two components of type M to compute p_y and p_z and a component to compute p, p_1, p_2 the relation

$$\Delta_{\bar{M}}(x, 2^{-n}) \leq |P| = \Delta_M(y, 2^{-n_1}) + \Delta_M(z, 2^{-n_2}) + c_p$$

with $c_p := |p| + |p_1| + |p_2|$. Using the relation $n = n_1 + n_2$ we get

$$n + 1 - \Delta_M(x, 2^{-n}) \geq (n_1 + 1 - \Delta_M(y, 2^{-n_1})) + (n_2 + 1 - \Delta_M(z, 2^{-n_2})) - (c_p + 1)$$

Under the assumption $x \in K_{\Delta_M}$ it follows, that there exists a $c \in \mathcal{N}$ such that $n + 1 - \Delta_M(x, 2^{-n}) < c$ for all n. It follows

$$n_1 + 1 - (\Delta_M(y, 2^{n_1}) < c + c_p + 1 \quad \text{and} \quad n_2 - \Delta_M(z, 2^{-n_2}) < c + c_p + 1$$

for all n_1 and all n_2. This means that $y, z \in K_{\Delta_M}$ as claimed by the first part of the theorem.

It remains to discuss the case $x \in K_{\Delta_M}^{o(n)}$. In this case it follows

$$n_1 - \Delta_M(y, 2^{-n_1}) = o(n) \quad \text{and} \quad n_2 - \Delta_M(z, 2^{-n_2}) = o(n)$$

For the proof of the theorem we need $o(n_1)$ and $o(n_2)$, respectively, on the right hand side of our equations. This follows for $n_1, n_2 = \Omega(n)$ as assumed in our theorem.

Lemma 10

$$\mu(K_{\Delta_M} - K_{\Delta_{\bar{M}}}) = 0$$

Proof: Analogously to the proof of $\mu(K_{\Delta_M}) = 1$, one proves $\mu(K_{\Delta_{\bar{M}}}) = 1$. It holds $K_{\Delta_{\bar{M}}} \subset K_{\Delta_M}$ because the machine $\bar{M}$ can simulate M by using the identical program p_x on one of the two submachines controlled by a program p independent from p_x. So the lengths of the two programs on M and $\bar{M}$ differ only by the constant $|p|$. It may be that an even shorter program configuration on $\bar{M}$ exists to compute an approximation of x of the same precision. But it cannot happen that a sequence relative to $\bar{M}$ is random and is not relative to M. The claim of the lemma follows.

4   Concluding Remarks

We have seen that the use of infinite computations and generalizations of the prefix approximation simplifies the theory. In no step we did really use assumptions on computability of the mappings in $h \in H$ or assumptions concerning the complexity of runtime of programs p to define Δ_M. Complexity aspects come in only in connection with the application of procedure technics or in programs to compute approximations on base of the coding theorem, which we applied in the proof of our last theorems. But this complexities are all on a very low level. Complexity hierarchies may play an important role in connection with the definition of Δ_M as C. P. Schnorr has proved, [3, 4]. Hierarchies may be generated too by networks of machines as we considered in a special case. But it is open if $K_{\Delta_M} \neq K_{\Delta_{\bar{M}}}$.

References

1. Kolmogorov, A.N.: Drei Zugänge zur Definition des Begriffs Informationsgehalt. Probl. Peredaci Inform. 1, 3–11 (1965) (in Russian)
2. Martin Löf, P.: The Definition of Random Sequences. Information and Control 8, 602–619 (1966)
3. Schnorr, C.-P.: Zufälligkeit und Wahrscheinlichkeit, eine algorithmische Begründung der Wahrscheinlichkeitstheorie. Lecture Notes in Mathematics, vol. 212, pp. 1–212. Springer, Heidelberg (1971)
4. Schnorr, C.-P.: Eine neue Charakterisierung der Zufälligkeit von Folgen, Habilitationsschrift zur Erlangung der Venia Legendi im Fach Mathematik der Universität des Saarlandes (1969)
5. Chaitin, G.I.: On the Length of Programs to Compute Finite Binary Sequences. J. Assoc. Comp. Machin. 13, 547–569 (1969)
6. Chaitin, G.I.: Algorithmic Information Theory. Cambridge University Press, Cambridge (1987)
7. Calude, C.: Theories of Computational Complexity. North-Holland, Amsterdam (1988)
8. von Mises, R.: Grundlagen der Wahrscheinlichkeitstheorie. Math. Zeitschrift 5, 5–99 (1910)
9. Li, M., Vitányi, P.: An Introduction to Kolmogorov Complexity and its Applications, pp. 1–546. Springer, Heidelberg (1993)
10. Hotz, G., Gamkrelidze, A., Gärtner, T.: Approximation of Arbitary Sequences by Computable Sequences - A new Approach to Chaitin-Kolmogorov-Complexity, 1–18 (unpublished, 2007); Gärtner T., Hotz, G.: Approximation von Folgen durch berechenbare Folgen - Eine neue Variante der Chaitin-Kolmogorov-Komplexität, Technischer Bericht A 01/02, März 2002, Fakultät für Mathematik und Informatik der Universität des Saarlandes, pp. 1–19
11. Chadzelek, T., Hotz, G.: Analytic Machines. Theoretical Computer Science 219, 151–167 (1999)
12. Hotz, G.: Algorithmische Informationstheorie, pp. 1–142. Teubner Texte zur Informatik, B. G. Teubner Verlag (1997)

Pervasive Theory of Memory

Ulan Degenbaev[1,*], Wolfgang J. Paul[1,**], and Norbert Schirmer[2,*]

[1] Saarland University, P.O. Box 15 11 50, 66041 Saarbrücken, Germany
[2] German Research Center for Artificial Intelligence (DFKI),
P.O. Box 15 11 50, 66041 Saarbrücken, Germany

Abstract. For many aspects of memory theoretical treatment already exists, in particular for: simple cache construction, store buffers and store buffer forwarding, cache coherence protocols, out of order access to memory, segmentation and paging, shared memory data structures (e.g. for locks) as well as for memory models of multi-threaded programming languages. It turns out that we have to unite all of these theories into a single theory if we wish to understand why parallel C compiled by an optimizing compiler runs correctly on a contemporary multi core processor. This pervasive theory of memory is outlined here.

1   Introduction

One subproject of the Verisoft-XT project[1] is to formally verify as big a portion as possible of the Microsoft Hyper-V virtualization product that is shipped as a component of Microsoft Windows Server 2008. This hypervisor is a multi-threaded C program with involved parallel algorithms and external assembler functions running in translated mode on contemporary multi core processors. The verification tool VCC [1] used to verify such programs is developed in parallel with the verification effort for the hypervisor and other programs. This paper is motivated by the question how one would prove the soundness of VCC. The rough road map is clear and was for instance followed with formal proofs in the former Verisoft project[2] [2]:

1. Define a semantics S for the subset C' of C used in the project. In the former Verisoft project big step and small step semantics for C'=C0 were used [3, 4, 5].
2. Show that the verification condition generator used is sound with respect to semantics S [5, 6].
3. Show that the compiler used correctly translates programs from C' to the instruction set architecture (ISA) of the processor used. In the Verisoft project a non optimizing compiler from C0 to the ISA of the VAMP processor [7, 8] was verified [3, 4].

* Work funded by the German Federal Ministry of Education and Research (BMBF) in the framework of the Verisoft XT project under grant 01 IS 07 008.
** Work partly funded by the German Federal Ministry of Education and Research (BMBF) in the framework of the Verisoft XT project under grant 01 IS 07 008.
[1] http://www.verisoftxt.de
[2] http://www.verisoft.de

S. Albers, H. Alt, and S. Näher (Eds.): Festschrift Mehlhorn, LNCS 5760, pp. 74–98, 2009.

4. in case one has doubts that the ISA in the manuals is the ISA realized by the hardware: show that the processor hardware correctly interprets the ISA [9].

In the context of the hypervisor effort we have to deviate from this road map due to the following difficulties:

1. Complexity of the processor: the documentation of the x64 ISA of contemporary multi core processors consists of thousands of pages [10, 11]. Of course the building plans of the processors are not public. Even if we had access to them they would be too complex to be completely verified using the present state of the art tools.
2. Memory model of the processor: modern processors use a weak shared memory model [12, 13, 14, 15]. A ten page white paper [16] is supplied to clarify this model beyond the thousands of pages of documentation.
3. Complexity of the compiler: in case of the hypervisor an optimizing Microsoft compiler (to whose source code we could gain access) translates multi-threaded C programs to the x64 ISA. This compiler is also too complex for present verification technology.
4. Compiler correctness: the theoretical treatment of compiler correctness for target architectures with a weak memory model is still a field of ongoing research [17, 18].

We proceed as follows: we first outline how to reverse engineer a memory system for processors which is consistent with the documentation [10, 11, 16] and with our ideas how to build processors [19, 20]. Section 3.2 gives simple sufficient conditions for store buffers (between processors and memory) to become invisible, namely in case of single processors and, trivially, in case of fenced memory transactions (a fenced transaction is only executed when the store buffer is empty). In the spirit of [7] Section 3.3 sketches very briefly how to show hardware correctness of a memory system consisting of a single cache and a main memory. In Section 3.3 we outline a proof of the corresponding result for a cache coherent shared memory. In order to obtain the result we later need, one has to combine three arguments: i) a classical transaction based correctness proof for cache coherence protocols, ii) its extension to compatible families of protocols as introduced in [21] and used in modern processors, and iii) a construction of the sequential order from the termination times of hardware transactions. In Section 3.4 we outline the arguments, why translated 'linear memory' is realized by multi level address translation. In Section 3.5 we reverse engineer a multi core processor with Tomasulo scheduler, memory management units, store buffers as well as coherency snooping as introduced in [22] and outline the correctness proof.

Assuming that we guessed the memory model correctly we then show in Section 3.6 how to initialize a contemporary multi core processor such that the hardware threads see the weak memory model derived previously in translated linear memory.

Finally we turn to the theory of compilation for multi-threaded C programs in weak memory models. Starting from a small step semantics for sequential programs we derive as a starting point an unrestricted naive parallel C semantics, which unfortunately we don't know how to compile into an efficient parallel assembler program. In Section 4.2 we review the correctness theorem from [3] for a non optimizing compiler for a sequential subset of C and then modify its statement (without proof) for optimizing compilers for multi-threading code; for a formal correctness proof of an optimizing compiler for a sequential subset of C see [23]. In the short Section 4.3 we sketch how to compile volatile variables such that in the compiled program they form a sequentially consistent portion of the weak memory. Using test and set operations on volatile variables we can implement locks which in turn permit to implement synchronized parallel C; this last step is explained in Section 4.4.

2   Notation

We denote the concatenation of bit strings $a \in \{0,1\}^n$ and $b \in \{0,1\}^m$ by $a \circ b$. For bits $x \in \{0,1\}$ and positive natural numbers $n \in \mathbb{N}^+$ we define inductively $x^1 = x$ and $x^n = x^{n-1} \circ x$. Thus, for instance $0^5 = 00000$ and $1^2 = 11$.

Overloading symbols like $+$, $\cdot$, and $<$ we will allow arithmetic on bit strings $a \in \{0,1\}^n$. In these cases arithmetic is binary modulo 2^n (with nonnegative representatives).

We model memories m as mappings from addresses a to byte values $m(a)$. For natural numbers d we denote by $m_d(a)$ the content of d consecutive memory cells (from right to left) starting at address a, so $m_d(a) = m(a + d - 1) \circ \cdots \circ m(a)$. We select ranges of a bit string by $x[hi{:}lo]$, e.g. $x[11{:}0]$ to select the 12 least significant bits of x.

3   Architecture Aspects

3.1   Sequential Memory

The state of a sequential memory with address range A and data range D is modeled by a function

$$m : A \to D$$

where $m(a)$ denotes the current content of memory cell with address a. We consider here three kinds of atomic memory transactions: read, write as well as test and set. We number transactions with indices $i \in \mathbb{N}_0$ and define the predicates

- $r(i)$: transaction i is a read,
- $w(i)$: transaction i is a write, and
- $ts(i)$: transaction i is test and set.

With each transaction i we associate an address $ad(i)$ and (input or output) data $data(i)$. We define the memory content before transaction i by m^i. The semantics of read, write and test and set transactions can then be defined by:

- $r(i) \rightarrow data(i) = m^i(ad(i)) \wedge m^{i+1}(a) = m^i(a),$

- $w(i) \rightarrow m^{i+1}(a) = \begin{cases} data(i) & \text{if } ad(i) = a, \\ m^i(a) & \text{otherwise, and} \end{cases}$

- $ts(i) \rightarrow data(i) = \begin{cases} 1 & \text{if } m^i(ad(i)) = 0, \\ 0 & \text{otherwise} \end{cases}$

$\wedge$

$$m^{i+1}(a) = \begin{cases} 1 & \text{if } ad(i) = a \wedge m^i(ad(i)) = 0, \\ m^i(a) & \text{otherwise.} \end{cases}$$

The predicate

$$W(a, i) \equiv \exists j < i : ad(j) = a \wedge (w(j) \vee (ts(j) \wedge data(j) = 1))$$

says that memory at address a has been written before transaction i. For such a and i we define the last transaction before transaction i that wrote to address a

$$last(a, i) = \max\{j < i : ad(j) = a \wedge (w(j) \vee (ts(j) \wedge data(j) = 1))\}.$$

Because $m^{j+1}(a) = m^j(a)$ for $j \in [last(a, i) + 1 : i - 1]$ one has

Lemma 1
$$m^i(a) = \begin{cases} m^{last(a,i)+1}(a) & \text{if } W(a, i), \\ m^0(a) & \text{otherwise} \end{cases}$$

and hence

$$r(i) \rightarrow data(i) = \begin{cases} data(last(ad(i), i) & \text{if } W(a, i) \\ m^0(ad(i)). \end{cases}$$

Observe that any system obeying the last equation defines a memory system, namely

$$m^i(a) = \begin{cases} data(last(a, i)) & \text{if } W(a, i), \\ m^0(a) & \text{otherwise.} \end{cases}$$

3.2   Store Buffers

A store buffer sb is a small queue between processor and memory m storing pending write transactions (see Fig. 1). We provide store buffer entries sbe with the following components:

- $sbe.ad$: the address of the write transaction,
- $sbe.data$: the data to be written, and
- the ghost component $sbe.index$: the index of the write transaction.[3]

---

[3] Recall that ghost components are not implemented and serve only for mathematical arguments.

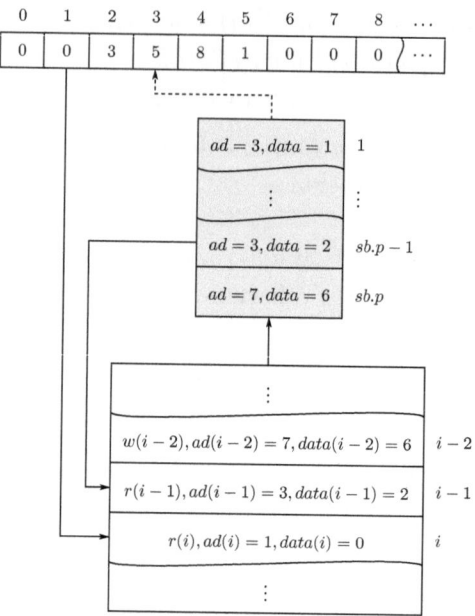

Fig. 1. Store buffer

We model a configuration of a store buffer as a pair $sb = (sb.p, sb.m)$ where $sb.p$ is the number of currently pending store requests and $sb.m$ maps the set $[1 : sb.p]$ to the set of store buffer entries. Initially the store buffer is empty

$$sb^0.p = 0.$$

If transaction i is a write request, its index, address and data are inserted at the end of the queue. Thus for the new configuration sb' we have

$$sb'.p = sb.p + 1$$
$$sb'.m(k) = sb.m(k) \text{ if } k < sb'.p$$
$$sb'.m(sb'.p).ad = ad(i)$$
$$sb'.m(sb'.p).data = data(i)$$
$$sb'.m(sb'.p).index = i.$$

If a write request is sent to the memory, it is deleted from the front of the queue

$$sb'.p = sb.p - 1$$
$$sb'.m(k) = sb.m(k + 1) \text{ for } 1 \leq k \leq sb'.p.$$

The store buffer stores requests in temporal order, i.e. we have

Invariant 1

$$k < k' \rightarrow sb.m(k).index < sb.m(k').index.$$

Predicate $hit(a, sb)$ signals that a write request with address a is in the store buffer:

$$hit(a, sb) \equiv \exists k \leq sb.p : sb.m(k).ad = a.$$

The entire system consists of

- the processor, and
- the memory system consisting of memory and store buffer.

A memory system step deletes the first store buffer entry and sends it to the memory. This maintains

Invariant 2. *Let $j = sb.m(1).index - 1$. Then*

$$m^i(a) = \begin{cases} data(last(a, j)) & : W(a, j) \\ m^0(a). \end{cases}$$

A processor step sends a transaction to the memory system. Write transactions are written into the store buffer. A test and set transaction causes the store buffer to be flushed before being executed. Read transactions $r(i)$ are answered using store buffer forwarding: in case of a store buffer hit $(hit(ad(i), sb)$ we determine the last store buffer entry which has a write request leading to the hit

$$k = max\{k' : sb.m(k').ad = ad(i)\}$$

and return the data in store buffer entry $sb.m(k')$. Otherwise we return data from memory

$$data(i) = \begin{cases} sb.m(k).data & hit(ad(i), sb) \\ m(ad(i)). \end{cases}$$

The invariants imply that the memory system behaves like a single memory:

Lemma 2

$$r(i) \rightarrow data(i) = \begin{cases} last(ad(i), i) & if\ W(ad(i), i), \\ m^0(ad(i)) & otherwise. \end{cases}$$

Thus store buffers are invisible in systems with a single processor. If several processors are connected with store buffers to a shared memory (Fig. 2) the store buffers are visible. Consider the following two threads where shared variables x and y are initially 0 and r1 and r2 are local variables stored in registers:

```
    x = 1;              y = 1;
    r1 = y;             r2 = x;
```

Thread 1	Thread 2

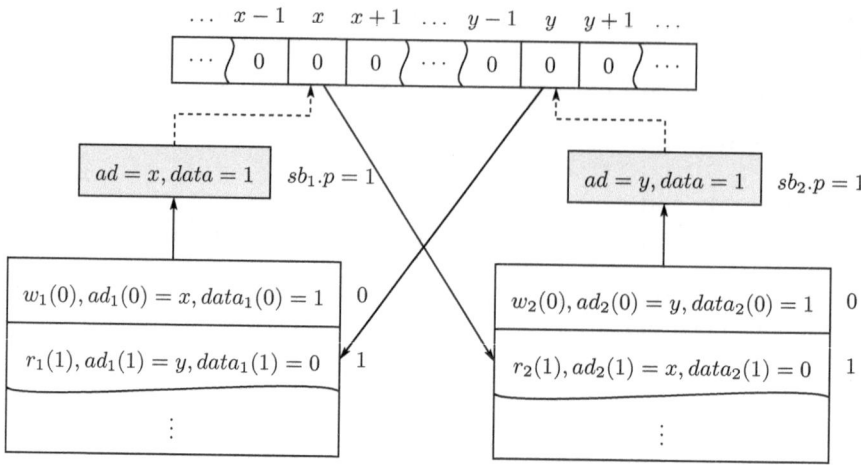

Fig. 2. Multiple store buffers

Can it happen that both r1 and r2 contain 0 in the end? On a sequentially consistent machine (for example a machine without store buffers), this cannot happen, since either the assignment to x happens before the assignment to y or vice versa and the program order is preserved within threads. However, if we take store buffers into account the outcome is valid. Consider the situation where both the assignment to x as well as to y are in the local store buffers, but have not yet emerged to the memory. Hence the read of y in the first thread still sees the value 0 and the read of x in the second thread also sees 0.

A brute force way to make store buffers invisible is to use fenced transactions.

Definition 1. *If t is a transaction, we denote by ft the corresponding fenced transaction. A fenced transaction from a processor is directly sent to the memory; it is only executed when its store buffer is empty.*

A trivial consequence is

Lemma 3. *If for a time interval T and an address range A all transactions during T with addresses in A are fenced, then the memory system behaves for this time interval and address range like a shared memory.*

3.3   Caches

Memories m are usually implemented by one or more levels of caches ca which are backed up by a main memory mm. Caches are small and fast memories; they can be implemented in many ways [24, 25, 26]. A uniform view on all kinds of cache constructions is provided by *abstract caches*. Decompose addresses $a \in A$ into (cache) line address $a.la$ and offset $a.\mathit{off}$:

$$a = a.la \circ a.\mathit{off}$$

and let LA be the set of line addresses. The configuration ca of an abstract cache is then simply specified by a pair of mappings

- $ca.valid : LA \to \{0,1\}$. A line la is present in the cache if $ca.valid(la) = 1$.
- $ca.data : A \to D$. Formally this is just an ordinary memory configuration with the full address range A, but values $ca.data(a)$ are only considered meaningful, if the corresponding cache line is valid, i.e. if $ca.valid(a.la) = 1$.

It is easy to see abstract caches are natural abstractions for the usual cache constructions; we show this for direct mapped caches. Consider the usual decomposition of addresses $a \in A \subseteq \{0,1\}^n$

$$a = a.tag \circ a.line \circ a.off$$

where $a.off \in \{0,1\}^o$ is the offset within a cache line, $a.line \in \{0,1\}^l$ is the local address of a cache line in the cache, $a.tag \in \{0,1\}^t$, and $o+l+t = n$. A direct mapped cache c has three memories all addressed by local line addresses $line$:

- $c.data(line) \in D^o$ is the cache line stored in the local cache line $line$,
- $c.valid(line) \in \{0,1\}$ indicates if the data in line $line$ is valid,
- $c.tag(line) \in \{0.1\}^t$ completes the local line address $line$ to a full line address.

The corresponding abstract cache $ca(c)$ can be defined by

$$ca(c).valid(tag \circ line) = 1 \leftrightarrow c.valid(line) = 1 \wedge c.tag(line) = tag$$
$$ca(c).data(tag \circ line \circ off) = c.data(line)[off].$$

In a correctness proof for a cache system on the hardware level one has – among others – to consider the following components of the hardware configuration h:

- the main memory $h.mm$, and
- in case of a direct mapped cache, the cache memories $h.c.data$, $h.c.tag$, $h.c.valid$.

From the direct mapped cache $h.c$ one abstracts the abstract cache $ca(h.c)$. From this one defines the (simulated) memory system $m(h)$ simulated by the hardware in configuration h as

$$m(h)(a) = \begin{cases} ca(h.c)(a) & \text{if } ca(h.c).valid(a.line) = 1, \\ h.mm(a) & \text{otherwise.} \end{cases}$$

A hardware correctness proof has also to consider the buses between the cache and main memory as well as the logic controlling transfers of cache lines between cache and main memory. Also hardware correctness proofs have to break read and write transactions etc. down to the cycle level. For memory transaction i one might have to consider start cycles $s(i)$ (a request signal is activated) and end cycles $e(i)$ (a busy signal is taken away or is not activated in the first place). A proof[4] that this memory system is simulated has to establish (among other things):

---

[4] For a non pipelined cache.

Lemma 4

1. *A read transaction i starting in cycle $s(i)$ returns in cycle $e(i)$ the data $m(h^{s(i)})(ad(i))$*
2. *A write transaction i starting in cycle $s(i)$ produces after cycle $e(i)$ the simulated memory*

$$m(h^{e(i)+1})(a) = \begin{cases} data(i) & \text{if } a = ad(i), \\ m(h^{s(i)})(a)) & \text{otherwise.} \end{cases}$$

The first formal correctness proofs for memory systems with caches that can be synthesized to running hardware are reported in [27, 7].

Cache Coherent Shared Memory. The vanilla implementation of shared memory for p processors $P(1), \ldots, P(p)$ is to connect each processor $P(i)$ with its own cache $ca(i)$ and to back up the caches $ca(i)$ with a main memory mm. The entire memory system is supposed to simulate a single sequentially consistent shared memory m [28]. In order to achieve this goal the caches observe each others' transactions via a special bus (this is called snooping) and run a cache coherence protocol. Instead of a single valid bit the caches use a set St of several states for each cache line in order to keep track what cache has what data and how these data match the data in main memory. In abstract caches we therefore replace function $ca.valid$ by a state function

$$ca.s : LA \to St.$$

A very common set of states introduced in [21] is

$$St = \{M, O, E, S, I\}.$$

For $s = ca.s(la)$ the intended meaning is

- $s = M$: the line is exclusive and modified. 'Modified' is not 'clean'.
- $s = O$: the line is shared and modified ('owned').
- $s = E$: the line is exclusive and clean. 'Clean' means that the data in the caches matches the data in main memory and 'exclusive' means that no other cache holds this line la; we formalize this below.
- $s = S$: the line is shared and clean. 'Shared' is not 'exclusive'.
- $s = I$: the line is invalid.

The caches have to maintain the following invariants about the cache states of each line.

Invariant 3. *Exclusive lines are only in one cache:*

$$ca(i).s(la) \in \{E, M\} \wedge i \neq j \to ca(j).s(la) = I.$$

Invariant 4. *Clean lines match data in main memory:*

$$ca(i).s(la) \in \{E, S\} \wedge a.la = la \to ca(i).data(a) = mm(a).$$

Invariant 5. *Lines la owned by different caches match:*

$$ca(i).s(la) = ca(j).s(la) = O \wedge a.la = la \rightarrow ca(i).data = ca(j).data.$$

The cache coherence protocol has to decide for each processor transaction whether to announce it on the special bus (in this case we call the transaction *public*) or not (we call the transaction *private*). Private processor transactions are: line invalidations on clean or shared data (the local state is changed to I), read hits (the local state stays the same), write hits on exclusive data (the new state is M), test and set hit if the cached data is $\neq 0$ (the local state stays the same). There is also a private transaction between a cache and main memory, namely if the cache flushes (writes to main memory and invalidates) a clean line. All other transactions are public.

If one views the processors as a distributed system delivering the transactions one by one[5] to their caches then for each of the common protocols it is very easy to show, that the invariants are maintained. Due to the (unrealistically simple) distributed system one can number the transactions $t(x)$ simply by the order in which they are sent to the memory system. The simulated memory m^x before transaction $t(x)$ is

$$m^x(a) = \begin{cases} ca^x(i).data(a) & \text{if } ca^x(i).s(a) \neq I \text{ for some } i, \\ mm^x(a) & \text{otherwise.} \end{cases} \tag{1}$$

Assuming that in the initial caches ca^0 all lines la are invalid ($ca^0(i).s(la) = I$) the invariants imply among other things

Lemma 5

1. $m^x(a)$ is is well defined by Equation 1,
2.

$$m^x(a) = \begin{cases} data(last(a, x)) & \text{if } W(ad(x), x), \\ mm^0(ad(x)) & \text{otherwise, and} \end{cases}$$

3. *if* $t(x)$ *is a read transaction, then the memory system returns data* $m^x(ad(x))$.

Variants of this lemma have been extensively studied and model checked. Producing at the hardware level a formal correctness proof for a cache coherent shared memory is still considered a major open problem. Indeed we are not aware of a paper and pencil proof for this problem. For such a proof one has to give a complete implementation, e.g. in the style of [19] or [7].

If we denote transaction j of processor i by $t(i, j)$ we have to consider the start cycles $s(i, j)$ and end cycles $e(i, j)$ of these transactions. Typical durations $e(i, j) - s(i, j) + 1$ might be 1 for read hits, 2 for exclusive write hits and many more cycles for public transactions. For a straightforward implementation of the transactions on a single shared data bus between caches and main memory and a single special bus one will be able to show

---

[5] Which is pointless; the very idea of shared memory is to parallelize transactions.

Lemma 6. *The following transactions do not overlap:*

- *public transactions,*
- *private transactions on the same processor,*
- *private and public transactions with the same address, and*
- *private writes and any transaction with the same address.*

The hardware version of Equation 1 of the memory $m(h)$ simulated by hardware h is

$$m(h)(a) = \begin{cases} ca(i,h).data(a) & \text{if } ca(i,h).s(a) \neq I \text{ for some } i, \\ h.mm & \text{otherwise.} \end{cases} \tag{2}$$

With the help of Lemma 6 one shows that the invariants hold for each cycle, and one gets

Lemma 7. $m(h)(a)$ *is well defined by Equation 2.*

In the spirit of [7] one can define a total order $O(i,j) \in \mathbb{N}$ for the transactions $t(i,j)$ using their end cycles $e(i,j)$: order transactions by their end cycles; order transactions with the same end cycle arbitrarily. Denote by

$$M(t) = max\{O(i,j) : e(i,j) \leq t\}$$

the largest index of a transaction that has completed until cycle t. Let z be the sequential index of a transaction $t(i,j)$ for some i and j, i.e. $z = O(i,j)$ resp. $(i,j) = O^{-1}(z)$.

We define predicate $W'(a,z)$ indicating that address a has been written by a transaction with sequential index $z' = O(i,j) < z$:[6]

$$W'(a,z) \equiv \exists z' < z : ad(O^{-1}(z')) = a \, \wedge$$
$$(w(O^{-1}(z')) \vee (ts(O^{-1}(z')) \wedge data(O^{-1}(z')) = 1)).$$

We define $last(a,z)$ as the last sequential index z' before z of a transaction writing to $ad(i,j)$:

$$last(a,z) = max\{z' < z : ad(O^{-1}(z')) = a \, \wedge$$
$$(w(O^{-1}(z')) \vee (ts(O^{-1}(z')) \wedge data(O^{-1}(z')) = 1))\}.$$

Lemma 8. *The hardware simulates a shared memory which is sequentially consistent with respect to ordering* $O(\ ,\ )$: *let* $z = M(t)$, *then*

1.

$$m(h^t)(a) = \begin{cases} data(last(a,z)) & \text{if } W'(a,z), \\ h.mm^0(a) & \text{otherwise,} \end{cases}$$

---

[6] We extend functions $ad(j)$, $data(j)$, $w(j)$, $r(j)$, $ts(j)$ defined in 3.1 to multiprocessor case as $ad(i,j)$, $data(i,j)$, $w(i,j)$, $r(i,j)$, $ts(i,j)$ to take additional parameter i – the index of a processor.

2. *a read transaction starting in cycle $s(i,j)$ returns in cycle $e(i,j)$ the data $m(h^{s(i,j)})(ad(i,j))$, and*

3.
$$W'(ad(i,j), O(i,j)) \rightarrow last(a, O(i,j)) < M(s(i,j)).$$

There is one further highly interesting and important property about cache coherence protocols which to the best of our knowledge have not received much theoretical treatment, namely compatibility within a family F of cache coherence protocols. If the special bus between caches helping to control the cache coherence protocol contains signals like the ones in the classical paper introducing the MOESI protocol [21], then one can specify with each memory access (i,j) the cache coherence protocol $mmode(i,j) \in F$ to be used for transaction i on processor j, and one gets

Lemma 9. *If a compatible family of cache coherence protocols is used in a memory system with caches and the memory mode used is specified separately for each transaction, then the memory system still simulates a sequentially consistent memory.*

Consistent families of cache coherence protocols are implemented in the processors of modern PCs. The result of the lemma is stated quite explicitly in [21]. We have seen neither a paper and pencil proof nor a model checked version of this important result.

3.4   Memory Management Units

Address Translation. We partition memory into pages; here we use the common page size $4K = 2^{12}$. We partition addresses a into page base addresses $a.ba$ and page offsets $a.pof \in \{0,1\}^{12}$, such that $a = a.ba \circ a.pof$. We denote a page with base address ba of memory m by $pg(m, ba)$. It consists of the $4K$ consecutive memory cells starting at address $pa \circ 0^{12}$:

$$pg(m, ba) = m_{4K}(ba \circ 0^{12}).$$

Let $V \subseteq A$ be a set of (virtual user) addresses to be translated by a memory management unit (MMU) and let

$$V.ba = \{a.ba : a \in V\}$$

be the set of page addresses in V. An abstract translation of address range V is specified by

- a translation function $T : V.ba \rightarrow A.ba$ specifying where to redirect memory accesses to pages in A, and
- rights functions $r : V.ba \rightarrow \{0,1\}$ specifying the access rights to pages. We consider $r = EXE$ (executable) and $r = RW$ (readable and writable).

For a processor running in translated mode a memory management unit has to redirect memory accesses to addresses $a \in V$ to $T(a.ba) \circ a.pof$ in the following situations

- if a comes from the program counter and $EXE(a.ba) = 1$,
- if a is the effective address of a write access and $RW(a.ba) = 1$, and
- if a is the effective address of a read access.

For all other addresses $a \in V$ and for all addresses $a \notin V$ a page fault has to be generated.

Page Tables are pages that are used to specify abstract translations. They commonly consist of page table entries *pte* occupying 4 bytes resp. one word in a page table. Within a page we can index the entries with page indices $px \in \{0, 1\}^{10}$. Thus we can define entry px of page table (with base address) ba as

$$pg(m, ba)[px] = pg(m, ba)_4(px \circ 0^2).$$

Intuitively multilevel address translation is done by traversing the graph G whose nodes are the page tables and whose edges are specified by a certain component *pte.ba* of the page table entries *pte*. But notice that the graph G is dynamic: it can be edited by the processor while the MMU traverses it. Page table entries *pte* usually have components like

- *pte.ba* the base address of the next page table or – at the last step of translation – a user page,
- the present bit *pte.p* indicating that the data of the entry is meaningful,
- rights bits *pte.r*, and
- possibly accessed and dirty bits *pte.acc* and *pte.d*; we will skip over them here most of the time.

Walks. Multi Level Address Translation is achieved by *walking* resp. traversing the page tables in K steps; common values are $K = 3$ or $K = 4$. The information gathered during the traversal is summarized in *walk*. Walk w has the following components:

- $w.vba$: the virtual base address to be translated,
- $w.ba$: the base address of the next page to be accessed,
- $w.r$ for the rights r: the logical AND of bits *pte.r* in the entries traversed so far, and
- $w.level \in [K : 0]$: the number of page tables that have yet to be traversed.

Walking starts with *initial walks* of level $w.level = K$. No rights have yet been restricted, so $w.r = 1$ for all r. The base address $w.ba$ of the page table where the traversal starts is stored as the *ba*-component of a processor register dedicated for this purpose; in *x64* processors this register is called *CR3*. Thus $w.ba = CR3.ba$. For the actual traversal we decompose base addresses ba of pages into K page indices $ba.px[i]$:

$$ba = ba.px[K] \circ ba.px[K-1] \circ \ldots \circ ba.px[1] = ba.px[K : 1].$$

The width of $px[i]$ depends on the size of a page table entry, which can be 4 or 8 bytes. In case of a 4-byte page table entry, there are $4K/4 = 1024$ entries per page table, and, therefore, the width of $px[i]$ is 10 bits. Respectively, for 8-byte page table entries the width $px[i]$ is 9 bits.

Extension of a level x walk w to a walk $wext(w)$ makes use of the level x page index $w.vba.px[x]$ of the address to be translated, the MMU accesses entry

$$pte = pg(m, w.ba)[w.vba.px[x]].$$

If it is not present ($pte.p = 0$) a page fault is generated. Otherwise, one sets

$$wext(w).ba = pte.ba$$
$$wext(w).r = w.r \land pte.r$$
$$wext(w).level = w.level - 1$$

The walk w is complete if level $w.level = 0$. The translated base address $w.ba$ of a complete walk is a translation for the walks virtual base address $w.vba$. An iterated walk extension $wext^x(w)$ is obtained in the obvious way by $wext^0(w) = w$ and $wext^{x+1}(w) = wext(wext^x(w))$

Translation Look Aside Buffers. Walking the page tables is slow as it requires many memory accesses. Therefore one collects translations $(w.vba, w.ba, w.r)$ found during the walking in a cache called the *translation look aside buffer* resp. the TLB. However, as processors do *not* keep this cache consistent with the page tables, it is the users responsibility to evict translations from the TLB, that should not be used any more. Translations (there can be several) for single virtual base addresses are evicted by so called *invlpg(vba)* instructions. There are also instructions for clearing the entire TLB.

Implementing an Abstract Translation. Suppose virtual address range V, translation function T, and rights functions r of an abstract translation are given. We construct a tree G of page tables such that walking G produces the translations prescribed by the abstract translation. Denote by

$$P_x = \{vba.px[K : x] : vba \in V.ba\} \text{ for } 2 \leq x \leq K + 1$$

the set of prefixes of the virtual base addresses formed by page indices from $K + 1$ down to 2. Note that $P_{K+1} = \epsilon$ is an empty prefix. Let

$$P = \bigcup_{x=2}^{K+1} P_x$$

be the set of all prefixes. For each prefix $p \in P$ we allocate in memory m a separate page table with base address $ptba(p)$, such that

$$\forall q \in P : q \neq p \Rightarrow ptba(q) \neq ptba(p).$$

The entries of the page table corresponding to prefix $vba.px[K : x]$ are defined by induction on x from the leaves ($x = 2$) to the root ($x = K + 1$). Let pte be the page table entry with index $vba.px[x-1]$ in the page table corresponding to prefix $vba.px[K:x]$, i.e.

$$pte = pg(m, ptba(vba.px[K:x]))[vba.px[x-1]].$$

For $x = 2$ set the present bits $pte.p = 1$ if $vba \in V.ba$, and for present entries set

$$pte.ba = T(vba)$$
$$pte.r = r(vba)$$

For $x > 2$ set $pte.p = 1$ if $vpa.px[K:x-1] \in P_{x-1}$, and for present entries set

$$pte.ba = ptba(vpa.px[K:x-1])$$
$$pte.r = 1$$

and point with special purpose register $CR3$ to the root of the tree obtained in this way

$$CR3 = ptba(\epsilon).$$

By induction on x one now easily shows

Lemma 10. *Let $vba \in V.ba$, let $pr = vba.px[K:x+1]$ and let w be a walk with $w.ba = vba, w.level = x, w.ba = ptba(pr)$ and $w.r = 1$. Then x-fold walk extension of w gives the desired translation:*

$$wext^x(w).ba = T(vba)$$
$$wext^x(w).r = r(vba).$$

A similar argument shows that initial walks with $w.vba \notin V.ba$ hit a not present entry at level y where

$$y = max\{x : vba.px[K:x] \notin P_x\}.$$

3.5 Out of Order Execution

Tomasulo schedulers as shown in Fig. 3 are the standard mechanism for the out of order execution of instructions in processors. Instructions are fetched in program order in the instruction fetch (IF) stage. They wait in the issue stage for a free reservation station (RS) of a functional unit capable of executing the instruction, and for a free slot in the reorder buffer (ROB). From the issue stage instructions proceed to a reservation station. At this point three things happen:

1. The instruction receives a *tag*; this is a local number for instructions issued but not written back. There are as many tags as places in the reorder buffer. The reorder buffer is usually implemented as a RAM implementing a queue that eventually holds the results (including the interrupts produced or sampled) of instructions; at issue time the instruction is inserted at the end of the queue. The natural *tag* to be used for an instruction is its RAM address in the ROB.

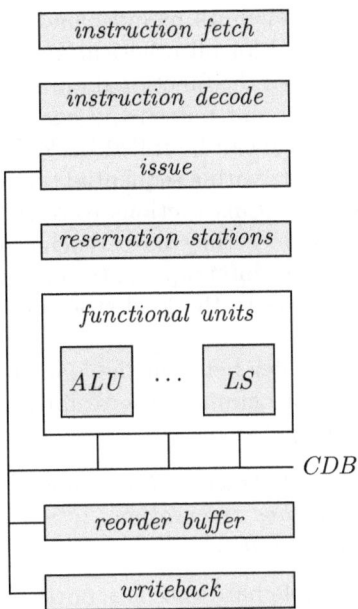

Fig. 3. Tomasulo scheduler

2. Register operands are looked up in the register files. If a register does not contain valid data (because an instruction writing the data is in flight), a *tag* field associated with the register contains the tag t of the last such instruction.
3. The results of such instructions with tag t are searched in the ROB and on the common data bus (CDB). If not all operands are found, the instructions producing the desired results are still in the reservation stations or the functional units. The reservation station snoops on the common data bus for the results of instructions with tags t occuring on the CDB. Once all operands are gathered and the functional unit can accept a new operand, instructions proceed to the functional unit, then later via the CDB to the ROB. They are written back when they are at the head of the queue implemented by the ROB. Thus, retirement of instruction is again in program order.

The classical correctness statement of out of order mechanisms then has the form

Lemma 11. *The mechanism of a Tomasulo scheduler (as shown in Fig. 3) preserves the sequential semantics of machine instructions that do not access memory; in these situations reservation stations and reorder buffer are invisible to the programmer.*

A (hopefully) reasonable paper and pencil proof can be found in [19]. There are numerous formal proofs for this result at various levels of detail. At the most detailed level the proof concerns synthesizable hardware [27, 20].

Load Store Units. If functional units include load store units LS accessing a memory system m (they should for all practical purposes!) a few extra precautions have to be taken: as long as the functional units do not produce irreversible results, an instruction that has not passed the head of the ROB and thus has not reached the write back stage can be rolled back. This permits to implement precise interrupts (i.e. interrupts with a sequential semantics). But write instructions in memory units (and read instructions to devices with read side effects) cannot easily be rolled back once they have reached the load store unit. One possible way to maintain precise interrupts is to send a write instruction (and a load instruction to an I/O port) to the load store unit only if it is at the head of the ROB.

One often inserts a store buffer between the load store unit and the memory system (see Fig. 4). Because of Lemma 2 the resulting memory system behaves like a single memory and one gets

Lemma 12. *The memory unit shown in Fig. 4 preserves the sequential semantics of machine instructions; thus reservation stations, reorder buffer and store buffer are invisible to the programmer.*

A formal proof for synthesizable hardware is reported in [27].

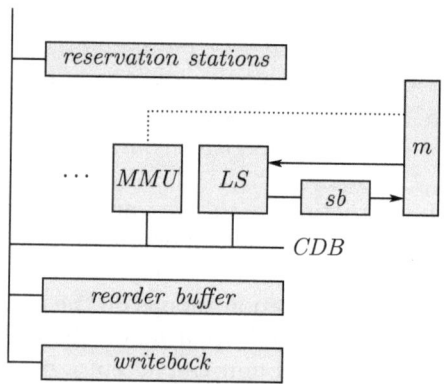

Fig. 4. Memory units

Memory Management Units. Intuitively, the control of a processor with a memory management unit has to split translated loads and stores into two microoperations:

i) find the address translation either by quickly looking it up in the TLB or by slowly walking the page tables, and
ii) perform the memory transaction using the translated address of step i) as an operand.

Tomasulo schedulers permit to implement this in a natural way. MMU and load store unit are separate functional units; if a translated memory transaction is fetched, two microinstructions are issued: one to the MMU computing the desired translation and a second one to the LS unit performing the actual access. For MMUs setting 'accessed' and 'dirty' bits accesses to the tables are potentially writing. Therefore a conservative implementation would perform the accesses only if the microinstruction computing the translation is at the head of the ROB (slowing down the process of page table walking even further). Notice that it is very natural, that a load instruction whose translation is already in the TLB overtakes a previous access whose address needs to be translated by walking. As a result the ROB entries of such load instructions contain kind of precomputed results; we deal with the problems arising from this in the multi processor case shortly.

The data path used by the MMU deserves some attention. One can provide a separate access path bypassing the store buffer from the MMU to the memory system m. Also one can forward results from the MMU directly to the LS unit. Note that due to the different access path into the memory system even in the case of a single processor Lemma 2 does not apply any more. Thus we get

Lemma 13. *The memory unit shown in Fig. 4 with MMU bypassing the store buffer almost preserves the sequential semantics of translated machine instructions: reservation stations and reorder buffer are invisible to the programmer, but the store buffer stays visible.*

Thus, a page table walk might miss a sequentially earlier page table update which is still in the store buffer.

Coherency Snoops. One would fear that in the multiprocessor case the programmer model becomes even more complicated, but in patents like [22] one finds counter measures. One of them is coherency snooping: the ROB entries of processor j holding (precomputed) results of load instructions (i, j) store also the translated address a and participate in the snooping protocol of the caches. If a write to address a is snooped on the cache of a different processor j' (sequentially earlier writes on processor j are handled by store buffer forwarding) the load instruction is either

- rolled back and repeated; this allows other processors to prevent the termination of the load instruction by repeated writes to address a, or
- the result of the load instruction is replaced by the data written by processor j' to address a; this gives the memory model provided by modern processors.

Lemma 14. *Suppose the memory unit shown in Fig. 4 is used with several processors connected to a cache coherent shared memory m and uses coherency snooping. Then locally sequential semantics is almost preserved; reservation stations and reorder buffers are invisible. Store buffers are visible.*

For the proof we use notation from section 3.3. Let t be the last cycle when a read transaction (i, j) is in the ROB before it is retired, and let $m(h^t)$ be the memory simulated by the memory system in cycle t as defined in Equation 2. The memory depends only on the write operations which have completed until cycle t; let $z = M(t)$. Write operations by the LS unit and the MMU are issued on each processor only when they are at the head of the ROB, i.e. they are issued in order. Thus memory $m(h^t)$ already corresponds to an in order execution on each processor and we have as before

$$m(h^t)(a) = \begin{cases} data(last(a, z)) & \text{if } W(a, z), \\ h.mm^0(a) & \text{otherwise.} \end{cases}$$

Let $t' \le t$ be the last cycle before t when the data for read transaction (i, j) was updated in the ROB, either by the execution of the load instruction or subsequently by the coherency snooping. Then the ROB writes back result $m(h^{t'})(ad(i, j))$ for transaction (i, j). Between cycles t' and t coherency snooping does not update the ROB for transaction (i, j), hence no write to address $ad(i, j)$ even started in this period and we have

$$\begin{aligned} m(h^{t'})(ad(i, j)) &= m(h^t)(ad(i, j)) \\ &= data(last(ad(i, j), z)). \end{aligned}$$

3.6   Initializing an x64 Processor

Figure 5 gives a a very schematic view of the instruction set architecture of contemporary PC processors as documented on about 3000 pages in [10] or only about 1500 pages in [11]. The major blocks are the processor core, MMU with TLB, memory system with main memory and caches, store buffers as well as the I/O devices which are accessed like the main memory. Multi core processors have several processor cores, MMUs and store buffers connected to one memory system and the devices.

Boxes labeled acc stand for memory 'access registers' holding addresses, data, etc. of memory transactions. The core contains numerous user registers R as well as numerous system registers; for us system register $CR3$ which serves as the origin of TLB walks is particularly important. The segmentation mechanism is a legacy feature going back to the x86 architecture. It can be made invisible by configuring the entire physical address space as a single segment with no restriction of rights. The memory can be accessed in many memory modes. At least one of them (UC – uncachable) completely bypasses the caches and thus makes the caches visible to the programmer. The good news is that the memory modes which do not bypass the caches are compatible. Thus, if no I/O devices are accessed and only compatible memory modes are used, then by Lemmas 9 and 14 the user sees a sequentially consistent physical memory PM and store buffers (see Fig. 6-a). If I/O devices are accessed in uncachable mode life is simple too. But if devices are accessed in memory modes using the cache, than the actual device access only takes place in case of cache misses. This even makes the states of the cache lines visible.

After a reset signal x64 machines are in a very simple operation mode where only a single processor is running and paging is switched off. Because paging is switched off the MMU is not visible. Because only one processor is running by Lemma 2 the store buffer is not visible. In this mode page tables as specified

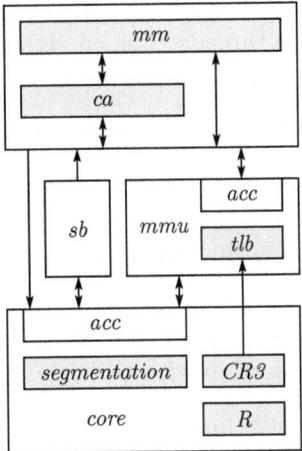

Fig. 5. x64 memory

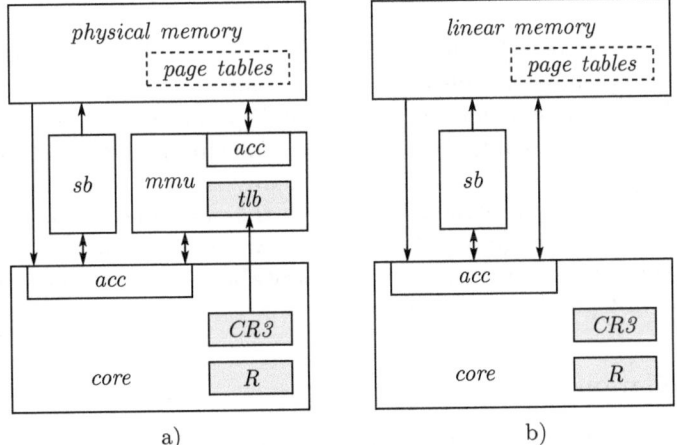

Fig. 6. x64 memory

in section 3.4 can be written into the physical memory (Fig. 6-a). If we turn on all processors, clear all TLBs and enable paging (i.e. we run the processors in translated mode, then by Lemma 10 an abstract translation is realized. In this mode the MMUs become invisible and the users see store buffers and a sequentially consistent shared 'linear' memory LM (Fig. 6-b).

4   Programming Language

4.1   Naive Parallel C Semantics

We are considering parallel C programs whose threads run in an interleaved fashion on multi core machines. The obvious approach to define the semantics of such programs is to start with the small step semantics of single threads and then to interleave the steps of the threads. The configuration c of a single threaded abstract C machine can be defined essentially using the following components:

- a program rest $c.pr$ consisting of a sequence of C instructions yet to be executed,
- a global memory $c.gm$,
- a heap memory $c.hm$, and
- a local memory stack $c.lms$ consisting of $c.rd$ (recursion depth) many memory frames.

For details of the particular semantics used in the Verisoft project[7] see [3, 4]. Generalizing this to a configuration with multiple threads i is straight forward:

- use for each thread i a local program rest $c.pr[i]$ and a local memory stack $c.lms[i]$,
- share the global memory $c.gm$ and the heap $c.hm$, and
- now interleave the (small step semantics) steps of the threads defined in this way.

There is no way to beat the elegance of this definition. Unfortunately we don't know how to implement it efficiently. For threads i let $p(i)$ be the program of thread i. Before running on a parallel machine C programs $p(i)$ are first compiled to a machine program $code(p(i)))$; even with the simplest non optimizing compiler a single small step semantics step is usually translated to several machine instructions. The hardware then interleaves the machine instructions instead of the steps of the C semantics. Hence if one wants to define efficiently implementable parallel C semantics one has to worry about the process of compilation, preferably by an optimizing compiler.

4.2   Compilation

The compiled programs $code(p(i))$ – running say in linear memory $LM(h)$ of a hardware configuration h – have to simulate the programs $p(i)$ of the C threads. For now we only sketch compiler correctness for a single thread. A simulation relation $consis(c, alloc, h)$ between C configurations c and hardware configurations h is defined with the help of an allocation function $alloc$. This functions maps elementary C variables x to 'allocated (linear) base addresses' $alloc(c, x)$. We define some typical properties of relation $consis$.

- For variables x let $asize(x)$ be the number of bytes allocated by the compiler for variable x; it depends only on the type of x. Let $va(c, x)$ be the value of variable x in configuration c. Then the $asize(x)$ bytes in linear memory $LM(h)$ should coincide with the C value of the variable

$$LM_{asize(x)}(alloc(c, x)) = va(c, x).$$

---

[7] http://www.verisoft.de

- Suppose p is a pointer; thus its value is another variable $va(c,p) = y$. Assume we have 8 byte addresses. Then the 8 bytes in linear memory following $alloc(c,p)$ should be the allocated base address of y

$$LM_8(alloc(c,p)) = alloc(c,y).$$

- For non optimizing compilers code is compiled statement by statement. The first statement of the program rest is $head(c.pr)$. It is translated to $code(head(c.pr))$. Let $start(code(head(c.pr)))$ be the address in linear memory where this piece of translated code is allocated. Then the program counter $h.pc$ should point there

$$h.pc = start(code(head(c.pr))).$$

Clearly, one needs to modify this condition for optimizing compilers.

For non optimizing compilers one obtains the following step by n-step simulation theorem

Lemma 15. *For every C computation c^0, c^1, ... there exist i) a hardware computation h^0, h^1, ..., ii) a sequence of step numbers s^0, s^1, ... and iii) a sequence of allocation functions $alloc^0$, $alloc^1$, ... such that*

$$consis(c^T, alloc^T, h^{s(T)})$$

holds for all T.

For a formal proof of this result see e.g. [4]. Optimizing compilers exploit the fact, that we do not really care for simulations to hold for every C step. It suffices if the relation holds for the 'visible' C steps T, for example when the program does I/O. Let us call these steps *I/O steps*. Then a possible correctness statement for an optimizing compiler would look like:

Lemma 16. *For every C computation c^0, c^1, ... there exist i) a hardware computation h^0, h^1, ..., ii) a sequence of step numbers s^0, s^1, ... and iii) a sequence of allocation functions $alloc^0$, $alloc^1$, ... such that*

$$consis(c^T, alloc^T, h^{s(T)})$$

holds for all I/O steps T.

For a formal correctness proof for an optimizing compiler (with respect to a big step semantics) see [23].

4.3   Volatile Variables

Compiler directives allow to declare shared variables x as volatile. Intuitively speaking this warns the compiler that these variables are shared and thus accesses to such variables should not be optimized to registers. In order to make compiler construction easy we syntactically restrict the use of volatile variables x. A thread can only perform assignments of the form

$$y = x \text{ or } x = y$$

where y is a thread local variable. We include any such assignment into the I/O steps and we implement any such assignment by a fenced read resp. write. By the trivial Lemma 3 then the store buffers in (Fig. 6-b) become invisible for transactions involving volatile variables and we obtain

Lemma 17. *The portion of memory allocated to volatile variables forms a sequentially consistent portion of linear memory LM.*

4.4   Synchronized Parallel C

Using test and set operations on volatile variables it is straightforward to implement locks using textbook shared memory algorithms [29]. Using locks one can exclusively reserve memory regions R of the shared C variables (e.g. certain data structures) temporarily to threads i, for certain intervals I of C-steps. During such intervals I thread i can do computations on region R like in ordinary sequential C computations: due to the locking no other thread accesses region R during interval I, thus the store buffers are by Lemma 2 invisible. However, at the end of interval I when the lock is released the compiler must guarantee that the updates of region R performed by thread i become visible to the other processors. If the compiler treats a lock release of thread i as an I/O step for the thread, then this is guaranteed by Lemma 16.

Currently we work on extensions of these basic programming disciplines and fencing policies for shared memory accesses to cover more programming idioms by our theory framework.

References

1. Cohen, E., Moskal, M., Schulte, W., Tobies, S.: A practical verification methodology for concurrent programs. Technical Report MSR-TR-2009-15, Microsoft Corp. (2009)
2. In der Rieden, T., Paul, W.J.: Beweisen als Ingenieurwissenschaft: Verbundprojekt Verisoft (2003–2007). In: Reuse, B., Vollmar, R. (eds.) Informatikforschung in Deutschland, pp. 321–326. Springer, Heidelberg (2008)
3. Leinenbach, D., Petrova, E.: Pervasive compiler verification – From verified programs to verified systems. In: 3rd intl Workshop on Systems Software Verification (SSV 2008). Electronic Notes in Theoretical Computer Science, vol. 217C, pp. 23–40. Elsevier Science B.V., Amsterdam (2008)
4. Leinenbach, D.C.: Compiler Verification in the Context of Pervasive System Verification. PhD thesis, Saarland University, Computer Science Department (2008)
5. Schirmer, N.: Verification of Sequential Imperative Programs in Isabelle/HOL. PhD thesis, Technical University of Munich (2006)

6. Schirmer, N.: A verification environment for sequential imperative programs in Isabelle/HOL. In: Baader, F., Voronkov, A. (eds.) LPAR 2004. LNCS, vol. 3452, pp. 398–414. Springer, Heidelberg (2005)
7. Beyer, S., Jacobi, C., Kroening, D., Leinenbach, D., Paul, W.: Putting it all together: Formal verification of the VAMP. International Journal on Software Tools for Technology Transfer 8(4-5), 411–430 (2006)
8. Dalinger, I., Hillebrand, M., Paul, W.: On the verification of memory management mechanisms. In: Borrione, D., Paul, W. (eds.) CHARME 2005. LNCS, vol. 3725, pp. 301–316. Springer, Heidelberg (2005)
9. Tverdyshev, S., Shadrin, A.: Formal verification of gate-level computer systems. In: Rozier, K.Y. (ed.) LFM 2008: Sixth NASA Langley Formal Methods Workshop. NASA Scientific and Technical Information (STI), NASA, pp. 56–58 (2008)
10. Intel Corporation: Intel 64 and IA-32 Architectures Software Developer's Manual: Volumes 1–3b (2009)
11. Advanced Micro Devices (AMD), Inc.: AMD64 Architecture Programmer's Manual: Volumes 1–3 (2006)
12. Adve, S.V., Gharachorloo, K.: Shared memory consistency models: A tutorial. IEEE Computer 29(12), 66–76 (1996)
13. Steinke, R.C., Nutt, G.J.: A unified theory of shared memory consistency. CoRR cs.DC/0208027 (2002)
14. Sarkar, S., Sewell, P., Nardelli, F.Z., Owens, S., Ridge, T., Braibant, T., Myreen, M.O., Alglave, J.: The semantics of x86-cc multiprocessor machine code. In: POPL 2009: Proceedings of the 36th annual ACM SIGPLAN-SIGACT symposium on Principles of programming languages, pp. 379–391. ACM, New York (2009)
15. Owens, S., Sarkar, S., Sewell, P.: A better x86 memory model: x86-tso. In: 22nd International Conference on Theorem Proving in Higher Order Logics (TPHOLs 2009). Springer, Heidelberg (to appear, 2009)
16. Intel: Intel 64 architecture memory ordering white paper. SKU 318147-001 (2007)
17. Midkiff, S.P., Lee, J., Padua, D.A.: A compiler for multiple memory models. Concurrency and Computation: Practice and Experience 16(2-3), 197–220 (2004)
18. Sevcík, J., Aspinall, D.: On validity of program transformations in the java memory model. In: Vitek, J. (ed.) ECOOP 2008. LNCS, vol. 5142, pp. 27–51. Springer, Heidelberg (2008)
19. Müller, S.M., Paul, W.J.: Computer Architecture: Complexity and Correctness. Springer, Heidelberg (2000)
20. Kröning, D.: Formal Verification of Pipelined Microprocessors. PhD thesis, Saarland University (2001)
21. Sweazey, P., Smith, A.J.: A class of compatible cache consistency protocols and their support by the ieee futurebus. In: ISCA 1986: Proceedings of the 13th annual international symposium on Computer architecture, pp. 414–423. IEEE Computer Society Press, Los Alamitos (1986)
22. Intel: Us patent 6687809 - maintaining processor ordering by checking load addresses of unretired load instructions against snooping store addresses (2004)
23. Leroy, X.: Formal certification of a compiler back-end, or: programming a compiler with a proof assistant. In: 33rd symposium Principles of Programming Languages, pp. 42–54. ACM Press, New York (2006)
24. Smith, A.J.: Cache memories. ACM Comput. Surv. 14(3), 473–530 (1982)
25. Stenström, P.: A survey of cache coherence schemes for multiprocessors. IEEE Computer 23(6), 12–24 (1990)

26. Pong, F., Dubois, M.: Verification techniques for cache coherence protocols. ACM Computing Surveys 29(1), 82–126 (1997)
27. Sawada, J., Hunt, W.A.: Results of the verification of a complex pipelined machine model. In: Pierre, L., Kropf, T. (eds.) CHARME 1999. LNCS, vol. 1703, pp. 313–316. Springer, Heidelberg (1999)
28. Lamport, L.: How to make a correct multiprocess program execute correctly on a multiprocessor. IEEE Trans. Comput. 46(7), 779–782 (1997)
29. Taubenfeld, G.: Synchronization Algorithms and Concurrent Programming. Pearson / Prentice Hall (2006)

Introducing Quasirandomness to Computer Science

Benjamin Doerr

Max-Planck-Institut für Informatik, 66123 Saarbrücken, Germany
http://www.mpi-inf.mpg.de/~doerr

Abstract. The paradigm of quasirandomness led to dramatic progress in different areas of mathematics, with the invention of quasi-Monte Carlo methods in numerical integration probably being the best known example. In the last two decades, discrete mathematics heavily used quasirandom ideas, leading, e.g., to notions like quasirandom graphs.

We feel that it is now time to exploit quasirandomness in computer science. As a first application, we propose and analyze a quasirandom analogue of the classical *randomized rumor spreading* protocol to disseminate information in networks.

1 Introduction

Randomized methods are well established both in mathematics and in computer science. Here are a few examples.

(i) *Monte Carlo integration:* We can approximate the integral of a function $f : [0,1]^d \to \mathbb{R}$ by the estimate $\frac{1}{n} \sum_{p \in P} f(p)$, where P is (multi-)set of n point chosen uniformly at random from $[0,1]^d$.

(ii) *Discrete mathematics:* To prove that for any $k \in \mathbb{N}$, there are graphs having girth $g(G) > k$ and chromatic number $\chi(G) > k$, Paul Erdős [Erd59] took a random graph on n vertices with each two vertices connected with probability $n^{-1+\frac{1}{2k}}$. With high probability, there are at most $n/2$ cycles of length k and shorter. Deleting an arbitrary vertex from each such cycle, we end up with a graph G' having girth $g(G') > k$. It is also not difficult to compute that $\chi(G') \geq n^{\frac{1}{2k}}/(6 \ln n)$. This shows that for n large enough, our construction yields a graph having both girth and chromatic number greater than k.

(iii) *Randomized algorithms:* There are so many randomized algorithms by now that it is hard to mention one without feeling guilty of neglecting another. The sorting algorithm Quicksort might be the most prominent example. Randomized primality tests only need time polylogarithmic in n to give a reasonable answer to the question whether a number n is a prime number or not. Random walks are the heart of many exploration algorithms. For example, they yield a simple $O(mn)$ time randomized algorithm for the *s-t*-connectivity problem in undirected graphs, that uses only a logarithmic amount of space [AKL+79]. Random sampling is an integral part of modern directions like property testing and sub-linear time algorithms.

S. Albers, H. Alt, and S. Näher (Eds.): Festschrift Mehlhorn, LNCS 5760, pp. 99–111, 2009.

For most of these examples, similar results avoiding the use of randomized methods exist. Often, they were found much later, they are more complicated, and inferior in further aspects. For example, it was only short ago that Reingold [Rei05] showed how to solve the undirected s-t-connectivity problem by a deterministic algorithm using logarithmic space. However, this algorithm is far from being practical. It is both difficult to implement and much less efficient.

For some examples, though, randomness turned out not to be necessary, and occasionally, the resulting deterministic approaches were much stronger. A very useful paradigm here is that of *quasirandomness*. In the following section we shall explain this concept and, using the historically early example of numerical integration, show how it naturally led to the very powerful quasi-Monte Carlo methods. For reasons of space, we will omit a discussion of quasirandomness in discrete mathematics, but continue in Sections 3 to 5 with what are now the first attempts to use quasirandomness in computer science. A number of open problems are outlined in Section 6.

2   Quasirandomness and Quasi-Monte Carlo Integration

In this section, we introduce the concept of quasirandomness via its most successful application, which is numerical integration. However, no knowledge in numerics is required for reading this section.

Numerical integration asks for estimating the value of an integral. Say we are given function $f : [0,1]^d \to \mathbb{R}$ from the d-dimensional unit cube into the real numbers. Often, the integral $I(f) := \int_{[0,1]^d} f(x)dx$ is hard to compute exactly. Hence, we are looking for a method of approximating its value.

An approach both simple and natural is *Monte Carlo integration*. For $n \in \mathbb{N}$ suitably large, we choose a multi-set P of n points uniformly from $[0,1]^d$. We estimate the value of the integral $I(f)$ by the average function value $E_P(f) := \frac{1}{n} \sum_{p \in P} f(p)$.

Naturally, we expect this value to be a reasonable approximation of $I(f)$, with the error $|E_P(f) - I(f)|$ being smaller the larger the number n of points is. We feel that this is natural, because a sufficiently large random point set should see small and large function values in a fair proportion.

In fact, what sounds natural can also be proven. The integration error can be bounded by an expression of order $O(n^{-1/2})$, hence it is nicely decreasing with growing sizes of the sample point set P. Note that here and in the following we treat the dimension d as a constant.

The paradigm of quasirandomness suggests that we do not stop at this point, but find out which property causes the random point set to be a good sample point set, and then try to find (not necessarily random) point sets that are particularly good in this respect.

For random sample points, it is easy to guess that they profit from being distributed evenly in the domain $[0,1]^d$. A closer analysis supports this guess and makes it precise. As a measure of uniformity, let us define the discrepancy of P to be

$$\mathrm{disc}(P) := \sup\{\,|\,|P \cap R| - n\,\mathrm{vol}(R)|\,|\,R \in \mathcal{R}\},$$

where $\mathcal{R} := \{[0, x[\mid x \in [0, 1]^d\}$ denotes the set of half-open axis-parallel rectangles in $[0, 1]^d$ with one corner being the origin and $\mathrm{vol}([0, x[) := \prod_{i=1}^{d} x_i$ denotes the volume of such a rectangle. Hence the discrepancy is a worst-case measure for how far the number of points of P contained in a rectangle R deviates from its fair value $n\,\mathrm{vol}(R)$.

The Koksma-Hlawka inequality, valid for any point set P, bounds the integration error via

$$|E_P(f) - I(f)| \le \tfrac{1}{n} \mathrm{disc}(P) V(f),$$

where $V(f)$ denotes the variation of f in the sense of Hardy and Krause (cf. e.g. [Mat99, p. 23]). We shall not define this variation here, since for a given function f this is a constant. From the (innocent looking, but deep) result that a random set of n points has an expected discrepancy of order $O(\sqrt{n})$, cf. [HNNW01], we obtain the previously stated bound on the integration error of Monte Carlo integration.

More importantly, we can read from the Koksma-Hlawka inequality that indeed random points are good *because* they are well distributed in the sense that they have small discrepancy. Following the quasirandom trail to its end, we should now try to find arbitrary point sets P having small discrepancy $\mathrm{disc}(P)$. Fortunately, such point sets exist. A number of explicit constructions are known that yield n-point sets having a discrepancy of $O(\log(n)^{d-1})$ only. Consequently, for these the Koksma-Hlawka inequality gives much better error guarantees. Also, a huge amount of experimental work shows that these point sets not only have a better error guarantee, but in fact are superior in many settings. Using such low-discrepancy point sets in numerical integration is called *quasi-Monte Carlo integration*.

3   Quasirandom Rumor Spreading

In this section, we demonstrate that quasirandom ideas can also be used in computer science. Our example will be a simple randomized protocol to disseminate information in networks. We shall first describe the random version and then develop a quasirandom one.

3.1   Random Rumor Spreading

In computer networks, the following task needs to be solved. One node of the network obtains some piece of information ("rumor") and needs to communicate it to all other nodes. This happens for example if we store copies of a database at each node of a network. If at some node an update is injected in the database, it has to be communicated to all other nodes. See [DGH+88, KDG03] for more details on where such problems occur.

Typically, as in the database synchronization setting, we are in the situation that updates may occur at arbitrary times at arbitrary nodes, and that we run a protocol that continuously tries to synchronize the databases. While keeping this setting in mind, for the analysis of this problem we shall assume that a single update occurs and we shall regard the process until it communicated this update to all other nodes.

We model the network topology via a graph $G = (V, E)$, that is, V represents the set of nodes and those nodes that can directly communicate with each other are connected via an edge $e \in E$. Unless G is a complete graph, we need intermediate nodes to help disseminating the rumor. We also want to use the help of such nodes to speed up the dissemination process. Our main aim is a fast spreading of the rumor. A second aim is robustness. By this we mean that the dissemination still works moderately well if some transmissions get lost, that is, do not reach their target due to all kinds of errors.

A very simple protocol achieving these aims surprisingly well is the following. The nodes act in a synchronized manner, that is, in rounds. In each round, each node that already knows the rumor contacts a random neighbor. If the neighbor does not yet know the rumor, it becomes informed in this round. This is known as the *randomized rumor spreading protocol*.

Though not a very elaborate protocol, randomized rumor spreading is highly efficient. Let G be a complete graph K_n, a d-dimensional hypercube Q_n ($n = 2^d$, $d \in \mathbb{N}$) or a random graph $G(n, p)$ with $p \geq (1 + \varepsilon) \ln(n)/n$, that is, a random graph $G = (V, E)$ on a fixed set of n vertices such that for all $u, v \in V$, $u \neq v$, independently we have $\Pr[(|\{u, v\} \in E) = p$. Then $O(\log n)$ rounds suffice to inform all nodes with high probability [FG85, FPRU90]. Here and in the following, "with high probability" shall mean with probability at least $1 - \frac{1}{n}$. Later, Elsässer and Sauerwald [ES07] extended this result to Cayley graphs, Sauerwald [Sau07] to expander graphs, and Berenbrink, Elsässer and Friedetzky to [BEF08] random regular graphs.

Clearly, $\log_2(n)$ rounds are necessary for any graph, simply because the number of informed nodes can at most double in each round. Hence all these results show the right order of magnitude.

For the complete graph, more accurate estimates for the broadcast time exist. Already the Frieze and Grimmet result [FG85] shows that $\ln(n) + \log_2(n) + o(\log(n))$ rounds suffice to inform all nodes with probability $1 - o(1)$. Pittel further reduces the error term to show that after $\ln(n) + \log_2(n) + \omega(1)$ rounds, all nodes are informed with probability $1 - o(1)$.

3.2   The Quasirandom Model

The paradigm of quasirandomness advises us to look for characteristic properties of the randomized rumor spreading process and then try to design a protocol that is particularly good with respect to these properties. One property of the randomized protocol that we might speculate to be the reason for its success is its *local fairness*. An informed vertex contacts its neighbors in a relativly balanced manner. If it became informed some time ago that is small compared to the number of its neighbors, then it will have contacted only few neighbors more than once. If it is informed for a long time, then it will have contacted each neighbor roughly equally often. Note that in the model where just a single message has to be distributed, the latter property is not useful.

Believing that this local fairness might be the reason for the success of randomized rumor spreading, we try to build a protocol that perfectionates local fairness. One way of doing so is the following. We equip each vertex with a cyclic permutation of its neighbors. This is the order in which it shall contact its neighbors. This clearly achieves local fairness in the above sense as well as possible. No neighbor is contacted a second time except after all neighbor were contacted once. Also, the numbers of times the different neighbors were contacted deviate by at most one.

Unfortunately, it is easy to see that we may design the cyclic permutations in a way that the protocol needs very long. On the complete graph, if all cyclic permutations have a particular vertex on the last position, then informing this vertex takes $n - 1$ rounds. To overcome such difficulties, we add a small grain of randomness to the protocol. We let each vertex choose its first addressee uniformly at random from its neighbors. All subsequent transmissions from this vertex are directed to its first addressee's successors in the cyclic list (in this order). This is what we shall call the *quasirandom rumor spreading protocol*. Clearly, the quasirandom protocol still has perfect local fairness.

Before analysing the quasirandom protocol, let us discuss it from the implementation point of view. From the theory perspective, we immediately see that it requires each vertex to store the permutation of its neighbors, which might need up to $\Theta(n \log n)$ bits. This was not necessary for the fully random model. However, we may assume that in most networks each vertex already has some list (array) of its neighbors, because the information of how to actually contact a neighbor has to be stored somewhere. Hence here the use of the lists does not increase the complexity. Rather, we might feel that the quasirandom protocol needs less resources. In particular, it needs much fewer random bits. This is nice if we feel that randomness is costly, and useful if we want to trace an actual run of the protocol.

The core question that needs to be answered, naturally, is if the quasirandom protocol works well even if we are not permitted to design the lists. Surprisingly, the answer is positive.

No matter how the lists present at each vertex look like, again $O(\log n)$ rounds suffice with high probability to inform all vertices of a complete graph K_n, a hypercube Q_n, an expander graph on n vertices or a random graph $G(n, p)$ with $p \geq (1+\varepsilon) \ln(n)/n$ [DFS08, DFS09]. Naturally, the lower bound of $\log_2(n)$ rounds presented above for the fully random model also holds for the quasirandom one. Hence again these bounds show the right order of magnitude.

For random graphs, we may even lower the edge probability p to $p = (\ln(n) + \omega(1))/n$. Then with probability $1 - o(1)$, the random graph is such that with high probability the quasirandom protocol independent of the starting point needs only $O(\log n)$ rounds. This is a notable advantage over the fully random model. Here, Feige et al. [FPRU90] showed that for $p = (\ln(n) + O(\log \log n))/n$, the random graph with probability $1 - o(1)$ is such that $\Theta(\log(n)^2)$ rounds are necessary to spread the rumor with high probability.

For the complete graph, we know even sharper bounds. In [ABD$^+$09], we showed that with probability $1 - o(1)$, the number of rounds needed to inform all vertices is $\log_2(n) + \ln_2(n) \pm o(\log(n))$. This was improved by Fountoulakis and Huber (private communication). They show that with probability $1 - o(1)$, the number S_n of rounds after which all vertices are informed, satisfies $\log_2(n) + \ln(n) - 4\ln(\ln(n)) \leq S_n \leq \log_2(n) + \ln(n) + \omega(1)$.

The bounds one can obtain for arbitrary graphs are also better for the quasirandom model. For the fully random model, it is known that $12n\ln(n)$ and $O(\Delta(G)(\operatorname{diam}(G) + \log(n)))$ rounds suffice to inform all vertices of an n-vertex graph G with high probability [FPRU90]. For the quasirandom model, it is easy to prove that after $2n - 3$ or $\Delta(G)\operatorname{diam}(G)$ rounds, with probability one all vertices are informed.

Besides the time needed to disseminate information to all nodes of a network, the robustness of the protocol is an important aspect. The fully randomized model, due to its high use of independent randomness is widely believed to be very robust. The only rigorous result in this direction is due to Elsässer and Sauerwald [ES06]. They showed that if each transmission independently with probability $1 - p$ fails to reach the target, then the time needed by the protocol increases by a factor of $O(1/p)$. In [DFS09], we show the same result for the quasirandom model.

Since the latter result does not regard possible constant factor differences between the two models, in current work with Anna Huber and Ariel Levavi (unpublished manuscript) we analyze the robustness of rumor spreading on the complete graph. We show that the quasirandom protocol informs all vertices of the complete graph in $\frac{1}{\log_2(1+p)}\log_2(n) + \frac{1}{p}\ln(n) + o(\log n)$ rounds. For the random model, we show a lower bound of the same magnitude. This demonstrates that the quasirandom protocol is as least as robust as the fully random one. It also shows that it is more robust than what one would expect at first, namely, that a fraction of p of the messages reaching their destination results in a run-time increase by a factor of $\frac{1}{p}$.

All results presented so far show that the quasirandom model achieves similar and rather superior results, while using a greatly reduced number of random bits. This raises the question if a further reduction of the number of random bits can be fruitful. This, however, is not true. In a model that can be seen as a reasonable extension of the quasirandom model to $\log_2(n) - b$ random bits per vertex, the time needed to inform all vertices of the complete graph can for unsuitable permutations go up to $\log_2(n) + \ln(n) + 2^{\Theta(b)} + o(\log n)$ rounds. We refer to [ABD$^+$09] for a precise statement of the result and thus avoid a lengthy discussion of how protocols with fewer random bits should look like.

4   Analyzing Quasirandomness

In this section, we shall give some insight in how to handle the reduced amount of randomness in the analysis. Clearly, with fewer independent random bits used in the quasirandom protocol, we have to deal with a huge number of dependencies. In particular, we cannot simply build on that we have a certain set of vertices informed after a certain time and continue our proof on this fact, but we have to regard the particular way how these vertices became informed. The fact that certain vertices are informed at a certain time or not, does yield statistical information on whether other vertices are informed and on where in the cyclic process of informing their neighbors they currently stand.

A second reason to be afraid of dependencies is the following. To prove statments holding with high probability, ususally so-called large-deviation bounds are employed. They assert that sums of independent random variables are strongly concentrated around their mean. If in the fully random model on the complete graph we have $n/2$ vertices informed and we expect them to inform roughly $cn := (\frac{1}{2}(1 - e^{-1/2})n$ new ones in the next round, then these bounds assert that we can be very sure to get almost that number. More precisely, the probability that less than $(1 - \varepsilon)cn$ vertices become informed, is less than $\exp(-\varepsilon^2 cn/2)$. Taking these two difficulties into account, it seems surprising that the above mentioned results could be proven.

In this section, we shall see that, in fact, coping with these dependencies needs not to be so difficult as it looks at first. Trying rather to communicate methods than deep results, we shall only regard a very simple problem, namely how to analyse quasirandom rumor spreading on the complete graph, and only to the extent of achieving the right order of magnitude. The main techniques used here, however, can be found in most other proofs on quasirandom rumor spreading as well. We shall therefore proof the following simple result.

Theorem 1. *With high probability, $O(\log n)$ rounds suffice to inform all vertices of the complete graph via the quasirandom rumor spreading model.*

The proof shall be completely self-contained apart from the following simple fact, which can be derived directly from Stirling's formula (see, e.g., [Rob55]) or more generally from Chernoff's inequality (see, e.g., [AS00]).

Lemma 1. *Let $X_1, \ldots, X_n$ be independent random variables uniformly distributed in $\{0, 1\}$. Then the probability that less than a quarter of them are zero, is less than $\exp(-n/8)$.*

Proof (of Theorem 1). For each vertex, fix the cyclic permutations of its neighbors used by the quasirandom model. For the ease of analysis, let us already now fix for each vertex the uniformly chosen random neighbor at which it will start informing its neighbors after it once itself becomes informed. Let u be any vertex. Let C be a constant chosen sufficiently large. We shall show that after $\Theta(C) \log n$ rounds, with probability $1 - n^{-\Theta(C)}$ all vertices are informed.

The proof relies on two key observations. The first is that we may *ignore* vertices, that is, assume that certain vertices stop informing their neighbors at a certain time. Clearly, this only makes the spreading of the information slower. More rigorously, by induction on the time t at which a vertex becomes informed in the original model, we easily see that no vertex becomes informed earlier in this weakened model. Consequently, the broadcast time of the weaker model cannot be smaller than the one of the original model. Note though that the order, in which vertices become informed, may change. This is why we fixed the random starting point at the beginning of the process.

The second observation, similar in nature, is that we may also assume that some vertices after becoming informed delay the start of their actions of informing neighbors. Again, this results in certain messages sent later than in the original model, and thus in vertices possibly becoming informed later, again slowing down the broadcasting process. We call this *delaying* the action of these vertices.

Clearly, we may use ignoring and delaying together and in arbitrary orders and ways. Any time bound proven in any such weakened model also holds in the true quasirandom model. We shall use both tools to form phases of the following kind. Vertices that become informed during a phase delay their action till the following phase. Then they (we shall call them *newly informed*) send out information, but all previously informed vertices are ignored (and again, freshly informed vertices are delayed). By this, we avoid most of the dependencies caused by the quasirandom model. Since the newly informed vertices have not participated in the broadcasting process so far, their random first addressees are stochastically independent of all events that happened so far. Since we ignore previously active vertices, a phase of length ℓ results in the newly informed vertices revealing their independent random starting points and contacting it and the $\ell - 1$ successors in the permutation.

We shall denote by I_t the set of vertices that are informed after round t. In consequence, $I_0 := \{u\}$ consists of the initially informed vertex. By N_t we shall denote vertices that are informed after round t, but have not yet sent out the rumor. Hence apart from $N_0 := \{u\}$, these are the vertices that became informed in the previous phase. We shall use this notation only for times t that are the end of some phase. Note that the phases as well as I_t and N_t depend on what ignoring and delaying assumptions we make. These will become clear in the course of the proof.

The first phase shall consist of $C \ln n$ rounds. Clearly, within this phase the initially informed vertex informes exactly $t_1 = C \ln n$ other vertices. They form N_{t_1}. No other vertices become informed due to our delaying assumption. Obviously, we have $|N_{t_1}| = C \ln n$ and $|I_{t_1}| = C \ln n + 1$.

Let us now convince ourselves of the following fact.

Fact: If at some time t we have $C \ln n \leq |N_t| < |I_t| \leq n/144$ and $|N_t| \geq \frac{1}{2}|I_t|$, then a single phase consisting of eight rounds with probability at least $1 - n^{-C/8}$ results in $|N_{t+8}| \geq 2|N_t|$ and $|N_{t+8}| \geq \frac{1}{2}|I_{t+8}|$.

For the proof of this fact, let $k := |N_t|$. Let $v_1, \ldots, v_k$ be an enumeration of N_t. Note that each vertex in N_t contacts a set of exactly eight other vertices (determined by the random first addressee and its cyclic permutation). They become informed (and hence end up in N_{t+8}), if they are not in I_t. A vertex may become contacted by several vertices in N_t. To avoid overcounting, let us call a vertex $v_i \in N_t$ *successful*, if it contacts exactly eight vertices that (i) are not in I_t and (ii) are not contacted by any of vertices $v_1, \ldots, v_{i-1}$.

Note that the total number of vertices that may become contacted in this phase together with the already informed vertices is at most $|I_t| + 8k < 9|I_t| \leq n/16$. Consequently, when analyzing whether vertex v_i is successful, at most that many vertices are already informed or were informed by vertices $v_1, \ldots, v_{i-1}$. These less than $n/16$ "bad" vertices determine altogether less than $n/2$ vertices having the property that they or one of their seven successors on v_i's permutation is bad. If the random starting first addressee of v_i is one of the other at least $n/2$ "nice" vertices, then v_i is successful. For the ease of argument, let us fix in some deterministic manner a set U_i of exactly $(n-1)/2$ nice vertices and call v_i successful only if its random first addressee is one of them. With this cosmetic operation, v_i is successful with probability $1/2$. Note that the events of the v_i being successful are independent, even though the set U_i are not.

By Lemma 1, with probability at least $1 - \exp(-k/8) \geq 1 - n^{-C/8}$, a quarter of the v_i are successful. This results in $N_{t+8} \geq \frac{1}{4}|N_t| \cdot 8 = 2|N_t|$ newly informed vertices. Note that, trivially, $|N_{t+8}| \geq \frac{1}{2}|N_{t+8}| + |N_t| \geq \frac{1}{2}(|N_{t+8}| + |I_t|) = \frac{1}{2}|I_{t+8}|$. This ends the proof of the fact.

From the state reached after the first phase, we may now repeatedly use the fact until a some time $t_2 \leq t_1 + 8\log_2(n)$ we have with probability $1 - \log_2(n)n^{-C/8}$ that $|I_{t_2}| > n/144$. This implies $|N_{t_2}| \geq n/288$. These newly informed vertices shall now inform all remaining vertices.

We regard a final phase of $k = 288(C+1)\ln(n)$ rounds. The probability that a fixed not yet informed vertex is contacted by a fixed vertex in N_{t_2} naturally is $k/(n-1)$. Hence the probability that none of the vertices in N_{t_2} contacts this vertex is $(1 - k/(n-1))^{|N_{t_2}|} \leq \exp(-k|N_{t_2}|/(n-1)) \leq \exp(-k/288) = n^{-C-1}$. Here we used the elementary estimate $1 + x \leq \exp(x)$ valid for all $x \in \mathbb{R}$. Now a simple union bound shows that with probability at least $1 - n^{-C}$, all uninformed vertices become contacted by vertices in N_{t_2}, and hence informed, in this last phase. $\square$

5   Experimental Results

The theoretical results we currently have indicate that the reduced amount of randomness in the quasirandom model does not reduce the performance of the broadcasting process. However, only in a few situations we were able to prove an advantage of the quasirandom model.

Since we suspect some advantage stemming from the even fairer way of contacting the neighbors, we implemented the protocol and ran a series of experiments. The complete discussion of their outcomes can be found in a forthcoming paper, preliminary results are described in [DFKS09]. Here, we shall only sketch some findings.

The results depicted in Fig. 1 show that the quasirandom model needs less time to inform all vertices. The advantage is small for the complete graph (which is no surprise given our very precise theoretical results), but becomes more visible for sparser graphs and is really striking for sparse random graphs. Note that

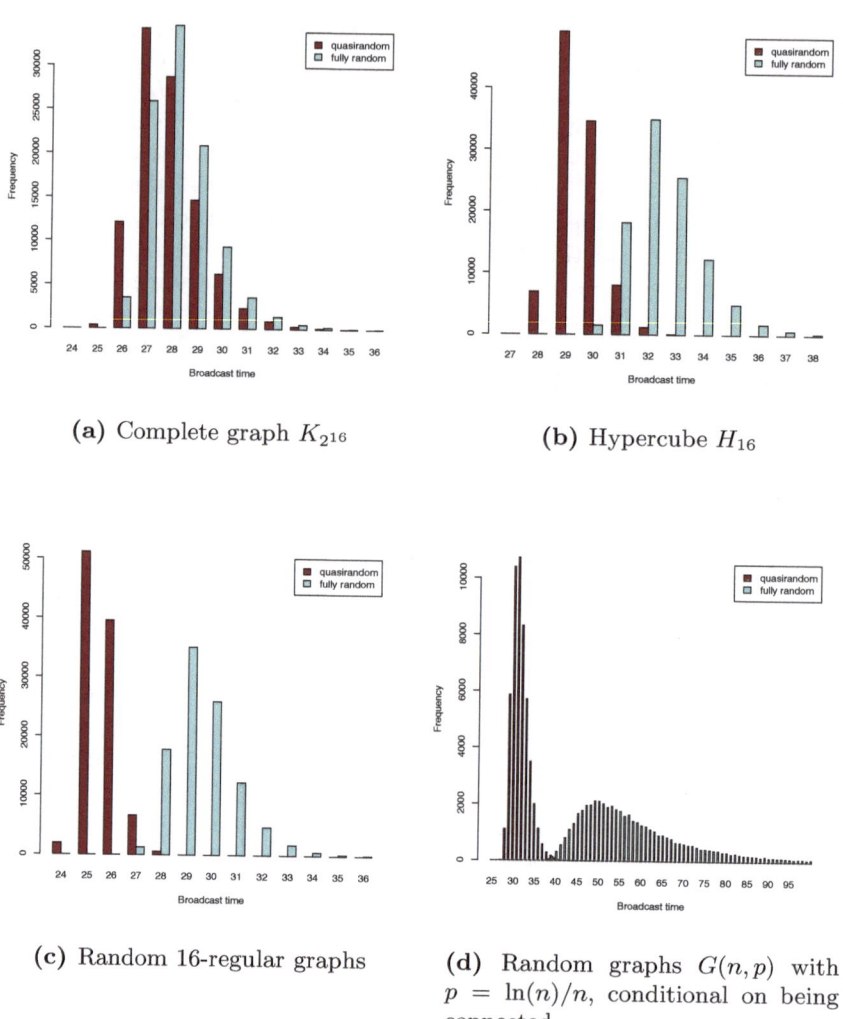

(a) Complete graph $K_{2^{16}}$

(b) Hypercube H_{16}

(c) Random 16-regular graphs

(d) Random graphs $G(n, p)$ with $p = \ln(n)/n$, conditional on being connected

Fig. 1. Empirical distributions of the broadcast times on four different graphs with $n = 2^{16}$ nodes. Thanks to Marvin Künnemann for producing these figures.

the quasirandom model not only is faster, but also the run-times are more concentrated. This means that the risk that the protocol takes longer than expected, is reduced here.

Concerning the robustness, we observed no significant difference between the two protocols.

We finally tried to understand if the particular lists chosen for the quasirandom protocol have an influence on the quality of the protocol. Again, for the graphs regarded so far, we could not see any significant difference. However, we could provoke some influence by regarding an example that has a stronger geometric flavor.

Let G_N be a 2-dimensional torus graph with vertex set $[N] \times [N]$ and each vertex having eight neighbors, namely in the four cardinal directions and the diagonals. In other words, the vertex (x, y) has the eight neighbors $(x + a, y + b)$, where $(a, b) \in \{-1, 0, 1\}^2 \setminus \{(0, 0)\}$ and addition is modulo N. Let us enumerate the directions $\{-1, 0, 1\}^2 \setminus \{(0, 0)\}$ in some counterclockwise manner, e.g., $d_1 = (1, 0)$, $d_2 = (1, 1)$, $d_3 = (0, 1)$, etc.

For $N = 2^6 = 64$, we experimentally determined the average times needed to broadcast a news to all vertices. If each vertex v has the canonical list $v + d_1, v + d_2, ...$, then we measures an average broadcast time of 84.57 rounds. This is even worse than the average number of 84.09 rounds needed by the fully random protocol. However, if we use a low-discrepancy order of the directions, things become much better. If each vertex serves its neighbors according to the order $(1, 5, 3, 7, 2, 6, 4, 8)$ of the directions, then the broadcast time drops to 77.10 rounds. A closer look (checking all possible orderings) in fact shows a strong correlation between the discrepancy of the sequence and the broadcast time.

6   Conclusion and Open Problems

In this paper, we gave a first application of quasirandomness in computer science. Our results show two important facts. (i) Quasirandom methods can successfully be used in computer science. For the rumor spreading example, we achieved moderate gains in the run-time via a model that is algorithmically even simpler than the fully random one. (ii) The second good news is that such quasirandom approaches can be analysed in spite of the dependencies naturally present in the random experiment. There is even the hope that general approaches like the delaying and ignoring concept exist, that allow to revert to classical methods from independent randomization.

From this work, a number of open questions arise.

– The results for random graphs, hypercubes and expander graph show that $O(\log n)$ rounds suffice with high probability for both the fully random and the quasirandom model. Our experimental investigation suggests that, in particular for hypercubes and moderately sparse random graphs ($p = (1 + \varepsilon) \ln(n)/n$), the quasirandom model is faster by a constant factor. Supporting this observation by a rigorous proof would be very desirable.

   Unfortunately, already for the fully random model the leading constant is only known for the complete graph. Hence to prove a constant factor advantage of the quasirandom model, one needs to prove good bounds for both models.

– An aspect not regarded in this paper is the number of messages that need to be sent. For the fully random model on the complete graph, Karp, Schindelhauer, Shenker and Vöcking [KSSV00] show that a suitable modification of the protocol can reduce the number of messages needed from $\Theta(n \log n)$ to an optimal $\Theta(n \log \log n)$. Similar improvements for random graphs and regular graphs have been obtained in [BEF08, Els06, ES08]. For the quasirandom protocol, nothing in this direction is published.

– Clearly the most general question is for which other computer science problems quasirandom ideas improve the existing approaches.

References

[ABD+09]   Angelopoulos, S., Bläser, M., Doerr, B., Fouz, M., Huber, A., Panagiotou, K.: Quasirandom Rumor Spreading: Tight Bounds and the Value of Random Bits (submitted, 2009)

[AKL+79]   Aleliunas, R., Karp, R., Lipton, R., Lovász, L., Rackoff, C.: Random walks, universal traversal sequences, and the complexity of maze problems. In: 20th IEEE Symposium on Foundations of Computer Science (FOCS), pp. 218–223 (1979)

[AS00]   Alon, N., Spencer, J.H.: The Probabilistic Method, 2nd edn. Wiley, Chichester (2000)

[BEF08]   Berenbrink, P., Elsässer, R., Friedetzky, T.: Efficient randomized broadcasting in random regular networks with applications in peer-to-peer systems. In: 27th ACM Symposium on Principles of Distributed Computing (PODC), pp. 155–164 (2008)

[DFKS09]   Doerr, B., Friedrich, T., Künnemann, M., Sauerwald, T.: Quasirandom rumor spreading: An experimental analysis. In: Proceedings of the 10th Workshop on Algorithm Engineering and Experiments (ALENEX), pp. 145–153 (2009)

[DFS08]   Doerr, B., Friedrich, T., Sauerwald, T.: Quasirandom rumor spreading. In: Proceedings of the 19th Annual ACM-SIAM Symposium on Discrete Algorithms (SODA), pp. 773–781 (2008)

[DFS09]   Doerr, B., Friedrich, T., Sauerwald, T.: Quasirandom rumor spreading: Expanders, push vs. pull, and robustness. In: Proceedings of the 36th International Colloquium on Automata, Languages and Programming (ICALP) (2009)

[DGH+88]   Demers, A.J., Greene, D.H., Hauser, C., Irish, W., Larson, J., Shenker, S., Sturgis, H.E., Swinehart, D.C., Terry, D.B.: Epidemic algorithms for replicated database maintenance. Operating Systems Review 22, 8–32 (1988)

[Els06]   Elsässer, R.: On the communication complexity of randomized broadcasting in random-like graphs. In: 18th ACM Symposium on Parallelism in Algorithms and Architectures (SPAA), pp. 148–157 (2006)

[Erd59]   Erdős, P.: Graph theory and probability. Canad. J. Math. 11, 34–38 (1959)

[ES06]   Elsässer, R., Sauerwald, T.: On the runtime and robustness of randomized broadcasting. In: Asano, T. (ed.) ISAAC 2006. LNCS, vol. 4288, pp. 349–358. Springer, Heidelberg (2006)

[ES07]   Elsässer, R., Sauerwald, T.: Broadcasting vs. Mixing and information dissemination on cayley graphs. In: Thomas, W., Weil, P. (eds.) STACS 2007. LNCS, vol. 4393, pp. 163–174. Springer, Heidelberg (2007)

[ES08]   Elsässer, R., Sauerwald, T.: On the power of memory in randomized broadcasting. In: 19th ACM-SIAM Symposium on Discrete Algorithms (SODA), pp. 218–227 (2008)

[FG85]   Frieze, A.M., Grimmett, G.R.: The shortest-path problem for graphs with random arc-lengths. Discrete Applied Mathematics 10, 57–77 (1985)

[FPRU90]   Feige, U., Peleg, D., Raghavan, P., Upfal, E.: Randomized broadcast in networks. Random Structures and Algorithms 1, 447–460 (1990)

[HNNW01]   Heinrich, S., Novak, E., Wasilkowski, G.W., Woźniakowski, H.: The inverse of the star-discrepancy depends linearly on the dimension. Acta Arithmetica 96, 279–302 (2001)

[KDG03]   Kempe, D., Dobra, A., Gehrke, J.: Gossip-based computation of aggregate information. In: 44th IEEE Symposium on Foundations of Computer Science (FOCS), pp. 482–491 (2003)

[KSSV00]   Karp, R., Schindelhauer, C., Shenker, S., Vöcking, B.: Randomized rumor spreading. In: 41st IEEE Symposium on Foundations of Computer Science (FOCS), pp. 565–574 (2000)

[Mat99]   Matoušek, J.: Geometric Discrepancy. Springer, Berlin (1999)

[Rei05]   Reingold, O.: Undirected ST-connectivity in log-space. In: Proceedings of the 37th ACM Symposium on Theory of Computing (STOC), pp. 376–385 (2005)

[Rob55]   Robbins, H.: A remark on Stirling's formula. American Mathematical Monthly 62, 26–29 (1955)

[Sau07]   Sauerwald, T.: On mixing and edge expansion properties in randomized broadcasting. In: Tokuyama, T. (ed.) ISAAC 2007. LNCS, vol. 4835, pp. 196–207. Springer, Heidelberg (2007)

Part II
Sorting and Searching

Reflections on Optimal and Nearly Optimal Binary Search Trees

J. Ian Munro

Cheriton School of Computer Science, University of Waterloo,
Waterloo, Ontario N2L 3G1, Canada
imunro@uwaterloo.ca

Abstract. We take a rather informal look at the development of techniques for finding optimal and near optimal binary search trees.[1] The point of view is both that of the "early" development of the field and as a set of wonderful examples for teaching algorithms.

1 Introduction

Kurt Mehlhorn's first publications on algorithms, as opposed to complexity[2] [3, 1], were on approximate solutions to the optimal[3] binary search tree problem. The problem is well known, given a set of values, the probability of a search for each of these values and the probabilty of a search for some value in each of the gaps between them, we are to find a binary search tree with the least possible expected search cost. We let $\{a_i : i = 1, n, a_i < a_{i+1}\}$ denote the set of values. $\boldsymbol{p} \overset{\text{def}}{=} \{p_i : i = 1, n\}$ denotes the probabilities of requests for a_i and $P = \sum_{i=1}^{n} p_i$. Similarly $\boldsymbol{q} \overset{\text{def}}{=} \{q_i : i = 0, n\}$ gives the probability of a request for a value between a_{i-1} and a_i, with the implicit notation that $a_0 = -\infty$ and $a_{n+1} = \infty$. Again, $Q = \sum_{i=0}^{n} q_i$). The solution is an easy dynamic programming exercise today, though not necessarily easy when it was discovered. A major shortcoming of the dynamic programming solution is its quadratic space requirement. This lead to work by several researchers including Mehlhorn on space efficient methods to find nearly optimal trees faster and with much less space. The problem also

---

[1] One of the joys of writing this paper was going through much of the work on the topic from the 1970's. I first thought of Kurt Mehlhorn's ICALP '75 paper [1], so I twirled around to get it from my bookshelf. I had the preceding one, from Saarbrücken, but not the 1975 Proceedings. So then I went for the journal (SICOMP) version, which was on the bookshelf, with a more than generous coating of dust which I blew off. Surviving the dust storm, I got down to rereading that paper and others, mostly recovered from the internet.

[2] I am counting the paper with Zvi Galil on monotone circuits [2] as "complexity".

[3] Almost all authors on the topic, including this one, refer to "optimal" binary search trees. Knuth [4, 5] uses the term "optimum" in a manner analogous to the distinction in combinatorics between "minimum", meaning a global minimum, and "minimal", meaning a local minimum. There "optimum" refers to a global best, whereas "optimal" may suggest a solution that is best only in some local context.

S. Albers, H. Alt, and S. Näher (Eds.): Festschrift Mehlhorn, LNCS 5760, pp. 115–120, 2009.

has several "relaxations". One can restrict attention to the case in which the probability of a successful search is 0 (i.e. $(p_i = 0 : i = 1, n)$). In another dimension one can relax the concern on the order of key values. These two constraints, combined, give the problem of finding a Huffman code [6].

2   Optimal Binary Search Trees

The notion of an optimal binary search tree seems to have arisen in a variety of contexts. For example, Knuth [5] cites the 1959 paper of Gilbert and Moore [7] in that respect. They were concerned primarily with the case $p_i = 0$ and developed a cubic algorithm. The "big step" came more than a decade later, by which time dynamic programming was more "mainstream". Knuth's solution [4] was the classic "3-loop" dynamic programming solution, like that for parsing arbitrary context free languages [8, 9] three or four years earlier. The twist, in terms of algorthmics was that Knuth's method took $O(n^2)$ time, though still $\Theta(n^2)$ space. As a basic dynamic programming solution, the idea is to find the optimal tree for every subrange of keys. This is done in increasing order by range size, so the optimal tree for the range (i, j), from the gap before key i to the gap after key j, has a root at some node k $(i \leq k \leq j)$ and the optimal subsolutions for ranges $(i, k-1)$ and $(k+1, j)$. Knuth's key observation was that, if one adds a key to the right of an interval, the root for that interval cannot move left (as we have nonnegative weights). Proving this is more subtle than one might expect, but given this fact one need only search for the root of interval (i, j) from the root of interval $(i, j-1)$ to the root of $(i-1, j)$. A quick calculation of the total runtime of the algorithm leads to a telescoping series and the conclusion that the algorithm runs in quadratic time. No progess has been made on significantly reducing the space requirement while still having a polynomial time algorithm that does not rely on conditions on the p_i and q_i values. Indeed this is even true if our goal is simply to determined whether a given tree is optimal. This remains as the most interesting open problem on the topic of optimal search trees.

At essentially the same time Hu and Tucker [10] gave a technique for the $p_i = 0$ case. It was improved and simplified by Garsia and Wachs [11] yielding a $\Theta(n \log n)$ time algorithm using only linear space. Note that if the weights are given in arbitrary order (or even left to right order) then under a "comparison and addition" model, $\Theta(n \log n)$ time is optimal.

3   Near Optimal Solutions and Analysis

A number of authors looked at approaches avoiding quadratic space. The work in Wayne Walker's Ph.D. thesis[4] [12] used the heuristic of roughly equalizing the weight on the left and right subtrees and then looking for a relatively large p_i

---

[4] Walker had an office just down the hall from me when we were graduate students at the University of Toronto. Indeed it was Walker and Gotlieb's work that introduced me to the topic.

value close by. A variety of tuning methods were tried and the approach worked well on experimental data. Curiously, or perhaps not, the analytic work on the cost of the optimal binary search tree was intertwined with that on approximate solutions. Paul Bayer, in his Master's thesis[5] [13], proved that (assuming $H > 2P/e$) $C_{opt} > H - P\lg(eH/2P)$. Here C_{opt} denotes the cost (expected number of key values inspected) of the optimal tree and H is the entropy of the discrete distribution: $H \stackrel{\text{def}}{=} \sum_{i=1}^{n} p_i \lg(1/p_i + \sum_{i=0}^{n} q_i \lg(1/q_i)$.

Bayer also gave an upper bound which was refined by Mehlhorn [14] to $H + 2 - P$. Note that the entropy is a function of the (multi) set of probabilities and independent of their order. Brian Allen[6] [15] showed that the bounds given are essentially as tight as possible and can almost swing from the lower bound to the upper bound by rearranging them. This holds even when H is almost as large as possible i.e. $\lg(n) - o(\lg(n))$.

Mehlhorn [3, 1, 14] formalized two top down approximation schemes, that were simplifications of the Walker and Gotlieb approach. Under one the root is chosen so that the left and right subtrees have total weight (probability of access) as close as possible. His second heuristic was to choose the root to minimize the maximum weight of either subtree. He and Bayer showed both produce trees in the range proven for the optimal tree [14, 13], though either can differ from the optimal by essentially $\lg H$ [15].

An $O(n \log n)$ time, linear space implementation of such an approximation scheme is straightforward. One creates an auxiliary array containing the sum of the probabilities up to each internal node. That is $w_i = \sum_{j=1}^{i-1} p_j + \sum_{j=0}^{i-1} q_j$. which itself can be computed in linear time as $w_i = w_{i-1} + p_{i-1} + q_{i-1}$. A binary search then suffices to determine the root of the tree. The recursion gives us a worst case behaviour of $O(n \log n)$ and indeed $O(n)$ if the root "tends to be in the middle of its range". Fredman [16] suggested an ingeneous trick to guarantee linear runtime. A single comparison with the value in the middle of the range tells us whether the desired element is in the first or second half. One then starts at the end of the half range where the prospective root is to be found and repeatedly doubles the offset from that location until the prospective root has been bracketed. A straightforward binary search between the last two locations inspected completes the search in time $O(\log d)$ where d is the distance of the prospective root from the nearest end of the interval. It is not hard to show the entire process of constructing the tree is performed in linear time.

Over the past 30 years there have been several small improvements and careful analyses of the approximability of the optimal binary search tree problem. These have culminated with the work of De Prisco and De Santos [17, 18] and that of Douïeb and Bose [19] which give very detailed bounds of the quality of approximation. As suggested earlier, it is natural to ask whether one can confirm that a tree is optimal in less time and space than required to find the tree.

---

[5] Bayer didn't ever publish his work in a refereed forum, neverthless his thesis was a key reference for many of us working on related problems. I still have a copy.

[6] Allen's thesis focussed primarily on self organizing binary search trees, a forerunner to splay trees. He was my first Ph.D. student.

4   Trees with No Ordering on Keys

The oldest version of the optimal binary "search" tree problem is, of course, that of finding a Huffman code [6]. Here we are to find the optimal binary code for a set of symbols with no constraints on the ordering of the values, but a "prefix free" constraint that no code can be a prefix of any other. In the terms we have used, the latter condition means that $P = 0$. Given the elements and probabilities sorted by probability, the tree (and so the codes) can be found in linear time by a well known greedy algorithm giving a bottom up construction.

The method maintains a forest of optimal solutions for disjoint subsets of the elements. Initially this forest is a set of n leaves, namely the given elements. The process proceeds by replacing the two trees of lowest probability with a new tree consisting of a new root and with children as the roots of the two trees being replaced. This is a wonderful example of a special case of a priority queue that is easily implemented in constant time per update. There are two kinds of trees: the original leaves (which we will assume are given in sorted order by probability) and the other trees (which are created in increasing order by probability value). We, therefore, have two sorted lists and perform an update by removing the two with lowest probability (from either list) and put a new tree at the end of the "subtree" list. Like all other versions of the problem, certain inputs can force us to discover the sorted order of the probabilities. Hence an $\Omega(n \log n)$ lower bound under a model permitting comparisons and additions if the values are not presented as a sorted list. This suggests another interesting open question. Suppose we are given a binary tree with probabilities assigned to the leaves, we ask whether this is a tree for an optimal prefix free code. In other words, is this the same tree as that formed by the Huffman algorithm, albeit with an arbitrary permutation of the leaves on each level? Prove or disprove that this recognition problem can be solved in linear time.

One can extend the Huffman code problem to have probabilities associated with internal nodes as well as leaves. Again the "left to right" ordering is up to the algorithm. Given p and q each in sorted order (by probability value) a "three list" priority queue, analogous to the "two list" method for the usual Huffman code, gives a solution in linear time. The two lightest "leaves" or "subtrees" become children of the lightest remaining "internal node". All three are removed from their respective lists and the new tree goes at the end of the "subtree" list, giving a linear time algorithm if the input is presented in sorted order.

5   Conclusion

We have revisited the development of techniques for optimal and near optimal binary search trees, most notably the advances of the early to middle 1970's. It was an era when many of our algorithmic methods were taking form. Indeed we have discussed several real gems of algorithm design that are wonderful classroom examples:

- Basic dynamic programming and Knuth's observation to make a "3 loop deep" program to run in quadratic time

- Clean and very accurate bounds on the performance of optimal search trees
- The appropriate choices of a "greedy metric" to produce amazingly good approximation methods
- A method to guarantee linear runtime for divide and conquer algorithms taking a binary search to perform the split
- A couple of special case priority queues taking constant time per operation

With this key work done 35 to 40 years ago, it is suprising that there are still a few very interesting open problems.

Like Kurt Mehlhorn's subsequent work on algorithms, the ideas are crisp; the proofs are clean; the approximations are amazingly good; and the algorithms are both practical and reasonable to implement.

References

1. Mehlhorn, K.: Best possible bounds for the weighted path length of optimum binary search trees. In: Brakhage, H. (ed.) GI-Fachtagung 1975. LNCS, vol. 33, pp. 31–41. Springer, Heidelberg (1975)
2. Mehlhorn, K., Galil, Z.: Monotone switching circuits and boolean matrix product. In: Becvár, J. (ed.) MFCS 1975. LNCS, vol. 32, pp. 315–319. Springer, Heidelberg (1975)
3. Mehlhorn, K.: Nearly optimal binary search trees. Acta Inf. 5, 287–295 (1975)
4. Knuth, D.E.: Optimum binary search trees. Acta Inf. 1, 14–25 (1971)
5. Knuth, D.E.: The Art of Computer Programming. Sorting and Searching, vol. III. Addison-Wesley, Reading (1973)
6. Huffman, D.A.: A method for the construction of minimum-redundanct codes. Proceedings of the IRE 40(9), 1098–1101 (1952)
7. Gilbert, E.N., Moore, E.F.: Variable-length binary encodings. The Bell System Technical Journal 38, 933–968 (1959)
8. Younger, D.H.: Context-free language processing in time n^3. In: FOCS, pp. 7–20. IEEE, Los Alamitos (1966)
9. Younger, D.H.: Recognition and parsing of context-free languages in time n^3. Information and Control 10(2), 189–208 (1967)
10. Hu, T.C., Tucker, A.C.: Optimal computer search trees and variable-length alphabetical codes. SIAM J. on Applied Math. 21(4), 514–532 (1971)
11. Garsia, A.M., Wachs, M.L.: A new algorithm for minimum cost binary trees. SIAM J. Comput. 6(4), 622–642 (1977)
12. Walker, W.A., Gotlieb, C.C.: A top-down algorithm for constructing nearly optimal lexicographical trees. In: Read, R.C. (ed.) Graph Theory and Computing, pp. 303–323 (1972)
13. Bayer, P.J.: Improved Bounds on the Costs of OPtimal and Balanced Binary Search Trees. MAC Technical Memorandum 69, ans M.Sc. Thesis, M.I.T. (1975)
14. Mehlhorn, K.: A best possible bound for the weighted path length of binary search trees. SIAM J. Comput. 6(2), 235–239 (1977)
15. Allen, B.: On the costs of optimal and near-optimal binary search trees. Acta Inf. 18, 255–263 (1982)
16. Fredman, M.L.: Two applications of a probabilistic search technique: Sorting x + y and building balanced search trees. In: STOC, pp. 240–244. ACM, New York (1975)

17. DePrisco, R., DeSantis, A.: On binary search trees. Inf. Process. Lett. 45(5), 249–253 (1993)
18. DePrisco, R., DeSantis, A.: New lower bounds on the cost of binary search trees. Theor. Comput. Sci. 156(1&2), 315–325 (1996)
19. Bose, P., Douïeb, K.: Efficient construction of near-optimal binary and multiway search trees. To appear Proc. Algorithms and Data Structures Symposium (WADS) (2009)

Some Results for Elementary Operations

Athanasios K. Tsakalidis

Department of Computer Engineering and Informatics,
University of Patras, 26501 Patras, Greece
tsak@cti.gr
www.tsakalidis.gr

Abstract. We present a number of results for elementary operations concerning the areas of data structures, computational geometry, graph algorithms and string algorithms. Especially, we focus on elementary operations like the dictionary operations, list manipulation, priority queues, temporal precedence, finger search, nearest common ancestors, negative cycle, 3-sided queries, rectangle enclosure, dominance searching, intersection queries, hidden line elimination and string manipulation.

1 Introduction

We consider a number of results derived in the last 25 years in the theory of efficient algorithms. For each of them, we present the main ideas, the main theorems, and we follow the path of their impact in the research community. We intentionally avoid technical descriptions and strictly mathematical definitions and we appoint them in a plausible manner. For each kind of the operations we offer a separate section with the respective references.

The model of computation we consider is the *Pointer Machine (PM-machine)*, the *Pure Pointer Machine (PPM-machine)*, and the *Random-Access Machine (RAM-machine)*. In a pointer machine, memory consists of a collection of *records*. Each record consists of a fixed number of *cells*. The cells have associated *types*, such as *pointer*, *integer*, *real*, and access to memory is only possible by "pointers". In other words, the memory is structured as directed graph with bounded out-degree. The edges of this graph can be changed during execution. Pointer machines correspond roughly to high-level programming languages without arrays. This PM-machine allows arithmetic capabilities on the content of data fields. In case that no arithmetic capabilities are allowed, we get the Pure Pointer Machine. The instruction set of the PPM-Machine consists of two types of instructions-pointer manipulation and control management. In contrast, the memory of a RAM consists of an array of cells. A cell is accessed through its address and hence address arithmetic is available. We assume in all the models that storage cells can hold arbitrary numbers and that the basic arithmetic and pointer operations take constant time. This is called the uniform cost assumption. All our machines are assumed to be deterministic.

Three types of complexity analysis are customary in the algorithms area: worst-case analysis, average-case analysis and amortized analysis. In a *worst-case analysis* worst-case bounds are derived for each single operation. This is

S. Albers, H. Alt, and S. Näher (Eds.): Festschrift Mehlhorn, LNCS 5760, pp. 121–133, 2009.

the most frequent type of analysis. In an *average-case analysis*, we postulate a probability distribution on the operations of the abstract data type and computes the expected cost of the operations under this probability assumption. In an *amortized analysis*, we study the worst-case cost of a sequence of operations.

The externalization of a known data structure is the transformation of this data structure in such a way, that it can handle the same operations in an efficient way in case that the file handled cannot be stored entirely in the main memory. In this case, we have a main memory and a disk partitioned in blocks of size B. In this model of computation, the efficiency of the data structure used is measured by the number of *I/O-operations*. An *I/O-operation* (or simply *I/O*) is the operation of reading (or writing) a block from (or into) disk.

2   Dictionary Operations

Data structuring is the study of *concrete implementation* of frequently occurring *abstract data types*. An abstract data type is a set together with a collection of operations on the elements of the set. In the data type *dictionary*, the set is the powerset S of a universe U and the operations are *insertions* and *deletions* of the elements and the *test of membership (access)*.

Note that the operations *insertion* and *deletion* are destructive, the old version of the set S is destroyed by the operations, and, in this case, the data structure used is called *ephemeral*. The nonedestructive version of the problem motivated the development *persistent data structures* [3]. We distinguish *partial* and *full* persistence. A data structure is partially persistent, if all versions can be accessed but only the newest version can be modified and fully persistent if every version can be both accessed and modified.

A new data structure called *Interpolation Search Tree (IST)* is presented in [8], which supports interpolation search and insertions and deletions. The amortized insertion and deletion cost is $O(\log n)$. The expected search time in a random file is $O(\log \log n)$. This is not only true for the uniform distribution but for a wide class of probability distributions. Informally, a distribution defined over an interval I is *smooth*, if the probability density over any subinterval of I does not exceed a specific bound, however small this subinterval is (i.e., the distribution does not contain sharp peaks).

Given two functions f_1 and f_2, a density function $\mu = \mu[a, b](x)$ is (f_1, f_2)-*smooth* ([2], [8]), if there exists a constant β, such that for all $c_1, c_2, c_3, a \le c_1 < c_2 < c_3 \le b$, and all integers n, it holds that we have that

$$\int_{c_2 - \frac{c_3 - c_1}{f_1(n)}}^{c_2} \mu[c_1, c_3](x)dx \le \frac{\beta \cdot f_2(n)}{n}$$

where $\mu[c_1, c_3](x) = 0$ for $x < c_1$ or $x > c_3$, and $\mu[c_1, c_3](x) = \mu(x)/p$ for $c_1 \le x \le c_3$, where $p = \int_{c_1}^{c_3} \mu(x)dx$.

Intuitively, function f_1 partitions an arbitrary subinterval $[c_1, c_3] \subseteq [a, b]$ into f_1 equal parts, each of length $\frac{c_3 - c_1}{f_1} = O(\frac{1}{f_1})$; that is, f_1 measures how fine is the partitioning of an arbitrary subinterval. Function f_2 guarantees that no part, of the f_1 possible, gets more probability mass than $\frac{\beta \cdot f_2}{n}$; that is, f_2 measures the sparseness of any subinterval $[c_2 - \frac{c_3 - c_1}{f_1}, c_2] \subseteq [c_1, c_3]$. The class of (f_1, f_2)-smooth distributions (for appropriate choices of f_1 and f_2) is a superset of both regular [11] and uniform classes of distributions, as well as of several non-uniform classes ([2],[5]). Actually, *any* probability distribution is $(f_1, \Theta(n))$-smooth, for a suitable choice of β.

Herewith it is worthwhile to note that the data structure used can reflect the distribution function in some places in order to guide the searching properly and to the fact that a random IST has subtrees of rootic size with high probability.

In [2], a technique is presented which extends the technique of [8] to a larger class of distributions and better bounds on searches and updates.

In [6], the *IS-Tree*, a dynamic data structure based on interpolation search is presented, which consumes worst case linear space and can be updated in $O(1)$ time worst case when the update position is given. Furthermore, the elements can be searched in $O(\log \log n)$ time expected with high probability, given that they are drawn from a $(n^\alpha, n^{1/2})$-smooth distribution, for constant $1/2 < \alpha < 1$. The worst case search time is $O(\log^2 n)$.

The externalization [10] of this data structure, called *ISB-tree*, was introduced in [4]. It supports search operations in $O(\log_B \log n)$ expected I/Os and update operations in $O(1)$ worst-case I/Os provided that the update position is given and B is the block size. The expected search bound holds with high probability, if the elements are drawn by a $(n/(\log \log n)^{1+\epsilon}, n^{1-\delta})$-smooth distribution, where $\epsilon > 0$ and $\delta = 1 - \frac{1}{B}$ are constants. The worst case search bound is $O(\log_B n)$ block transfers.

AVL-trees were introduced by Adel'son-Velskii and Landis in 1962 [1]. A binary search tree is AVL if the height of the subtrees at each node differ by at most one. In [7] we analyse the amortized behavior of AVL-trees under a sequence of insertions. We show that the total rebalancing cost for a sequence of n arbitrary insertions is at most $2.618n$. For random insertions, the bound is improved to $2.26n$. We show that the probability that t or more balance changes are required decreases exponentially with t. Mixed insertions and deletions do not have amortized constant complexity. In [9] is shown that the total rebalancing cost for a sequence of only arbitrary deletions is $1.618n$.

References

1. Adel'son-Velskii, G.M., Landis, E.M.: An Algorithm for the Organization of Information. Dokl. Akad. Nauk SSSR 146, 263–266 (1962) (in Russian); English translation in Soviet. Math. 3, 1259–1262 (1962)
2. Andersson, A., Mattson, C.: Dynamic Interpolation Search in $o(\log \log n)$ Time. In: Lingas, A., Carlsson, S., Karlsson, R. (eds.) ICALP 1993. LNCS, vol. 700, pp. 15–27. Springer, Heidelberg (1993)
3. Driscoll, J.R., Sarnak, N., Sleator, D.D., Tarjan, R.E.: Making Data Structures Persistent. J. Comput. System Sci. 28, 86–124 (1989)
4. Kaporis, A., Makris, C., Mavritsakis, G., Sioutas, S., Tsakalidis, A., Tsichlas, K., Zaroliagis, C.: ISB-tree: A new indexing scheme with efficient expected behaviour. In: Deng, X., Du, D.-Z. (eds.) ISAAC 2005. LNCS, vol. 3827, pp. 318–327. Springer, Heidelberg (2005)

5. Kaporis, A., Makris, C., Sioutas, S., Tsakalidis, A., Tsichlas, K., Zaroliagis, C.: Improved Bounds for Finger Search on a RAM. In: Di Battista, G., Zwick, U. (eds.) ESA 2003. LNCS, vol. 2832, pp. 325–336. Springer, Heidelberg (2003)
6. Kaporis, A., Makris, C., Sioutas, S., Tsakalidis, A., Tsichlas, K., Zaroliagis, C.: Dynamic Interpolation Search Revisited. In: Bugliesi, M., Preneel, B., Sassone, V., Wegener, I. (eds.) ICALP 2006. LNCS, vol. 4051, pp. 382–394. Springer, Heidelberg (2006)
7. Mehlhorn, K., Tsakalidis, A.K.: An Amortized Analysis of Insertions into AVL-Trees. SIAM J. Comput. 15(1), 22–33 (1986)
8. Mehlhorn, K., Tsakalidis, A.: Dynamic Interpolation Search. Journal of the ACM 40(3), 621–634 (1993)
9. Tsakalidis, A.K.: Rebalancing Operations for Deletions in AVL-Trees. RAIRO Inform. Theor. 19(4), 323–329 (1985)
10. Vitter, J.: External Memory Algorithms and Data Structures: dealing with massive data. ACM Computing Surveys 33(2), 209–271 (2001)
11. Willard, D.E.: Searching unindexed and nonuniformly generated files in log log N time. SIAM J. Comput. 14(4), 1013–1029 (1985)

3   List Manipulation

In [10], we give a representation for linked lists which allows to efficiently *insert* and *delete* objects in the list and to quickly determine the order of two list elements. The basic data structure, called an *indexed BB[a]-tree* ([9], [2]), allows to do n insertions and deletions in $O(n \log n)$ steps and determine the order in constant time, assuming that the locations of the elements worked at are given. The improved algorithm does n insertions and deletions in $O(n)$ steps and determines the order in constant time. An application of this provides an algorithm which determines the *ancestor relationship* of two given nodes in a dynamic tree structure of bounded degree in time $O(1)$ and performs n arbitrary insertions and deletions at given positions in time $O(n)$ using *linear space*.

The amortized analysis of our algorithm is substantiallly based on the *weight−property* of BB[a]-trees, which is proved in [2]. The weight-property can be stated as follows: A node of weight w (i.e. w descendants) participates in only $O(n/w)$ structural changes of the tree when a sequence of n insertions and deletions is processed. This result improves the bounds given in [6], where only insertions were allowed. In [7], two algorithms are given. The first algorithm matches the $O(1)$ amortized time per operations of [10] and is simpler. The second algorithm permits all operations in $O(1)$ *worst-case* time. In [1], simpler solutions are given that match the bounds of [7].

The results of [10] are well used in the theory of persistent data structures [3–5, 8].

References

1. Bender, M.A., Cole, R., Demaine, E.D., Farach-Colton, M., Zito, J.: Two Simplified Algorithms for Maintaining Order in a List. In: Möhring, R.H., Raman, R. (eds.) ESA 2002. LNCS, vol. 2461, pp. 219–223. Springer, Heidelberg (2002)
2. Blum, N., Mehlhorn, K.: On the average number of Rebalancing Operations in weight-balanced trees. Theor. Comput. Sci. 11, 303–320 (1980)

3. Buchsbaum, A., Tarjan, R.E.: Confluently Persistent Deques via Data Structural Bootstrapping. In: Proc. of the fourth annual ACM-SIAM Symposium on Discrete Algorithms, pp. 155–164 (1993)
4. Dietz, P.F.: Fully Persistent Arrays (Extended Abstract). In: Proc. of the Workshop on Algorithma and Data Structures, pp. 67–74 (1989)
5. Dietz, P.F.: A Space Efficient Variant of Path Copying for Partially Persistent Sorted Sets. J. Comput. System Sci. 53, 148–152 (1996)
6. Dietz, P.F.: Maintaining Order in a Linked List. In: Proc. 14th ACM STOC, pp. 122–127 (1982)
7. Dietz, P.F., Sleator, D.D.: Two Algorithms for Maintaining Order in a List. In: Proc. 19th ACM STOC, pp. 365–372 (1987)
8. Driscoll, J.R., Sarnak, N., Sleator, D.D., Tarjan, R.E.: Making Data structures Persistent. J. Comput. System Sci. 28, 82–124 (1989)
9. Nievergelt, I., Reingold, E.M.: Binary Search Tress of Bounded Balance. SIAM J. Comput. 2, 33–43 (1973)
10. Tsakalidis, A.K.: Maintaining Order in a generalized Linked List. Acta Informatica 21(1), 101–112 (1984)

4 Priority Queues

A min-max priority queue is a priority queue that structures the elements with respect to the maximum element as well the minimum element. In [2], we present a simple and efficient implementation of a min-max priority queue, reflected *min-max priority queues*. The main merits of our construction are threefold. First, the space utilization of the reflected min-max heaps is much better than the naive solution of putting two heaps back-to-back [3]. Second, the methods applied in this structure can be easily used to transform ordinary priority queues into min-max priority queues. Third, when considering only the setting of min-max priority queues, we support merging in constant worst-case time which is a clear improvement over the best worst-case bound achieved [1].

References

1. Hoyer, P.: A General Technique for Implementation of Efficient Priority Queues. In: Proc. of the 3rd Israel Symposium on Theory of Computing systems (ISTCS) (1995)
2. Makris, C., Tsakalidis, A., Tsichlas, K.: Reflected Min-Max Heaps. Inf. Proc. Letters 86, 209–214 (2003)
3. Williams, J.W.J.: Algorithm 232. Comm. ACM 7(6), 347–348 (1964)

5 Temporal Precedence

In this section we refer to the *Temporal Precedence Problem* on PPM-machine. This problem asks for the design of a data structure, maintaining a set of stored elements and supporting the following two operations: *insert* and *precedes*. The Operation *insert(a)* introduces a new element a in the structure, while the operation *precedes(a, b)* returns true iff element a was inserted before element b

temporally. In [4], a solution is provided to the problem with worst-case time complexity $O(\log \log n)$ per operation and $O(n \log \log n)$ space, where n is the number of elements inserted. It was demonstrated that the *precedes* operation has a lower bound of $\Omega(\log \log n)$ for the *Pure Pointer Machine* model of computation. In [1] two simple solutions are presented with *linear* space and worst-case *constant* insertion time. In addition, two algorithms are described that can handle the *precedes*(a, b) operation in $O(\log \log d)$ time, where d is the temporal distance between the elements a and b. In [2], solutions are given, which match the same time and space bounds in simpler manner. The *Temporal Precedence Problem* is related to a very concrete problem that arises in parallel implementation of logic programming languages [3]. More specifically, in the And-Parallelism problem the problem of correct binding and assignment of variables can be reduced to the *insert* and *precedes* operations of the Temporal Precedence problem.

References

1. Brodal, G.S., Makris, C., Sioutas, S., Tsakalidis, A., Tsichlas, K.: Optimal Solution for the Temporal Precedence Problem. Algorithmica 33(4), 494–510 (2002)
2. Pontelli, E., Ranjan, D.: A Simple Optimal Solution for the Temporal Precedence Problem on Pure Pointer Machines. Theory of Computing Systems 38, 115–130 (2005)
3. Pontelli, E., Ranjan, D., Gupta, G.: On the Complexity of Parallel Implementation of Logic Programs. In: Ramesh, S., Sivakumar, G. (eds.) FST TCS 1997. LNCS, vol. 1346, pp. 123–137. Springer, Heidelberg (1997)
4. Ranjan, D., Pontelli, E., Gupta, G., Longpre, L.: The Temporal Precedence problem. Algorithmica 28(3), 288–306 (2000)

6   Finger Search

Finger search trees represent ordered lists into which pointers can be maintained, called fingers, from which searches can start; the time for a search, insertion or deletion is $O(\log d)$, where d is the number of items between the search starting point and the accessed item. The $O(\log d)$ bound can either be achieved in the amortized sence [3, 10] or in the worst case [9, 11, 14–16]. Finger search leads to optimal algorithms for the basic operations *union, intersection, difference*, which can support the development of efficient algorithms for sorting presorted files ([8], [7]) and for locally adaptive data schemes [2].

In [4], a general solution is proposed for the persistence problem. The authors develop simple, systematic efficient techniques for making different linked data structures persistent. They show first that if an ephemeral structure has nodes of bounded in-degree, then the structure can be made partially persistent at an amortized space cost of $O(1)$ per update step and a constant-factor increase in the amortized cost of access and update operations. Second, they present a method which can make a linked stucture of bounded in-degree fully persistent at an amortized time and space cost of $O(1)$ per update step and a worst-case time of $O(1)$ per access step. Finally, the authors present a partial persistent implementation of balanced search tree with worst-case time per operation of $O(\log n)$ and an amortized space cost of $O(1)$ per insertion or deletion. Combining this result with a *delayed updating technique* of Tsakalidis [16], we obtain a fully persistent form of balanced search trees with the same time and space bounds as in the partially persistent case. The technique employed in [4] is strongly related to *fractional cascading*. This relation can be used to support a *forget* operation which permits to explicitly delete versions and thus improves the space requirement [12].

The results of [15, 16] are well used in the theory of the persistent data structures [4, 5, 13] and in multidimensional search [6].

In [1], a new finger search tree is developed with worst-case constant update time in the PM-machine. This was a major problem in the field of Data Structures and was tantalizigly open for over 25 years [9], while many attempts by researchers were made to solve it. The result is a consequence of the innovative mechanism that guides the rebalancing operations, combined with incremental multiple splitting and fusion techniques over nodes.

References

1. Brodal, G.S., Lagogiannis, G., Makris, C., Tsakalidis, A.K., Tsihlas, K.: Optimal Finger Search Trees in the Pointer Machime. J. of Comput. System Sci. 67, 381–418 (2003)
2. Bentley, J.L., Sleator, D.D., Taran, R.E., Wei, V.K.: A locally adaptive Data Compression Scheme. Com. of ACM, 320–330 (1986)
3. Brown, M.R., Tarjan, R.E.: Design and Analysis of a Data Structure for representing Sorted Lists. SIAM J. Comput. 9, 594–614 (1980)
4. Driscoll, J.R., Sarnak, N., Sleator, D.D., Tarjan, R.E.: Making Data structures Persistent. J. Comput. System Sci. 28, 82–124 (1989)
5. Driscoll, J.R., Sleator, D.D., Tarjan, R.E.: Full Persistent Lists with Catenation. J. ACM 41(5), 943–959 (1994)
6. Duch, A., Martinez, C.: Improving the Performance of Multidimensional Search using Fingers. ACM Journal of Expermental Algorithmics 10, Art. no 2.4, 1–23 (2005)
7. Elmasry, A.: Adaptive Sorting with AVL Trees. IFIP TCS, 307–316 (2004)
8. Estivill-Castro, V., Wood, D.: A survey of Adaptive Sorting Algorithms. ACM Computing Surveys 24, 441–475 (1992)
9. Guibas, L.J., McCreight, E.M., Plass, M.F., Roberts, J.R.: A new represenation for Linear Lists. In: Proc. 9th Ann. ACM Symp. on Theory of Computing, pp. 49–60 (1978)
10. Huddleston, S., Mehlhorn, K.: A new Data Structure foe representing Sorted Lists. Acta Inform. 17, 157–184 (1982)
11. Kosaraju, S.R.: Localized Search in Sorted Lists. In: Proc. 13th Ann. ACM Symp. on Theory of Computing, pp. 62–69 (1981)
12. Mehlhorn, K., Näher, S., Uhrig, C.: Deleting Versions in Persistent Data Structures. Tech. Report, Univ. of Saarland, Saarbrücken (1989)
13. Sarnak, N., Tarjan, R.E.: Planar Point Location using Persistent Search Trees. Comm. ACM 29, 669–679 (1986)
14. Tarjan, R.E.: Private communication (1982)

15. Tsakalidis, A.K.: AVL-trees for Lícalized Search. Inform. and Control 67(1-3), 173–194 (1985)
16. Tsakalidis, A.K.: A Simple Implementation for Localized Search. In: Proc. WG 1985, Internat. Workshop on Graph-theoretical Concepts in Computer Science, pp. 363–374. Trauner-Verlag (1985); An optimal Implementation for localized seach. Tech. Rep. A84/06, Fachbereich Angewandte Mathematik and Informatik. Universität des Saarlandes, Saarbrücken, West Germany (1984)

7   Nearest Common Ancestors

Considering an arbitrary tree, the problem is to compute the nearest common ancestor of two given nodes x and y, denoted by $nca(x, y)$, which is defined as the lowest common node of the two paths from node x and node y to the root of the tree. In [1], a RAM-algorithm is presented running in $O(n)$ preprocessing time, $O(n)$ space and answering a query in $O(1)$ time. In [2], a PM-algorithm is given which requires $O(n \log \log n)$ preprocessing time, $O(n \log \log n)$ space and $O(\log \log n)$ *optimal* query time. In [1], the optimality of this query time is proved, and it is claimed that the algorithm of [2] can be modified to run on a PM-machine in linear time and space. Another *optimal* PM-algorithm with $O(n)$ preprocessing time, $O(n)$ space and $O(\log \log n)$ query time is described in [3].

Considering the dynamic case of one arbitrary tree, where the tree can be updated by insertions on the leaves or deletions of nodes, we get in [4] a PM-algorithm which needs $O(n)$ space, performs m arbitrary insertions on an initially empty tree in time $O(m)$, and allows to determine the nearest common ancestor of nodes x and y in time $O(\log(min\{depth(x), depth(y)\}) + a(k, k))$, where the second term is amortized over the k queries and $depth(x)$ is the distance from the node x to the root, and $a(k, k)$ is the inverse Ackerman function.

References

1. Harel, D., Tarjan, R.E.: Fast Algorithms for finding Nearest Common Ancestors. SIAM J. Comp. 13, 338–355 (1984)
2. van Leeuwen, J.: Finding Lowest Common Ancestors in less than Logarithmic Time. Unpublished report, Amherst, NY (1976)
3. van Leeuwen, J., Tsakalidis, A.K.: An optimal Pointer Machine Algorithm for Nearest Common Ancestors. Tech. Report, UU-CS-88-17, dept. of Computer Science, Univ. of Utrecht, Utrecht (1988)
4. Tsakalidis, A.K.: The Nearest Common Ancestor in a Dynamic Tree. Acta Informatica 25, 37–54 (1988)

8   Negative Cycle

The negative cycle problem is the problem of finding a negative length cycle in a directed graph with positive and negative edge-costs or proving that there are none. In [2] is shown that a negative cycle in a directed weighted graph with n

nodes and e edges can be computed in $O(n+e)$ time and $O(n+e)$ space. Assuming that the input of the algorithm is a weighted random digraph, it is proved in [1] that its average time complexity for dense graphs lies between $O(n \log n)$ and $O(min\{n^2/log^2 n, e\})$, the exact value depending on the probability with which an edge is present in the random graph, and for sparse random graph is $\Theta(n^2)$.

References

1. Spirakis, R., Tsakalidis, A.: A very fast practical algorithm for finding a negative cycle in a digraph. In: Kott, L. (ed.) ICALP 1986. LNCS, vol. 226, pp. 397–406. Springer, Heidelberg (1986)
2. Tsakalidis, A.: Finding a Negative Cycle in a Directed Graph. Techn. Report A85/05, Angewandte Mathematik und Informatik, FB-10, Univ. des Saarlandes, Saarbrücken (1985)

9   Three-Sided Queries

Let S be a set of n points in a two-dimensional space. A three-sided range query takes as arguments three coordinates x_1, x_2, y_1 and reports the set K of all points (x, y) of S with $x_1 \leq x \leq x_2$ and $y \leq y_1$. In [2] we consider 3-sided range queries on n points for a universe of $[N] \times \Re$, where N is the set of integers $\{0, ..., N-1\}$. We achieve $O(\log \log n + k)$ time, usings $O(N+n)$ space and preprocessing time, where k denotes the size of the output. This was later improved in [1] to $O(k)$ time, but with expected linear preprocessing time. The only dynamic sublogarithmic bounds on this problem can be found in [3], where it is attained $O\left(\frac{\log n}{\log \log n}\right)$ worst case or $O(\log n)$ randomized update time and $O\left(\frac{\log n}{\log \log n} + k\right)$ query time in linear space.

References

1. Alstrup, S., Brodal, G.S., Rauhe, T.: New Data Structures for Orthogonal Range Searching. In: FOCS 2000, pp. 198–207 (2000)
2. Fries, O., Mehlhorn, K., Näher, S., Tsakalidis, A.K.: A loglogn Data structure for Three-Sided Range Queries. Inf. Process. Lett. 25(4), 269–273 (1987)
3. Willard, D.E.: Examining Computational Geometry, Van Emde Boas Trees and Hashing from the perspective of the Fussion Tree. SIAM J. Comp. 29(3), 1030–1049 (2000)

10   Rectangle Enclosure

We consider two versions of the rectangle enclosure problem. Given a set S of rectangles in the plane, in the first version we report all the rectangles, which enclose a given query rectangle. In [1], a solution is given for the first version of the problem generalized in d-space, which needs $O(\log^{2d-2} \log \log n + k)$ query time, where k is the size of the answer in the case that the data structure used is static. In the dynamic case, the query time is $O(\log^{2d-1} n + k)$, and an update operation costs $O(\log^{2d-1} n)$. In both cases, the space used is $O(n \log^{2d-2} n)$. The query time in the static case is improved to $O(\log^{2d-2} n + k)$ in [3].

In the second version, we report all the pairs of the rectangles (R, R'), where $R, R' \in S$ and R' encloses R. In [6] a solution was given that needs $O(n)$ space and runs in $O(n \log^2 n + k)$ time. It has been an open problem for more than ten years how the $O(n \log^2 n)$ term of the reporting time could be reduced. In [4], an algorithm was given that solved this problem in $O(n + k)$ space and $O(n \log n \log \log n + k \log \log n)$ time. In [2] a subroutine of the previous algorithm is modified using persistence and periodic rebuilding of list structures and the space required is reduced to linear, while retaining the same time complexity. In [5], a simple solution is presented with the same time and space bounds.

References

1. Bistiolas, V., Sofotassios, D., Tsakalidis, A.: Computing Rectangle Enclosures. Comput. Geometry: Thery & Appl. 2(6), 301–308 (1993)
2. Bozanis, R., Kitsios, N., Makris, C., Tsakalidis, A.: The Space-Optimal Version of a known Rectangle Enclosure Reporting Algorithm. Inf. Process. Letters 61, 37–41 (1997)
3. Bozanis, P., Kitsios, N., Makris, C., Tsakalidis, A.: New Results on Intersection Query Problems. Computer J. 40(1), 22–29 (1997)
4. Gupta, P., Janardan, R., Smid, M., Dasgupta, B.: The Rectangle Enclosure and Point-Dominance Problems. Intern. J. Comput. Geom. Applic. 7(5), 437–457 (1997)
5. Lagogiannis, G., Makris, C., Tsakalidis, A.: A new Algorithm for Rectangle Enclosure Reporting. Inf. Process. Letters 72, 177–182 (1999)
6. Lee, D.T., Preperata, F.P.: An improved Algorithm for the Rectangle Enclosure Problem. J. Algorithms 3, 218–224 (1982)

11   Dominance Searching

In [1], several data structures are presented for the 3-dominance searching problem: store a set S of n points in $\Re^3$ in a data structure, such that the points in S dominating a query point can be reported efficiently. We say that a point $p = (p_1, p_2, p_3)$ dominates a point $q = (q_1, q_2, q_3)$, if and only if $p_i \geq q_i$ for all $1 \leq i \leq 3$ and $p \neq q$. All our data structures use *linear* space. The first data structure works for the restricted case where the coordinates of the points in S and of the query points are integers in the range $[0, N - 1]$. In this case, we achieve a query time of $O\left((\log \log N)^2 \log \log \log N + k \log \log N\right)$, where k is the number of answers to the query. The second and third data structure both work for the unrestricted case, where the coordinates are arbitrary reals. We achieve $O(\log n \log \log n + k)$ query time for pointer machines and $O(\log n + k)$ query time for random access machines. These results are improved in [2].

References

1. Makris, C., Tsakalidis, A.: Algorithms for three-dimensional Dominance Searching in linear space. Inf. Process. Letters 66, 277–283 (1998)
2. Afshani, P.: On dominance reporting in 3D. In: Halperin, D., Mehlhorn, K. (eds.) ESA 2008. LNCS, vol. 5193, pp. 41–51. Springer, Heidelberg (2008)

12   Intersection Queries

Generalized intersection searching problems are a class of problems that constitute an extension of their standard counterparts. In such problems, we are given a set of collored objects and we want to report or count the distinct collors of the oblects intersected by a query object. Many solutions have appeared for both iso-oriented and non-oriented objects. In [1, 2], it is shown how to improve the bounds of several generalized inresection searching problems as well as how to obtain upper bounds for some problems like arbitrary line segment and generalized triangle stabbing, which were not treated before.

In [3], efficient solutions are given for the following problems: the Static d-dimensional rectangle enclosure problem, with $O(n \log^{2d-2} n)$ space and $O(\log^{2d-2} n + k)$ query time, the generalized c-oriented polygonal intersection searching, with $O(n \log^2 n)$ space and $O(\log n + k)$ query time, the generalized rectangular point enclosure problem, with $O(n \log n)$ space and $O(\log n + k)$ and the two-dimensional dominance searching problem with respect to a set of obstacle points, with $O(n \log n)$ space and $O(\log n + k)$ query time.

References

1. Bozanis, P., Kitsios, N., Makris, C., Tsakalidis, A.: New upper Bounds gor Generalized Intersection Searching Problems. In: Fülöp, Z., Gecseg, F. (eds.) ICALP 1995. LNCS, vol. 944, pp. 464–474. Springer, Heidelberg (1995)
2. Bozanis, P., Kitsios, N., Makris, C., Tsakalidis, A.: Red-Blue Inrersection Reporting for Objects of Non-Constant Size. Computer J. 39(6), 5410546 (1996)
3. Bozanis, P., Kitsios, N., Makris, C., Tsakalidis, A.: New Results on Intersection Query Problems. Computer J. 40(1), 22–29 (1997)

13   Hidden Line Elimination

In a hidden line (or Surface) elimination problem we are given a set of objects in 3D space and a view-point and ask for the parts of the objects that are visible from the viewpoint. In a hidden line problem, the reported parts are line segments while in hidden surface problem, they are regions of surfaces. In [3], an algorithm is presented with optimal $O(n)$ space and worst case time $O(n \log n + k \log(n^2/k))$. In [4], a simple intersection sensitive algorithm is presented which solves the hidden line elimination problem in optimal $O(n)$ space and time complexity of $O((n + I) \log n)$, where I is the number of the intersections of the edges on the projection plane. An extension of this algorithm can solve the hidden surface removal problem in $O((n + I) \log n)$ time and $O(n + k)$ space, where k is the output size.

We consider the following problem as defined in [1]. Given a set of n isothetic rectangles in 3D space determine the subset of rectangles, that are not completely hidden. In [2], we present an optimal algorithm for this problem that runs in $O(n \log n)$ time and $O(n)$ space. Our results are an improvement over the one in [1] by a logarithmic factor in storage and are achieved by using a different approach. An analogous approach gives non-trivial solutions for other kinds of objects, too.

References

1. Grove, E.F., Murali, T.M., Vitter, J.S.: The object complexity model for hidden elimnination. Internat. J. Comput. Geom. Appl. 9, 207–217 (1999)
2. Kitsios, N., Makris, C., Sioutas, S., Tsakalidis, A., Tsaknakis, J., Vasiliadis, B.: An Optimal Algotithm for reporting Visible Rectangles. Inf. Process. Letters, 283–288 (2002)
3. Kitsios, N., Tsakalidis, A.: Space-Optimal Hidden Line Elimination for Rectangles. Inf. Process. Letters 60, 195–200 (1996)
4. Kitsios, N., Tsakalidis, A.: Space Reduction and an Extension for a Hidden Line Elimination Algorithm. Comp. Geom: Theory & Applic., 397–404 (1996)

14   String Manipulation

In [3], we consider several new versions of approximate string matching with gaps. The main characteristic of these new versions is the existence of gaps in the matching of a given pattern in a text. Algorithms are devised for each version, and their time and space complexities are stated. These specific versions of approximate string matching have various applications in computerized music analysis. In [5], we describe algorithms for computing typical regularities in strings that contain *don't care symbols*. We show also how our algorithms can be used to compute other string regularities, specifically the covers of both ordinary and circular strings.

Biological weighted sequences are used extensively in molecular biology as profiles for protein families, in the representation of binding sites and often for the representation of sequences produced by shotgun sequencing strategy. In [4], we introduce the *Weighted Suffix Tree*, an efficient data structure for computing string regularities in weighted sequences for molecular data. Repetitions, pattern matching and regularities in biological weighted sequences are also considered in [2]. In [6], we present algorithms for the *Motif Identification Problem* in Biological Weighted Sequences. The first algorithm extracts *repeated motifs*, the second algorithm extracts *common motifs* and the third alorithm extracts *maximal pairs*.

A multirepeat in a string is a substring that appears a predefined number of times. A multirepeat is maximal if it cannot be extended either to the right or to the left and produce a multirepeat. In [1], we present algorithms for two different versions of the problem of finding maximal multirepeats in a set of strings. In the first version we consider the case of arbitrary gaps and in the second version the case that the gap is bounded in a small range.

References

1. Bakalis, A., Iliopoulos, C.S., Makris, C., Sioutas, S., Theodoridis, E., Tsakalidis, A., Tsichlas, K.: Locating Maximal Multirepeats in Multiple Strings under various Constraints. Comput. J. 50(2), 178–185 (2006)
2. Christodoulakis, M., Iliopoulos, C., Mouchard, L., Tsakalidis, A., Tsichlas, K.: Computations of repetitions and Regularities of Biological Weighted Sequences. J. Comput. Biology 13(6), 1214–1231 (2006)
3. Crochemore, M., Iliopoulos, C., Makris, C., Rytter, W., Tsakalidis, A., Tsichlas, K.: Approximate String matching with Gaps. Nordic J. Comput. 9(1), 54–65 (2002)
4. Iliopoulos, C., Makris, C., Panagis, Y., Perdikuri, K., Theodoridis, E., Tsakalidis, A.: The Weighted Suffix Tree: An efficient Data Structure for handling Molecular Weighted Sequences and its Applications. Fundamenta Informaticae 71, 259–277 (2006)
5. Iliopoulos, C., Mohamed, M., Mouchard, L., Perdikuri, K., Smyth, W.F., Tsakalidis, A.: String Regularities with don't Cares. Nordic J. Comput. 10(1), 40–51 (2003)
6. Iliopoulos, C., Perdikuri, K., Theodoridis, E., Tsakalidis, A., Tsichlas, K.: Algorithms for extracting Motifs from Biological Weighted Sequences. J. Discrete Algorithms 5, 229–242 (2007)

Acknowledgement

We have presented some results derived since 1983 in collaboration with my students, my colleagues and my advisor Kurt Mehlhorn. I met Kurt in 1976. He was the youngest Professor in Germany and I was the oldest undergraduate student in our department. I was honored to live in his *scientific neighborhood* for 13 years as undergraduate student, as graduate student and as collaborator. Kurt was and remains the Professor, the Researcher and the Head of the Family. In 1979 I made Kurt's portait. I am grateful for everything that he taught me. I would like to say in public "Kurt, ich danke Dir, dass ich Dich in meinem Leben getroffen habe" (Kurt, I thank You that I have met You in my life).

Maintaining Ideally Distributed Random Search Trees without Extra Space

Raimund Seidel

Universität des Saarlandes
FR 6.2 Informatik,
Postfach 15 11 50
D-66041 Saarbrücken
Germany
rseidel@cs.uni-saarland.de
http://www-tcs.cs.uni-saarland.de/seidel

Abstract. We consider the problem of maintaing a random binary search tree under insertions and deletions under the conditions that (i) no extra permanent storage space be used besides the tree itself, and (ii) that at any point in time the tree be perfectly random, meaning that it is drawn from the ideal binary search tree distribution. We present a simple solution to this problem with an expected deletion time of $O(\log n)$ and expected insertion time of $O(\log^2 n)$ time.

Keywords: Binary Search Trees, Randomized Data Structures.

1 Introduction

Binary search trees are a basic data structure for storing dictionaries with keys from an ordered universe. They have been known since the early days of computers, see [7, p. 446] for a short survey. Early on it was noticed that if a set S of n keys is inserted in random order into an initially empty binary search tree using the standard leaf-insertion procedure, then *in expectation* the resulting tree is very well behaved in the sense that for every key the expected search time is logarithmic, and even the expected maximum search time of all keys is logarithmic. However, in 1975 Knott [5] noted that this random insertion order assumption was not maintained by the usual deletion algorithms and their variants. This led to a flurry of theoretical and experimental studies in the following years (see for instance [6, 8, 3]). Their results however were only partial or inconclusive.

The problem of "maintaining randomness" was only resolved when researches abandoned the assumption of randomness of the input and instead adopted the point of view that randomness be generated by the update algorithms themselves. This made probabilistic assumptions about input distributions unnecessary, an undeniable advantage, as such assumptions are hard to justify and their validity difficult to check.

Aragon and Seidel [1, 13] were the first to propose such a randomized binary search tree maintenance scheme. Shortly after, Bent and Dricsoll [2] proposed a somewhat different scheme which was later rediscovered by Martínez

S. Albers, H. Alt, and S. Näher (Eds.): Festschrift Mehlhorn, LNCS 5760, pp. 134–142, 2009.

and Roura [9]. For further related work see [12] and [4] although the latter is already subsumed by [1].

These proposed structures in their natural realizations maintain an ideally distributed binary search tree, meaning that at any point in time the tree looks as if it had been generated by inserting its keys in random order into an initially empty tree. However, in order to achieve this while keeping updates fast, the proposed structures have to store some additional "balance information" with the nodes of the tree: in the case of [13] it is random number (essentially a random virtual insertion time) and in the case of [2, 9] it is the subtree size. Seidel and Aragon in [13] also show that you can forego storing this "random virtual insertion time" and use a hash value of the key instead, which can be always recomputed when needed. If these hash values are sufficiently random, then logarithmic expected update times can still be achieved, however the trees that arise will not be ideally distributed any more in the sense described above.

In this note we show that it is possible to maintain ideally distributed random binary search trees without using any extra storage so that the expected deletion time is $O(\log n)$ and the expected insertion time is $O(\log^2 n)$. Our method is conceptually very simple and should be easy to implement. It remains to be seen whether its actual performance will be comparable to the hash based method mentioned above when space really is at a premium.

This paper was written on the occasion of *Kurt Mehlhorn's* 60th birthday. As far as I know Kurt has never worked on this particular problem. But this paper should have some references to his work. Fortunately Kurt has worked on many problems. Here [11] is a piece of work close to the subject matter considered in this paper.

Happy Birthday, Kurt!

2   Random Search Trees

There are at least three equivalent models of generating the ideal random binary search tree distribution. Assume we are dealing with a set A of n distinct keys $a_1 < a_2 < \cdots < a_n$.

The random insertion model: Pick a permutation π of $\{1, \ldots, n\}$ uniformly at random. Insert the keys in the order $a_{\pi(1)}, a_{\pi(2)}, \ldots, a_{\pi(n)}$ into an initially empty tree using the standard leaf insertion algorithm (see for instance Section 3.3.1 of Mehlhorn's book [10]).

The random root model: Pick one of the n keys uniformly at random, say a_r. Make it the root of the tree. Its left and right subtrees will be recursively built random search trees for the sets $\{a_1, \ldots, a_{r-1}\}$ and $\{a_{r+1}, \ldots, a_n\}$, respectively. (Of course the empty set is represented by the empty tree.)

The priority model: Independently draw n numbers $p_1, \ldots, p_n$ from some continuous distribution, say the uniform distribution on the real interval $[0, 1]$. Produce the *treap* for the set of pairs $\{(a_1, p_1), \ldots, (a_n, p_n)\}$. This "treap" stores a pair (a_i, p_i) at each node, and it is in symmetric order with respect to the keys a_i and a max-heap with respect to the priorities p_i. The resulting tree is unique if the keys are distinct (true by assumption) and the priorities are distinct (true with probability 1). See [13].

These three models are equivalent in the sense they produce the same distribution on n-node binary trees. This can be readily seen by noting that both the priority model and the random insertion model can be viewed as particular realizations of the random root model with particular mechanisms for choosing a random root.

The priority model underlies the work of Aragon and Seidel [13], the random root model is the basis for the work of Bent and Driscoll [2] and of Martínez and Roura. The random insertion model underlies the work of Heyer [4], however that work relies fundamentally on the random input assumption and cannot deal with arbitrary update sequences. We will ignore it for the rest of this paper.

The random root model implies that the probability P_T that a particular tree arises is exactly $\Pi_{v \in T}(1/s_T(v))$, where $s_T(v)$ is the size of the subtree of T rooted at v However, this probability will be of little concern in our discussions.

It is interesting to elucidate the sample spaces underlying the three models: In the case of the random insertion model it is simply Π_n the set of all n-permutations with uniform probability. In the case of the random priority model it is the real n-cube $[0, 1]^n$ with uniform distribution. In the case of the random root model, the sample space is complicated. It is essentially the set of all n-node binary trees itself, with the non-uniform probability distribution as indicated above.

Thus the sample space for the random root model is very complicated in comparison to the sample space for the random priority model. As a result of this the random root model appears to be less well suited for simple derivations of various quantities that arise in the analysis of algorithms on random binary trees, such as expected depth of a node, expected size of a subtree, expected spine length, or expected costs of updates assuming "expensive" rotations. Thus it is no coincidence that [9] refers to previous works for the analysis of such quantities. Unfortunately the given references seem to provide the claimed analyses only partially, if at all.

3   Updating Random Search Trees

The relative merits of the random priority model versus the random root model also show when dealing with updates of random search trees. Consider the random priority model first. Let $S = \{(a_1, p_1), \ldots, (a_n, p_n)\}$ be a set of n key and random priority pairs, and let S' be obtained from S by removing or adding one such pair. The treap T for S and the treap T' for S' are both well defined and unique, and by definition are drawn from ideal random search tree distributions. How T' is obtained from T is really irrelevant. Any algorithm can be used, though preferably a fast one. However, the priorities p_i definitely have to be available in some form.

The situation is quite different in the random root model. Let $A = \{a_1, \ldots, a_n\}$ be a set of n keys and let A' be obtained from A by removing or adding one key. The tree T' for A' resulting from updating the tree T for A must satisfy the random root model, assuming that T did so. This means that whether such a distributionally correct transformation happens depends on the update algorithm and its random choices. Most tree update algorithms will not do the job, and a main part of [9] is the proof that their update algorithms actually do perform distributionally correct transformations.

The update algorithms of [9] happen to be functionally the same as the fast update algorithms given in [13]. It is an interesting philosophical question whether this is just a coincidence or has to be the case.

In the following we give generic versions of the update algorithms, show how they are realized in an efficient manner if random priorities are stored (the random priority model), if subtree sizes are stored (the random root model), and finally, as main result of the paper, if no additional values are stored at all.

> **procedure** GENERIC-DELETE (a : key, T : tree)
>   **if** $T = tnull$ **then** **return**
>   **if** $a < T {\rightarrow} key$ **then** GENERIC-DELETE($a, T {\rightarrow} lchild$)
>   **else** **if** $a > T {\rightarrow} key$ **then** GENERIC-DELETE($a, T {\rightarrow} rchild$)
>   **else** ROOT-DELETE(T)
> **procedure** ROOT-DELETE(T : tree)
>   **if** IS-LEAF(T) **then** $T \leftarrow null$
>   **else** **if** $\boxed{\textbf{Lchild-Wins}(T)}$ **then** ROTATE-RIGHT(T)
>   ROOT-DELETE($T {\rightarrow} rchild$)
>   **else** ROTATE-LEFT(T)
>   ROOT-DELETE($T {\rightarrow} lchild$)

> **procedure** GENERIC-INSERT(x : item, T : tree)
>   **if** IS-NULL(T) **then** $T \leftarrow$ NEWNODE(x)
>   **else** **if** $\boxed{\textbf{newroot}(x, T)}$ **then** ROOT-INSERT(x, T)
>   **else** **if** $x.key < T {\rightarrow} key$ **then** GENERIC-INSERT($x, T {\rightarrow} lchild$)
>   **else** GENERIC-INSERT($x, T {\rightarrow} rchild$)
> **procedure** ROOT-INSERT(x : item, T : tree)
>   **if** IS-NULL(T) **then** $T \leftarrow$ NEWNODE(x)
>   **else** **if** $x.key < T {\rightarrow} key$ **then** ROOT-INSERT($x, T {\rightarrow} lchild$)
>   ROTATE-RIGHT(T)
>   **else** ROOT-INSERT($x, T {\rightarrow} rchild$)
>   ROTATE-LEFT(T)

These procedures assume call-by-reference semantics. ROTATE-RIGHT and ROTATE-LEFT are the usual tree rotation routines, see e.g. Section 3.5.1 of Mehlhorn's book [10].

3.1 Deletion

The GENERIC-DELETE routine detailed above first finds the the node storing the key a. This node is the root of the subtree T. You now need to remove the root of T and join the two subtrees into a single tree that is to replace T. The procedure ROOT-DELETE achieves just that by recursively rotating the root down into leaf position and then clipping it off. The total cost, ignoring the cost of **Lchild-Wins**(), is proportional to the length of one root-leaf path, which in expectation is logarithmic. The only question remaining in this process is whether to rotate the root left or right. In the generic procedure this is dictated by the framed predicate **Lchild-Wins**(T). This is the only place where the various models cause differences in the implementation.

In the random priority model, as realized in [13], the respective random priority is stored with every node. Overall the tree needs to be a max-heap with respect to these priorities. Thus if the root is removed from T the new root must be the child with larger priority. Thus **Lchild-Wins**(T) is realized by the simple predicate

$$(\ T{\rightarrow}lchild{\rightarrow}priority \ ) \ > \ (\ T{\rightarrow}rchild{\rightarrow}priority).$$

Note that a positive outcome to this predicate means that the largest of all the priorities in the left and right subtrees of T is in the left subtree. Since the priorities are independent random variables this happens with probablity $\ell/(\ell + r)$, where ℓ and r are the number of nodes in the left and right subtree of T respectively. In the natural realization of the random root model, as in [2, 9] the subtree size is stored with each node. Thus **Lchild-Wins**(T) is realized by a coinflip that comes up 1 with probability

$$(T \rightarrow lchild{\rightarrow}size \ )/(\ T{\rightarrow}lchild{\rightarrow}size \ + \ T{\rightarrow}rchild{\rightarrow}size \ ).$$

How can you proceed if neither *priority* nor *size* are stored with each node of the tree? The suprisingly straightforward answer, already noted in Section 6 of [13], is the following: use the random coinflip method of the random root model and, when needed, simply compute the size of a subtree by traversing it and counting its nodes. The cost for **Lchild-Wins**(T), implemented this way, is then proportional to the size of subtree T, which is about to be rotated. Thus the cost can be assigned to the rotation, and fortunately Theorem 3.1 of [13] assures that the expected deletion time is still logarithmic if the rotation cost is proportional to the size of the rotated subtree.

3.2 Insertion

The GENERIC-INSERT routine detailed above first checks whether the new item to be inserted is to be made the new root of the tree. This is dictated by the outcome of the framed test **newroot**(x, T), which is the only place in the procedure where the various models cause different implementations. If the outcome of the test is negative, then the item is recursively inserted in the appropriate subtree. If the outcome is positive, i.e. if the new item is to be made the new root, then the tree T needs to be split on the key of the new item and the resulting two trees are made children of the new root. In our implementation this is achieved via a simple recursive method involving rotations, that is tantamount to inserting the new item in the appropriate leaf position and then rotating it up into root position. The total cost, ignoring the cost of **newroot**(), will be proportional to the length of one root-leaf path, which in expectation is logarithmic.

How can the test **newroot**(x, T) be realized in the different models? In the random priority model, as realized in [13], a random priority is stored with every node of the tree, and the tree is a max-heap with respect to these priorities. The new item x to be inserted is endowed with a new random priority $x.p$. The test **newroot**(x, T) is simply performed by

$$x.p \; > \; T{\rightarrow}priority.$$

Notice that this test has a positive outcome with probability $1/(t+1)$, where t is the size of tree T, since a positive outcome means that $x.p$ is larger than all t priorities stored in T. Of course this is exactly the probability required by the random root model. The natural implementation of that model, as realized in [2, 9] stores with each node or the tree the size of its subtree. Thus **newroot**(x, T) is realized by a coinflip that comes up 1 with probability

$$1\big/(1{+}T{\rightarrow}size)\,.$$

How can you proceed if neither *priority* nor *size* are stored with each node of the tree? The approach taken in the case of deletions, namely determining *size* when needed by enumerating the nodes of the tree, is not really feasible. The insertion routine proceeds top-down. Already the first call to **newroot**(x, T) would cause the entire tree to be traversed and hence incur linear time. The important observation is that we are not interested in *size* itself, but we want to flip a coin with bias $1/(1 + size)$. Here is a simple way of achieving just that.

The abstract problem we are facing is as follows: We are given a set M of unknown size m that we can iterate over at constant cost per element. We want to flip a coin that comes up 1 with probability $1/(1{+}m)$. Let D be some continuous probability distribution. Draw a random number p from D. Now iterate over M and for every element enumerated draw a new independent random number from D. If this number is bigger than p then stop and return 0. If this does not happen vor any element of M then return 1.

Note that this method returns 1 iff p happens to be the largest of the $1 + m$ random numbers drawn independent from D. Of course this happens exactly with probability $1/(1{+}m)$, as required. What is the expected time taken by this method? Clearly this is proportional to the expectation of the random variable K that counts how often a random number is drawn from D. For $i > m + 1$ we have $\Pr[K \geq i] = 0$. For $1 \leq i \leq m + 1$ we have $\Pr[K \geq i] = 1/i$ since the procedure continues beyond the i-th draw form D iff p is the largest of the first i numbers drawn. Thus we get

$$\mathrm{Ex}[K] = \sum_{i \geq 1} \Pr[K \geq i] = \sum_{1 \leq i \leq m+1} 1/i = H_{m+1} \approx \log m\,.$$

So this implementation of **newroot**(x, T) takes $O(\log m)$ expected time, where m is the number of nodes in T. It needs no additionally stored information, but just needs to be able to enumerate the nodes of the tree T at constant cost per node, which is easy to do using any of the standard tree traversal algorithms.

How often is **newroot**( ) invoked during the insertion of x? This will be exactly the depth of x in the resulting tree, which in expectation is at most $1 + 2\log(n + 1) = O(\log n)$ (see e.g. Theorem 4.1 of [13]), where n is the size of the original tree.

Since **newroot**(x, T) is only invoked on subtrees of size at most n we immediately get that the overall expected time necessary for all these invocations is $O(\log^2 n)$, and since the remaining expected time for the insertion operation overall is logarithmic we get that the expected insertion time is $O(\log^2 n)$.

4   Possible Improvements?

We have presented ideal distribution preserving update methods for random search trees that require no extra space at all. For deletion the expected running time is $O(\log n)$ whereas for insertion it is $O(\log^2 n)$. This is a somewhat unusual situation since normally, if one of the two update operations turns out to be more costly, then it is the deletion. Are there possibilities of improving the insertion operation or its analysis? We briefly give, admittedly handwavy, arguments why three fairly obvious approaches for improvement will not work.

The first approach would be to simply improve the analysis. After all, using n as an upper bound to the sizes of all subtrees to which **newroot**( ) is applied seems quite generous. But it is quite clear that with high probability, say the first $(\log n)/10$ subtrees considered all have size at least, say $n^{1/4}$. This implies that the expected overall cost of all the **newroot**( ) invocations will be $\Omega(\log^2 n)$. It is an interesting problem to determine the *exact* expected number of random numbers drawn during our insertion routine. Presumably this will depend on the key rank of the inserted item x in the resulting tree.

The second approach would be to alter the insertion procedure so that it does not proceed top-down, but bottom-up, along the lines of the insertion procedure for treaps that first inserts the new item in leaf position (ignoring priorities) and then rotates the item back up until the heap conditions on the priorities are re-established. It turns out that the expected number of rotations is less than 2 (see Theorem 3.1 of [13]). But it seems that this method can·only be made to work correctly if explicit random priorities are available. Even if subtree sizes are available but they are only revealed going along a path *towards* the root, there is no clear strategy where to stop the up-rotations, even if the length of the remaining path to the root is known. This makes it unlikely that a strategy exists that has no subtree sizes available at all. (This observation also exhibits a clear difference between random search trees implemented with priorities as in [13] and with subtree sizes as in [2, 9]. The priority based implementation allows insertions at handles (a pointer to the predecessor or successor of the key to be inserted) in constant expected time, whereas the subtree-size based implementation needs logarithmic expected time, since it must proceed top-down.

The third approach, finally, suggest to reuse the work done by previous **newroot()** invocations. Assume that $T_1, \ldots, T_\ell$ is a sequence of subtrees on which **newroot()** invocations are performed. Note that for $1 < i \le \ell$ tree T_i is a subtree of T_{i-1} and hence it is contained in all trees T_j with $j < i$. When we enumerate the nodes of T_1 we first enumerate the nodes of T_ℓ, then the remaining nodes of $T_{\ell-1}$, then the remaining nodes of $T_{\ell-2}$, and so on. Assume that **newroot**(x, T_1) stops after k nodes have been enumerated. These nodes are also in T_2, and therefore at the invocation of **newroot**(x, T_2) you could take advantage of this. The details are not particularly difficult, but we don't need to discuss them, because of the following observation: Assume T_ℓ, and hence all T_i have size at least m. Then the expected maximum time taken by one of the ℓ invocations **newroot**(x, T_2) will be $\Omega(\ell \cdot \log m)$. From this an $\Omega(\log^2 n)$ lower bound follows for any sort of reuse approach.

The $\Omega(\ell \cdot \log m)$ bound can be proven as follows. Consider ℓ independent random variables $K_1, \ldots, K_\ell$, all with the same distribution as the random variable K discussed above. Consider the random variable $X = \max\{K_1, \ldots, K_\ell\}$. Since $\Pr[K_j \ge i] = 1/i$, we get that the probability that each K_j, and hence their maximum, is at least i is given by $1 - (1 - 1/i)^\ell$. Thus we get

$$\mathrm{Ex}[X] = \sum_{1 \le i \le m+1} \Pr[X \ge i] \ge \sum_{\ell < i \le m} \left(1 - (1 - \frac{1}{i})^\ell\right) \ge \sum_{\ell < i \le m} \frac{\ell}{2i} = \ell(H_m - H_\ell)/2.$$

This uses the inequality $(1 - 1/i)^\ell \le 1 - \ell/(2i)$, which holds for $i \ge \ell$.

5   A Riddle at the End

The random variable K discussed above was considered in the context of a finite set M. It is interesting to see what happens if M is infinite.

Consider the following process: A random number p is drawn from some continuous probability distribution D. Following this you draw independently further random numbers from the same distribution until you get one that is larger than p. What is the expected number of random numbers that you draw?

References

1. Aragon, C.R., Seidel, R.G.: Randomized Search Trees. In: Proc. IEEE Symp. FOCS, pp. 540–545 (1989)
2. Bent, S.W., Driscoll, J.R.: Randomly balanced search trees (manuscript, 1991)
3. Eppinger, J.L.: An Empirical Study of Insertion and Deletion in Binary Search Trees. Commun. ACM 26(9), 663–669 (1983)
4. Heyer, M.: Randomness Preserving Deletions on Special Binary Search Trees. M.Sc. Thesis, Dept. of Computer Science, Nat. Univ. of Ireland, Cork (2005)
5. Knott, G.D.: Deletions in Binary Storage Trees. PhD Thesis, Computer Science Dept., Stanford Univ. (1975)
6. Jonassen, A.T., Knuth, D.E.: A Trivial Algorithm Whose Analysis Isn't. J. Comput. Syst. Sci. 16(3), 301–322 (1978)

7. Knuth, D.E.: The Art of Computer Programming. Sorting and Searching, vol. III. Addison-Wesley, Reading (1973)
8. Knuth, D.E.: Deletions That Preserve Randomness. IEEE Trans. Software Eng. 3(5), 351–359 (1977)
9. Martínez, C., Roura, S.: Randomized Binary Search Trees. J. ACM 45(2), 288–323 (1998)
10. Mehlhorn, K.: Data Structures and Algorithms 3. Springer, Heidelberg (1984)
11. Mehlhorn, K.: A Partial Analysis of Height-Balanced Trees under Random Insertions and Deletions. SIAM J. on Comput. 11(4), 748–760 (1982)
12. Pugh, W.: Skip Lists: A Probabilistic Alternative to Balanced Trees. Commun. ACM 33(6), 668–676 (1990)
13. Seidel, R., Aragon, C.R.: Randomized Search Trees. Algorithmica 16(4/5), 464–497 (1996)

A Pictorial Description of
Cole's Parallel Merge Sort

Torben Hagerup

Institut für Informatik, Universität Augsburg, 86135 Augsburg, Germany
`hagerup@informatik.uni-augsburg.de`

Abstract. A largely pictorial description is given of a variant of an ingenious parallel sorting algorithm due to Richard Cole. The new description strives to achieve greater simplicity by exploiting symmetries that were not explicit in the original exposition and that can be conveyed nicely with pictures. Not paying attention to constant factors allows an additional slight simplification of the algorithm.

1 Introduction

In 1988 Richard Cole published two sorting algorithms for the parallel random-access machine or PRAM, a model of computation that comprises consecutively numbered processors with lock-step access to a shared memory [2]. One algorithm is for the concurrent-read exclusive-write or CREW variant of the PRAM, while the other algorithm works on the more restrictive exclusive-read exclusive-write or EREW PRAM. Neither PRAM variant allows writing to the same memory cell in the same step by several processors. The CREW PRAM allows reading from the same memory cell in the same step by several processors, while the EREW PRAM does not.

Both algorithms sort n items using n processors, $O(\log n)$ time and $O(n)$ space, which is optimal, up to a constant factor, as concerns the running time, the time-processor product, and the space. The existence of PRAM algorithms with these characteristics was already implied earlier by the so-called AKS network of Ajtai, Komlós and Szemerédi [1] and its descendants, but PRAM algorithms derived in this manner are deemed impractical due to their complexity and their large constant factors.

Both of Cole's algorithms are based on the natural paradigm of merging in a binary tree. If the merging at each level of the tree is completed before the merging at the level above it starts, the total sorting time will be $\Omega(\log n \log \log n)$ on the CREW PRAM and $\Omega((\log n)^2)$ on the EREW PRAM. In order to reduce the time to $O(\log n)$, Cole developed clever schemes for pipelining the merges. In general, the merging at a level of the tree starts before the merging at the level below it has completed, in a sense using small samples of the full set of items as a "scaffolding" that allows items arriving later to be put in place more speedily.

In the case of the algorithm for the CREW PRAM, working out the details of the idea expressed in the previous paragraph leads to a complete algorithm

S. Albers, H. Alt, and S. Näher (Eds.): Festschrift Mehlhorn, LNCS 5760, pp. 143–157, 2009.

in a relatively straightforward manner. On the EREW PRAM, however, where simultaneous reading is not allowed, things become more involved. For reasons elucidated by Cole, it is necessary to send scaffolding information not only up, but also down the tree. The result is a rather intricate pattern of interacting data streams that move any which way through the tree. Although Cole's exposition is admirable, there are many facts to be kept in mind simultaneously and many somewhat tedious details to verify.

This work aims at a description of Cole's sorting algorithm for the EREW PRAM that is simpler and easier to verify. One starting point is the realization that although the merge tree is clearly a rooted tree, in that the information ascends from the leaves to the root, much is to be gained in simplicity from ignoring this fact to the extent possible and considering the tree as a free (i.e., unrooted) tree. The nodes in the tree can be made to treat all of their incident edges in a uniform way. In fact, it is natural to associate computational steps not with nodes, but with edges, and to let all edges execute the same procedure in each of a number of identical stages. This lends a pleasing symmetry to the algorithm that is particularly useful when it is presented pictorially—a central part of the algorithm can be viewed as a game about drawing arrows according to certain simple rules, and one immediately notices facts whose verification at the textual level requires a certain effort and is probably less reliable.

2   Preliminaries

Consider the task of sorting elements of a universe U according to a total order $<$ on U. The word *item* will be used to denote an element of U. Let $-\infty$ and ∞ be symbolic quantities such that $-\infty < x < \infty$ for all items x.

For every integer $k \geq 1$, if a set A consists of the items $x_1, \ldots, x_m$ and $x_1 < \cdots < x_m$, a *k-interval* of A is a set of the form $\{x \in U \mid x_i \leq x < x_{i+k}\}$, where $0 \leq i \leq m + 1 - k$, $x_0 = -\infty$, and $x_{m+1} = \infty$. When A and B are finite sets of items, we will say that A is a *9-cover* of B if no 1-interval of A contains more than 9 elements of B and, more generally, that A is *dense* in B if, for every integer $k \geq 1$, no k-interval of A contains more than $3k + 6$ elements of B.

The *rank* of an item x in a finite set A of items is the number $|\{y \in A : y \leq x\}|$ of items in A smaller than or equal to x. For every integer $c \geq 1$, let the *c-sample* of a finite set A of items be the subset of those items in A whose rank in A is a multiple of c. Define a *regular sample* to be either a 1-sample (i.e., a copy) or a 3-sample.

We shall need the following technical result, essentially due to Cole.

Lemma 1. *Let A, B, A' and B' be finite sets of items such that A and B as well as A' and B' are disjoint, A is dense in A', and B is dense in B'. Then the 3-sample of $A \cup B$ is dense in the 3-sample of $A' \cup B'$.*

Proof. Let S and S' be the 3-samples of $A \cup B$ and of $A' \cup B'$, respectively. Let I be a k-interval of S for some integer $k \geq 1$ and take $k_A = |A \cap I|$ and $k_B = |B \cap I|$. Since S is the 3-sample of $A \cup B$, $k_A + k_B \leq 3k$. If $\overline{k}_A$ and $\overline{k}_B$

are the numbers of intervals of A and of B, respectively, intersected by I, then $\overline{k}_A \leq k_A + 1$ and $\overline{k}_B \leq k_B + 1$. Because A and B are dense in A' and B', respectively, $|(A' \cup B') \cap I| \leq 3(\overline{k}_A + 2) + 3(\overline{k}_B + 2) \leq 3(k_A + k_B + 6) \leq 3(3k + 6)$. But then $|S' \cap I| \leq \lceil 3(3k + 6)/3 \rceil = 3k + 6$.

Without this being repeated on every occasion, the following lemmas assume every set of items manipulated by an algorithm to be stored compactly in a sorted array. Moreover, when it is stated that a task can be carried out in constant time with a certain number of processors, every processor assigned to the computation is supposed to know beforehand the rank of its own number in the set of all numbers of processors assigned to the computation and the size and starting address of every array that holds part of the input or is to receive part of the output. The space needed in addition to that taken up by the input and output is constant per processor.

When A and B are sets of items, the *ranking* of A in B is a function that maps every item in A to its rank in B. With A represented in a sorted array as described above, the ranking of A in B is represented in an array with the same index set as that of A. We say that A is *ranked* in B if the ranking of A in B is available. The *cross-ranking* of A and B consists of the ranking of A in B and the ranking of B in A, and we will say that A and B are *cross-ranked* or that A is *cross-ranked* with B if the cross-ranking of A and B is available. As observed by Cole, a shorthand for denoting rankings is convenient, especially in a pictorial representation. The ranking of A in B and the cross-ranking of A and B will be denoted by $A \longrightarrow B$ and $A \longleftrightarrow B$, respectively. When A is a 9-cover of B, we may express this additional fact by writing the ranking of A in B as $A \longmapsto B$.

The four lemmas below deal with simple ranking problems. They are illustrated in Fig. 1 and will be referred to using the short names indicated in parentheses. In Fig. 1, the meaning of the implication arrow $\Rightarrow$ is that, given the rankings to the left of the arrow, the rankings to its right can be computed in constant time with as many processors as the total size of the sets on which rankings are computed. Technically, when a ranking $A \longrightarrow B$ is to be produced and either A or B is empty, we assume that no computation is required (so that zero processors suffice).

Lemma 2 (subset rule). *Let A, B and C be sets of items such that A and B are disjoint and assume that the rankings of A in B and of $A \cup B$ in C are available. Then the ranking of A in C can be computed in constant time with $|A|$ processors.*

Proof. For each $x \in A$, add the ranks of x in A (trivially available) and in B (available by assumption) to obtain the rank of x in $A \cup B$. Then look up the rank of x in C and store it in an output array.

Lemma 3 (union rule). *Let A, B and C be pairwise disjoint sets of items, every two of which are cross-ranked. Then the cross-ranking of $A \cup B$ and C can be computed in constant time with $|A| + |B| + |C|$ processors.*

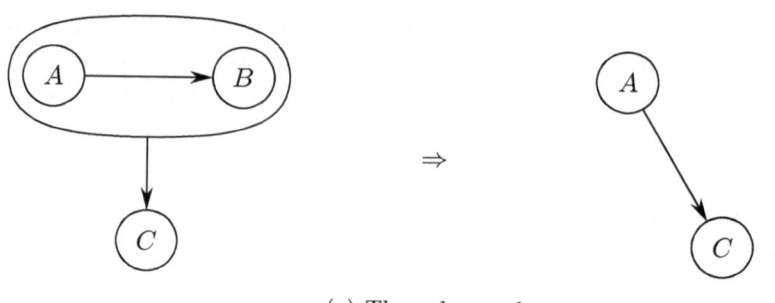

(a) The subset rule.

(b) The union rule.

(c) The sample rule. S is a regular sample of B.

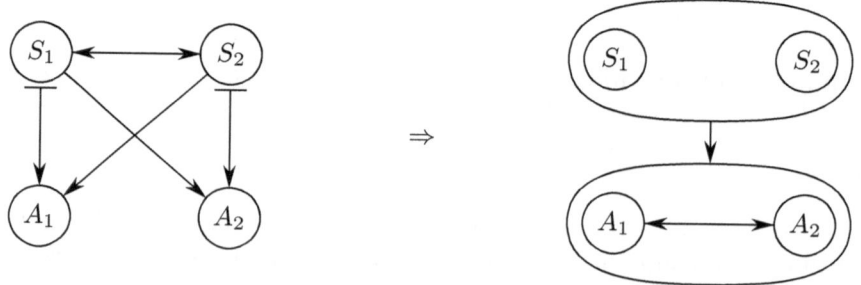

(d) The cross rule. S_i is a 9-cover of A_i, for $i = 1, 2$.

Fig. 1. Simple rules for deriving rankings from other rankings

Proof. For each $x \in A$, obtain the rank q of x in $A \cup B$ as in the previous proof and copy the rank of x in C to position q of an output array; proceed analogously for each $x \in B$. This computes the ranking of $A \cup B$ in C. To obtain the rank in $A \cup B$ of each $x \in C$, add the ranks of x in A and B.

Lemma 4 (sample rule). *Let A and B be finite disjoint sets of items and let S be a regular sample of B. If A is ranked in B, then A can be ranked in S in constant time with $|A|$ processors. If B is ranked in A, then S can be ranked in A in constant time with $|S|$ processors.*

Proof. S is a c-sample of B for a $c \in \{1,3\}$ that can easily be determined— except if B is empty—by testing whether $|S| = |B|$. If an item in A has rank q in B, its rank in S is $\lfloor q/c \rfloor$. If an item in S has rank q in S, its rank in A is that of the item in B whose rank in B is cq.

Lemma 5 (cross rule). *Let S_1, S_2, A_1 and A_2 be sets of items such that S_1 and S_2 as well as A_1 and A_2 are disjoint, each of S_1 and S_2 is ranked in each of A_1 and A_2, S_1 and S_2 are cross-ranked, and S_i is a 9-cover of A_i, for $i = 1, 2$. Then, in constant time and with $|S_1| + |S_2|$ processors, we can cross-rank A_1 and A_2, form $S_1 \cup S_2$ and $A_1 \cup A_2$, and rank $S_1 \cup S_2$ in $A_1 \cup A_2$.*

Proof. If we associate a processor with each item in S_1 and S_2, each such processor can obtain the rank in $S_1 \cup S_2$ of its associated item x by adding the ranks of x in S_1 and S_2. In constant time, we can therefore form $S_1 \cup S_2$ and associate with each item $x \in S_1 \cup S_2$ a processor that knows the ranks of x in S_1 and S_2 as well as whether x came from S_1 or S_2. Suppose that $x \neq \max(S_1 \cup S_2)$, so that x has a successor x' in $S_1 \cup S_2$. By pretending to be associated with the successor, if any, of x in its original set (S_1 or S_2), the processor associated with x can easily discover whether x' came from the same set as x. From this it can deduce the ranks of x' in S_1 and S_2 and then, for $i = 1, 2$, look up the rank q_i of x in A_i and the rank q_i' of x' in A_i. It proceeds to read the elements in A_i of ranks $q_i, \ldots, q_i' - 1$, for $i = 1, 2$, to merge the corresponding sequences, each of which contains at most 10 items, and to place the resulting sequence in an output array starting in the $(q_1 + q_2)$th position. This computes $A_1 \cup A_2$, except for the at most 18 smallest and the at most 20 largest items, which are easily handled by the processors associated with the smallest and largest items in $S_1 \cup S_2$, and the cross-ranking of A_1 and A_2 can be obtained as a by-product. Finally, the rank in $A_1 \cup A_2$ of each item in $S_1 \cup S_2$ is found as the sum of its ranks in A_1 and A_2.

3   High-Level Description

Suppose that the task at hand is to sort $n \geq 2$ pairwise distinct items $x_1, \ldots, x_n$. Let T be an undirected free tree whose internal nodes are all of degree 3 and with exactly $n + 1$ leaves $r, v_1, \ldots, v_n$. Replace each undirected edge $\{u, v\}$ in T by the two directed edges (u, v) and (v, u) and let $G = (V, E)$ be the resulting directed graph. With $L = \{r, v_1, \ldots, v_n\}$, let $E_1 = \{(u, v) \in E \mid u \in L\}$ and $E_3 = E \setminus E_1$ be the set of edges in E out of leaves and out of inner nodes, respectively. For each $e = (u, v) \in E$, we denote by $\tilde{e}$ the reverse edge (v, u). Moreover, for each edge $e = (v, w) \in E$, an *immediate predecessor* of e is an edge in E of the form (u, v) with $u \neq w$. For each $e \in E$, let $L(e) = \{x_i \mid e$ lies on a simple path in G from v_i to $r\}$. We will call an edge $e \in E$ *upward* if $|L(e)| > 0$, and *downward* otherwise. The *height* of an upward edge $e \in E$ is the length of a longest simple path in G whose last edge is e. This terminology corresponds to imagining r placed as a root at the top of T and defining the height of an upward edge as one more than the usual height of its lower endpoint. In this view, for each upward edge $e = (u, v)$, $L(e) = \{x_i \mid v_i$ is a descendant of $u\}$.

The algorithm to be described works in $2d$ *stages*, numbered $1, \ldots, 2d$, where d is the diameter of T. Before and after every stage, the algorithm stores for each $e \in E$ three sets of items, $A'(e)$, $S(e)$ and $S'(e)$, each of which is represented in a sorted array. All of these sets are initially empty. At a high level of abstraction, each of the $2d$ stages *processes* each edge $e \in E$ by executing the following steps:

1. If $|A'(e)| = |L(e)| > 0$, then set $c := 1$; otherwise set $c := 3$.
2. If $e \in E_1$, then set $A'(e) := L(e)$. If $e \in E_3$, compute $A'(e) := S'(e_1) \cup S'(e_2)$, where e_1 and e_2 are the two immediate predecessors of e.
3. Let $S'(e)$ be the c-sample of $A'(e)$.

In each stage, informally, each edge $e \in E_3$ fetches samples from its immediate predecessors, forms their union and provides its own sample of the union. If e is upward and had collected all items in $L(e)$ already in the previous stage, it passes them all on; otherwise its sample is a 3-sample.

If the execution of $A'(e) := S'(e_1) \cup S'(e_2)$ is thought of as moving a copy of each item in $S'(e_1)$ or $S'(e_2)$ across e, then the set of edges across which copies of a particular item x_i are moved span a subgraph of G without length-2 cycles, and therefore an outtree. It follows that whenever the algorithm forms the union of two sets of items, the two sets are disjoint.

Let us say that an edge $e \in E$ is *complete* in a stage if the relation $|A'(e)| = |L(e)| > 0$ holds at the beginning of that stage. By induction on h, one can show that an upward edge of height h is complete in a stage t if and only if $t \geq 2h$. For the basis, an upward edge e of height 1 sets $A'(e) := L(e)$ in stage 1 and has $A'(e) = L(e)$ forever after. Assume now that $h \geq 2$ and that the claim holds for all upward edges of height at most $h - 1$ and consider an upward edge e of height h with immediate predecessors e_1 and e_2. The relations $S'(e_1) = L(e_1)$ and $S'(e_2) = L(e_2)$ hold at the beginning of stage t if and only if $t \geq 2(h - 1) + 1$, by induction, and therefore e is complete in stage t if and only if $t \geq 2h$, as desired. Since the edge e_r entering r is of height at most d, it follows that $L(e_r) = \{x_1, \ldots, x_n\}$ can be obtained in sorted form as $A'(e_r)$ at the end (or, in fact, at the beginning) of the last stage. Thus the algorithm computes the desired result.

If an upward edge $e \in E$ is complete for the first time in a stage t, the set $S'(e)$ is the 3-sample of $L(e)$ in stage $t - 1$ and is $L(e)$ itself in stage t and in every later stage. Let us therefore call an upward edge of height h *active* in a stage t exactly if $t \leq 2h$. A downward edge e is defined to be active exactly when $\tilde{e}$ is. Let E^* be the set of active edges. The following lemma is instrumental in bounding the resource requirements of the algorithm.

Lemma 6. *At the beginning and end of every stage, $\sum_{e \in E^*} |A'(e)| \leq 10n$.*

Proof. Let us call the reciprocal of the number c computed as part of the processing of an edge $e \in E$ in a particular stage the *sampling density* of e in that stage. Imagine each item not as a discrete entity, but as a commodity that can be present in arbitrary amounts. Moreover, imagine that the c-sample computed in step 3 of the algorithm does not contain selected items, but rather includes $1/c$ of the amount of each item present in $A'(e)$. Since a c-sample of a set A never includes more than $|A|/c$ items, the total amount of items present in a set manipulated by the algorithm according to this fictitious accounting is an upper bound on the number of items present in the set in the actual execution.

Fix an item x_i. A positive amount of x_i can be present in $A'(e)$ for an active edge e only if G contains a simple path p that starts at v_i and has e as its last edge, and then the amount of x_i in $A'(e)$ at the end of a stage t is upper-bounded by the product of the sampling densities in stage t of the edges on p other than e. All edges preceding the first active edge e' on p must be upward, and therefore common to all relevant paths p. Moreover, all edges on p after e' have sampling density $1/3$, and the same is true of e' unless e' is upward. Therefore the amount of x_i present in $A'(e)$, summed over all active edges e, is at most $1 + 3\sum_{j=0}^{\infty}(2/3)^j = 10$ (see Fig. 2). The lemma follows by summation over all n items x_i.

4   The Execution of a Stage

The detailed description of a single stage is where pictures will be most useful. Nodes in T are drawn as polygons, two such polygons sharing a corner exactly if the two corresponding nodes are adjacent. More specifically, nodes in T of degree 3 and degree 1 are drawn as triangles and as thirds of triangles, respectively, and Fig. 3(a) shows conventions that will be used throughout for drawing the sets stored by the algorithm between stages. Note, in particular, that the sets associated with an edge $e = (u, v)$ are shown inside the polygon representing the node v that e enters.

As mentioned in the introduction, an efficient execution of the algorithm hinges on the availability of suitable "scaffolding". Before and after every stage, the algorithm stores the following scaffolding information:

A. For each $e \in E_3$, the cross-ranking $S(e_1) \leftrightarrow S(e_2)$ of $S(e_1)$ and $S(e_2)$, where e_1 and e_2 are the two immediate predecessors of e.
B. For each $e \in E$, the ranking $S(e) \mapsto S'(e)$ of $S(e)$ in $S'(e)$.
C. For each $e \in E$, the cross-ranking $S'(\tilde{e}) \leftrightarrow A'(e)$ of $S'(\tilde{e})$ and $A'(e)$.

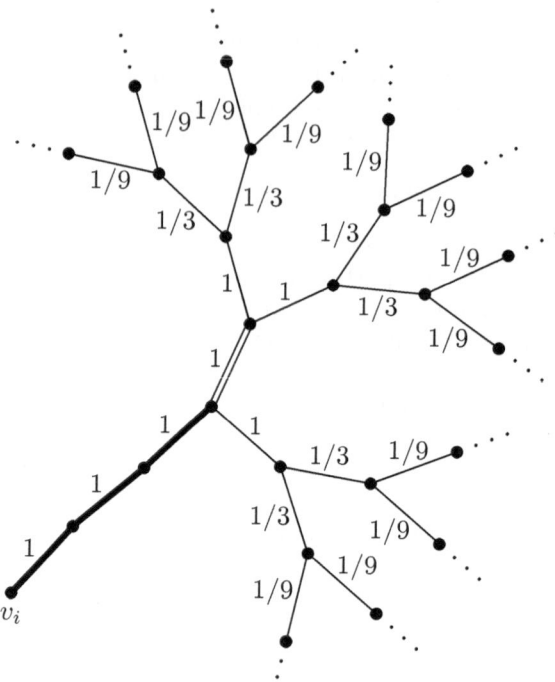

Fig. 2. The tree of paths p from v_i to edges e with positive amounts of x_i in $A'(e)$. A thick edge has sampling density 1. If it is also black, it is inactive. Each edge e is labeled with the maximum possible amount of x_i present in $A'(e)$.

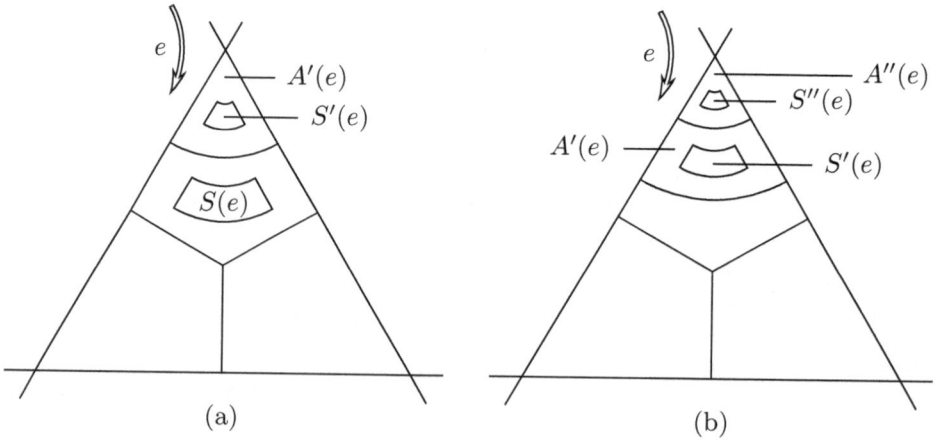

Fig. 3. The pictorial representation of sets manipulated by the algorithm for each edge $e \in E$. (a): Between stages. (b): During the processing of e. The sets $S(e)$, $S'(e)$ and $S''(e)$ are indicated only in figures for which they are of relevance.

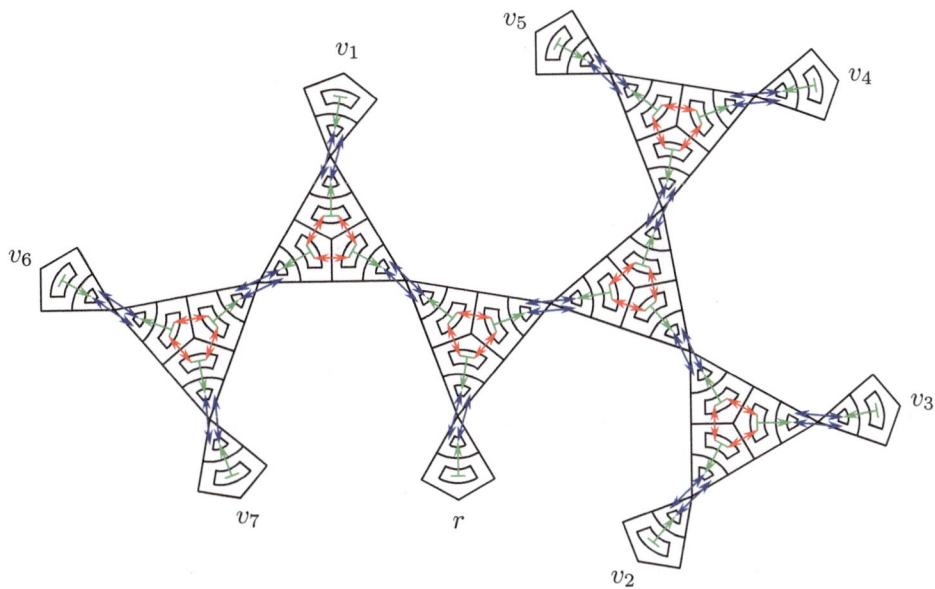

Fig. 4. An example tree T with the rankings available at the beginning of each stage

Fig. 4 shows an example tree T with the rankings A–C available at the beginning of each stage. The following color coding of the rankings will be used throughout: A: red; B: green; C: blue. We will consider the availability of these rankings as invariants with the same names A–C. As anticipated in the short-hand above, invariant B includes the fact that for each $e \in E$, $S(e)$ is a 9-cover of $S'(e)$.

Two additional invariants that hold before and after every stage are formulated below. The first of these is illustrated in Fig. 5, while the other invariant is implicit already in the drawing conventions of Fig. 3.

D. For each $e \in E_3$, $A'(e) = S(e_1) \cup S(e_2)$, where e_1 and e_2 are the immediate predecessors of e.
E. For each $e \in E$, $S'(e)$ is a regular sample of $A'(e)$.

Before the first stage, invariants A–E are trivially satisfied, since all relevant sets are empty.

At a more detailed level, the processing of each edge $e \in E$ in each stage is refined as follows:

1. If $|A'(e)| = |L(e)| > 0$, then set $c := 1$; otherwise set $c := 3$.
2. If $e \in E_1$, then set $A''(e) := L(e)$ and rank $A'(e)$ in $A''(e)$. Otherwise, with e_1 and e_2 taken to be the two immediate predecessors of e, cross-rank $S'(e_1)$ and $S'(e_2)$, set $A''(e) := S'(e_1) \cup S'(e_2)$ and rank $A'(e)$ in $A''(e)$.
3. Let $S''(e)$ be the c-sample of $A''(e)$.

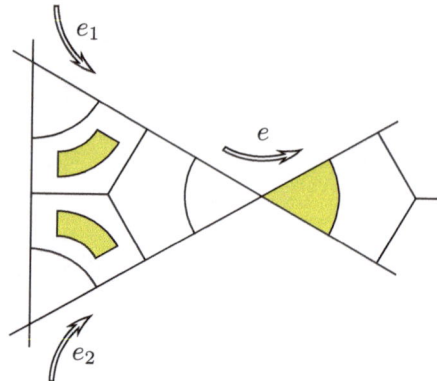

Fig. 5. Invariant D: The union of the two yellow sets in the same triangle is the third yellow set

4. Rank $S'(e)$ in $S''(e)$.
5. Cross-rank $S''(\bar{e})$ and $A''(e)$.
6. Set $A'(e) := A''(e)$, $S(e) := S'(e)$ and $S'(e) := S''(e)$.

Steps 1–6 above are easily seen to have the same net effect on $A'(e)$ and $S'(e)$ as steps 1–3 of the high-level description. The sets $A''(e)$ and $S''(e)$ can be thought of as "the new values" of $A'(e)$ and $S'(e)$, respectively, just as $S(e)$ is the value of $S'(e)$ from the previous stage. Fig. 3(b) shows the conventions used for drawing the sets associated with an edge e during the processing of e. One may imagine new sets "sprouting" in the corners of triangles.

Invariants D and E hold at the end of every stage, as an immediate consequence of the computation carried out in that stage. Therefore they always hold outside of step 6. The following lemma proves that the "cover part" of invariant B also holds outside of step 6.

Lemma 7. *At the end of every stage, $S(e)$ is dense in $S'(e)$ for every $e \in E$.*

Proof. By induction on the stage number t. The claim is trivial for $e \in E_1$ and for $t = 1$. For $e \in E_3$ and $t \geq 2$, consider the situation just before the execution of step 6 in stage t. Invariants D and E show that with e_1 and e_2 taken to be the two immediate predecessors of e, $S'(e)$ is a regular sample of $S(e_1) \cup S(e_2)$, whereas $S''(e)$ is a regular sample of $S'(e_1) \cup S'(e_2)$. More precisely, if e is not complete in stage t, $S'(e)$ and $S''(e)$ are the 3-samples of $S(e_1) \cup S(e_2)$ and of $S'(e_1) \cup S'(e_2)$, respectively. By induction, $S(e_i)$ is dense in $S'(e_i)$, for $i = 1, 2$, so Lemma 1 shows that $S'(e)$ is indeed dense in $S''(e)$. And if e is complete in stage t, $S'(e)$ is a regular sample of $L(e) = S''(e)$ and therefore clearly dense in $S''(e)$.

Steps 1, 3 and 6 are trivial. The execution of the other steps is described below. An alternative, essentially stand-alone description is provided by Figs. 6–9 and their captions.

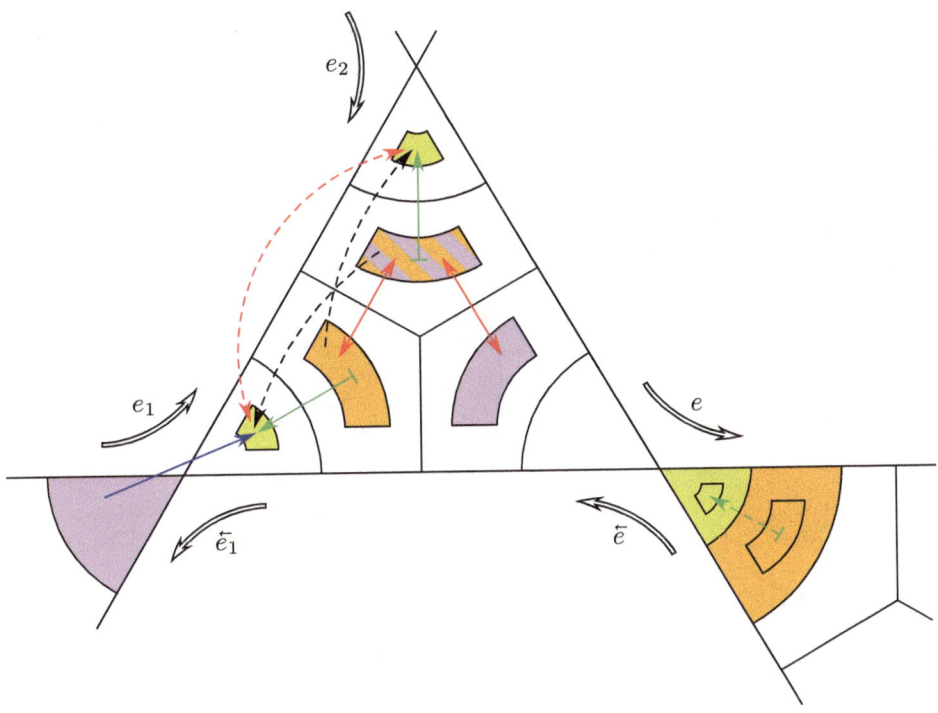

Fig. 6. The execution of steps 2 and 4 for an edge $e \in E_3$. The solid arrows are available at the start of the phase. Invariant D, applied to the pink sets, the rightmost solid red arrows, the blue arrow and the subset rule allow the drawing of the downward-pointing dashed black arrow. The other dashed black arrow follows by symmetry. The cross rule now allows the drawing of the dashed red arrows and, by the sample rule and invariant D, applied to the orange sets and to the yellow sets, the dashed green arrow.

2. The necessary computation is trivial if $e \in E_1$, so consider the case $e \in E_3$. In the following two sentences, various invariants are applied to $\bar{e}_1$ rather than to e. By invariant C, $A'(\bar{e}_1)$ is ranked in $S'(e_1)$. But by invariant D, $A'(\bar{e}_1) = S(e_2) \cup S(\bar{e})$, and $S(e_2)$ and $S(\bar{e})$ are cross-ranked by invariant A, so the subset rule allows us to rank $S(e_2)$ in $S'(e_1)$. By symmetry, we can rank $S(e_1)$ in $S'(e_2)$. Moreover, by invariants A and B, we have the rankings of $S(e_i)$ in $S'(e_i)$, for $i = 1, 2$, as well as the cross-ranking of $S(e_1)$ and $S(e_2)$. The cross rule now implies that we can cross-rank $S'(e_1)$ and $S'(e_2)$, merge the two sets to obtain $A''(e) = S'(e_1) \cup S'(e_2)$, and rank $A'(e) = S(e_1) \cup S(e_2)$ (invariant D) in $A''(e)$ (see Fig. 6).

4. By Invariant E, $S'(e)$ is a regular sample of $A'(e)$. Since $S''(e)$ is clearly a regular sample of $A''(e)$ and $A'(e)$ was ranked in $A''(e)$ in step 2, it suffices to appeal to both parts of the sample rule.

5. Our first goal will be to cross-rank $S'(\bar{e})$ with $A''(e)$ and, by both parts of the sample rule, with $S''(e)$. Assume first that $e \in E_1$. There is nothing to do in stage 1, since $S'(\bar{e})$ is empty. In every later stage, we have $A''(e) =$

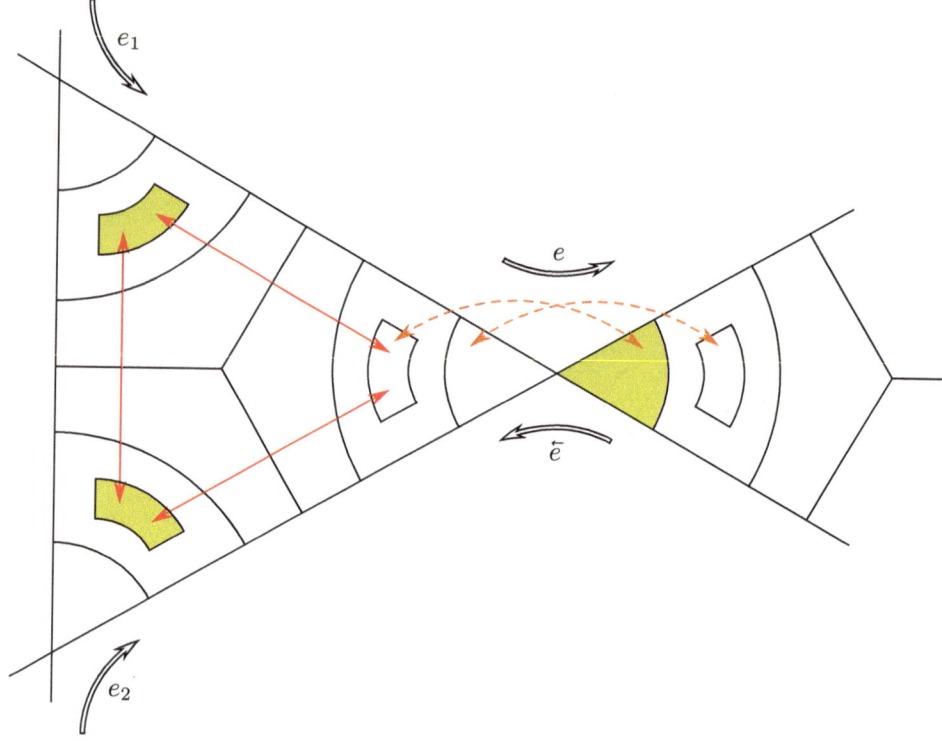

Fig. 7. The first part of the execution of step 5 for an edge $e \in E_3$. The red arrows were computed in step 2 (Fig. 6). Invariant D, applied to the yellow sets, the red arrows and the union rule allow the drawing of the leftmost pair of dashed orange arrows. The other dashed orange arrows follow by symmetry.

$A'(e)$, so the desired ranking is available, according to invariant C. Assume now that $e \in E_3$. As is easy to see by symmetry, step 2 cross-ranked every two of $S'(\bar{e})$, $S'(e_1)$ and $S'(e_2)$, where e_1 and e_2 are the two immediate predecessors of e. Therefore, by the union rule, we can cross-rank $S'(\bar{e})$ and $A''(e) = S'(e_1) \cup S'(e_2)$ (see Fig. 7).

By symmetry, we also have the rank of $S'(e)$ in $S''(\bar{e})$. From step 4, we have the ranks of $S'(e)$ in $S''(e)$ and, by symmetry, of $S'(\bar{e})$ in $S''(\bar{e})$. By invariant E, $S'(e)$ is a regular sample of $A'(e)$, so invariant C and both parts of the sample rule show that we can cross-rank $S'(e)$ and $S'(\bar{e})$. Now, by the cross rule and invariant B, applied at the end of the stage, we can rank $S''(e)$ in $S''(\bar{e})$ (see Fig. 8).

Since $S''(e)$ is a regular sample of $A''(e)$, it is a 9-cover of $A''(e)$, and we can trivially rank $S''(e)$ in $A''(e)$. At this point, we have ranked each of $S''(e)$ and $S'(\bar{e})$ in each of $S''(\bar{e})$ and $A''(e)$, and we have the cross-ranking of $S''(e)$ and $S'(\bar{e})$. Therefore, by the cross rule, we can obtain the cross-ranking of $S''(\bar{e})$ and $A''(e)$ (see Fig. 9).

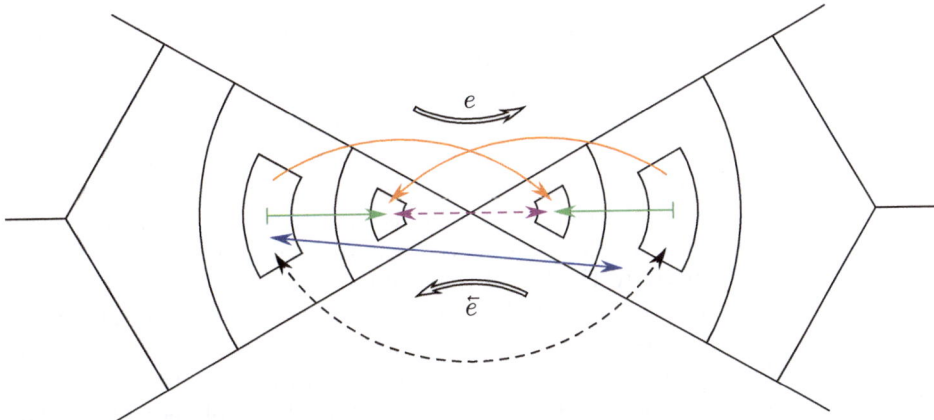

Fig. 8. The execution of step 5 continued. The orange arrows derive with the sample rule from those drawn in the previous figure. The black arrows follow in the same way from the blue arrows, whose presence is guaranteed by invariant C. The green arrows were drawn in step 4 (Fig. 6). The cross rule now allows the drawing of the magenta arrows.

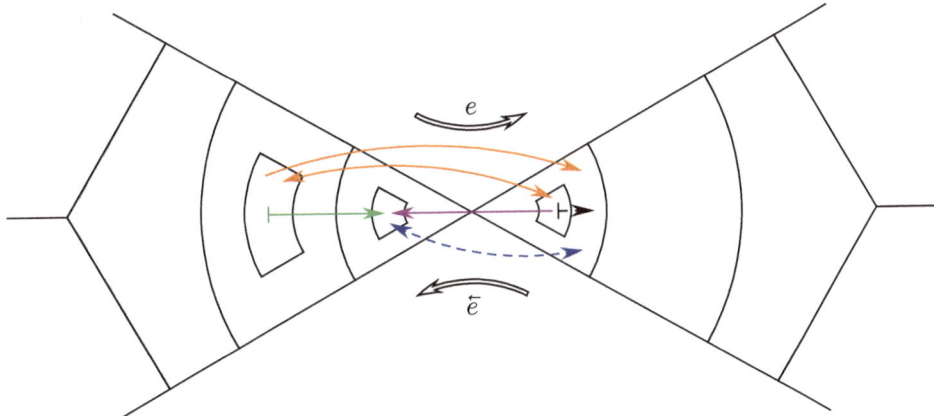

Fig. 9. The third and final part of the execution of step 5. The magenta arrow was drawn in the previous figure. The orange arrows derive with the sample rule from those drawn in Fig. 7. The green arrow was copied from the previous figure, and the black arrow is trivial. The cross rule now allows the drawing of the blue arrows.

The rankings required by invariants A, B and C for the next stage are computed in steps 2, 4 and 5, respectively. Therefore invariants A–E hold at the beginning and end of every stage.

5   Detailed Implementation

This section spells out the nitty-gritty remaining details of the algorithm. Recall the following standard method of allocating consecutively numbered resource units such as processors or memory cells to jobs $J_1, \ldots, J_m$: If J_i needs a_i resource units, for $i = 1, \ldots, m$, compute the prefix sums $s_0, \ldots, s_m$, where $s_i = \sum_{j=1}^{i} a_j$ for $i = 0, \ldots, m$, and assign to J_i the resource units numbered $b + s_{i-1}, \ldots, b + s_i - 1$, for $i = 1, \ldots, m$, where b is the number of the first available resource unit. The prefix sums $s_0, \ldots, s_m$ can be computed in $O(\log m)$ time with $O(m)$ processors by means of a balanced binary tree with $a_1, \ldots, a_m$ as its leaves: In a bottom-up sweep over the tree, each node learns the sum of the leaves in the maximal subtree below it, and in a subsequent top-down sweep, it learns the sum of the leaves strictly to the left of that subtree.

For each $e \in E$, let us say that the sets $A'(e)$, $S(e)$, $S'(e)$, $A''(e)$ and $S''(e)$ are in the *custody* of e. For each $e \in E$, the total size of the sets in the custody of e just before the execution of step 6 in a stage t in which e is active is bounded by a constant times the size of $A'(e)$ at the beginning or end of one of the stages $t - 1$ and t. Indeed, $A''(e)$ is just $A'(e)$ at the end of stage t, $S'(e)$ and $S''(e)$ are subsets of $A'(e)$ and $A''(e)$, respectively (invariant E), and $S(e)$ is empty or equals $S'(e)$ at the beginning of the previous stage. Therefore, by Lemma 6, the total size of the sets in the custody of active edges, as well as of the rankings computed for these sets, is $O(n)$ at all times.

Sets in the custody of inactive upward edges never again change, and sets in the custody of inactive downward edges cannot influence the output of the algorithm. Therefore it is not necessary to associate processors with inactive edges. It is not necessary to allocate space for sets in the custody of inactive edges either, except when such sets are read during the processing of an active edge. This can happen only in steps 2 and 5 of the processing of an active edge e with immediate predecessors e_1 and e_2, where $S'(e_1)$ and $S'(e_2)$ are read. To cope with this exception, when an edge e becomes inactive, the custody of $S'(e)$ is transferred to those active edges of which e is an immediate predecessor, each of which stores a copy of $S'(e)$ together with any rankings computed for $S'(e)$. The total space requirements remain $O(n)$.

By associating a processor with each edge in E and carrying out a "dry run" of the sorting algorithm in which sets of items are replaced by their sizes, merging of (disjoint) sets is replaced by addition of their sizes, etc., it is possible, in $O(d)$ time, to compute for each $e \in E$ and for $t = 1, \ldots, 2d$ the total space needed for the sets in the custody of e in stage t. (To prevent this computation from needing $\Theta(dn)$ space, it is preceded by an even more rudimentary computation that records for each edge e only when a set in the custody of e becomes nonempty for the first time and when e becomes inactive, so that space proportional to the number of intervening stages can be allocated to e.) Now, for $t = 1, \ldots, 2d$, the allocation of space to edges in stage t can be planned by computing prefix sums in the manner described in the beginning of the section. Since $|E| = O(n)$, the $2d$ independent prefix-sums computations can be carried out in a pipelined fashion in $O(d + \log n)$ total time with $O(n)$ processors.

If we allocate one processor per memory cell ever used by the algorithm and intersperse these memory cells with information about the sizes and starting addresses of relevant arrays, it is clear from Lemmas 2–5 that each stage can be executed in constant time. The available processors can also effectuate any necessary custody transfers, as discussed above, and copy sets that are to survive from one stage to the next between their old and new locations in memory. This takes place between stages and needs constant time per stage.

So far, the algorithm uses $O(n)$ processors, $O(d + \log n)$ time and $O(n)$ space. By letting each physical processor simulate a constant number of virtual processors, we can reduce the processor count to exactly n, and a proper choice of the tree T ensures that $d = O(\log n)$. This reproves Cole's original result: The algorithm sorts n items using n processors, $O(\log n)$ time and $O(n)$ space.

6   Comparison with Cole's Description

Cole's sets $UP(v)$, $SUP(v)$ and $OLDSUP(v)$ correspond to what is here called $A'(e)$, $S'(e)$ and $S(e)$, respectively, where e is the edge from v to its parent. Similarly, $DOWN(v)$, $SDOWN(v)$ and $OLDSDOWN(v)$ correspond to $A'(\tilde{e})$, $S'(\tilde{e})$ and $S(\tilde{e})$. Cole's assumptions (a) and (c) correspond to our invariant A, for e and for $\tilde{e}$, (b) and (d) correspond to B, and (f) and (g) correspond to C, while assumption (e) is not used here.

In Cole's variant of the algorithm, a node uses the sampling densities $1/4, \ldots,$ $1/4, 1/2, 1$ over the stages in which it is active. We may express this by saying that the algorithm adheres to the *sampling regime* $(4, 2)$ (with an implicit 1 at the end). The algorithm in fact works correctly with any sampling regime of the form $(z_0, \ldots, z_l)$, where $z_0, \ldots, z_l$ are integers with $z_1, \ldots, z_l > 1$ and $z_0 > 2$ (the latter condition ensures that $\sum_{j=0}^{\infty} (2/z_0)^j < \infty$; cf. the proof of Lemma 6). Cole proposes an even more general alternative, namely to use sampling densities $1/2$ and $1/4$ at alternate levels of the tree. Here the sampling regime (3) was chosen as a simplest possibility.

References

1. Ajtai, M., Komlós, J., Szemerédi, E.: An $O(n \log n)$ sorting network. In: 15th Annual ACM Symposium on Theory of Computing (STOC 1983), pp. 1–9 (1983)
2. Cole, R.: Parallel merge sort. SIAM J. Comput. 17, 770–785 (1988)

Self-matched Patterns, Golomb Rulers, and Sequence Reconstruction

Franco P. Preparata

Computer Science Department, Brown University, 115 Waterman Street, Providence, RI 02912-1910, USA

Abstract. The reconstruction of an unknown target sequence from the collection of its subsequences is a problem originating in computational biology (sequencing by hybridization), but with combinatorial significance in its own right. Retracing the history of the topic, this paper explores the power of the subsequence pattern in extracting information from the target, and proposes a new class of effective patterns, named self-matched, characterized by their autocorrelation function.

1 Introduction

The motivation for the topic of this note originates in computational biology. As a potential alternative to traditional methods for nucleic acids sequencing based on gel electrophoresis, about two decades ago [1, 2, 4, 5, 6] a significantly novel approach was proposed. Resorting to the emerging microarray technology and exploiting Watson-Crick nucleotide complementarity (hybridization), the objective was to obtain by a single laboratory experiment *all* substrings of a specified length of a target sequence with no positional information (the so-called *sequence spectrum*). Sequencing would then be reduced to the reconstruction of the sequence from its substrings. Clearly, the sequencing-by-hybridization (SBH) technology is a complex biochemical-combinatorial interaction, and these two facets are strongly interdependent. The modeling of the combinatorial aspects must closely reflect the biochemical constraints, and in turn the features of the reconstruction algorithms may substantially affect the nature of the laboratory experiments. In any case, the present goal is not to present anew a critical analysis of SBH, carried out elsewhere [8]; instead, our objective is to analyze the process of extracting information for the reconstruction of a target sequence. Although reconstruction pertains to sequences over an arbitrary finite alphabet, due to the immediate origin of the problem and for concreteness of presentation, with no loss of generality we shall make explicit reference to the 4-letter alphabet of nucleic acids, although the analysis applies to sequences over any finite alphabet.

A crucial component of the approach is the "probing pattern", i.e., the format of the subsequences for which the probing of the target is carried out. In the simplest form (the most natural, and, chronologically, the earliest one proposed [4]), the probing pattern is a string of k symbols (contiguous or solid

S. Albers, H. Alt, and S. Näher (Eds.): Festschrift Mehlhorn, LNCS 5760, pp. 158–169, 2009.

probes). The strong interdependence between two strings used consecutively in the reconstruction (they overlap in $k-1$ of their k symbols) casts suspicion on their effectiveness in extracting information, and, in fact, a detailed formal analysis confirms this expectation. Some alternatives to the contiguous probing pattern were proposed, typically through the insertion of one or two gaps of "don't care" positions in the probing pattern [6] [1].

Despite the modest performance improvements, the adoption of "gaps" signaled an important transition from the notion of "strings" to the notion of "subsequences" conforming to an arbitrary pattern. Complying with standard terminology, such subsequences will be frequently referred to as "probes". In this context, a probing scheme was subsequently proposed [7] which was shown to come very close to achieving the full potential of the approach from a combinatorial-algorithmic viewpoint (this point will be addressed in the Concluding Remarks at the end of this note). Such scheme exhibited a highly structured conformation: denoting with a "1" a natural base, and with a "0" a universal base, the pattern was represented by the binary sequence $1^s(0^{s-1}1)^r$, for integers r and s.

As it turns out, those structured gapped patterns are closely related to more general high-performing probing schemes, referred to here as *self-matched* patterns, discussed in Section 3.

2   Sequence Reconstruction

Sequence reconstruction is an incremental process, adding one symbol at each step, conventionally from left to right. Sequence reconstruction is easily described when the probing pattern has the form 1^k (a string of length k), since, in this case, it can be pictured as the traversal of a directed graph (refer to Figure 1), of which nodes correspond to sequence $(k-1)$-tuples and arcs to sequence k-tuples.

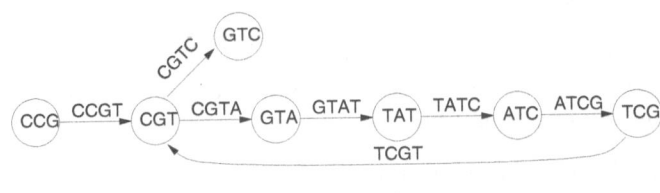

sequence:**CCGTATCGTC**

Fig. 1. Reconstruction graph for sequence CCGTATCGTC and probing pattern 1111

This graph is clearly Eulerian (only two vertices, the initial and final ones, have odd degree) and sequence reconstruction corresponds to tracing an Eulerian path on the graph. The reconstruction is unambiguous if and only if there is a unique Eulerian path, and a necessary condition for the existence of distinct Eulerian paths is the occurrence of repeated $(k-1)$-tuples along the sequence.

---

[1] The realization of a "don't care" nucleotide is theoretically effected by so-called universal bases [3], exhibiting nonspecific hybridization to the four natural bases.

A general feature of sequence reconstruction is that performance is to be measured by the length m of sequences that can be unambiguously reconstructed. This estimate, therefore, is to be made with reference to an ensemble of random sequences, which is normally taken as the set of sequences of a given length with independent identically distributed symbols. The estimate, in turn, must be qualified by its confidence level. Essentially, we always deal with reconstructions with a prescribed confidence level.

When the probing pattern is a string, as in the above example, and there are two or more probes in the spectrum differing only in their rightmost symbol, then each of the paths issuing from a repeated $(k-1)$-tuple corresponds to an actual segment of the target. However, when we consider more general gapped patterns of the form $1(0 \vee 1)^{l-2}1$ (of length l with $k < l$ symbols equal to 1), the event of multiple paths emerging from a node does not necessarily correspond to repeated $(l-1)$-tuples, but only to identical subsequences of $k-1$ symbols and length $l-1$ occurring in the sequence; therefore, only one of these paths is guaranteed to be continued (the "correct" path). The other ("spurious") paths are continued to the extent that the spectrum contains subsequences supporting their extension: such subsequences are referred to as *fooling probes* (because they fool the reconstruction process). The whole rationale of the gapped pattern approach is that the probability that a spurious path be mistaken as the correct one decays rapidly with the length of its segment being extended beyond the branching.

Therefore, a branching with gapless probes reveals an actual segment of the target sequence (there are no fooling probes in this case); a branching with gapped probes is normally attributed to fooling probes, because k is significantly smaller than l. Moreover, branching events are likely to occur, because, when gapped probes are used, the target sequence is meant to be significantly longer and, therefore, its spectrum is likely to contain fooling probes.

It follows that a key aspect of the approach is the design of probing patterns achieving this two-fold objective:

1. Minimize the probability that a spurious path is deterministically extended.
2. Minimize the average length of the extended spurious paths, in order to reduce the computational burden.

To gain insight into the relation between the probing pattern and the stated objectives, it is appropriate to analyze the mechanism by which sequence reconstruction fails.

The overall process can be pictured as the construction of a tree whose nodes are sequence symbols and whose edges describe the successor relation. A subset of these nodes have outdegree > 1 (branching nodes). The subtree rooted at a branching node contains the correct path and (at least) one competing (spurious) path. Normally, the spurious path is expectedly of short length, because the absence of an extending fooling probe in the spectrum causes its termination. Referring to two paths issuing from a branching node, there are situations (failures) when the spurious path can be deterministically extended, and they can be classified into two classes:

1. **Mode 1.** The two paths issuing from the branching node are identical except for their initial (branching) symbol, and therefore the two alternatives cannot be distinguished. This failure is caused by k fooling probes which agree with the correct path *except* in the branching position. These probes occur scattered along the target sequence, with possible overlaps (the spectrum provides no positional information). This failure mode is illustrated in the following example:

Example 1. For probing pattern 100100111, suppose the algorithm detects the following situation (symbols at the branching position are shown within brackets):

```
...A C G A G T C C T [G] A G T G A T A T A T ... correct path
                     [T] A G T G A T A T A T ... spurious path


        C * * G * * C T [T]
          G * * T * * T [T] A
            A * * C * * [T] A G
              C * * [T] * * T G A
                [T] * * T * * T A T
```

The last five rows illustrate the relevant fooling probes.

2. **Mode 2.** The two competing paths do not agree, except, partially, in their respective prefixes. This situation is caused by a string of length $l-1$ (called *supporting segment*), whose prefix of length $J \le l-1$ coincides with a suffix of the correct path up to the branching position. The branching of the alternative path and possible disagreements of its prefix of length $l-1-J$ with the homologous prefix of the correct path beyond the branching are supported by fooling probes. Once the spurious path has incorporated the supporting segment, its deterministic extension is guaranteed.

This more complex failure mode is illustrated by the following example:

Example 2. For probing pattern 100100111, suppose the algorithm detects the following situation (relevant disagreements within brackets):

```
...A C G A G T C C T [G] A G T [G] A T A T A T ... correct path
                     [T] A G T [A] A T C T G G ... spurious path


    1   C * * G * * C T T
    2     G * * T * * T T A
    3       A * * C * * T A G
    4         C * * T * * T A A
    5       T * * T * * G T A
```

Here the pair [G][T] is the ambiguous branching, the top path represents the correct extension. In italics is the length-8 supporting segment CTTAGTAA, which establishes the spurious path shown at the bottom. This segment occurs *elsewhere* in the sequence. The last five rows illustrate the relevant fooling probes. The branching disagreement [G-T] is compensated by probes 1-4, and disagreement [G-A] is compensated by probes 4 and 5.

The analysis of Mode 1 is relatively simple, since the occurrence of a Mode 1 event is due to k fooling probes scattered along the sequence. The analysis of Mode 2 is significantly more complex, due to the interplay between two actual subsequences of the target, and has been carried out in some detail elsewhere [8]. However, the two modes, although inherently different, appear to share the common feature that a substantial number of fooling probes are needed to support the branching. Therefore, in what follows we shall consider Mode 1, where exactly k fooling probes are needed to support the branching in the case of failure.

We observe incidentally that, although the notion of fooling probe is conceptually extraneous to the analysis of the original string-based probing approach, even this method can be viewed within the same optics. In fact, in such case a branching occurs when there are two identical $(k-1)$-tuples along the sequence: clearly one of the two emerging paths is determined by probes occurring *elsewhere* (corresponding to a "jump" in the reconstruction), i.e., these probes can be legitimately called "fooling" probes. The problem is that these probes are not due to random events (scattered occurrences along the sequence), but are guaranteed to exist since, due to the chosen probing pattern 1^k, they all occur at the same site.

Returning to the analysis of Mode 1, given l and k, how does the choice of the probing pattern affect the probability of occurrence of a collection of k fooling probes causing failure?

A useful tool is the notion of "alignment matrix" of a failure event. Such event is determined by a collection of k fooling probes positionally aligned with respect to the position of branching. The alignment matrix is a $k \times (2l-1)$ binary table, whose rows and columns correspond respectively to probes and positions; an entry is either 1 or "." (a visual alternative for binary 0) depending upon whether a position is sampled or not by a probe.

Example 3: For probing pattern 1001101001 we have the following alignment matrix:

```
1 . . 1 1 . 1 . . 1 . . . . . . . . .
. . . 1 . . 1 1 . 1 . . 1 . . . . . .
. . . . . 1 . . 1 1 . 1 . . 1 . . . .
. . . . . . 1 . . 1 1 . 1 . . 1 . . .
. . . . . . . . . 1 . . 1 1 . 1 . . 1
```

Note that each row of the alignment matrix is a shift of the probing pattern, so that the central column is $[1, 1, \ldots, 1]^T$.

In general, the k fooling probes may be completely disjoint or partially overlapping, i.e., they may occur at $u \le k$ *sites*. Therefore, for any fixed value of u (number of sites), the assignment of probes to sites is a surjection from a k-set to a u-set and the number of such assignments is given by a Stirling number of the second kind $S_{k,u}$.

Given a specific assignment of h probes to u sites, site i produces a constraint of ν_i sequence symbols, where ν_i is the weight of the bit-by-bit OR of the probes assigned to i. To a first approximation, if m is the length of the sequence, the assignment yields a term

$$\binom{m}{u} 4^{-(\nu_1+\ldots+\nu_u)},$$

i.e., a monomial of degree u in m, expressing the probability of failure occurring at a specific step of the sequence reconstruction. Since there are m sequence positions where the failure branching may occur, we shall obtain a monomial in m of degree $u+1$.

In principle, the analysis can be carried out for each of the $\sum_{u=1}^{k} S_{k,u}$ probe-to-site assignments (this number is known as Bell Number $B(k)$), and is, in general, a very tedious task even for small values of l and k. The end-result is a polynomial $q_\pi(m)$ in the variable m of degree $k+1$, which characterizes the performance of pattern π in sequence reconstruction. This polynomial has the general form

$$q_\pi(m) = 3m(c_1 m + \ldots + c_{k-1} m^{k-1} + c_k m^k) \tag{1}$$

where the factor $3m$ is due to the position where the failure branching occurs (there are 3 choices for each sequence position). The coefficient c_k is determined completely by the single k-site assignment (when all sites are distinct), and is obviously independent of the pattern for a given weight k, while all other coefficients of $q_\pi(m)$ depend upon the number of constrained symbols for different probe/site assignments, and are strongly pattern-dependent (through row overlap).

To gain insight into the structure of $q_\pi(m)$, we refer to the alignment matrix of the probing pattern π. Introducing the formal variable z we can represent π as the polynomial (with binary coefficients)

$$p(z) = \sum_{i=1}^{k} z^{j_i}$$

with $j_1 = 0$ and $j_k = l - 1$. In this representation the i-th row of the alignment matrix is the polynomial $p(z)z^{l-1-j_i}$. If we add the matrix entries column-by-column, we obtain the polynomial

$$R(z) = p(z) + z^{l-1-j_{k-1}}p(z) + z^{l-1-j_{k-2}}p(z) + \ldots + z^{l-1-j_1}p(z)$$

$$= p(z)\left(z^{l-1}\left(\frac{1}{z^{j_k}} + \frac{1}{z^{j_{k-1}}} + \ldots + \frac{1}{z^{j_1}}\right)\right)$$

$$= p(z)\left(z^{l-1}p\left(\frac{1}{z}\right)\right) = p(z)p^*(z)$$

where $p^*(z)$ is the polynomial obtained from $p(z)$ by reversing the order of its coefficients.

Polynomial $R(z) = \sum_{j=0}^{2l-2} R_j z^j$ is normally called the *autocorrelation function* of pattern $p(z)$. Coefficient R_{l-1} has value k and for the off-center terms the following symmetry holds: $R_j = R_{2l-2-j}$.

3   On the Relation between the Autocorrelation Function and the Probability of Reconstruction Failure

We shall show below that the autocorrelation function of a pattern is a major indicator of the effectiveness of the pattern in extracting information from a target sequence.

3.1   Golomb Rulers

We noted above that, with the exception of c_k, the coefficients of polynomial $q_\pi(m)$ (formula (1)) are pattern-dependent (a probe/site assignment determines specific row-disjunctions in the alignment matrix).

However, if the autocorrelation function is such $R_j \leq 1$ for any off-center value R_j (i.e. for $j \neq l-1$), then the only possible row overlaps occur in correspondence with the central column in position $l-1$, and the coefficients of $q_\pi(m)$ attain their minimum values for a given k. Patterns with this property are known in the literature as *Golomb rulers*[2]. It is immediate that the length l of a Golomb ruler of weight k is bounded below by $\binom{k}{2}+1$, because each off-center column in the alignment matrix may contain at most a single 1[3]. Since any two rows of the alignment matrix of a Golomb ruler overlap only in the central column, all u-site probe/site assignments yield identical, minimal contributions to the coefficient c_u. The value of this contribution is $4^{k-u}/4^{k^2}$, because each assignment leaves the smallest overall number $k-u$ of unconstrained symbols. Recalling that there are $S_{k,u}$ such assignments, the smallest value c_u^* of c_u is therefore $S_{k,u}4^{k-u}/4^{k^2}$. By $q^*(m)$ we shall denote the polynomial $q_\pi(m)$ for a Golomb pattern of weight k, i.e.,

$$q^*(m) = 3m(c_1^*m + \ldots + c_{k-1}^*m^{k-1} + c_k^*m^k).$$

The polynomial $q^*(m)$ can be used to define the gold standard against which weight-k patterns can be compared. Specifically, for a conventional value ϵ of probability of failure (here and hereafter we assume $\epsilon = 0.1$), the equation

$$q^*(m_k^*) = \epsilon$$

defines the length m_k^* of the sequences that can be reconstructed with confidence $1 - \epsilon$. We now develop an estimate of the value m_k^*. It can be verified that

$$\frac{c_u^*m^u}{c_k^*m^k} = \frac{S_{k,u}4^{k-u}}{S_{k,k}}m^{u-k} = S_{k,u}\left(\frac{4}{m}\right)^{k-u}, \tag{2}$$

---

[2] Golomb rulers ave been extensively studied in connection with the design of radar pulses.

[3] However, this lower bound is met only for $k \leq 4$.

Using a convenient upper bound to $S_{k,u}$ of the form[4]

$$S_{k,u} < u^{2(k-u)}$$

we obtain

$$S_{k,u}\left(\frac{4}{m^*}\right)^{k-u} < \left(\frac{u^2}{4^{k-3}}\right)^{k-u}.$$

The latter result shows that only for $u \geq k-1$ the contribution of lower-degree terms should not be neglected, for usual values of k. In fact, it can be verified that, for the usual values $k = 8$ and $m > 6,000$, the two highest degree terms account for a fraction of the value of $q^*(m)$ larger than 0.9999. For this reason we shall neglect in our analysis all other terms except the two of the highest degree. Within this approximation, using formula (2),

$$q^*(m) \approx 3m(c^*_{k-1}m^{k-1} + c^*_k m^k) = 3m^{k+1}c^*_k(1 + \frac{c^*_{k-1}}{mc^*_k})$$

$$= 3m^{k+1}\frac{1}{4^{k2}}\left(1 + \binom{k}{2}\frac{4^{k-(k-1)}}{m}\right)$$

because $S_{k,k-1} = \binom{k}{2}$. If we assume for the moment that $m^*_k > 4^{k-2}$, we have that $\binom{k}{2}\frac{4}{m^*} << 1$, and we may use the simpler approximation

$$m^*_k \approx \left(\frac{4\epsilon}{3}\right)^{\frac{1}{k+1}} 4^{k-1} \approx 0.79 \cdot 4^{k-1}$$

(For example, for $k = 8$ and $\epsilon = 0.1$ we obtain $m^*_k = 13,085$. We also verify that $m^*_k = 0.79 \cdot 4^{k-1} > 4^{k-2}$.)

3.2 Self-matched Patterns

However, the comparative evaluation of different patterns must be done for a fixed choice of $[l, k]$, since this pair is a measure of the cost of a probing scheme

---

[4] Using the well-known recurrence

$$S_{u+s,u} = S_{u-1+s,u-1} + uS_{u-1+s,u}$$

we can obtain a (very conservative) upper-bound to $S_{u+s,u}$ of the form

$$U_{u+s,u} = u^{2s}$$

Indeed, $U_{u,u} = 1$ (since $s = 0$), thus satisfying the boundary condition $S_{u,u} = 1$. Moreover, $U_{1+s,1} = 1$, thus satisfying the boundary condition $S_{1+s,1} = 1$. Finally, $U_{u+s,u}$ satisfies the above recurrence relation, because

$$U_{u+1+s,u+1} = (u+1)^{2s} > u^{2s} + (u+1)^{2s-1} > u^{2s} + u\cdot(u+1)^{2(s-1)} = U_{u+s,u} + uU_{u+s,u+1}$$

(note that, in general, l is expected to be substantially smaller than $\binom{k}{2} + 1$, a lower bound to the achievable length of a weight-k Golomb ruler).

If for a chosen pattern any off-center term R_j is larger than 1, this results in an increase of the values of some coefficient c_u for $u < k$ and consequently $q_\pi(m)$ will achieve the target threshold ϵ for a value of $m < m_k^*$.

We shall now consider the patterns for which $R_j \leq 2$ for $j \neq l - 1$, and call these patterns *self-matched*[5]. Clearly, $R_j = 1$ means that in column j only one of the aligned probes has a 1; analogously, $R_j = 2$ means that in column j there are exactly two aligned probes with a 1 (pairwise overlapping). Since $R_j \leq 2$, no two aligned probes share more than one position beside the central position $l - 1$. Conversely, each $R_j = 2$ is due to a specific pair of aligned probes.

When considering the $\binom{k}{2}$ mappings of k probes to $k - 1$ sites, there is a single site to which any 2 probes are assigned. Therefore, a pair of probes, corresponding to a value 2 in the autocorrelation function, will affect the coefficient c_{k-1} exactly once. The modification is that 2 rather than just one symbols remain unconstrained, resulting in a contribution of

$$\frac{4^2}{4^{k^2}} - \frac{4}{4^{k^2}}$$

to coefficient c_{k-1} with respect to its value for the case of Golomb rulers. As a consequence (our approximation of) $q_\pi(m)$ increases by $3m^k s_2 12/4^{k^2}$ and m decreases by

$$\frac{s_2 \cdot 12}{k + 1}$$

units with respect to the value m_k^*. In addition, for a given length l, a *compact* self-matched pattern is defined as one for which all off-center terms $R_j \in \{1, 2\}$. Such compact pattern has the smallest value of $s_2 = (\binom{k}{2} - l + 1)$. For example, a $[20, 8]$ compact self-matched pattern has a decrease of about 24 units with respect to the optimum $m_8^* \approx 13,085$. This illustrates how close the performance of a self-matched pattern is to that of an equal-weight Golomb ruler.

Although self-matched patterns can be readily discovered by computer search, a synthesis technique is presented as an appendix.

4   Concluding Remarks

In Section 2 we have contrasted the adoptions of solid and gapped probes in sequence reconstruction. We have outlined that reconstruction failure in the case of solid probes occurs because of the non-uniqueness of Eulerian paths in the graph determined by the sequence $(k - 1)$-tuples. We have then discussed the mechanisms of reconstruction failure when adopting gapped probes, and it may now appear surprising that the notion of Eulerian path does not seem to

---

[5] This denotation relates to their suitability for being used as signals in matched-filter radar detection (peaked autocorrelation function).

play any role. In reality, only Failure Mode 1 is the new phenomenon due to the appearance of fooling probes; the equally important Failure Mode 2 can be likened to the branching occurring with solid probes.

Of course, the non-uniqueness of Eulerian paths is a legitimate cause of failure also for gapped probes. An important feature, however, is that gapped-probe reconstruction operates as if the probe length is $l > k$ (typically, $l > 2k$). The emergence of non-unique Eulerian paths depends upon the sequence length, so that reliably reconstructible sequences are of length $O(2^l)$. However, such length is well beyond the range of gapped-probe methods. In fact, as the target sequence length grows, the spectrum becomes increasingly denser, and when the probability of finding *any* probe equal $1/4$, i.e., the *a priori* probability of any symbol, sequence reconstruction becomes a runaway branching process, and failure is due to computational reasons. This critical bound is approximately $1.15 \cdot 4^{k-1}$, and note that $4^{k-1} \ll 2^l$. In addition, self-matched probes (and the structured gapped probes) achieve with 90% confidence a reconstruction length $m \approx 0.794^{k-1}$, which is very close to the computational upper bound.

References

1. Bains, W., Smith, G.C.: A novel method for DNA sequence determination. Jour. of Theoretical Biology 135, 303–307 (1988)
2. Drmanac, R., Labat, I., Bruckner, I., Crkvenjakov, R.: Sequencing of megabase plus DNA by hybridization. Genomics 4, 114–128 (1989)
3. Loakes, D., Brown, D.M.: 5-Nitroindole as a universal base analogue. Nucleic Acids Research 22(20), 4039–4043 (1994)
4. Lysov, Y.P., Florentiev, V.L., Khorlin, A.A., Khrapko, K.R., Shih, V.V., Mirzabekov, A.D.: Sequencing by hybridization via oligonucleotides. A novel method. Dokl. Acad. Sci. USSR 303, 1508–1511 (1988)
5. Pevzner, P.A.: l-tuple DNA sequencing: computer analysis. Journ. Biomolecul. Struct. & Dynamics 7(1), 63–73 (1989)
6. Pevzner, P.A., Lipshutz, R.J.: Towards DNA-sequencing by hybridization. In: Privara, I., Ružička, P., Rovan, B. (eds.) MFCS 1994. LNCS, vol. 841, pp. 143–258. Springer, Heidelberg (1994)
7. Preparata, F.P., Upfal, E.: Sequencing-by-Hybridization at the information-theory bound: An optimal algorithm. Journal of Computational Biology 7(3/4), 621–630 (2000)
8. Preparata, F.P., Upfal, E., Heath, S.A.: Sequence reconstruction from nucleic acid micro-array data. In: Nunnally, B. (ed.) Analytic techniques for DNA sequencing, pp. 177–193. M. Dekker, New York (2005)

Appendix: A Synthesis Technique of Self-matched Patterns

A pattern is a monic polynomial $p(z)$ of degree $l - 1$. For some integer $s < l - 1$ we may express $p(z)$ in the following form:

$$p(z) = p_0(z) + z^s p_1(z)$$

where both $p_0(z)$ and $p_1(z)$ are monic polynomials, of respective degrees $d_0 < s$ and $d_1 = l - s$, with $d_0 + d_1 \leq l - 2$. The autocorrelation $R_p(z)$ of $p(z)$ has the expression:

$$R_p(z) = (p_0(z) + z^s p_1(z)) z^{l-1} \left(p_0 \left(\frac{1}{z} \right) + z^{-s} p_1 \left(\frac{1}{z} \right) \right)$$

$$= z^{l-1-d_0} p_0(z) z^{d_0} p_0 \left(\frac{1}{z} \right) + z^{l-1-d_1} p_1(z) z^{d_1} p_1 \left(\frac{1}{z} \right)$$

$$+ z^{l-1-s} p_0(z) p_1 \left(\frac{1}{z} \right) + z^{l-1+s} p_0 \left(\frac{1}{z} \right) p_1(z)$$

$$= z^{l-1-d_0} R_{p_0}(z) + z^{l-1-d_1} R_{p_1}(z) + z^{l-1-s} p_0(z) p_1 \left(\frac{1}{z} \right) + z^{l-1+s} p_0 \left(\frac{1}{z} \right) p_1(z).$$

If we define $g_{01}(z) = p_0(z) p_1(1/z) z^{d_1}$ (a monic polynomial of degree $d_0 + d_1$), we have $p_1(z) p_0(1/z) z^{d_0} = z^{d_0+d_1} g_{01}(1/z) = g_{10}(z)$ (also a monic polynomial of degree $d_0 + d_1$, obtained by reversing the order of the coefficients of $g_{01}(z)$) and we can write:

$$R_p(z) = z^{l-1-d_0} R_{p_0}(z) + z^{l-1-d_1} R_{p_1}(z) + z^{l-1-s-d_1} g_{01}(z) + z^{l-1+s-d_0} g_{10}(z).$$

This equation is the key to our construction technique.

Since $R_p(z)$, an autocorrelation function, has the property $R_{p,l-1-i} = R_{p,l-1+i}$, it is sufficient to consider the coefficients of the terms of degree $\geq l - 1$. (Note that $z^{l-1-s-d_1} g_{01}(z)$ and $z^{l-1+s-d_0} g_{10}(z)$ do not interfere, since the rightmost term of the former has degree $l - 1 - s + d_0 i < l - 1$ and the leftmost term of the latter has degree $l - 1 + s - d_0 > l - 1$, by the hypothesis that $d_0 < s$.) In other words, we factor from $R_p(z)$ the power z^{l-1} and retain only the terms of degree ≥ 0 (thereby dropping polynomial $g_{01}(z)$). Such construction yields the polynomial

$$\hat{R}_p(z) = \hat{R}_{p_0}(z) + \hat{R}_{p_1}(z) + z^{s-d_0} g_{10}(z)$$

where we have denoted the truncated polynomials with a "hat" diacritic.

A first obvious necessary condition is that none of the above three terms violate the condition of a self-matched pattern, i.e., none of their coefficients pertaining to degrees ≥ 1 exceeds the value 2.

Sufficient conditions are that

1. The sum $\hat{R}_{p_0} + \hat{R}_{p_1}$, also verifies the stated condition on its coefficients (in which case we say that the two patterns are *compatible*).
2. The parameter s has been chosen as the smallest value for which the polynomial $\hat{R}_{p_0} + \hat{R}_{p_1} + z^{s-d_0} g_{10}(z)$ does not violate the coefficient condition.

Example 3: We express polynomials as strings of coefficients. Let $p_0 = 1101$ and $p_1 = 100011$ (so that $d_0 = 3$ and $d_1 = 5$). Next we obtain $R_{p_0} = 1113111$ ($\hat{R}_{p_0} = 3111$) and $R_{p_1} = 11001310011$ ($\hat{R}_{p_1} = 310011$), so that both p_0 and p_1 are self-matched patterns (they are, actually, Golomb rulers and, therefore, automatically compatible), thereby obtaining $\hat{R}_{p_0} + \hat{R}_{p_1} = 621111$.

Finally, $g_{10} = 101111121$ so that to avoid violation of the property g_{10} must be shifted 2 positions to the right (yielding $s - d_0 = 2$, or $s = 5$) to obtain for $\hat{R}(z)$ the sequence

$$
\begin{array}{l}
2\,1\,1\,1 \\
3\,1\,0\,0\,1\,1 \\
\underline{1\,0\,1\,1\,1\,1\,1\,2\,1} \\
\\
6\,2\,2\,1\,2\,2\,1\,1\,1\,2\,1
\end{array}
$$

corresponding to the pattern $p = 11010100011$.

Part III
Combinatorial Optimization with Applications

Algorithms for Energy Saving*

Susanne Albers**

Department of Computer Science, University of Freiburg
Georges Koehler Allee 79, 79110 Freiburg, Germany
`salbers@informatik.uni-freiburg.de`

Abstract. Energy has become a scarce and expensive resource. There
is a growing awareness in society that energy saving is a critical issue.
This paper surveys algorithmic solutions to reduce energy consumption
in computing environments. We focus on the system and device level.
More specifically, we study power-down mechanisms as well as dynamic
speed scaling techniques in modern microprocessors.

Keywords: Dynamic speed scaling, power-down mechanisms, scheduling, competitive analysis, probabilistic analysis, approximation
algorithms.

1 Introduction

With increasing CPU clock speeds and higher levels of integration in processors,
memories and controllers, power consumption has become a major concern in
computer system design over the past years. Power dissipation is critical in bat-
tery operated mobile computing devices that have proliferated in recent years. In
these devices, obviously, the amount of available energy is severely limited. Moreover, power consumption is a major concern in desktop computers and servers.
Electricity costs impose a substantial strain on the budget of data and computing centers, where servers and, in particular, CPUs account for 50–60% of the
energy consumption. In fact, Google engineers, maintaining thousands of servers,
recently warned that if power consumption continues to grow, power costs can
easily overtake hardware costs by a large margin [11]. In addition to cost, energy
dissipation causes thermal problems. Most of the consumed energy is converted
into heat, resulting in wear and reduced reliability of hardware components.

For these reasons, there has recently been considerable research interest in
the design and analysis of *energy-efficient algorithms* that reduce the energy
consumption while minimizing compromise to service. This survey focuses on
energy saving mechanisms on the system and device level. In this context, there
are basically two techniques to save energy.

---

* An extended and modified version of this survey, aiming at a different audience, will
appear in the *Communications of the ACM*.
** Work supported by a Gottfried Wilhelm Leibniz Award of the German Research
Foundation.

S. Albers, H. Alt, and S. Näher (Eds.): Festschrift Mehlhorn, LNCS 5760, pp. 173–186, 2009.

(1) *Power-down mechanisms*: When a system is idle, it can be transitioned into low-power standby or sleep states. This technique is well-known and widely used to save energy. One has to find out when to shut down a system, taking into account that a transition back to the active mode requires extra energy.

(2) *Speed scaling*: Microprocessors currently sold by chip makers such as AMD and Intel are able to operate at variable speed. The higher the speed, the higher the power consumption is. The goal is to save energy by utilizing the full speed/frequency spectrum of a processor and applying low speeds whenever possible.

The power management problems described above are *online problems* in that a system is usually not aware of future events. A power-down mechanism, during an idle period, usually has no information when the period ends. Is it worthwhile to move to a lower-power state and benefit from the reduced energy consumption, given that the system must finally be powered up again at a cost to the active mode? A speed scaling algorithm typically does not know future jobs. Should lower speed levels be used at the expense of delaying the service of tasks that may arrive in the near future?

Despite the handicap of not knowing the future, an online strategy should achieve a provably good performance. Here we resort to *competitive analysis* [29], where an online algorithm *ALG* is compared to an optimal offline algorithm *OPT* that knows the entire future and can compute an optimal solution. Online algorithm *ALG* is called *c-competitive* if, for every input, the total energy consumption of *ALG* is at most c times that of *OPT*.

In this survey we first present the most important results known for power-down mechanisms. Then we address dynamic speed scaling algorithms.

2   Power-Down Mechanisms

Power-down mechanisms are a common technique to save energy. We encounter them on an every day basis. The display of our desktop turns off after some period of inactivity. Our laptop transitions to a standby or hibernate mode if it has been idle for a while. In these settings, there usually exist idleness thresholds that specify the length of time after which a system is powered down. From an algorithmic point of view, we would like to design strategies that determine such thresholds and perform well relative to the optimum.

Formally, we are given a device that always resides in one of several states. In addition to the active state, there can be several standby and sleep modes. These states have individual power consumption rates. The energy incurred in transitioning the system from a higher-power to a lower-power state is usually negligible. However, a power-up operation consumes a significant amount of energy. Over time the device experiences an alternating sequence of active and idle periods. During active periods, the system must reside in the active mode to perform the required tasks. During idle periods, the system may be moved to lower-power states. An algorithm has to decide when to perform the transitions and to which states to move. The goal is to minimize the total energy consumption. As the energy consumption during the active periods is fixed, assuming that prescribed tasks have to be performed, we concentrate on energy minimization in the idle intervals. In fact, we focus on any idle period and optimize the energy consumption in any such time window.

In the following we will first study systems that consist of two states only. Then we will address systems with multiple states. We stress that we consider the minimization of energy. We ignore the delay that arises when a system is transitioned from a lower-power to a higher-power state.

2.1   Systems with Two States

Consider a two-state system that may reside in an active state or in a sleep state. We assume without loss of generality that the power consumption rate in the active state is 1, i.e. the system consumes one energy unit per time unit. The power consumption rate in the sleep mode is 0. The results we present in the following generalize to arbitrary consumption rates. Suppose that β, $\beta > 0$, energy units are required to transition the system from the sleep state to the active state. The energy of transitioning from the active to the sleep state is assumed to be 0. If this is not the case, we can simply fold the corresponding energy into the cost of β incurred in the next power-up operation. The system experiences an idle period whose length T is initially unknown.

We first observe that an optimal offline algorithm OPT, knowing T in advance, is simple to formulate. If the value of T, counted in time units, is smaller than the value of β, OPT remains in the active state throughout the idle period. If T is at least β, OPT transitions to the sleep state right at the beginning of the idle period and powers up to the active state at the end of the period.

The following deterministic online algorithm mimics the behavior of OPT.

Algorithm ALG-D: In an idle period, remain in the active state first. After β time units, if the period has not ended yet, transition to the sleep state.

Theorem 1. *ALG-D is 2-competitive and this is the smallest competitiveness a deterministic online algorithm can achieve.*

Proof. We first analyze $ALG\text{-}D$ and consider two cases. If the value of T is smaller than the value of β, then $ALG\text{-}D$ consumes T units of energy during the idle interval and this is in fact equal to the consumption of OPT. If T is at least β, then $ALG\text{-}D$ first consumes β energy units to remain in the active state. An additional power-up cost of β is incurred at the end of the idle interval. Hence, $ALG\text{-}D$'s total cost is 2β, while OPT incurs a cost of β for the power-up operation at the end of the idle period.

We next verify that no deterministic online algorithm can achieve a competitive ratio smaller than 2. If an algorithm transitions to the sleep state after exactly t time units, then in idle period of length t it incurs a cost of $t + \beta$ while OPT pays $\min\{t, \beta\}$ only.   □

We remark that power management in two-state systems corresponds to the famous ski-rental problem, a cornerstone problem in the theory of online algorithms, see e.g. [16].

Interestingly, it is possible to beat the competitiveness of 2 using randomization. A randomized algorithm transitions to the sleep state according to a probability density function $p(t)$. The probability that the system powers down during the first t_0 time units of an idle period is $\int_0^{t_0} p(t)dt$. Karlin et al. [20] determined the best probability distribution.

Algorithm ALG-R: Transition to the sleep state according to the probability density function

$$p(t) = \begin{cases} \frac{1}{(e-1)\beta}e^{t/\beta} & 0 \leq t \leq \beta \\ 0 & \text{otherwise.} \end{cases}$$

Theorem 2. [20] *ALG-R achieves a competitive ratio of $\frac{e}{e-1}$, and this is the smallest competitive ratio a randomized strategy can achieve.*

Here $e \approx 2.71$ is the Eulerian number and hence $\frac{e}{e-1} \approx 1.58$, which is considerably below the deterministic bound of 2.

From a practical point of view, it is also instructive to study stochastic settings where the length of idle periods is governed by probability distributions. In practice, short periods might occur more frequently. Probability distributions can also model specific situations where either very short or very long idle periods are more likely to occur, compared to periods of medium length. Of course, such a probability distribution may not be known in advance but can be learned over time.

Let $Q = (q(T))_{0 \leq T < \infty}$ be a fixed probability distribution on the length T of idle periods. For any $t \geq 0$, the deterministic algorithm ALG_t that always powers down after exactly t time units incurs an expected cost of

$$E[ALG_t(Q)] = \int_0^t Tq(T)dT + (t+\beta)\int_t^\infty q(T)dT \qquad (1)$$

on idle periods generated according to Q. Karlin et al. [20] proposed the following strategy to handle probabilistic settings.

Algorithm ALG-P: Given a fixed Q, let A_Q^* be the deterministic algorithm ALG_t that minimizes (1).

Theorem 3. [20] *For any Q, the expected energy consumption of ALG-P is at most $\frac{e}{e-1}$ times the expected optimum consumption.*

2.2 Systems with Multiple States

Many modern devices do not have only one but several low-power states. Specifications of such systems are given, for instance, in the Advanced Configuration and Power Management Interface (ACPI) that establishes industry-standard interfaces enabling power management and thermal management of mobile, desktop

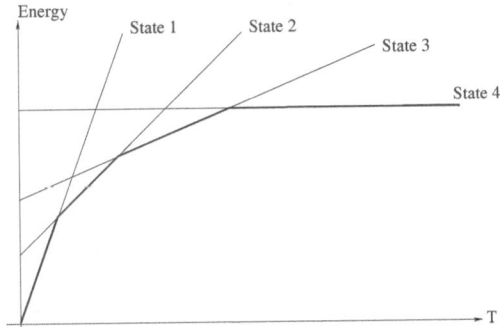

Fig. 1. Illustration of the optimum cost in a four-state system

and server platforms. A description of the ACPI power management architecture built into Microsoft Windows operating systems can be found at [1].

Consider a system with ℓ states $s_1, \ldots, s_\ell$. Let r_i be the power consumption rate of s_i. We number the states such that $r_1 > \ldots > r_\ell$. Hence s_1 is the active state and s_ℓ represents the state with lowest energy consumption. Let β_i be the energy required to transition the system from s_i to the active state s_1. We assume again that transitions from higher-power to lower-power states incur 0 cost because the corresponding energy is usually negligible. The goal is to construct a state transition schedule minimizing the total energy consumption in an idle period.

Irani et al. [17] presented online and offline algorithms. They first observe that the total energy incurred by an optimal offline algorithm OPT in an idle period of length T is given by

$$OPT(T) = \min_{1 \le i \le \ell} \{r_i T + \beta_i\}.$$

Hence, OPT chooses the state that allows for the smallest total cost consisting of energy consumption in the period and final power-up cost. Interestingly, the optimal cost has a simple graphical representation, see Figure 1. If we consider all linear functions $f_i(t) = r_i t + \beta_i$, then the optimum energy consumption is given by the lower envelope of the arrangement of lines.

We can use this lower envelope to guide an online algorithm which state to use at any time. Let $S_{OPT}(t)$ denote the state used by OPT in an idle period of total length t, i.e. $S_{OPT}(t)$ is the state $\arg\min_{1 \le i \le \ell} \{r_i t + \beta_i\}$. The following algorithm, proposed in [17], traverses the state sequence as suggested by the optimum offline algorithm.

Algorithm Lower-Envelope: In an idle period, at any time t, use state $S_{OPT}(t)$.

Intuitively, over time, *Lower-Envelope* visits the states represented by the lower envelope of the functions $f_i(t)$. If currently in state s_{i-1}, the strategy transitions to the next state s_i at time t_i, where t_i is the solution to the equation

$r_{i-1}t + \beta_{i-1} = r_i t + \beta_i$. Here we assume that states whose functions do not occur on the lower envelope, at any time, are discarded. Note that the algorithm is a generalization of ALG-D for two-state systems.

Theorem 4. [17] *Lower-Envelope is 2-competitive.*

The competitiveness of 2 is the smallest ratio achievable by a deterministic online algorithm.

Irani et al. [17] also studied the setting where the length of idle periods is generated by a probability distribution $Q = (q(T))_{0 \leq T < \infty}$. They determined the time t_i when an online strategy should move from state s_{i-1} to s_i, $2 \leq i \leq \ell$. Let t_i be the time t that minimizes

$$\int_0^t r_{i-1} T q(T) dT + \int_t^\infty q(T)(r_{i-1}t + (T - t)r_i + \beta_i - \beta_{i-1}) dT.$$

Intuitively, the above expression is the expected cost of a deterministic algorithm ALG_t that powers down after t time units, assuming that only states s_{i-1} and s_i are available.

Algorithm ALG-P(ℓ): Change states at the transition times $t_2, \ldots, t_\ell$ defined above.

Note that ALG-$P(\ell)$ is a generalization of ALG-P for two-state systems.

Theorem 5. [17] *For any fixed probability distribution Q, the expected energy consumption of ALG-$P(\ell)$ is at most $\frac{e}{e-1}$ times the expected optimum consumption.*

Furthermore, Irani et al. presented an approach how to learn an initially unknown Q. They combined the approach with ALG-$P(\ell)$ and performed experimental tests for an IBM mobile hard drive with four power states. It shows that the combined scheme achieves low energy consumption close to the optimum and usually outperforms many single-value prediction algorithms.

Augustine et al. [4] investigate generalized multi-state systems in which the state transition energies may take arbitrary values. Let $\beta_{ij} \geq 0$ be the energy required to transition from s_i to s_j, $1 \leq i, j \leq \ell$. Augustine et al. demonstrate that *Lower-Envelope* can be generalized and achieves a competitiveness of $3 + 2\sqrt{2} \approx 5.8$. This ratio holds for any state system. Better bounds are possible for specific system. Augustine et al. devise a strategy that, for a given system S, achieves a competitive ratio of $c^* + \epsilon$, for any $\epsilon > 0$, where c^* is the best competitiveness possible for S. Finally, the authors consider stochastic settings and develop optimal state transition times.

3   Dynamic Speed Scaling

Many modern microprocessor can run at variable speed. Examples are the Intel SpeedStep and the AMD processor PowerNow. High speeds result in higher performance but also high energy consumption. Lower speeds save energy but performance degrades. If the processor runs at speed s, then the required power is s^α, where $\alpha > 1$ is a constant. In practical application α is usually in the range between 2 and 3. The well-known cube-root rule states that the power is proportional to s^3. Obviously, energy consumption is power integrated over time. The goal is to dynamically set the speed of a processor so as to minimize energy consumption, while still providing a desired quality of service.

Dynamic speed scaling leads to many challenging scheduling problems. At any time a scheduler has to decide not only which job to execute but also which speed to use. Consequently, there has been considerable research interest in the design and analysis of efficient scheduling algorithms. This section reviews the most important results developed over the past years. We first address scheduling problems with hard job deadlines. Then we consider the minimization of response times and other objectives.

In general, two scenarios are of interest. In the offline setting all jobs to be processed are known in advance. In the online setting jobs arrive over time and an algorithm, at any time, has to make scheduling decisions without knowledge of any future jobs. Recall that an online algorithm ALG is c-competitive if, for every input, the objective function value (typically the energy consumption) of ALG is within c times the value of an optimal solution.

3.1   Scheduling with Deadlines

In a seminal paper, initiating the algorithmic study of speed scaling, Yao, Demers and Shenker [30] investigated a scheduling problem with strict job deadlines. At this point this framework is by far the most extensively studied algorithmic speed scaling problem.

Consider n jobs $J_1, \ldots, J_n$ that have to be processed on a variable-speed processor. Each job J_i is specified by a release time r_i, a deadline d_i and a processing volume w_i. The release time and the deadline mark the time interval in which the job must be executed. The processing volume is the amount of work that must be done to complete the job. The processing time of a job depends on the speed. If J_i is executed at constant speed s, it takes w_i/s time units to finish the job. Preemption of jobs is allowed, i.e. the processing of a job may be suspended and resumed later. The goal is to construct a feasible schedule minimizing the total energy consumption.

The framework by Yao et al. assumes that there is no upper bound on the maximum processor speed. Hence there always exists a feasible schedule satisfying all job deadlines. Furthermore, it is assumed that a continuous spectrum of speeds is available. We will discuss later how to relax these assumptions.

Fundamental Algorithms. Yao, Demers and Shenker [30] first study the offline setting and develop an algorithm for computing optimal solutions, minimizing total energy consumption. The strategy is known as YDS referring to the initials of the authors. The algorithm proceeds in series of iterations. In each iteration, a time interval of maximum density is identified and a corresponding partial schedule is constructed. The *density* Δ_I of a time interval $I = [t, t']$ is the total work to be completed in I divided by the length of I. More precisely, let S_I be the set of jobs J_i that must be processed in I, i.e. that satisfy $[r_i, d_i] \subseteq I$. Then

$$\Delta_I = \frac{1}{|I|} \sum_{J_i \in S_i} w_i.$$

Intuitively, Δ_I is the minimum average speed necessary to complete all jobs that must be scheduled in I.

Algorithm *YDS* repeatedly determines the interval I of maximum density. In such an interval I the algorithm schedules the jobs of S_I at speed Δ_I using the *Earliest Deadline First (EDF)* policy, i.e. among the available unfinished jobs the one with the earliest deadline is executed. Then *YDS* removes the set S_I as well as the time interval I from the problem instance. More specifically, for any unscheduled job J_i with $d_i \in I$, the new deadline time is set to $d_i := t$. For any unscheduled J_i with $r_i \in I$, the new release time is $r_i := t'$. Time interval I is discarded. We give a summary of the algorithm in pseudo-code.

Algorithm YDS: Initially $\mathcal{J} := \{J_1, \ldots, J_n\}$. While $\mathcal{J} \neq \emptyset$, execute the following two steps. (1) Determine the interval I of maximum density. In I process the jobs of S_I at speed Δ_I according to *EDF*. (2) Set $\mathcal{J} := \mathcal{J} \setminus S_I$. Remove I from the time horizon and update the release times and deadlines of unscheduled jobs accordingly.

Theorem 6. [30] *For any job instance, YDS computes an optimal schedule minimizing the total energy consumption.*

Obviously, when identifying intervals of maximum density, *YDS* only has to consider intervals whose boundaries are equal to the release times and deadlines of the jobs. A straightforward implementation of the algorithm has a running time of $O(n^3)$. Li et al. [25] showed that the time can be reduced to $O(n^2 \log n)$. Further improvements are possible if the job execution intervals form a tree structure [24].

Yao et al. [30] also devised two elegant online algorithms, called *Average Rate* and *Optimal Available*. Whenever a new job J_i arrives at time r_i, its deadline d_i and processing volume w_i are known. For any incoming job J_i, *Average Rate* considers the *density* $\delta_i = w_i/(d_i - r_i)$, which is the minimum average speed necessary to complete the job in time if no other jobs were present. At any time t the speed $s(t)$ is set to the accumulated density of jobs active at time t. A job J_i is *active at time* t if $t \in [r_i, d_i]$. Available jobs are scheduled according to the *EDF* policy.

Algorithm Average Rate: At any time t use a speed of $s(t) = \sum_{J_i : t \in [r_i, d_i]} \delta_i$. Available unfinished jobs are scheduled using *EDF*.

Yao et al. [30] analyzed *Average Rate* and proved an upper bound on the competitiveness.

Theorem 7. [30] *The competitive ratio of Average Rate is at most $2^{\alpha-1}\alpha^\alpha$, for any $\alpha \geq 2$.*

Bansal et al. [5] showed that the analysis is essentially tight by providing a nearly matching lower bound.

Theorem 8. [5] *The competitive ratio of Average Rate is at least* $((2-\delta)\alpha)^\alpha/2$, *where* δ *is a function of* α *that approaches zero as* α *tends to infinity.*

The second strategy *Optimal Available* is computationally more expensive than *Average Rate*. It always computes an optimal schedule for the currently available work load. This can be done using *YDS*.

Algorithm Optimal Available: Whenever a new job arrives, compute an optimal schedule for the currently available unfinished jobs.

Bansal, Kimbrel and Pruhs [8] gave a comprehensive analysis of the above algorithm and proved the following result.

Theorem 9. [8] *The competitive ratio of Optimal Available is exactly* α^α.

The above theorem implies that in terms of competitiveness, *Optimal Available* is better than *Average Rate*. Bansal et al. [8] also presented a new online algorithm, called *BKP* according to the initials of the authors, that approximates the optimal speeds of *YDS* by considering interval densities. For times t, t_1 and t_2 with $t_1 < t \leq t_2$, let $w(t, t_1, t_2)$ be the total processing volume of jobs that are active at time t, have a release time of at least t_1 and a deadline of at most t_2.

Algorithm BKP: At any time t use a speed of

$$s(t) = \max_{t' > t} \frac{w(t, et - (e-1)t', t')}{t' - t}.$$

Available unfinished jobs are processed using *EDF*.

Theorem 10. [8] *Algorithm BKP achieves a competitive ratio of* $2(\frac{\alpha}{\alpha-1})^\alpha e^\alpha$.

For large values of α, the competitiveness of *BKP* is better than that of *Optimal Available*.

All the above online algorithms attain constant competitive ratios that depend on α but no other other problem parameter. The dependence on α is exponential. For small values of α, which occur in practice, the competitive ratios are reasonably small. A result by Bansal et al. [8] implies that the exponential dependence on α is inherent to the problem.

Theorem 11. [8] *Any randomized online algorithm has a competitiveness of at least* $\Omega((4/3)^\alpha)$.

Refinements. *Bounded speed:* The problem setting considered so far assumes a continuous, unbounded spectrum of speeds. However, in practice only a finite set of discrete speed levels $s_1 < s_2 < \ldots < s_d$ is available. The offline algorithm *YDS* can be adapted easily to handle feasible job instances, i.e. inputs for which feasible schedules exist using the restricted set of speeds. Note that feasibility can be checked easily by always using the maximum speed s_d and scheduling available jobs according to the *EDF* policy. Given a feasible job instance, the modification of *YDS* is as follows. We first construct the schedule according to *YDS*. For each identified interval I of maximum density we approximate the desired speed Δ_I by the two adjacent speed levels $s_k < \Delta_I < s_{k+1}$. Speed s_{k+1} is used first for some δ time units and s_k is used for the last $|I| - \delta$ time units in I, where δ is chosen such that the total work completed in I is equal to the original amount of $|I|\Delta_I$. An algorithm with an improved running time of $O(dn \log n)$ was presented by Li and Yao [26].

If the given job instance is not feasible, the situation is more delicate. In this case it is impossible to complete all the jobs. The goal is to design algorithms that achieve good *throughput*, which is the total processing volume of jobs finished by their deadline, and at the same time optimize energy consumption. Papers [6, 13] present algorithms that even work online. At any time the strategies maintain a pool of jobs they intend to complete. Newly arriving jobs may be admitted to this pool. If the pool contains too large a processing volume, jobs are expelled such that the throughput is not diminished significantly. The algorithm by Bansal et al. [6] is 4-competitive in terms of throughput and constant competitive with respect to energy consumption.

Temperature minimization: High processor speeds lead to high temperatures which impair a processor's reliability and lifetime. Bansal et al. [8] consider the minimization of the maximum temperature that arises during processing. They assume that cooling follows Newton's law, which states that the rate of cooling of a body is proportional to the difference in temperature between the body and the environment. Bansal et al. [8] show that algorithms *YDS* and *BKP* have favorable properties. For any jobs sequence, the maximum temperature is within a constant factor of the minimum possible maximum temperature, for any cooling parameter a device may have.

Sleep states: Irani et al. [19] investigate an extended problem setting where a variable-speed processor may be transitioned into a sleep state. In the sleep state, the energy consumption is 0 while in the active state even at speed 0 some non-negative amount of energy is consumed. Hence [19] combines speed scaling with power-down mechanisms. In the standard setting without sleep state, algorithms tend to use low speed levels subject to release time and deadline constraints. In contrast, in the setting with sleep state it can be beneficial to speed up a job so as to generate idle times in which the processor can be transitioned to the sleep mode. Irani et al. [19] develop online and offline algorithms for this extended setting. Baptiste et al. [10] and Demaine et al. [15] also study scheduling problems where a processor may be set asleep, albeit in a setting without speed scaling.

3.2 Minimizing Response Time

A classical objective in scheduling is the minimization of response times. A user releasing a task to a system would like to receive feedback, say the result of a computation, as quickly as possible. User satisfaction often depends on how fast a device reacts. Unfortunately, response time minimization and energy minimization are contradicting objectives. To achieve fast response times a system must usually use high processor speeds, which lead to high energy consumption. On the other hand, to save energy low speeds should be used, which result in high response times. Hence one has to find ways to integrate both objectives.

Consider n jobs $J_1, \ldots, J_n$ that have to be scheduled on a variable-speed processor. Each job J_i is specified by a release time r_i and a processing volume w_i. When a job arrives, its processing volume is known. Preemption of jobs is allowed. In the scheduling literature, response time is referred to as *flow time*. The flow time f_i of a job J_i is the length of the time interval between release time and completion time of the job. We seek schedules minimizing the total flow time $\sum_{i=1}^{n} f_i$.

Limited energy: Pruhs et al. [27] assume that a fixed energy volume E is given and the goal is to minimize the total flow time of the jobs. The authors consider unit-size jobs, i.e. all jobs have the same processing volume, and study the offline scenario where all the jobs are known in advance. Pruhs et al. [27] show that optimal schedules can be computed in polynomial time. However, in this framework with a limited energy volume it is hard to construct good online algorithms. If future jobs are unknown, it is unclear how much energy to invest for the currently available tasks.

Energy plus flow times: Albers and Fujiwara [2] proposed another approach to integrate energy and flow time minimization. They consider a combined objective function that simply adds the two costs. Let E denote the energy consumption of a schedule. We wish to minimize $g = E + \sum_{i=1}^{n} f_i$. Albers and Fujiwara concentrate on unit-size jobs and show that optimal offline schedules can be constructed in polynomial time using a dynamic programming approach. In fact the algorithm can also be used to minimize the total flow time of jobs given a fixed energy volume.

Most of [2] is concerned with the online setting where jobs arrive over time. Albers and Fujiwara present a simple online strategy that processes jobs in batches and achieves a constant competitive ratio. Batched processing allows one to make scheduling decisions, which are computationally expensive, only every once in a while. This is certainly an advantage in low-power computing environments. Nonetheless, Albers and Fujiwara conjectured that the following algorithm achieves a better performance with respect to the minimization of g: At any time, if there are ℓ active jobs, use speed $\sqrt[\alpha]{\ell}$. A job is active if it has been released but is still unfinished. This algorithm and variants thereof have been the subject of extensive analyses [6, 7, 9, 23], not only for unit-size but also for arbitrary size jobs. Moreover, unweighted and weighted flow times have been considered.

The currently best result is due to Bansal et al. [7]. They modify the above algorithm slightly by using a speed of $\sqrt[\alpha]{\ell + 1}$ whenever ℓ jobs are active. Inspired by a paper of Lam et al. [23] they apply the *Shortest Remaining Processing Time*

(SRPT) policy to the available jobs. More precisely, among the active jobs, the one with the least remaining work is scheduled.

Algorithm Job Count: At any time if there are $\ell \geq 1$ active jobs, use speed $\sqrt[\alpha]{\ell+1}$. If no job is available, use speed 0. Always schedule the job with the least remaining unfinished work.

Theorem 12. [7] *Algorithm* Job Count *is 3-competitive for arbitrary size jobs.*

Further work considering the weighted flow time in objective function g can be found in [7, 9]. Moreover, [6, 23] propose algorithms for the setting that there is an upper bound on the maximum processor speed.

All the above results assume that when a job arrives, its processing volume is known. Papers [14, 23] investigate the harder case that this information is not available.

3.3   Extensions and Other Objectives

Parallel processors: The results presented so far address single-processor architectures. However, energy consumption is also a major concern in multi-processor environments. Currently, relatively few results are known. Albers et al. [3] investigate deadline-based scheduling on m identical parallel processors. The goal is to minimize the total energy on all the machines. The authors first settle the complexity of the offline problem by showing that computing optimal schedules is NP-hard, even for unit-size jobs. Hence, unless $P \neq NP$, optimal solutions can not be computed efficiently. Albers et al. [3] then develop polynomial time offline algorithms that achieve constant factor approximations, i.e. for any input the consumed energy is within a constant factor of the true optimum. They also devise online algorithms attaining constant competitive ratios. Lam et al. [21] study deadline-based scheduling on two speed-bounded processors. They present a strategy that is constant competitive in terms of throughput maximization and energy minimization.

Bunde [12] investigates flow time minimization in multi-processor environments, given a fixed energy volume. He presents hardness results as well as approximation guarantees for unit-size jobs. Lam et al. [22] consider the objective function of minimizing energy plus flow times. They design online algorithms achieving constant competitive ratios.

Makespan minimization: Another basic objective function in scheduling is makespan minimization, i.e. the minimization of the point in time when the entire schedule ends. Bunde [12] assumes that jobs arrive over time and develops algorithms for single and multi-processor environments. Pruhs et al. [28] consider tasks having precedence constraints defined between them. They devise algorithms for parallel processors given a fixed energy volume.

4   Conlusions

This article has surveyed algorithmic approaches to save energy. Another survey on algorithmic problems in power management was written by Irani and

Pruhs [18]. The past months have witnessed considerable research activity and it is conceivable that energy conservation from an algorithmic point of view will continue to be an active area of investigation. Many open problems remain. With respect to power-down mechanisms, for instance, it would be interesting to design strategies that take into account the latency that arises when a system is transitioned from a sleep state to the active state. As for speed scaling techniques, we need a better understanding of strategies for multi-processor environments as multi-core architectures become more and more common not only in servers but also in desktops and laptops.

References

1. http://www.microsoft.com/whdc/system/pnppwr/powermgmt/default.mspx
2. Albers, S., Fujiwara, H.: Energy-efficient algorithms for flow time minimization. ACM Transactions on Algorithms 3 (2007)
3. Albers, S., Müller, F., Schmelzer, S.: Speed scaling on parallel processors. In: Proc. 19th ACM Symposium on Parallelism in Algorithms and Architectures, pp. 289–298 (2007)
4. Augustine, J., Irani, S., Swamy, C.: Optimal power-down strategies. SIAM Journal on Computing 37, 1499–1516 (2008)
5. Bansal, N., Bunde, D.P., Chan, H.-L., Pruhs, K.: Average rate speed scaling. In: Laber, E.S., Bornstein, C., Nogueira, L.T., Faria, L. (eds.) LATIN 2008. LNCS, vol. 4957, pp. 240–251. Springer, Heidelberg (2008)
6. Bansal, N., Chan, H.-L., Lam, T.-W., Lee, L.-K.: Scheduling for speed bounded processors. In: Aceto, L., Damgård, I., Goldberg, L.A., Halldórsson, M.M., Ingólfsdóttir, A., Walukiewicz, I. (eds.) ICALP 2008, Part I. LNCS, vol. 5125, pp. 409–420. Springer, Heidelberg (2008)
7. Bansal, N., Chan, H.-L., Pruhs, K.: Speed scaling with an arbitrary power function. In: Proc. 20th ACM-SIAM Symposium on Discrete Algorithm, pp. 693–701 (2009)
8. Bansal, N., Kimbrel, T., Pruhs, K.: Speed scaling to manage energy and temperature. Journal of the ACM 54 (2007)
9. Bansal, N., Pruhs, K., Stein, C.: Speed scaling for weighted flow time. In: Proc. 18th Annual ACM-SIAM Symposium on Discrete Algorithms, pp. 805–813 (2007)
10. Baptiste, P., Chrobak, M., Dürr, C.: Polynomial time algorithms for minimum energy scheduling. In: Arge, L., Hoffmann, M., Welzl, E. (eds.) ESA 2007. LNCS, vol. 4698, pp. 136–150. Springer, Heidelberg (2007)
11. Barroso, L.A.: The price of performance. ACM Queue 3 (2005)
12. Bunde, D.P.: Power-aware scheduling for makespan and flow. In: Proc. 18th Annual ACM Symposiun on Parallel Algorithms and Architectures, pp. 190–196 (2006)
13. Chan, H.-L., Chan, W.-T., Lam, T.-W., Lee, K.-L., Mak, K.-S., Wong, P.W.H.: Energy efficient online deadline scheduling. In: Proc. 18th Annual ACM-SIAM Symposium on Discrete Algorithms, pp. 795–804 (2007)
14. Chan, H.-L., Edmonds, J., Lam, T.-W., Lee, L.-K., Marchetti-Spaccamela, A., Pruhs, K.: Nonclairvoyant speed scaling for flow and energy. In: Proc. 26th International Symposium on Theoretical Aspects of Computer Science, pp. 255–264 (2009)
15. Demaine, E.D., Ghodsi, M., Hajiaghayi, M.T., Sayedi-Roshkhar, A.S., Zadimoghaddam, M.: Scheduling to minimize gaps and power consumption. In: Proc. 19th Annual ACM Symposium on Parallel Algorithms and Architectures, pp. 46–54 (2007)

16. Irani, S., Karlin, A.R.: Online computation. In: Hochbaum, D. (ed.) Approximation Algorithms for NP-Hard Problems, pp. 521–564. PWS Publishing Company (1997)
17. Irani, S., Shukla, S.K., Gupta, R.K.: Online strategies for dynamic power management in systems with multiple power-saving states. ACM Transaction in Embedded Computing Systems 2, 325–346 (2003)
18. Irani, S., Pruhs, K.: Algorithmic problems in power management. SIGACT News 36, 63–76 (2005)
19. Irani, S., Shukla, S.K., Gupta, R.: Algorithms for power savings. ACM Transactions on Algorithms 3 (2007)
20. Karlin, A.R., Manasse, M.S., McGeoch, L.A., Owicki, S.S.: Competitive randomized algorithms for nonuniform problems. Algorithmica 11, 542–571 (1994)
21. Lam, T.-W., Lee, L.-K., To, I.K.K., Wong, P.W.H.: Energy efficient deadline scheduling in two processor systems. In: Tokuyama, T. (ed.) ISAAC 2007. LNCS, vol. 4835, pp. 476–487. Springer, Heidelberg (2007)
22. Lam, T.-W., Lee, L.-K., To, I.K.-K., Wong, P.W.H.: Competitive non-migratory scheduling for flow time and energy. In: Proc. 20th Annual ACM Symposium on Parallel Algorithms and Architectures, pp. 256–264 (2008)
23. Lam, T.-W., Lee, L.-K., To, I.K.K., Wong, P.W.H.: Speed scaling functions for flow time scheduling based on active job count. In: Halperin, D., Mehlhorn, K. (eds.) ESA 2008. LNCS, vol. 5193, pp. 647–659. Springer, Heidelberg (2008)
24. Li, M., Liu, B.J., Yao, F.F.: Min-energy voltage allocation for tree-structured tasks. Journal on Combintorial Optimization 11, 305–319 (2006)
25. Li, M., Yao, A.C., Yao, F.F.: Discrete and continuous min-energy schedules for variable voltage processors. Proc. National Academy of Sciences USA 103, 3983–3987 (2006)
26. Li, M., Yao, F.F.: An efficient algorithm for computing optimal discrete voltage schedules. SIAM Journal on Computing 35, 658–671 (2005)
27. Pruhs, K., Uthaisombut, P., Woeginger, G.J.: Getting the best response for your erg. ACM Transactions on Algorithms 4 (2008)
28. Pruhs, K., van Stee, R., Uthaisombut, P.: Speed scaling of tasks with precedence constraints. Theory of Computing Systems 43, 67–80 (2008)
29. Sleator, D.D., Tarjan, R.E.: Amortized efficiency of list update and paging rules. Communcations of the ACM 28, 202–208 (1985)
30. Yao, F.F., Demers, A.J., Shenker, S.: A scheduling model for reduced CPU energy. In: Proc. 36th IEEE Symposium on Foundations of Computer Science, pp. 374–382 (1995)

Minimizing Average Flow-Time

Naveen Garg

Indian Institute of Technology Delhi

Abstract. This article looks at the problem of scheduling jobs on multiple machines both in the online and offline settings. It attempts to identify the key ideas in recent work on this problem for different machine models.

1 Introduction

The task of scheduling jobs on multiple machines so as to optimize performance, has long been studied in the Optimization and Computer Science communities for it provides a rich collection of interesting problems which have led to the development of interesting algorithmic tools and techniques. One important measure of the quality of a schedule, which is particularly relevant in the online setting is the average flow time of the schedule. The flow time of a job is the total time it remains in the system, which is the same as the difference in its completion and release times. The average flow time of a schedule is just the average of the flow times of the jobs. By minimizing the average flow time one is minimizing the average time a job spends in the system. This can also be viewed as minimizing the L_1-norm of the flow times of the jobs. When fairness is an issue, one could consider minimizing the L_p-norm, $p \geq 1$ of the flow times. An alternative would be to minimize the maximum *stretch* of a job where the stretch of a job is the ratio of its flow time to its processing time. In this article we will only be concerned with minimizing the average flow time.

In scheduling when we talk of multiple machines, we have multiple options. If the machines are identical and a job can go on any machine then we have the setting of Parallel Scheduling. Suppose the machines have differing speeds so that a machine i with twice the speed as machine i' can finish the same job in half the time as required by i'. We refer to this setting as Related Scheduling. The most general setting is that of Unrelated Scheduling where for each job j and machine i we are given the time required by machine i to process job j, say p_{ij} and these times could be completely arbitrary and unrelated to each other. A setting which is intermediate in complexity between Unrelated Scheduling and Related Scheduling is that of Parallel Scheduling with Subset Constraints; here a job j can be processed only on a specified subset of machines, say S_j, and its processing time is identical on all machines in S_j.

The problem of minimizing average flow time can be considered both in the online and offline scenarios. In the offline case, we have complete knowledge of the processing time and release time of every job. In the online scenario, the

S. Albers, H. Alt, and S. Näher (Eds.): Festschrift Mehlhorn, LNCS 5760, pp. 187–198, 2009.

processing time of a job is known only on its release; this is also referred to as the *clairvoyant setting*. A more challenging scenario in the online case is the *non-clairvoyant* setting where we get to know the processing time of a job only when it completes.

For some of the above problems we will be able to develop algorithms which have the property of *immediate dispatch*. This implies that the scheduling algorithm can decide which machine to schedule a job j on, the moment j is released. Note that this decision — which machine to schedule a job on — is the crux of the problem. This is because, once we know what jobs are to be scheduled on a particular machine we can determine the optimum schedule for this machine using the *shortest remaining processing time* (SRPT) rule. On the other hand, for some other problems we will be able to develop competitive algorithms only when we permit *resource augmentation*. This means that our algorithm has more resources, in particular, it has machines with higher speeds, than the optimum algorithm.

One key technique we employ in our algorithms is that of Linear Programming. We use a time indexed formulation for the flow time problem and then suitably round the fractional program obtained.

2   Previous Work

The special case of this problem when all machines are identical has received considerable attention [1, 2, 14]. There has been recent progress on the problem where machines can have different speeds [11, 10], or when a job can go only on a subset of machines [12].

For the problem of minimizing total flow time in the offline setting, we know matching lower and upper bounds on the approximation ratio for all machine models [13] except for the setting of unrelated machines. Here while a lower bound of $\Omega(\log P)$ is known there is no upper bound known for this problem. For the subset setting in [12] we develop an algorithm that maps the problem to computing an unsplittable flow in a suitably defined graph.

When each machine is provided a small additional speed (say ϵ) then we [6] show that one can get a $O(1 + 1/\epsilon)^2$ competitive algorithm for minimizing weighted flow time even on unrelated machines. This result relies on an interesting potential function. For the non-clairvoyant setting it is known that without speed augmentation we cannot get a bounded competitive ratio even for single machines [15]. With speed augmentation, competitive algorithms are known only for single and parallel machines.

A set of flow time problems where there is a large gap in our understanding is the setting of weighted flow time. For single machines, the problem is NP-hard and a $1 + \epsilon$ approximation is achievable in quasi-polynomial time [7]. For single machines, in the online setting, an $O(\min(\log W, \log^2 P))$ upper bound [5] and an $\Omega(\sqrt{\min(\log W, \log \log P)})$ lower bound [3] on the competitive ratio is known. No results are known for minimizing weighted flow time on multiple machines in the offline setting, besides the quasi-polynomial-result which also extends to a constant number of machines. In the online setting a lower bound of $\Omega(\min(W^{1/2}, P^{1/2}, (n/m)^{1/4}))$ is known [8] on the competitive ratio of any online algorithm for minimizing weighted flow time on 2 or more machines.

3   An LP Formulation for Flow Time

We first develop the LP for the **Parallel Scheduling** problem. There are m machines and n jobs. Job j has size p_j and is released at time r_j. For the purposes of this formulation we will assume that p_j and r_j are integers.

Our formulation will allow migratory schedules (where a job can migrate from one machine to another) as feasible solutions. In fact, it turns out to be convenient to even allow such schedules which process the same job simultaneously over multiple machines. For each job j, machine i and time t, we have a variable $x_{i,j,t}$ which denotes the fraction of the interval $[t, t+1]$ on machine i for which job j is processed.

$$\text{minimize } \sum_j \sum_i \sum_t x_{i,j,t} \cdot \left(\frac{t-r_j}{p_j} + \frac{1}{2} \right)$$

subject to

for all machines i and time t	$\sum_j x_{i,j,t} \leq 1$
for all jobs j	$\sum_i \sum_t x_{i,j,t} = p_j$
for all jobs j, machines i, and time $t < r_j$	$x_{i,j,t} = 0$
for all jobs j, machines i, time t	$x_{i,j,t} \in [0,1]$

The first constraint refers to the fact that a machine can process at most one job at any point of time. The second constraint says that job j gets completed in the schedule while the third constraint captures the simple fact that we cannot process a job before its release date. It should be clear that any feasible solution gives rise to a schedule where jobs can migrate across machines and may even get processed simultaneously on different machines. The only non-trivial part of the LP is the objective function.

Given a feasible solution x to the LP let $f_j(x) = \sum_i \sum_t x_{i,j,t} \cdot \left(\frac{t-r_j}{p_j} + \frac{1}{2} \right)$ denote the contribution of job j to the objective. We call this quantity the *fractional flow time* of j. Let $F_j(S)$ denote the (integral) flow time of job j in a schedule S. Let x_S be a feasible solution to the LP corresponding to the schedule S. To justify the choice of the objective function we would like to argue that for any schedule S and all jobs j, $f_j(x_S) \leq F_j(S)$. Let j be scheduled from time $a + r_j$ to time $a + r_j + p_j$ on machine i in schedule S. Note that the fractional flow time of this job equals

$$\frac{p_j}{2} + a + \sum_{i=0}^{p_j - 1} \frac{i}{p_j} \leq a + p_j.$$

Since $a + p_j$ is the flow time of this job the fractional flow time is only smaller than the actual flow time. If schedule S was such that the job j was not done contiguously on one machine but split into smaller pieces and done across many machines, even then we would have this property. Therefore the sum of the fractional flow times is at most the total flow time and so the optimum solution to the LP is a lower bound on OPT.

Why have we chosen to define fractional flow time of a job j as above and not, for instance, as $f'_j(x) = \sum_i \sum_t x_{i,j,t} \cdot \frac{t-r_j}{p_j}$? Consider the following simple example which establishes a large gap between the fractional flow time as defined here and the (integral) flow time. Job j is released at time 0 and has a processing time of m. Clearly the minimum flow time is m. However, a fractional solution could schedule 1 unit of j simultaneously on the m machines so that the fractional flow time of j is now 1. Note that with the original definition of fractional flow time j would have a fractional flow time of $m/2 + 1$.

This alternate definition of fractional flow time would work if we could ensure that in our LP solution no job is simultaneously processed on multiple machines. This for instance could be achieved by adding the constraint $\sum_i x_{i,j,t} \leq 1$ for every job j and time t, to the LP. However, it turns out to be more convenient to work with the earlier definition of fractional flow time and that is the one we follow here.

Note that

$$\sum_j f_j(x) = \sum_j f'_j(x) + \sum_{i,j,t} x_{i,j,t}/2. \tag{1}$$

For the case of Parallel Scheduling, the last term equals $\sum_j p_j/2$ and hence minimizing $\sum_j f_j(x)$ is the same as minimizing $\sum_j f'_j(x)$. Let

$$r_x(t) = \sum_{j:r_j \leq t} \sum_i \sum_{t' \geq t} \frac{x_{i,j,t'}}{p_j}$$

denote the total unfinished fraction of jobs at time t in the solution x. It is easy to see that $\sum_j f'_j(x)$ equals $\sum_t r_x(t)$ and this gives us a handle on minimizing total fractional flow time. If we had only one machine the optimum solution to the LP would, at time t, schedule that unfinished job with the smallest processing time; we call this the *shortest job first* (SJF) rule. The same idea applies for multiple parallel machines. At each time t, we would schedule the smallest unfinished job, simultaneously on all the machines. If the smallest job has only αm, $\alpha < 1$, units of processing left, then this job will occupy a fraction α of the time slot on each of the m machines. The SJF rule is followed to determine what job(s) to schedule in the remainder of the time slot.

What happens when the machines are not identical. To address this we will first have to modify our LP formulation. Related Scheduling is a special case of Unrelated Scheduling where the processing time of job j on machine i, $p_{ij} = p_j/s_i$ where s_i is the speed of machine i. Hence, we will write a formulation only for the Unrelated Scheduling problem. Note that the fraction of the processing requirement of job j which is done on machine i equals $\sum_t x_{i,j,t}/p_{ij}$. For each job j we replace the second constraint by

$$\sum_i \sum_t \frac{x_{i,j,t}}{p_{ij}} = 1 \tag{2}$$

The fractional flow time of a job j is now defined as

$$f_j(x) = \sum_i \sum_t x_{i,j,t} \cdot \left(\frac{t - r_j}{p_{ij}} + \frac{1}{2} \right)$$

and the objective function equals $\sum_j f_j(x)$. As before, let

$$f_j'(x) = \sum_i \sum_t x_{i,j,t} \cdot \frac{t - r_j}{p_{ij}}$$

and so equation 1 continues to hold.

The quantity $\sum_{i,j,t} x_{i,j,t}$ equals the total processing time of the schedule and unlike the Parallel Scheduling case, this is not a constant for Related Scheduling. For the Related Scheduling problem Let $m_x(t) = \sum_i \sum_j x_{i,j,t}$ be the number of machines used in an optimum solution x in the time slot $[t, t+1]$. This quantity need not be an integer and if it were, say 7.2, then in this time period solution x would use the fastest 7 machines fully and the next fastest machine to an extent of 0.2.

Note that $\sum_t m_x(t)$ equals the total processing time of the schedule. If we knew $m_x(t)$ for all times t, then minimizing total fractional flow time would be the same as minimizing $\sum_j f_j'(x)$. As is the Parallel Scheduling problem, the latter would be minimized if we schedule, in the time slot $[t, t+1]$, the unfinished job with the shortest processing time, simultaneously, on the $m_x(t)$ fastest machines. Thus knowledge of $m_x(t)$ for all times t is sufficient to determine the optimal solution x. We will see later that this observation would lead to a 2-competitive online algorithm for minimizing total fractional flow time on related machines.

For the case of Unrelated Scheduling, and even the special case of Parallel Scheduling with Subset Constraints, there is no simple way of minimizing $\sum_j f_j'(x)$. We will later show an example that will help us better understand the difficulty in doing this minimization. When we have faster machines than the optimum algorithm — the setting of resource augmentation — we shall be able to do this minimization.

4   Rounding the LP Solution

Let x be the optimum solution to the LP for the Parallel Scheduling problem. Note that

$$\sum_j f_j'(x) = \sum_{i,j,t} \frac{t \cdot x_{i,j,t}}{p_j} - \sum_j r_j.$$

We first assume that all jobs have processing times of the from $2^i, i \geq 0$. If $p_j = 2^k$ we say job j is of class k. Let $x_{i,t}(k)$ be the total fraction of slot $[t, t+1]$ on machine i occupied by jobs of class k in solution x. Note that $\sum_k x_{i,t}(k) \leq 1$. For ease of presentation we will subdivide this slot on machine i into smaller slots, one for each class. The jobs of the original slot are also partitioned into these smaller slots so that all jobs of the same class are in the same smaller slot which is also labeled with that class. The *volume* of a smaller slot is just the total fraction of the jobs it contains.

We now use this subdivision of slots to construct another solution to the LP, y, with the property that every job is scheduled on only one machine. Further, in y a job of class k is scheduled only in slots of class k. This second property ensures that the fractional flow time of y is at most that of x.

The total volume of class k slots is the same on each machine. This implies that in solution y, for every class k, we should have the same number of jobs on each machine. This naturally suggests an algorithm which distributes jobs of each class in a round robin manner on the machines. The solution y is then obtained by considering jobs in increasing order of their release times and scheduling each job on the machine assigned to it, in the earliest available slots of that class.

The above procedure might leave some slots empty as there might be no job of that class which could be scheduled there. However, the total volume of class k slots which remain empty on a machine cannot exceed 2^k. This implies that on each machine we might not be able to schedule at most one class k job in the class k slots available to us. Further, suppose we had 7 jobs of class k and 5 machines. Then the total volume of class k slots on each machine is $1.4 \cdot 2^k$. On 2 of these 5 machines the round robin algorithm would schedule 2 jobs of class k and so for one of these jobs enough slots would not be available. So in all, up to 2 jobs of class k might remain unscheduled on each machine.

Let $R_{i,k}$ be the set of unscheduled jobs of class k on machine i and let $R = \cup_{i,k} R_{i,k}$. We let y denote the solution only for the jobs not in R. It is clear that the fractional flow time of y is no more than the fractional flow time of x. We first bound the (integral) flow time of jobs not in R in terms of the fractional flow time of the solution y. To do this we schedule the jobs not in R on the same machine as they are scheduled in solution y, but using the SJF rule. Since we know that SJF is optimal for fractional flow time, this solution, say z, has fractional flow time at most that of y. The schedule z has the nice property that a job can be interrupted only by another job of lower class. This property implies that the total flow time of jobs of class k is at most the total fractional flow time of these jobs plus the total processing time of the schedule z which is at most $T = \sum_j p_j$. Hence the total flow time of jobs in z is at most the fractional flow time of z plus KT, where K is the number of classes.

We now schedule the jobs of $R_{i,k}$ on machine i in the first available empty slots in z after their release. Hence each job of $R_{i,k}$ can have a flow time of at most the total processing time of jobs scheduled on i. Thus the total flow time of all jobs in R is at most $2KT$. Putting everything together, we get that the total flow time of all jobs is at most the fractional flow time of x plus $3KT$ which is at most OPT $+ 3KT$.

How do we get around the assumption that job sizes are of the kind 2^i. Note that rounding job sizes so that they are of this form can increase the flow time by an unbounded amount. We get around this difficulty by rounding down the job sizes, to the nearest power of 2, in the objective function, but leaving them, as is, in the constraints. Here, we make the assumption that the smallest processing time is at least 1, and this is no loss of generality. Note that the optimum solution to this new LP is at most twice the optimum solution to the original LP. Further since the constraints remain the same, the set of feasible solutions is identical in the two LPs. A job whose processing time is between 2^k and 2^{k+1} is considered to be of class k for applying the above rounding procedure.

Our analysis would now differ at a few places. First note that the objective function now equals

$$\sum_{i,j,t} \frac{t \cdot x_{i,j,t}}{\lfloor p_j \rfloor} - \sum_j r_j \frac{p_j}{\lceil p_j \rceil}$$

and since the second term is a constant, minimizing the objective is the same as minimizing the first term. Since in the solution y we schedule class k jobs only in class k slots, the value of the first term for solution y is no more than the value of this term for solution x. The volume of class k slots that remain empty on a machine in the solution y might now be 2^{k+1} so that up to 2 jobs might remain unscheduled. This together with the additional job of class k that might not get scheduled on machine i due to insufficient volume of class k slots implies that the size of $R_{i,k}$ might be as large as 3. Hence the total flow time of the jobs in R is at most $3KT$ and this implies that the total flow time is no more than $2\text{OPT} + 4KT$. If the largest processing time is P then the number of classes, K, is at most $\log P$. Further since the optimum flow time is at least as large as T, we get a $(2 + 4 \log P)$-approximation.

5   Rounding the LP Solution for Related Machines

We renumber the machines so that $s_1 \geq s_2 \geq \cdots \geq s_m$. Let x be an optimum solution to the linear program. We will assume that the job sizes are powers of 2. Once again, note that

$$\sum_j f'_j(x) = \sum_{i,j,t} \frac{t \cdot x_{i,j,t}}{p_j} - \sum_j r_j.$$

As in the case of **Parallel Scheduling** we will use the solution x to subdivide slots and assign them a class. However, now we define the volume of a slot on machine i as the amount of processing that can be done in this slot. Thus the volume of a slot $[t, t+1]$ on machine i is s_i.

How should we now assign jobs to machines? We would once again like to have the property that for each class, k, and machine, i, we have at most a constant number of class k jobs that or not scheduled in the class k slots on machine i. However, one additional property we have to ensure is that the total processing time of our schedule is not much more than the total processing time of the solution x. One possible assignment is the following: for any class k and integer i, $1 \leq i \leq m$, if the total volume of class k jobs on machines i through m is $\alpha 2^k$ then the number of jobs assigned to these machines is $\lfloor \alpha \rfloor$. This assignment has the property that for any k, the number of class k jobs assigned to a machine is at most one more than the number of class k jobs that can fit into the class

k slots on that machine. Further, for any i the total processing time of class k jobs assigned to machines i through m in this assignment is less than the total processing time contributed by the class k slots on machines i through m in the solution x. Hence the processing time of this assignment is less than the processing time of solution x.

Motivated by this assignment, we consider the following online assignment of jobs to machines. Initially each slot is unmarked. At any time, when we encounter a class k slot on machine i we schedule an unfinished class k job already assigned to this machine in this slot. If there is no such job then we check to see if the total volume of unmarked slots of class k on machines i though m exceeds 2^k. If so we assign a class k job to this machine and schedule it in this slot. Unmarked class k slots of total volume 2^k on machines i through m are now marked. While marking slots we always consider machines in increasing order of their index.

This procedure would first schedule a class k job on machine 1, then on machine 2 and so on. When the job, scheduled on machine i finishes, we would have 2^k unmarked slots on machine i and so will schedule another job on it. Thus, once a class k job is scheduled on machine i, no class k slot on this machine would go idle. This implies that the total volume of empty class k slots is at most $m2^k$. The last job scheduled on each machine might not find enough slots of the same class to finish; in this case we use the empty slots to complete the job. However, we might still not have finished scheduling all the jobs as we would still have some unmarked slots. Using the same procedure as before, we schedule a class k job on machine i if the total volume of unmarked jobs on machines i to m exceeds 2^k. This step would ensure that all jobs finish and that each machine would have at most 2 jobs of class k which would not be completely scheduled in the class k slots.

In the above description we seem to require knowledge of when the schedule finishes on each machine. We get around this by using the empty slots in solution x to schedule jobs in solution y. Note that there might be empty slots in the middle of the schedule. We schedule the lowest class job that can be scheduled in an empty slot using the above procedure and only if no job can be scheduled do we leave the slot empty. As we did for Parallel Scheduling, we use the solution y only for deciding what job to assign to each machine. Once this is known we just follow the SJF rule to schedule the jobs on the machines; let this be solution z. Once again we have that the total flow time of all jobs in solution z is no more than the fractional flow time of x plus $2KT$ where T is now the processing time of schedule z which is the same as the processing time of y. In constructing y we moved jobs from slower machines in solution x to faster machines and hence the processing time of y is less than that of x. The second term of the objective function of the LP equals half the processing time and hence $T \leq 2\mathsf{OPT}$. This implies that the flow time of z is at most $(1 + 4\log P)\mathsf{OPT}$. We can get around the assumption that job sizes are powers of 2, by applying the same rounding trick that we used for the Parallel Scheduling problem.

Given an optimal solution x to the linear program we have shown how to round it in an online manner to get a schedule z with flow time at most $O(\log P)$ times the optimum. To get an online algorithm for the Related Scheduling problem we also need to show how to compute x in an online manner. Recall that the key stumbling block in computing x is to determine $m_x(t)$ which is the number of machines in use at time t.

We now discuss a recent result on minimizing flow time and energy and show how it solves our problem of computing an optimal x online. We are given a machine which can run on any speed $s \le S$. When the machine runs on speed s it finishes s units of processing in 1 unit of time. Let $P : [0..S] \to \Re^+$ be the power function, where $P(s)$ is the amount of power consumed when the machine runs on speed s. Jobs arrive online and we need to determine at each instant what speed to run the machine on so as to minimize the sum of the fractional flow time and the energy consumed. For a job j the fractional flow time is equal to $f'_j(x)$, where x is the schedule constructed. Hence, once we have determined the speed schedule, we should follow SJF to minimize the total fractional flow time. Bansal et al. [4] give a 2-competitive online algorithm for minimizing fractional flow time and energy for any convex, continuous, power function.

We would like to view $m_x(t)/2$ as the power consumed at time t. If $m_x(t)$ is, say 3.2, then we view this as a single machine running at speed $s_1 + s_2 + s_3 + (0.2)s_4$ and consuming 1.6 units of power. Hence by minimizing fractional flow time plus energy we would have obtained an optimal solution to the linear program. The 2-competitive algorithm online algorithm of Bansal et al. thus gives an online algorithm for computing a solution x' to the LP which has value at most twice the optimum. This together with the online algorithm for rounding an LP solution yields an $O(\log P)$-competitive algorithm for minimizing flow time on unrelated machines.

6 Rounding the LP Solution for Parallel Scheduling with Subset Constraints

Let S_j denote the subset of machines on which job j can be scheduled and let p_j, r_j be the processing time and release time of j. Let x be an optimal solution to the LP. Note that the second term in the objective function is now equal to $\sum_j p_j/2$ and hence minimizing the total fractional flow time for this problem is the same as minimizing $\sum_j f'_j(x)$.

However, it is not possible to minimize $\sum_j f'_j(x)$ in an online manner. Consider the following example on three machines. All jobs have processing time 1 and are of two kinds; jobs of type a can be scheduled on machines 1 and 2 while jobs of type b can go to machines 2 and 3. At each time, t, $0 \le t \le T - 1$, 2 jobs of each type are released. Note that at each time instant only 3 units of processing can be done and so at time T we would have T jobs left. Suppose at least $T/2$ of these are of type a. Then from time T to $T + L$, $L >> T$, we release 2 jobs of type a at each time. Since at every instant in the interval $[T..(T + L)]$ we have at least $T/2$ unfinished jobs the quantity $\sum_j f'_j(x)$ for this schedule is at least $TL/2$. The optimum schedule for this instance would finish all type a jobs by time T and schedule the remaining type b jobs from time T to $2T$. Hence the flow time of this schedule is $O(T^2 + L)$ which implies that the competitive ratio of every online algorithm is $\Omega(T)$.

We will therefore develop an offline algorithm to round the LP solution x. We again assume that all processing times of powers of 2. We construct a directed graph $G = (V, E)$ with a source vertex s, a vertex $v(j)$ for job j and one vertex $v(i, j, k)$ for machine i, job j and class k. We number jobs by increasing release times with the first job numbered 0. Vertex $v(i, j, k)$ has exactly one incoming edge from vertex $v(i, j - 1, k)$, for $j \geq 0$ and from vertex s for $j = 0$. Vertex $v(j)$ has edges from all vertices $v(i, j, k)$ where job j is of class k and $i \in S_j$. We will now construct a flow with s as the source and vertices $v(j)$, $0 \leq j < n$ as the sink. The flow on the edge from $v(i, j, k)$ to $v(j)$ equals $\sum_t x_{i,j,t}/2^k$ and is the fraction of job j done on machine i. Thus the total flow into each sink vertex is 1. The flow on the edges not incident to the sink is given by conservation. Note that the flow on the edge $(v(i, j - 1, k), v(i, j, k))$, say f, equals the total fraction, processed on machine i, of the class k jobs released at or after r_j. This implies that the volume of class k slots after time r_j on machine i is at least $f2^k$.

We now use the single source unsplittable flow algorithm of Dinitz et al. [9] to compute a flow f' such that the 1 unit of flow that goes from s to $v(j)$ flows along a single path. [9] show that this can be done so that for all edges, $e \in E$, $f'(e) \leq f(e) + 1$. The unsplittable flow f' naturally determines an assignment of jobs to machines; if the flow reaching $v(j)$ does so along the edge $v(i, j, k)$ then job j is assigned to machine i. Once we have the assignment of jobs to machines, we can compute a schedule y by considering jobs in order of their release times and scheduling a job of class k in the earliest available class k slots on the machine it is assigned to.

Would some jobs remain unscheduled because there are no slots available for them? If, for each class, k we add 2^k class k slots on each machine then no job will remain unassigned. This is because for any machine i, the number of class k jobs which have release time greater than or equal to r_j and which are assigned to i is at most $f + 1$ where f is the original flow on the edge $(v(i, j - 1, k), v(i, j, k))$. Further, the volume of class k slots on machine i after time r_j in the solution x is at least as large as $f2^k$. By adding 2^k jobs of class k at the end on each machine we ensure that for each j the volume of class k slots available after time r_j is at least as large as the total volume of jobs assigned to this machine which have release time at least r_j. This implies that all class k jobs can be scheduled in the available slots. Equivalently, if we did not add the extra slots, at most one job of each class will remain unscheduled on each machine.

The rest of the development is the same as for the case of Parallel Scheduling. Our final schedule z would have a flow time bounded by $\text{OPT} + 2KT$ where $T = \sum_j p_j \leq \text{OPT}$. This gives a $(1 + 2 \log P)$-approximation for minimizing flow time for the Parallel Scheduling with Subset Constraints problem.

7 Minimizing Flow Time on Unrelated Machines

The rounding algorithm we discussed for Parallel Scheduling with Subset Constraints also applies to Unrelated Scheduling when all processing times p_{ij} are powers of 2;

the only additional thing that needs to be taken care of is that the processing time does not increase much. When the processing times are not powers of 2 then by treating each distinct processing time as a class, we can still apply the algorithm but now the approximation ratio is O(number of distinct processing times). The technique of rounding down processing times to powers of 2, in the objective function and leaving them as such in the constraints does not apply in the setting of unrelated machines. This is because we now have

$$\sum_j f_j'(x) = \sum_{i,j,t} \frac{t \cdot x_{i,j,t}}{\lfloor p_{ij} \rfloor} - \sum_j r_j \sum_i \frac{\sum_t x_{i,j,t}}{\lfloor p_{ij} \rfloor}$$

and the second term is no more a constant. If the second term were a constant, for instance when $p_{ij} = \alpha \lceil p_{ij} \rceil$ then we could ignore the second term and obtain a schedule, y, in which each job is scheduled on only one machine by rearranging jobs of a class in slots of the same class. When we did this the contribution of the first term in solution y is no more than its contribution in solution x. However, now y might be such that

$$\sum_j r_j \sum_i \frac{\sum_t y_{i,j,t}}{\lfloor p_{ij} \rfloor} < \sum_j r_j \sum_i \frac{\sum_t x_{i,j,t}}{\lfloor p_{ij} \rfloor}$$

and so we will not be able to argue any more that the fractional flow time of y is less than that of x. In fact we do not know of any approximation algorithm for Unrelated Scheduling.

8   Online Algorithm for Unrelated Machines with Speed Augmentation

When the online algorithm is given ϵ more speed than the offline algorithm, we can obtain a $O((1 + \epsilon^{-1})^2)$-competitive algorithm. Note that in the example showing that no online algorithm can be competitive on unrelated machines, we had argued that any online algorithm would accumulate $T/2$ jobs by time T which the algorithm cannot get rid off for a long time $L >> T$ and this led to a flow time of $\Omega(LT)$. However, if the machines of the online algorithm had ϵ more speed then in roughly $T\epsilon^{-1}$ time units it would be able to get rid off all these extra jobs accumulated.

Our online algorithm is actually very natural. When a job arrives it is assigned to the machine where it would lead to the smallest increase inflow time. To determine the increase in flow time on machine i due to the arrival of a job, we maintain the set of jobs that are assigned to i but not finished yet. Since these jobs will be scheduled on i using the SJF rule, the increase in flow time due to the addition of an arriving job j to this set can be easily computed.

The analysis of this algorithm relies on a potential function which shows that the fractional flow time of our solution is less than $1 + \epsilon^{-1}$ times the fractional flow time of a non-migratory optimum, when we allow the online algorithm an additional speed ϵ. Here the fractional flow time of a job, j, is as defined by the function $f_j'(x)$. If we give our machines an additional ϵ speed then (integral) flow time is no more than $1 + \epsilon^{-1}$ times this fractional flow time. Putting these together we get an $O(1+\epsilon^{-1})^2$-competitive algorithm for scheduling on unrelated machines which have ϵ more speed than the machines of the offline algorithm.

Acknowledgments. Much of the work discussed in this article was done jointly with Amit Kumar, V.N. Muralidhara and Jivitej Chaddha.

References

1. Avrahami, N., Azar, Y.: Minimizing total flow time and total completion time with immediate dispatching. In: SPAA, pp. 11–18. ACM, New York (2003)
2. Awerbuch, B., Azar, Y., Leonardi, S., Regev, O.: Minimizing the flow time without migration. In: STOC, pp. 198–205 (1999)
3. Bansal, N., Chan, H.-L.: Weighted flow time does not admit o(1)-competitive algorithms. In: SODA, pp. 1238–1244 (2009)
4. Bansal, N., Chan, H.-L., Pruhs, K.: Speed scaling with an arbitrary power function. In: SODA 2009: Proceedings of the Nineteenth Annual ACM -SIAM Symposium on Discrete Algorithms, Philadelphia, PA, USA, pp. 693–701. Society for Industrial and Applied Mathematics (2009)
5. Bansal, N., Dhamdhere, K.: Minimizing weighted flow time. ACM Transactions on Algorithms 3(4) (2007)
6. Chadha, J., Garg, N., Kumar, A., Muralidhara, V.N.: A competitive algorithm for minimizing weighted flow time on unrelated machines with speed augmentation. In: STOC (2009)
7. Chekuri, C., Khanna, S.: Approximation schemes for preemptive weighted flow time. In: STOC, pp. 297–305 (2002)
8. Chekuri, C., Khanna, S., Zhu, A.: Algorithms for minimizing weighted flow time. In: STOC, pp. 84–93 (2001)
9. Dinitz, Y., Garg, N., Goemans, M.: On the single-source unsplittable flow problem. In: FOCS 1998: Proceedings of the 39th Annual Symposium on Foundations of Computer Science, Washington, DC, USA, p. 290. IEEE Computer Society, Los Alamitos (1998)
10. Garg, N., Kumar, A.: Better algorithms for minimizing average flow-time on related machines. In: Bugliesi, M., Preneel, B., Sassone, V., Wegener, I. (eds.) ICALP 2006. LNCS, vol. 4051, pp. 181–190. Springer, Heidelberg (2006)
11. Garg, N., Kumar, A.: Minimizing average flow time on related machines. In: STOC, pp. 730–738 (2006)
12. Garg, N., Kumar, A.: Minimizing average flow-time: Upper and lower bounds. In: FOCS, pp. 603–613 (2007)
13. Garg, N., Kumar, A., Muralidhara, V.N.: Minimizing total flow-time: The unrelated case. In: Hong, S.-H., Nagamochi, H., Fukunaga, T. (eds.) ISAAC 2008. LNCS, vol. 5369, pp. 424–435. Springer, Heidelberg (2008)
14. Leonardi, S., Raz, D.: Approximating total flow time on parallel machines. In: STOC, pp. 110–119 (1997)
15. Motwani, R., Phillips, S., Torng, E.: Non-clairvoyant scheduling. Theoretical Computer Science 130(1), 17–47 (1994)

Integer Linear Programming in Computational Biology

Ernst Althaus[1], Gunnar W. Klau[2], Oliver Kohlbacher[3], Hans-Peter Lenhof[4,*], and Knut Reinert[5]

[1] Department of Computer Science, Johannes-Gutenberg-Universität Mainz
[2] Life Sciences Group, Centrum Wiskunde & Informatica (CWI), Amsterdam
[3] Center for Bioinformatics Tübingen, Eberhard-Karls-Universität Tübingen
[4] Center for Bioinformatics Saar, Saarland University
[5] Institute for Computer Science, Freie Universität Berlin
`lenhof@bioinf.uni-sb.de`

Abstract. Computational molecular biology (bioinformatics) is a young research field that is rich in NP-hard optimization problems. The problem instances encountered are often huge and comprise thousands of variables. Since their introduction into the field of bioinformatics in 1997, integer linear programming (ILP) techniques have been successfully applied to many optimization problems. These approaches have added much momentum to development and progress in related areas. In particular, ILP-based approaches have become a standard optimization technique in bioinformatics. In this review, we present applications of ILP-based techniques developed by members and former members of Kurt Mehlhorn's group. These techniques were introduced to bioinformatics in a series of papers and popularized by demonstration of their effectiveness and potential.

1 Introduction

Computational molecular biology (or bioinformatics) is a young research field that develops computational approaches for scientific problems arising in the life sciences. At the outset, the methodological focus was on the development of efficient algorithms, data structures, and optimization techniques that were able to deal with the data arising in life science applications. Due to the development of high-throughput methods for biomedical data analysis and the rise of systems biology, statistical learning approaches have gained in importance during the last decade. However, statistical analyses and simulations of physiological and pathological processes have raised a large number of novel hard optimization problems. Since the true nature of most of the biological processes being investigated is still obscure, they have often been modeled in the simplest possible way as linear processes. Since the majority of these optimization problems are discrete ones, it is not surprising that integer linear programming (ILP)

---

[*] Corresponding Author.

S. Albers, H. Alt, and S. Näher (Eds.): Festschrift Mehlhorn, LNCS 5760, pp. 199–218, 2009.
© Springer-Verlag Berlin Heidelberg 2009

based approaches have gained in importance and that ILP-based approaches like branch-and-cut (B&C) algorithms and Lagrangian relaxations (LR), etc., have become standard optimization techniques in bioinformatics only 10 years after having been applied in the field for the first time.

The first papers in bioinformatics that are indirectly linked to ILP-based techniques presented complexity results and reductions of original bioinformatics problems to standard combinatorial optimization problems. For example, Alizadeh, Karp, Weisser, and Zweig [1] proved in 1994 that a particular physical mapping problem could be reformulated as a traveling salesman problem. The first ILP formulations for bioinformatics problems together with novel B&C approaches were published in 1997. Reinert, Lenhof, Mutzel, Mehlhorn, and Kececioglu [2] proposed a B&C algorithm for the Maximum Weight Trace problem, a particular version of the multiple alignment problem. Christof, Jünger, Kececioglu, Mutzel, and Reinelt [3] published a B&C approach for the physical mapping problem. The first application of ILP-based approaches to RNA sequence-structure alignment was suggested by Lenhof, Reinert, and Vingron [4] in 1998. In 2000, the first B&C approach was applied to a problem in structural bioinformatics, the so-called side chain optimization problem. This approach is due to Althaus, Kohlbacher, Lenhof, and Müller [5]. In 2004, Klau, Rahmann, Schliep, Vingron, and Reinert [6] applied the first ILP-based approach to the problem of designing microarrays. The first application in the area of systems biology and biological networks was published by Dittrich, Klau, Rosenwald, Dandekar, and Müller [7] in 2008. The above listing of central publications underlines the fact that Kurt Mehlhorn's group has made substantial contributions in this area, leading to the popularization of these techniques in bioinformatics. Actually, the first publication traces back to an idea of Kurt Mehlhorn who, in 1996, proposed applying ILP-based techniques to the challenging optimization problems his bioinformatics subgroup dealt with. In the meantime, these techniques have been exploited in almost all areas of computational molecular biology and a large number of papers have been published applying ILP-based approaches to bioinformatics problems. As a consequence of this development, first reviews and annotated bibliographies with ILP-based approaches and mathematical programming as the main focus have been published. For example, in 2008, Guiseppe Lancia [8] presented a very informative review of mathematical programming in computational molecular biology that also gives a comprehensive overview of ILP-based approaches in this field.

In this paper, we will focus on approaches that have been suggested and published by members and former members of Kurt Mehlhorn's group. We will summarize the contributions in a form similar to an annotated bibliography without going into technical details and briefly sketch further research stimulated by the original papers. The application areas range from sequence analysis (Section 2), structural bioinformatics (Section 3), and probe design for microarray experiments (Section 4) to computational systems biology (Section 5) and vaccine design (Section 6). Finally, we will give a short summary and outlook in Section 7.

2   Sequence Analysis

2.1   Multiple Sequence Alignment

DNA encodes the blueprints of all living organisms and carry the instructions for building proteins and non-coding RNA molecules. Like DNA, proteins and RNA are linear polymers composed of basic building blocks (amino acids, nucleotides). Hence, they can be represented by sequences of letters, where each letter represents one building block. While DNA sequences contain the instructions for building proteins and RNA molecules, the amino acid sequences of proteins define their 3D structures and, thereby, their functionality. Thus, it is not surprising that the comparison of sequences plays a crucial role in the life sciences and that the development of algorithms and methods for studying and analyzing molecular sequences has been one of the major goals at the rising of computational molecular biology. These so-called sequence alignment methods aim at exhibiting the commonalities and differences of a given set of sequences by calculating a kind of two-dimensional matrix where each row represents a sequence and the columns exhibit their common patterns and also their differences. To array the letters in a suitable way, a single operation, the insertion of so-called gaps, usually represented by the '-' symbol is allowed. The insertion of a gap at a certain position in a sequence shifts all letters of the corresponding suffix of the sequence to the right, i.e., it increases the column numbers of the suffix letters by the number of inserted '-'s. Table 1 depicts a multiple sequence alignment and a typical three-dimensional structure (here, myoglobin) is shown in Fig. 1.

To assess the quality of alignments, a large number of scoring functions has been suggested that, in turn, lead to the definition of optimization problems of

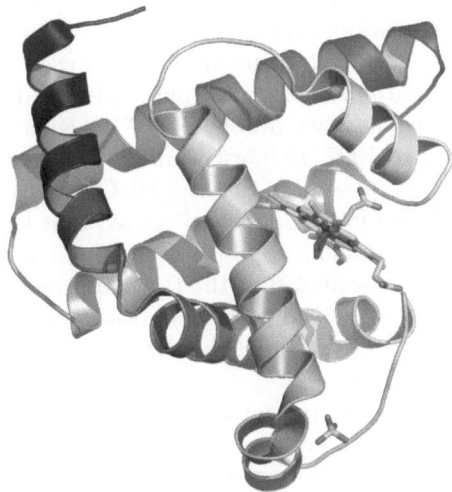

Fig. 1. A 3D model showing the helical domains of myoglobin

Table 1. Multiple sequence alignment of six globin sequences: Human hemoglobin subunit alpha (UniProt accession: P69905), human hemoglobin subunit beta (P68871), horse hemoglobin subunit alpha (P01958), horse hemoglobin subunit beta (P02062), sperm whale myoglobin (P02185) and European yellow lupin leghemoglobin-2 (P02240)

```
HBA_HUMAN    -MVLSPADKTNVKAAWGKVGAHAGEYGAEALERMFLSFPTTKTYFPHF-DLSH
HBB_HUMAN    MVHLTPEEKSAVTALWGKV--NVDEVGGEALGRLLVVYPWTQRFFESFGDLST
HBA_HORSE    -MVLSAADKTNVKAAWSKVGGHAGEYGAEALERMFLGFPTTKTYFPHF-DLSH
HBB_HORSE    -VQLSGEEKAAVLALWDKV--NEEEVGGEALGRLLVVYPWTQRFFDSFGDLSN
MYG_PHYCA    -MVLSEGEWQLVLHVWAKVEADVAGHGQDILIRLFKSHPETLEKFDRFKHLKT
LGB2_LUPLU   MGALTESQAALVKSSWEEFNANIPKHTHRFFILVLEIAPAAKDLFSFLKGTSE

HBA_HUMAN    -----GSAQVKGHGKKVADALTNAVAHVDD---M--PNALSALSDLHAHKLRVD
HBB_HUMAN    PDAVMGNPKVKAHGKKVLGAFSDGLAHLDN---L--KGTFATLSELHCDKLHVD
HBA_HORSE    -----GSAQVKAHGKKVGDALTLAVGHLDN---L--PGALSNLSDLHAHKLRVD
HBB_HORSE    PGAVMGNPKVKAHGKKVLHSFGEGVHHLDN---L--KGTFAALSELHCDKLHVD
MYG_PHYCA    EAEMKASEDLKKHGVTVLTALGAILKKKGH---H--EAELKPLAQSHATKHKIP
LGB2_LUPLU   VPQ--NNPELQAHAGKVFKLVYEAAIQLQVTGVVVTDATLKNLGSVHVSK-GVA

HBA_HUMAN    PVNFKLLSHCLLVTLAAHLPAEFTPAVHASLDKFLASVSTVLTSKYR------
HBB_HUMAN    PENFRLLGNVLVCVLAHHFGKEFTPPVQAAYQKVVAGVANALAHKYH------
HBA_HORSE    PVNFKLLSHCLLSTLAVHLPNDFTPAVHASLDKFLSSVSTVLTSKYR------
HBB_HORSE    PENFRLLGNVLVVVLARHFGKDFTPELQASYQKVVAGVANALAHKYH------
MYG_PHYCA    IKYLEFISEAIIHVLHSRHPGDFGADAQGAMNKALELFRKDIAAKYKELGYQG
LGB2_LUPLU   DAHFPVVKEAILKTIKEVVGAKWSEELNSAWTIAYDELAIVIKKEMNDAA---
```

the following form: Given a set of sequences and a scoring function, calculate an alignment of the sequences that is optimal with respect to the scoring function. Besides heuristics, dynamic programming approaches have been developed to solve these problems. Unfortunately, the runtime and the storage requirements of these dynamic programming approaches grow exponentially with the number of sequences and the corresponding optimization problems are NP-hard. Therefore, calculating the optimal multiple alignment of more than 10 sequences was, and still is, a real challenge, but in many applications users would like to compare dozens of sequences. A variant of the multiple sequence alignment (MSA) problem is the Maximum Weight Trace problem (MWT), which has been introduced by John Kececioglu [9]. In 1997, Reinert et al. [2] proposed an ILP formulation for the MWT problem and presented the first B&C approach. They derived several classes of facet-defining inequalities and proved that for all but one class the corresponding separation problem can be solved in polynomial time. Moreover, first experimental results indicated that the B&C approach is able to solve problem instances that cannot be solved with state-of-the-art dynamic programming approaches. Lenhof et al. [10] extended this B&C approach to a segment-based multiple alignment algorithm.

In 2002, Althaus et al. [11] proposed a more general ILP formulation for multiple alignment with arbitrary gap costs and presented a B&C approach to solve the resulting ILPs. The new B&C approach was the first algorithm that could deal with truly affine gap costs. The quality of the approach has been evaluated using the BAliBase database. The evaluation showed that the approach produces high-quality alignments comparable or even superior to the best programs developed so far.

In 2006, Althaus et al. [12] published a paper where they discuss the most important facet-defining inequalities of the solution polyhedron of their ILP approach. They proved that the three (exponentially large) classes of natural valid inequalities considered in the B&C approach are both facet-defining for the convex hull of integer solutions and separable in polynomial time. Experimental results for several benchmark instances again demonstrated that this B&C approach outperforms other leading tools.

Given this ILP formulation, Althaus and Canzar [13, 14] propose a Lagrangian relaxation approach by dualizing all inequalities involving more than two sequences into the objective function. The resulting Lagrangian subproblem is a pairwise sequence alignment problem with some additional cost terms stemming from the multipliers of the dualized constraints. It can be solved by an extension of the known dynamic programming approach.

Fischetti et al. [15] studied the minimum routing cost tree (MRCT) problem. They studied IP models for this problem and compared several algorithms. Furthermore, they applied the proposed techniques to the sum-of-pairs (SP) multiple sequence alignment problem. Here, they considered and modified the alignment technique proposed by Dan Gusfield [16]. The alignment algorithm is an approximation technique that has a performance ratio of $2 - (2/n)$, i.e., it guarantees that the calculated alignment has a score smaller than or equal to $2 - (2/n)$ times the optimal score.

The question of which scoring function is best for which alignment, or which scoring function is suitable for a specific alignment problem or a given family of sequences are key challenges in this area. To solve this problem, it is useful to look at the so-called inverse multiple alignment problem, in which a set of high-quality alignments manually curated by experts is given and the goal is to identify the best scoring function that favors these alignments, or at least similar ones. Kececioglu and Kim [17] formulated the inverse multiple alignment problem as a linear program and proposed cutting plane techniques to solve the corresponding LP.

2.2 Sequence-Structure Alignments (RNA)

RNA is another important biomolecule. Until very recently, the central dogma of molecular biology was that DNA is transcribed into its working copy RNA, and RNA in turn is translated into proteins, the actual functional units in the cell. In the last few years, however, it became evident that RNA itself is able to trigger or inhibit important functions in the cell [18], tremendously increasing the interest in the study of RNA molecules. From an algorithmic point of view, the sequence alignment algorithms for DNA still apply to RNA sequences, the only difference being that the four-letter alphabet contains a U instead of the T. It has been shown, however, that the sequence alone does not carry all information necessary to compute reliable alignments. An RNA sequence folds

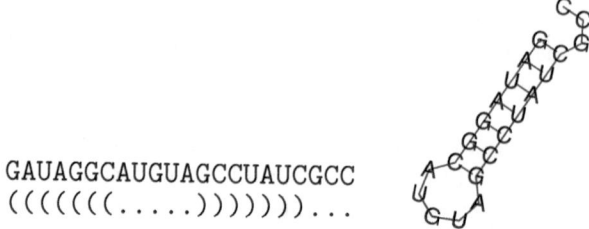

GAUAGGCAUGUAGCCUAUCGCC
(((((((. . . . .))))))) . . .

Fig. 2. Two ways to depict an RNA sequence and corresponding secondary structure. Left: the bracket notation in which pairing brackets indicate base pairs. Right: an alternative way to represent the structure using a graph.

back onto itself and can form hydrogen bonds between pairs of G-C, A-U, and G-U pairs. These bonds lead to the distinctive *secondary structures* of an RNA sequence. Figures 2 and 3 show common representations of small toy examples of RNA sequences together with their secondary structure. In the course of evolution, RNA sequences mutate at a much higher rate than the structure that they are forming, following the structure-function paradigm. RNA molecules with different sequences but the same or similar secondary structure are likely to belong to the same functional family, in which the secondary structure is conserved by selective pressure. This means that the computation of reliable alignments should take structural information into account. Figure 4 shows an example of two possible alignments of two RNA sequences and structures, where the first maximizes the structural similarity and the second maximizes the sequence similarity. Figure 4 also contains a so-called *pseudoknot* depicted by the red line crossing the other lines in the secondary structure. Pseudoknots do occur naturally in some classes of RNA families. Their presence or absence in the corresponding computational models plays an important role in the computational complexity of the corresponding optimization problems. Allowing pseudoknots makes the problems computationally hard [19]. Hence, most approaches assume a pseudoknot-free, *nested* structure as their input. A nested structure can be drawn as an outer-planar graph in its circular representation (see Fig. 3 on the right side for an illustration): Nested structures allow a straightforward decomposition of the entire structure into smaller substructures leading to polynomial time algorithms based on the principle of *dynamic programming*. In addition, it is well known that the multiple alignment problem is NP-hard [20] even without considering secondary structure. Considering the above introductory discussion, the aim is to solve the *sequence-structure alignment* problem: Given two or more RNA sequences, we want to find an optimal multiple sequence-structure alignment.

Lenhof et al. [4] gave a graph-based model that they used to define an integer linear program, which they then solved using the branch-and-cut principle. They align RNA sequences with known structure to those of unknown structure by maximizing the sequence and structure score. Their approach allows handling

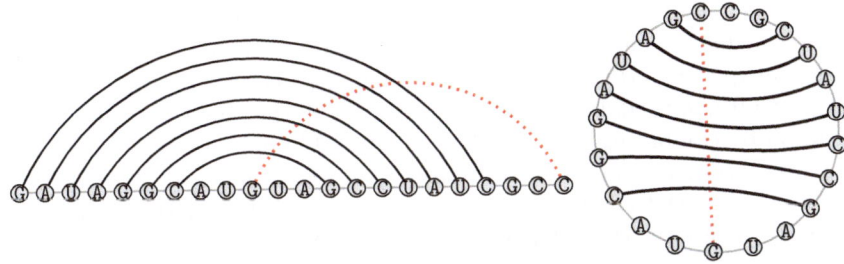

Fig. 3. Graph-based representations of RNA structures. The left side shows the standard graph representation, whereas on the right side a circular graph representation is given. Adding the dotted red edge yields a pseudoknot, i.e., crossing base pairs, in the secondary structure.

pseudoknots and is able to tackle problem instances with a sequence length of approximately 1400 bases. However, for many problem instances of that size, their algorithm already requires a prohibitively large amount of resources. Lancia et al. developed a branch-and-cut algorithm [21] that is similar to [4] for the related problem of aligning contact maps. In subsequent work [22], Lancia and Caprara introduced Lagrangian relaxation to the field of computational biology: Their formulation is based on previous work in the field of quadratic programming problems like the Quadratic Knapsack Problem [23] or the Quadratic Assignment Problem [24]. In [25] Bauer and Klau adapted the Lagrangian relaxation formulation to the problem of aligning two RNA structures: Their implementation yields an algorithm that is an order of magnitude faster than the algorithm from [4] for solving the same instances with respect to the same objective function. Along these lines, Bauer, Klau, and Reinert [26] described an initial integer linear programming formulation for solving multiple RNA structures simultaneously. Althaus et al. (see Section 2.1) presented a formulation for aligning multiple sequences with arbitrary gap costs which also contains extensive polyhedral studies about facet-defining inequalities. Finally, Bauer et al. presented in [27] a graph-based model that unified the formulations given in [26] and the work [12] on pure sequence alignment described in the previous section for the

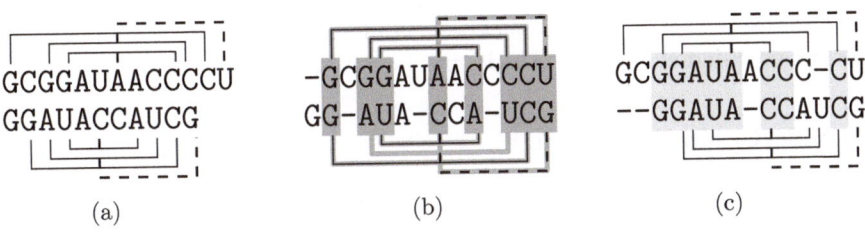

Fig. 4. a) Two RNA sequences with their corresponding secondary structure, b) the alignment that maximizes sequence and structure score (in gray), and c) the alignment maximizing sequence score alone (in light gray)

simultaneous alignment of multiple RNA structures. They concentrated on a sound mathematical description of the approach and provided a first formulation for multiple structural RNA alignments including arbitrary gap costs in the graph-based framework. In a companion paper [28], they focused on the heuristic application by combining the efficiently computable pairwise alignment method with a standard consistency-based multiple alignment method [29]. This results in a state-of-the-art tool since the comparison with other methods proved once more that the application of ILP-based optimization techniques can result in very efficient tools.

3   Structural Bioinformatics

Structural bioinformatics, an area rich in optimization problems, is primarily concerned with problems related to protein structure, for example, the prediction of protein structures from their sequence, the prediction of interactions between proteins (docking), or the design of new protein structures. These problems can be formulated as optimization problems, however, the objective functions are often complex. Typically, they involve the optimization of an energy function based on physical interactions in high-dimensional spaces spanned by the coordinates of the protein atoms.

Proteins are linear polymers composed of 20 basic building blocks called amino acids. These 20 amino acids share a common chemical structure (the backbone), however, they differ in the so-called side chains, functional groups defining an amino acids physical and functional properties. While proteins can be described as a sequence of amino acids on the most coarse level, they have rich three-dimensional structure, which is responsible for a protein's function. Each protein folds into its characteristic three-dimensional or tertiary structure, which is determined by the amino acid sequence. The prediction of the 3D structure from the primary structure is a very difficult task. It has been one of the key problems in bioinformatics for the last decades: Given the primary structure of a protein, determine its three-dimensional structure. If structural information of a related protein with known 3D structure is used, the problem is denoted as threading. The structure prediction problem can be split into two subproblems. Hereby, the prediction of the protein's fold – the structure of the backbone without its side-chains – is the first problem. Consequently, the optimization or positioning of its side-chains given a rigid backbone conformation is the remaining second subproblem called side-chain positioning (see Figure 5).

3.1   Side-Chain Positioning

A well-studied and frequently occurring subproblem is the optimal placement of the side chains of the amino acids forming a protein. This task can be formulated as a discrete optimization problem by using a rotamer library that represents the conformational space of each side chain by a discrete set of conformations (rotamers). The respective combinatorial optimization problem is denoted as

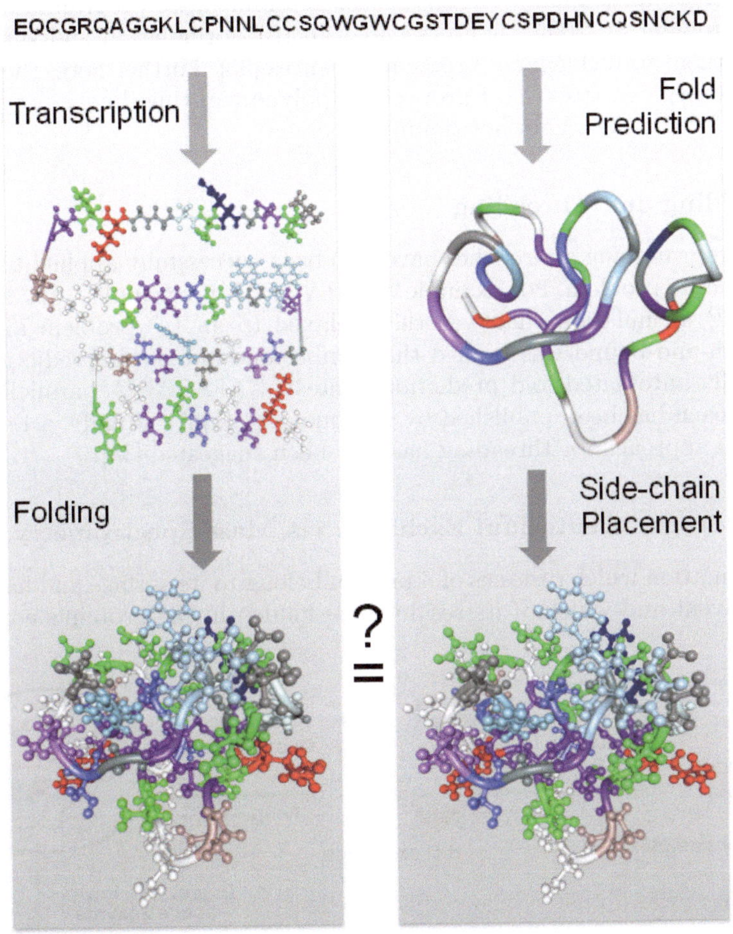
EQCGRQAGGKLCPNNLCCSQWGWCGSTDEYCSPDHNCQSNCKD
Transcription
Fold Prediction
Folding
Side-chain Placement
?
=

Fig. 5. The synthesis of proteins entails the translation of mRNA into a linear sequence of amino acids. These amino acid chains fold into a three-dimensional structure. Most structure prediction approaches first try to determine the fold of the backbone. The more detailed amino acid side-chain positions can then be predicted by an additional step, the side-chain placement or positioning. If these two problems are solved correctly, the resulting 3D structure will be close to the experimentally observed protein structure.

the global minimum energy conformation (GMEC) problem, because we have to identify the set of rotamers that build the conformation with minimal energy. To solve this problem, many different algorithms have been proposed, among them different ILP approaches. The first ILP formulation and B&C approach is due to Althaus et al. [5, 30]. They applied the side chain optimization problem to protein-protein docking. A quite similar ILP formulation was derived by Eriksson et al. [31].

Chazelle et al. [32] first suggested using semidefinite programming for the GMEC problem. In 2005, they presented an ILP formulation and showed that it allows them to tackle large problem instances [33]. Furthermore, they relaxed the integrality constraint and proposed a polynomial-time linear programming heuristic for the side chain positioning problem.

3.2   Folding and Threading

Integer programming approaches have also been successfully applied to the protein threading problem. For example, the RAPTOR program [34, 35, 36] relying on an ILP formulation that was then relaxed to an LP problem and solved via branch-and-bound was ranked the number one individual prediction server at the fully-automated fold prediction contest, CAFASP3. A parallelization of this approach has been published by Andonov et al. [37]. Finally, a Lagrangian relaxation approach for threading has also been suggested [38].

3.3   Hydrogen-Deuterium Exchange via Mass Spectrometry

The information which residues of a protein belong to its surface and have access to the solvent and which of its residues are hidden in the protein's core can be

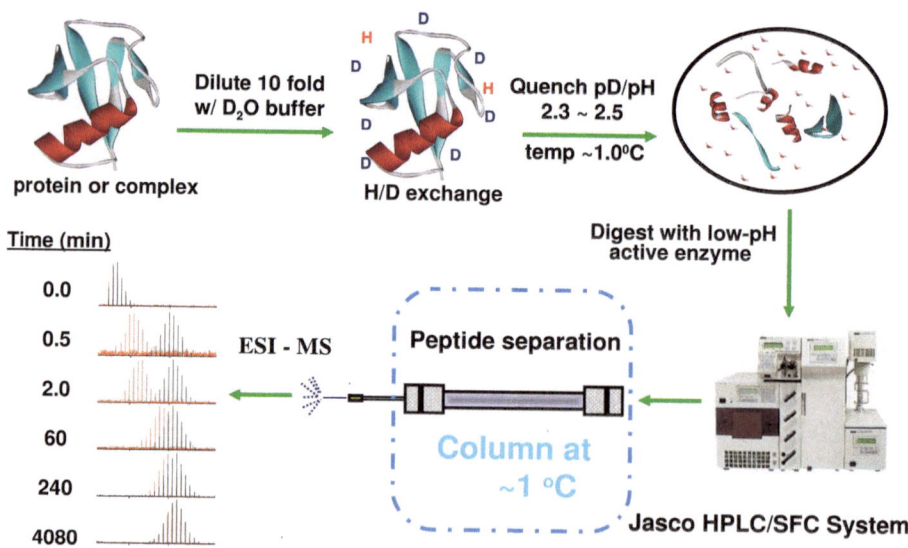

Fig. 6. A schematic view of the HDX experiment. The protein is diluted in a D_2O buffer. At different time points, it is then digested and analyzed with a mass spectrometer. After a separation of the different digested fragments, we obtain their average mass uptakes at the time points. All steps in the HDX experiment are automated and performed by a CTC PAL robot: sample dilution, mixing, quench/digestion, timing and HPLC injection.

used, e.g., to evaluate predicted 3D models and to eliminate false positive predictions. Solution-phase hydrogen/deuterium exchange (HDX) with high-resolution mass analysis permits identification of the solvent access and contact surfaces in a protein/protein complex. Measuring HDX via mass spectrometry exhibits the deuterium uptake of the peptic fragments that are obtained by digestion (see Figure 6 for an illustration of the experiment). The evaluation of this data is done by first analyzing the individual fragments with state-of-the-art methods which reveal cumulative exchange rates of the fragments. From these, it is possible to infer exchange rates of single amino acids or small parts of the protein. These inferences are mainly done by manual inspection, which requires a large amount of time and is vulnerable to errors [39].

To explain the experiment, we simplify the biochemistry. Given a protein consisting of n amino-acids, where each amino-acid has its *exchange rate* k_i. Dilution of the protein in D_2O will cause that a specific hydrogen of the amino acid will exchange to a deuterium, resulting in an increase of the mass by one neutron mass. The expectation that the hydrogen is exchanged at time t is $1 - e^{k_i t}$. Using mass-spectrometry, we are able to measure the average mass uptake of peptic fragments (parts of the protein) at different time points. Given exchange rates $k_1, \ldots, k_n$, the expected uptake of a peptic fragment from amino-acid i to j after time t is $\sum_{l=i}^{j}(1 - e^{k_l t})$. Our problem is to find exchange rates $k_1, \ldots, k_n$ so that the expected and the measured mass uptakes are as similar as possible. This is a non-linear curve-fitting problem.

Althaus et al. [40] provided the first automatic methods to solve this problem, i.e., they proposed an algorithm to obtain exchange rates below the level of digested fragments. They applied two different approaches to solve this problem. In [40] they used known methods to analyze single digested fragments. Given a fragment from i to j and the mass uptake of this fragment over time, these methods return $i - j + 1$ exchange rates, which are most likely the exchange rates

No.	Amine Acid Sequence	Rates
	G L S D G E W Q Q V L N V W G K V E A D I A G H G A E V L	s m f
1	←——→	15 8 5
2	←————————————————————→	7 2 1
3	←——————————————→	5 2 1
4	←——————————————————————————————→	12 1 4
5	←————————→	5 1 1
6	←——→	11 1 3
7	←——————————→	4 1 1
8	←————————→	3 1 0
9	←————————————————→	7 1 0

Fig. 7. We show the cumulative exchange rates of several overlapping fragments partitioned into three classes (slow, medium, fast). We want to automatically draw conclusions on the exchange rates of single amino acids, like that the second D has to have medium exchange rate concluded from the fragments 3 and 5.

within the fragment. This cumulative exchange rate information is computed for all fragments. The calculated exchange rates are partitioned into three classes. Then, we compute the assignment of exchange rates to amino acids such that the number of errors is minimized (see Figure 7).

In [41] Althaus et al. extended this approach by computing exchange rates directly from the deuterium uptake. Given a model for the exchange process (like the $(1 - e^{kt})$ from the simplified explanation of the experiment) and deuterium uptake for the fragments over time, the problem of finding the exchange rates for the amino acids can be formulated as a non-linear curve fitting problem. As the number of parameters of this curve fitting problem is very high, standard tools to estimate parameters are not applicable. By discretizing the exchange rates to a set of candidates, Althaus et al. were able to formulate an integer linear program whose solution provides the optimal solution to the curve fitting problem with respect to the discretization and which is nevertheless solvable in reasonable time.

4   Probe Design for Microarray Experiments

Microarray technology has become a widely used analytical technique in the life sciences, providing a cost-efficient way to determine levels of specified RNA or DNA molecules in a biological sample. Typically, one measures the amount of gene expression in a cell by observing hybridization of mRNA to different probes on a microarray, each probe targeting a specific gene. A distinct and likewise important application, arising, for example, in medicine, environmental sciences, industrial quality control, or biothreat reduction, is the identification of biological agents in a sample. This wide range of applications leads to the same methodological problem: To determine the presence or absence of *targets*—such as viruses or bacteria—in a biological sample.

Given a collection of genetic sequences of targets, one faces the challenge of finding short oligonucleotides, the *probes*, which allow the detection of targets in a sample by hybridization experiments. The experiments are conducted using either unique or non-unique probes, and the problem at hand is to compute a minimal design, i.e., to select a minimal set of probes from a larger set of probe candidates that enables us to infer the targets in the sample from the hybridization results. When testing for more than one target in the sample, a relevant problem in practice, the design of an optimal probe set becomes NP-hard for the case of non-unique probes.

The example in Table 2 illustrates the problem: It consists of four targets $t_1, \ldots, t_4$ and nine candidate probes $p_1, \ldots, p_9$. Hybridization between targets and probes is characterized by an incidence matrix. Assume first that we know that the sample contains at most one target. The goal is to select a minimal set of probes that allows us to infer the presence of a single target. In this example, it is sufficient to use probes p_1, p_2, and p_3 for detecting the presence of a single target (e.g., for target t_2 probes p_1 and p_3 hybridize, while p_2 does not). Minimizing the number of probes is a reasonable objective function, since it is proportional

Table 2. A small example of a target-probe incidence matrix

	p_1	p_2	p_3	p_4	p_5	p_6	p_7	p_8	p_9
t_1	1	1	1	0	1	1	0	0	0
t_2	1	0	1	1	0	0	1	1	0
t_3	0	1	1	1	0	1	1	0	1
t_4	0	1	0	0	1	0	1	1	1

to the cost of the experiment. In the example, $\{p_1, p_2, p_3\}$ is an optimal choice. The smaller choice $\{p_4, p_8\}$, however, does not allow distinguishing between the empty target set and the set $\{t_1\}$.

Now assume that target t_2 and target t_3 are *simultaneously* present in the sample, but none of the remaining targets are. In this case all three probes p_1, p_2, and p_3 hybridize. This case cannot be distinguished from the case where only t_1 is present. As a remedy, we could take all the probes $p_1, \dots, p_9$. This is, however, not necessary. With p_1, p_4, p_5, and p_9, all target sets of cardinality ≤ 2 can be distinguished, with the exception of $\{t_1, t_3\}$ vs. $\{t_2, t_4\}$. Adding probe p_8 to the design allows making this last distinction as well. It is clear that taking all probes results in the best possible separation between all subsets. However, we can often achieve the same quality with a substantially smaller number of probes.

In addition to the difficulty illustrated above, the problem is aggravated by the presence of errors. Usually, the false positive error rate f_1, i.e., the rate at which the experiment reports a hybridization when there is none, and the false negative rate f_0, i.e., the rate at which the experiment fails to report a hybridization, are up to 5%. As a remedy, it is customary to build some redundancy into the design; e.g., demanding that two targets be separated by more than one probe and that each target hybridize to more than one probe.

In [42], Klau, Rahmann, Schliep, Vingron, and Reinert have presented the first approach for selecting a minimal probe set respecting the redundancy properties for the case of non-unique probes in the presence of a small number of multiple targets in the sample. Their approach is based on an ILP formulation and a branch-and-cut algorithm. Experimental results on real and artificial data show that the exact ILP approach significantly reduces the number of probes needed as compared to traditional greedy method, while preserving the decoding capabilities of existing approaches. In [43], the authors have extended their work and have presented an additional, more elegant ILP formulation.

5   Computational Systems Biology

In systems biology, biological processes are often modeled as networks. Hence, the construction and analysis of large biological networks, such as protein-protein interaction (PPI), metabolic, gene-regulatory, and signal transduction networks, have become major research topics. The increasing quantity of available data creates the need for automated analysis methods to better understand cellular processes, network organization, evolutionary changes, and disease mechanisms.

Well-established microarray technologies provide a wealth of information on gene expression in various tissues and under diverse experimental conditions. Integrating, e.g., protein-protein interaction and gene expression data generates a meaningful biological context in terms of functional association for differentially expressed genes.

5.1 Network-Based Disease Bioinformatics

Frequently, large scale medical expression profiling studies investigate many experimental conditions simultaneously, thereby generating multiple measures of significance, typically expressed in the form of p-values. Especially in tumor biology, expression profiling has become a widely-used technique for the classification of different tumors and tumor subtypes. Furthermore, in the clinical context, various patient-associated data is available that—in conjunction with expression data—provides valuable information on the influence of specific genes on disease-specific pathophysiology. In particular, the analysis of survival data allows establishing gene expression signatures to make predictions about the prognosis and to assess the disease relevance of certain genes. The cellular function of an individual gene cannot be understood on the level of isolated components alone, but needs to be studied in the context of its interplay with other gene products. The combined analysis of expression profiles and protein-protein interaction data thus allows the detection of previously unknown disregulated modules in interaction networks not recognizable by the analysis of pathways defined *a priori*.

To identify interaction modules in this setting it is necessary to devise firstly an adequate scoring function on networks and secondly an algorithm to find maximally-scoring subnetworks. This problem, as stated by Ideker et al. [44], has been proven to be NP-hard. Traditionally, it has been approached via heuristics. Some of these often computationally demanding algorithms tend to deliver large high-scoring networks, which may be difficult to interpret.

Dittrich, Klau, Rosenwald, Dandekar, and Müller [7] have presented a novel, ILP-based approach for this problem, which uses its relation to the well-studied prize-collecting Steiner tree problem (PCST). Their method utilizes a modular scoring function, based on signal-noise decomposition implemented as a mixture model that permits the smooth integration of multivariate p-values derived from various sources. Given the resulting protein scores, a branch-and-cut algorithm, originally developed for PCST, delivers provably optimal and suboptimal solutions to the maximal-scoring subgraph problem. The resulting subnetwork size can be controlled by an adjustment parameter that is statistically interpretable as false discovery rate. Figure 8 shows an optimal subnetwork that has been computed based on microarray and survival data collected in a study involving around 200 patients suffering from two different lymphoma subtypes. The ILP-based algorithm was able to discover biologically meaningful disregulated modules that included and extended modules that are well-known for the pathogenesis of the two tumor subtypes. Moreover, a direct comparison with a widely-used heuristic approach on simulated data clearly demonstrated the

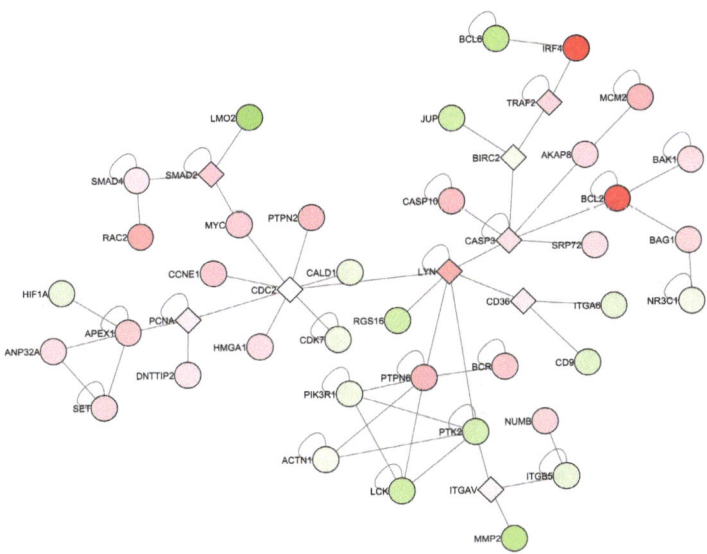

Fig. 8. Example of a potentially disregulated functional module showing differences between the ABC and GCB lymphoma subtypes. This optimal subnetwork has been detected using a score based on a gene-wise two-sided t-test for differential expression and a univariate Cox regression hazard model. The derived subnetwork captures the differentially expressed interaction modules associated with the increased malignancy of the ABC subtype. Coloring is according to the fold change where red denotes an overexpression in ABC and green in GCB. Diamond nodes represent negative scoring genes additionally included in the optimal solution.

shortcomings of the heuristics. Surprisingly, the exact algorithm was also able to compute the optimal solutions much faster than the heuristics needed to generate their significantly worse solutions.

5.2 Comparative Network Analysis

Based on the assumption that evolutionary conservation implies functional significance, comparative network analysis may help to elucidate protein pathways and interactions. Moreover, it facilitates the generation, investigation, and validation of hypotheses about the underlying networks, and transfer functional annotations. In addition to component-based comparative approaches, network alignments provide the means to study conserved network topology such as common pathways and more complex network motifs. Yet, unlike in classical sequence alignment, the comparison of networks becomes computationally challenging already in the pairwise case; as most meaningful assumptions instantly lead to NP-hard problems.

Recently, many heuristic approaches have been proposed for aligning two or multiple networks. In [45], Klau has introduced the *maximum structural matching* formulation as a basis for an exact approach to global pairwise network

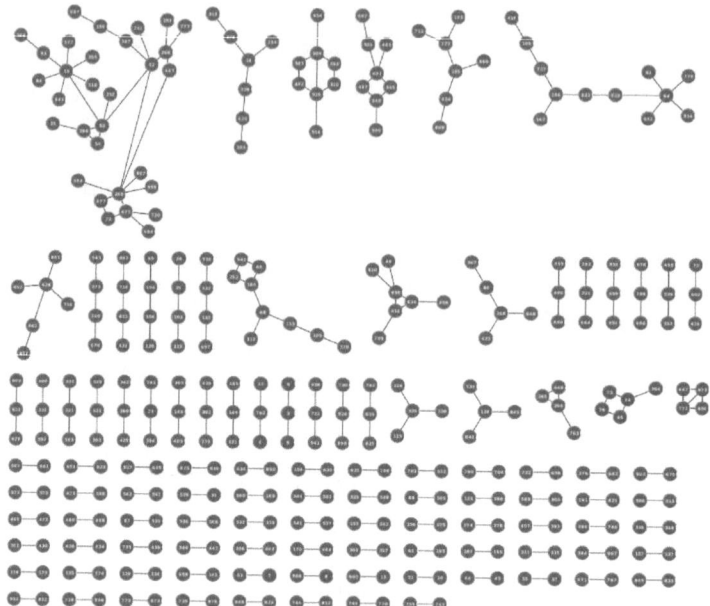

Fig. 9. Maximum common subnetworks with respect to the number of conserved interactions between *Rattus norvegicus* and *Mus musculus* protein-protein interaction networks

alignment. He reformulates the problem in terms of an ILP and proposes a Lagrangian relaxation approach based on this formulation. The approach is inspired by the work on RNA sequence-structure alignment and contact map overlap described in Section 2.2. Figure 9 shows an example of a provably maximum common subnetwork between two protein-protein interaction networks from rat and mouse.

6   Vaccine Design

The human immune system is one of the most complex biological systems. It has evolved to protect the host from invading pathogens (viruses, bacteria, etc.) and also to recognize aberrant cells (e.g., in the context of cancer). Vaccines are one of the most successful therapeutic strategies known. They activate the immune system to recognize and remove invading pathogens or cancer cells and thus prevent infection or aid in cancer therapy. Over the last decade, tailor-made vaccines based on individual peptides (short fragments of proteins) have attracted considerable interest, in particular, because they also facilitate a personalized therapy. These peptide-based or epitope-based vaccines trigger an immune response by confronting the immune system with peptides derived from proteins originating from the pathogen or from cancer-specific proteins. The peptides bind to major histocompatibility complex (MHC) molecules and thus initiate an immune response. However, there are many allelic variants of MHC molecules, meaning that different patients typically bind different repertoires of peptides. Due to economic and regulatory issues, one cannot simply add all immunogenic peptides to such a peptide mix. Hence, it is crucial to identify the optimal set of peptides for a vaccine, given constraints such as MHC allele frequencies in the target population, peptide mutation rates, and maximum number of selected peptides. The selection of these peptides results in an interesting optimization problem. Toussaint et al. [46, 47] were able to formulate the vaccine design problem as an ILP. Most immunological requirements can be integrated in an elegant fashion as constraints and the resulting problems can then be efficiently solved. The ILP approach is the first method to solve this problem to optimality and it outperforms previous (heuristic) approaches with respect to overall immunogenicity.

7   Summary

The preceding discussion of ILP-based approaches highlights how Kurt Mehlhorn's group has made substantial contributions to bioinformatics research through the introduction of ILP techniques to numerous different problem areas. The work initiated by his group was often seminal and the research issues that were raised remained active topics for several years emphasizing their relevance. Half of the papers cited in this review article are co-authored by former members of Kurt Mehlhorn's group, although the review covers many of bioinformatics' core problems, especially in the area of sequence analysis, structural bioinformatics, probe design for microarray experiments, computational systems biology, and vaccine design. It stands to reason that ILP techniques are now established in bioinformatics. Bioinformatics is still a young discipline and – as Donald Knuth put it – "biology easily has 500 years of exciting problems to work on" [48]. We believe that combinatorial optimization techniques will play a key role in tackling some of these problems in the future.

Acknowledgements

The authors are grateful to Anne Dehof for providing Fig. 5.

References

1. Alizadeh, F., Karp, R., Weisser, D., Zweig, G.: Physical mapping of chromosomes using unique probes. In: Proceedings of the Annual ACM-SIAM Symposium on Discrete Algorithms (SODA 1994), pp. 489–500 (1994)
2. Reinert, K., Lenhof, H., Mutzel, P., Mehlhorn, K., Kececioglu, J.: A branch-and-cut algorithm for multiple sequence alignment. In: Proceedings of the First Annual International Conference on Computational Molecular Biology (RECOMB 1997), pp. 241–249 (1997)

3. Christof, T., Jünger, M., Kececioglu, J., Mutzel, P., Reinelt, G.: A branch-and-cut approach to physical mapping with end-probes. In: Proceedings of the First Annual International Conference on Computational Molecular Biology (RECOMB 1997), pp. 84–92 (1997)

4. Lenhof, H.P., Reinert, K., Vingron, M.: A polyhedral approach to RNA sequence structure alignment. In: Proceedings of the Second Annual International Conference on Computational Molecular Biology (RECOMB 1998), pp. 153–162 (1998)

5. Althaus, E., Kohlbacher, O., Lenhof, H., Müller, P.: A combinatorial approach to protein docking with flexible side-chains. In: Proceedings of the Second Annual International Conference on Computational Molecular Biology (RECOMB 2000), pp. 15–24 (2000)

6. Klau, G., Rahmann, S., Schliep, A., Vingron, M., Reinert, K.: Optimal robust non-unique probe selection using integer linear programming. Bioinformatics 20, i186–i193 (2004)

7. Dittrich, M., Klau, G.W., Rosenwald, A., Dandekar, T., Müller, T.: Identifying functional modules in protein-protein interaction networks: an integrated exact approach. Bioinformatics 24(13), i223 (2008)

8. Lancia, G.: Mathematical programming in computational biology: an annotated bibliography. Algorithms 1(2), 100–129 (2008)

9. Kececioglu, J.: The maximum weight trace problem in multiple sequence alignment. In: Apostolico, A., Crochemore, M., Galil, Z., Manber, U. (eds.) CPM 1993. LNCS, vol. 684, pp. 106–119. Springer, Heidelberg (1993)

10. Lenhof, H.P., Morgenstern, B., Reinert, K.: An exact solution for the segment-to-segment multiple sequence alignment problem. Bioinformatics 15(3), 203–210 (1999)

11. Althaus, E., Caprara, A., Lenhof, H.P., Reinert, K.: Multiple sequence alignment with arbitrary gap costs: Computing an optimal solution using polyhedral combinatorics. In: Proceedings of the 1st European Conference on Computational Biology (ECCB 2002), pp. 4–16 (2002)

12. Althaus, E., Caprara, A., Lenhof, H.P., Reinert, K.: A branch-and-cut algorithm for multiple sequence alignment. Mathematical Programming 105, 387–425 (2006)

13. Althaus, E., Canzar, S.: A Lagrangian relaxation approach for the multiple sequence alignment problem. In: Dress, A.W.M., Xu, Y., Zhu, B. (eds.) COCOA 2007. LNCS, vol. 4616, pp. 267–278. Springer, Heidelberg (2007)

14. Althaus, E., Canzar, S.: A Lagrangian relaxation approach for the multiple sequence alignment problem. J. Combinat. Opt. 16(2), 127–154 (2008)

15. Fischetti, M., Lancia, G., Serafini, P.: Exact algorithms for minimum routing cost trees. Networks 39, 161–173 (2002)

16. Gusfield, D.: Efficient methods for multiple sequence alignment with guaranteed error bounds. Bulletin of Mathematical Biology 55, 141–154 (1993)

17. Kececioglu, J., Kim, E.: Simple and fast inverse alignment. In: Apostolico, A., Guerra, C., Istrail, S., Pevzner, P.A., Waterman, M. (eds.) RECOMB 2006. LNCS (LNBI), vol. 3909, pp. 441–455. Springer, Heidelberg (2006)

18. Mattick, J.S.: The functional genomics of noncoding RNA. Science 309(5740), 1527–1528 (2005)

19. Goldman, D., Istrail, S., Papadimitriou, C.H.: Algorithmic aspects of protein structure similarity. In: FOCS, pp. 512–522 (1999)

20. Wang, L., Jiang, T.: On the complexity of multiple sequence alignment. J. Comput. Biol. 1(4), 337–348 (1994)

21. Lancia, G., Carr, R., Walenz, B., Istrail, S.: 101 optimal PDB structure alignments: a branch-and-cut algorithm for the maximum contact map overlap problem. In: Proc. of the Fifth Annual International Conference on Computational Biology, pp. 193–202. ACM Press, New York (2001)
22. Caprara, A., Lancia, G.: Structural Alignment of Large-Size Proteins via Lagrangian Relaxation. In: Proc. of RECOMB 2002, pp. 100–108. ACM Press, New York (2002)
23. Caprara, A., Pisinger, D., Toth, P.: Exact solution of the quadratic knapsack problem. Informs J. on Computing 11(2), 125–137 (1999)
24. Carraresi, P., Malucelli, F.: A reformulation scheme and new lower bounds for the quadratic assignment problem. In: Quadratic Assignment and Related Topics. DIMACS Series in Discrete Mathematics and Theoretical Computer Science, pp. 147–160. AMS Bookstore (1994)
25. Bauer, M., Klau, G.W.: Structural Alignment of Two RNA Sequences with Lagrangian Relaxation. In: Fleischer, R., Trippen, G. (eds.) ISAAC 2004. LNCS, vol. 3341, pp. 113–123. Springer, Heidelberg (2004)
26. Bauer, M., Klau, G.W., Reinert, K.: Multiple structural RNA alignment with Lagrangian relaxation. In: Casadio, R., Myers, G. (eds.) WABI 2005. LNCS (LNBI), vol. 3692, pp. 303–314. Springer, Heidelberg (2005)
27. Bauer, M., Klau, G.W., Reinert, K.: An exact mathematical programming approach to multiple RNA sequence-structure alignment. Algorithmic Operations Research (2008); Special Issue on Biology, Medicine, and Health Care
28. Bauer, M., Klau, G.W., Reinert, K.: Accurate multiple sequence-structure alignment of RNA sequences using combinatorial optimization. BMC Bioinformatics 8(1), 271 (2007)
29. Notredame, C., Higgins, D.G., Heringa, J.: T-Coffee: A novel method for fast and accurate multiple sequence alignment. Journal of Molecular Biology 302, 205–217 (2000)
30. Althaus, E., Kohlbacher, O., Lenhof, H.P., Müller, P.: A combinatorial approach to protein docking with flexible side-chains. J. Comput. Biol. 9(4), 597–612 (2002)
31. Eriksson, O., Zhou, Y., Elofsson, A.: Side-chain positioning as an integer programming problem. In: Gascuel, O., Moret, B.M.E. (eds.) WABI 2001. LNCS, vol. 2149, pp. 128–141. Springer, Heidelberg (2001)
32. Chazelle, B., Kingsford, C., Singh, M.: A semidefinite programming approach to side chain positioning with new rounding strategies. Informs J. Comput. 16, 380–392 (2004)
33. Kingsford, C., Chazelle, B., Singh, M.: Solving and analyzing side-chain positioning problems using linear and integer programming. Bioinformatics 21, 1028–1039 (2005)
34. Xu, J., Li, M., Kim, D., Xu, Y.: RAPTOR: optimal protein threading by linear programming. J. Bioinformatics Comput. Biol. 1(1), 95–117 (2003)
35. Xu, J., Li, M.: Assessment of RAPTOR's linear programming approach in CAFASP3. Proteins 53(suppl. 6), 579–584 (2003)
36. Xu, J., Li, M., Xu, Y.: Protein threading by linear programming: theoretical analysis and computational results. J. Combinat. Opt. 8(4), 403–418 (2004)
37. Andonov, R., Balev, S., Yanev, N.: Protein threading: From mathematical models to parallel implementations. Informs J. Comput. 16(4), 393–405 (2004)
38. Veber, P., Yanev, N., Andonov, R., Poirriez, V.: Optimal protein threading by cost-splitting. In: Casadio, R., Myers, G. (eds.) WABI 2005. LNCS (LNBI), vol. 3692, pp. 365–375. Springer, Heidelberg (2005)

39. Zhang, Z., Post, C.B., Smith, D.L.: Amide hydrogen exchange determined by mass spectrometry: application to rabbit muscle aldolase. Biochemistry 35, 779–791 (1996)
40. Althaus, E., Canzar, S., Emmett, M.R., Karrenbauer, A., Marshall, A.G., Meyer-Baese, A., Zhang, H.: Computing H/D-exchange speeds of single residues from data of peptic fragments. In: Proceedings of the 23rd Annual ACM Symposium on Applied Computing, Fortaleza, Ceará, Brazil (2008)
41. Althaus, E., Canzar, S., Ehrler, C., Emmett, M.R., Karrenbauer, A., Marshall, A.G., Meyer-Bäse, A., Tipton, J., Zhang, H.: Discrete fitting of hydrogen-deuterium-exchange-data of overlapping fragments. In: The 2009 International Conference on Bioinformatics & Computational Biology (in press, 2009) (accepted for publication)
42. Klau, G.W., Rahmann, S., Schliep, A., Vingron, M., Reinert, K.: Optimal robust non-unique probe selection using integer linear programming. In: Proceedings of the Twelfth International Conference on Intelligent Systems for Molecular Biology (ISMB 2004), pp. 186–193 (2004)
43. Klau, G.W., Rahmann, S., Schliep, A., Vingron, M., Reinert, K.: Integer linear programming approaches for non-unique probe selection. Discrete Applied Mathematics 155, 840–856 (2007)
44. Ideker, T., Ozier, O., Schwikowski, B., Siegel, A.F.: Discovering regulatory and signalling circuits in molecular interaction networks. Bioinformatics 18(suppl. 1), S233–S240 (2002)
45. Klau, G.W.: A new graph-based method for pairwise global network alignment. BMC Bioinformatics 10(suppl. 1), S59 (2009)
46. Toussaint, N.C., Dönnes, P., Kohlbacher, O.: A mathematical framework for the selection of an optimal set of peptides for epitope-based vaccines. PLoS Comput. Biol. 4(12), e1000246 (2008)
47. Toussaint, N.C., Kohlbacher, O.: OptiTope – A web server for the selection of an optimal set of peptides for epitope-based vaccines. Nucl. Acids Res. (in press, 2009)
48. Knuth, D.E.: Donald Knuth – Computer Literacy Bookshops Interview (1993), http://tex.loria.fr/historique/interviews/knuth-clb1993.html

Via Detours to I/O-Efficient Shortest Paths

Ulrich Meyer*

Institute for Computer Science,
Goethe University,
60325 Frankfurt/Main, Germany
umeyer@cs.uni-frankfurt.de
http://www.uli-meyer.de

Abstract. During the winter semester 1996/1997 Kurt Mehlhorn gave a lecture series on *algorithms for very large data sets*. Towards the end he covered graph traversal problems like finding shortest paths on sparse graphs in external-memory. Kurt concluded the topic stating that there may not be much hope to solve these basic problems I/O-efficiently. The author, just about to finish his master's studies those days, took this as a challenge. The following paper reviews some detours, dead-ends, and happy ends of the author's still ongoing research on external-memory graph traversal.

1 Introduction

Shortest path problems are among the most fundamental and also the most commonly encountered graph problems, both in themselves and as subproblems in more complex settings [3]. Besides obvious applications like preparing travel time and distance charts [28], shortest path computations are frequently needed in telecommunications and transportation industries [53], where messages or vehicles must be sent between two geographical locations as quickly or as cheaply as possible. Other examples are complex traffic flow simulations and planning tools [28], which rely on solving a large number of individual shortest path problems.

One of the most commonly encountered subtypes is the Single-Source Shortest-Path (SSSP) version: let $G = (V, E)$ be a graph with $|V|$ nodes and $|E|$ edges, let s be a distinguished vertex of the graph, and c be a function assigning a non-negative real *weight* to each edge of G. The objective of the SSSP is to compute, for each vertex v reachable from s, the weight dist(v) of a minimum-weight ("shortest") path from s to v; the weight of a path is the sum of the weights of its edges.

Breadth-First Search (BFS) [50] can be seen as the unweighted version of SSSP; it decomposes a graph into levels where level i comprises all nodes that can be reached from the source via i edges. The BFS numbers also impose an order on the nodes within the levels. BFS has been widely used since the late 1950's; for example, it is an ingredient of the classical separator algorithm for planar graphs [33].

* Partially supported by the DFG grant ME 3250/1-1, and by MADALGO - Center for Massive Data Algorithmics, a Center of the Danish National Research Foundation.

S. Albers, H. Alt, and S. Näher (Eds.): Festschrift Mehlhorn, LNCS 5760, pp. 219–232, 2009.

We are mainly interested in the case where the input graphs for BFS and SSSP are too big to fit in the main-memory of the computing device. The world-wide-web for example can be looked upon as a massive graph where each webpage is a node and the hyperlink from one page to another is an edge between the nodes corresponding to those pages. As of April 2009, it is estimated that the indexed web contains at least 20 billion webpages [18].

We consider the commonly accepted external-memory (EM) model of Aggarwal and Vitter [2]. It assumes a two level memory hierarchy with faster internal memory having a capacity to store M vertices/edges. In an I/O operation, one block of data, which can store B vertices/edges is transferred between disk and internal memory. The measure of performance of an algorithm is the number of I/Os it performs. The number of I/Os needed to read N contiguous items from disk is $\mathrm{scan}(N) = \Theta(N/B)$. The number of I/Os required to sort N items is $\mathrm{sort}(N) = \Theta((N/B)\log_{M/B}(N/B))$. For all realistic values of N, B, and M, $\mathrm{scan}(N) < \mathrm{sort}(N) \ll N$. A large number of results have appeared in the area of external-memory computing. Vitter provides the most recent and extensive overview [56].

2   State of the Art in 1997 and Main Problems

When Kurt Mehlhorn covered **EM** shortest path in his lecture series on *algorithms for very large data sets* in the winter semester 1996/97 all known BFS and SSSP approaches took $\Omega(|V|)$ I/Os for general graphs in the worst-case. Using simple calculations Kurt impressively demonstrated that for typical main-memory sizes (already in 1997) if $|V|$ exceeds M by a small constant factor, the expected time to perform the required I/Os would clearly break all practical limits. At that time, an I/O operation on a high-end hard disk was accounted with about 10 milliseconds and in spite of increased throughput these values have not decreased tremendously ever since. However, there is a recent trend to replace hard disks by flash memory devices, which can perform *read*-I/Os in just 0.1 milliseconds. On the other hand, for huge inputs like the world-wide-web graph mentioned before, $|V| = 2 \cdot 10^{10}$ I/Os would still mean nearly one *month* of I/O waiting time.

Before explaining the main difficulties of external-memory graph traversal we first shortly review internal-memory (**IM**) algorithms [16] for BFS and SSSP. They typically visit the vertices of the input graph G in a one-by-one fashion; appropriate candidate nodes for the next vertex to be visited are kept in some data-structure Q (a FIFO-queue for BFS and a priority-queue for SSSP). After a vertex v is extracted from Q, the adjacency list of v, i.e., the set of neighbors of v in G, is examined in order to update Q: unvisited neighboring nodes are inserted into Q; for SSSP the priorities of nodes already in Q may be updated. The short description above already contains the main difficulties for I/O-efficient graph-traversal algorithms:

(a)   Unstructured indexed access to adjacency lists.
(b)   Remembering visited nodes.
(c)   (The lack of efficient) Decrease_Key operations in **EM** priority-queues.

The impact of (a) depends on the sizes of the adjacency lists; if a list contains k edges then it takes $\Theta(1 + k/B)$ I/Os to retrieve all its edges. This is efficient if $k = \Omega(B)$, but wasteful if, e.g., $k = \mathcal{O}(1)$.

Problem (b) can be partially overcome by solving the graph problems *in phases* [13]: a dictionary DI of maximum capacity $|\text{DI}| < M$ is kept in internal memory; DI serves to remember visited nodes. Whenever the capacity of DI is exhausted, the algorithms make a pass through the external graph representation: all edges pointing to visited nodes are discarded, and the remaining edges are compacted into new adjacency lists. Then DI is emptied, and a new phase starts by visiting the next element of Q. This phase-approach explored in [13] is most efficient if the quotient $|V|/|\text{DI}|$ is small; $\mathcal{O}(\lceil |V|/|\text{DI}| \rceil \cdot \text{scan}(|V|+|E|))$ I/Os are needed in total to perform all graph compactions. Additionally, $\mathcal{O}(|V|+|E|)$ operations are performed on Q.

As for SSSP, problem (c) is less severe if (b) is resolved by the phase-approach: instead of actually performing Decrease_Key operations, several priorities may be kept for each node in the external priority-queue; after a node v is dequeued for the first time (with the smallest key) any further appearance of v in Q will be ignored. In order to make this work, superfluous elements still kept in the **EM** data structure of Q are marked obsolete right before DI is emptied at the end of a phase; the marking can be done by scanning Q. Thus, plugging in the respective I/O-bounds yields $\mathcal{O}(|V| + \lceil |V|/M \rceil \cdot \text{scan}(|V| + |E|))$ I/Os for BFS and $\mathcal{O}(|V| + \lceil |V|/M \rceil \cdot \text{scan}(|V| + |E|) + \text{sort}(|E|))$ I/Os for SSSP.

For undirected graphs, another possibility is to solve (b) and (c) by applying extra bookkeeping and extra data structures like the I/O-efficient tournament tree of Kumar and Schwabe [32]. In that case the graph traversal can be done in *one* phase, even if $n \gg M$. The respective I/O-bounds are $\mathcal{O}(|V|+\text{sort}(|V|+|E|))$ I/Os for BFS and $\mathcal{O}(|V| + \lceil |V|/M \rceil \cdot \text{scan}(|V| + |E|) + \text{sort}(|E|))$ I/Os for SSSP.

Problems (b) and (c) usually disappear in the semi-external memory (**SEM**) setting where it is assumed that $M = c \cdot |V| < |E|$ for some appropriately chosen positive constant c: e.g., the **SEM** model may allow to keep a boolean array for (b) in internal memory; similarly, a node priority queue with Decrease_Key operation for (c) could reside completely in **IM**.

3   First Attempts: Nice Results – Just Not for EM

In our first attempts to improve the known **EM** graph traversal algorithms we tried to apply some kind of parallel simulation. The parallel random access machine [31] (PRAM) is one of the most widely studied abstract models of a parallel computer. A PRAM consists of p independent processors and a shared memory, which these processors can synchronously access in unit time. The performance of PRAM algorithms is usually described by the two parameters *time* (assuming an unlimited number of available PUs) and *work* (the total number of operations needed). A fast and efficient parallel algorithm minimizes both time and work; ideally the work is asymptotic to the sequential complexity of the problem.

A PRAM simulation [13] translates one step of the parallel algorithm into a constant number of global scanning and sorting operations. This scheme has been used successfully to yield efficient solutions for **EM** list-ranking – a problem that previously suffered from a high I/O penalties due to unstructured accesses to adjacency lists just like BFS and SSSP. However, the simulation requires a fast and work-efficient parallel algorithm for the problem at hand. Unfortunately, the parallel SSSP problem has so far resisted solutions that are fast and work-efficient at the same time: no PRAM algorithm is known that terminates with $\mathcal{O}(|V| \cdot \log |V| + |E|)$ work (the sequential complexity of Dijkstra's algorithm [20] with Fibonacci heaps [23]) and sublinear running time for arbitrary graphs with nonnegative edge weights.

Conservative Parallelization. Instead of becoming frustrated with worst-case efficiency we decided to explore the average-case behavior of Dijkstra's algorithm first. In fact together with Kurt we found a number of criteria [17, 41] which can be used to identify vertices in the priority queue whose tentative distances are already final and which therefore can be removed from the priority queue earlier than in Dijkstra's algorithm – and in parallel. Unfortunately, it turned out that even under favorable side-conditions (random graphs with random edge weights) our best criteria still required $\Omega(|V|^{1/3})$ parallel phases: too many for standard PRAM simulation, which would have resulted in an $\Omega(|V|^{1/3} \cdot \text{sort}(|V| + |E|))$ I/O approach, thus taking $\Omega(|V|)$ I/Os again for sufficiently large values of $|V|$.

Aggressive Parallelization. Motivated by our parallelization of Dijkstra's algorithm mentioned above we kept on searching for improved work-efficient SSSP parallelizations, still hoping to find the link to external-memory. A major step forward – as we thought then – was to shift from conservative to aggressive node selection (yielding a label-correcting approach). Indeed the parallel running time for SSSP on random graphs with random edge weights could be reduced to $\mathcal{O}(\log^2 n)$ using just linear work on average. Also, for other interesting graph classes (like graphs modeling the WWW, telephone calls or social networks) the theoretical bounds could be improved [40]. A recent experimental study [34] by Madduri and co-workers also demonstrated the practical relevance for one of our parallel SSSP algorithms, the Δ-stepping [49]. Still – all the parallel results we obtained [17, 39, 40, 45, 49] suffered from heavy dependence on the graph diameter. Thus, once more, there was little hope to transform any of these approaches toward I/O-efficient SSSP for general graphs.

Average-Case Efficient Sequential SSSP. A nice feature of our parallel algorithms was that on certain graph classes they provably took only $\mathcal{O}(|V| + |E|)$ work (instead of $\mathcal{O}(|V| \cdot \log |V| + |E|)$) in the context of random edge weights. Thus, running them sequentially should beat Dijkstra's algorithm for example. Consequently we began the search for better average-case efficient sequential SSSP approaches. Eventually [48] we presented both label-setting and label-correcting algorithms that solve the SSSP problem on such arbitrary directed graphs with random edge weights in time $\mathcal{O}(|V| + |E|)$ on the average. For independent random edge weights, the average-case time-bound can also be obtained with high probability.

Our new SSSP algorithms do not use exact priority queues, but simple hierarchical bucket structures with adaptive splitting instead: The label-setting version aims to split the current bucket until a single vertex remains in it, whereas the label-correcting algorithm adapts the width of the current bucket to the maximum degree of the vertices contained in it. We also came up with constructive existence proofs for graph classes with random edge weights on which several previous SSSP algorithms are forced to run in superlinear time on average.

Our initial results triggered further research by Goldberg [25] and Hagerup [29] who gave alternative SSSP algorithms that achieve linear average-case time, too. In particular, Goldberg also demonstrated the practical advantages of average-case efficient SSSP computations [26, 27].

4   First Steps toward General EM Graph Traversal

While being busy with sequential and parallel SSSP algorithms we still tried a couple of other directions in order to improve **EM** graph traversal on general graphs.

For semi-external depth-first search (DFS) we proposed a heuristic [54] that maintains a tentative DFS forest which is modified by I/O-efficiently scanning non-tree edges to reduce the number of cross edges. It processes batches of $\mathcal{O}(|V|)$ edges with internal DFS and then only replaces an edge in the tentative forest if necessary. Further improvements are obtained by applying node and edge reduction heuristics. For a large suite of data sets (including real world data) our heuristic ran between 10 and 200 times faster than the best known alternative. Our **SEM** DFS implementation played a crucial role in a network analysis tool [21] for the Webgraph.

On the theoretical side we investigated external-memory DFS for undirected planar graphs [10]. Based on the ideas behind a previous parallel algorithm [55] we obtained the first $o(|V|)$-I/O solution; our approach required $\mathcal{O}(\text{sort}(|V|) \cdot \log|V|)$ I/Os. The result was later superseded by Maheshwari and Zeh [36] who showed that planar DFS (and BFS & SSSP) can be solved using just $\mathcal{O}(\text{sort}(|V|))$ I/Os.

Another theoretical contribution was a space-efficient internal-memory dictionary data-structure [22], which is helpful in the context of the phase approach discussed in Section 2. It offers a smooth transition between bit-vector and hash-table representation and yields better constants than previous approaches.

5   I/O-Efficient Breadth-First Search

Only after the year 2000 we found appropriate means to speed-up **EM** BFS for general undirected graphs. The main idea was to use a preprocessing phase to restructure the adjacency lists of the graph representation. It should group the vertices of the input graph into clusters of small diameter in G and store the adjacency lists of the nodes in a cluster contiguously on the disk, such that many related adjacency lists can be accessed by few I/Os during the actual BFS computation. The latter needs to be appropriately modified in order to exploit that the adjacency lists loaded from a cluster are useful to create the next BFS levels.

Overlapping Clusters. Our first attempt [38] worked with overlapping clusters, one for each graph vertex covering its complete neighborhood up to some maximum distance k. These neighborhoods could be constructed using $\mathcal{O}(k)$ sorting and scanning phases. Within the subsequent BFS computation it was then sufficient to only load the k-neighborhood of carefully selected BFS levels using unstructured access to the respective clusters of all vertices in these levels. For general graphs with (small) maximum node degree d a total I/O bound of $\mathcal{O}(\frac{|V|}{\gamma \cdot \log_d B} + \text{sort}(|V| \cdot B^\gamma))$ could be obtained by choosing k maximal with $k \cdot d^k \leq B^\gamma$ for some arbitrary constant $0 < \gamma \leq 1/2$. Unfortunately, the approach needed $\mathcal{O}(|V| \cdot B^\gamma)$ space. Thus, both the applicability (due to space-blowup) and the absolute I/O savings were still rather limited.

Disjoint Clusters. In our follow-up approach [37] (MM_BFS) – this time together with Kurt – we developed two **EM** clustering schemes that produce *disjoint* clusters of adjacency-lists.

The *Euler tour clustering method* first builds a spanning tree T for G. Then it constructs an Euler tour around T. The tour is broken at the source node and the elements of the resulting list are then stored in consecutive order using an external memory list-ranking algorithm. Thereafter, the Euler tour is chopped into *chunks* of $\mu = \mathcal{O}(\sqrt{B})$ nodes. Duplicates are removed in a way that each node only remains in the first chunk it originally occurs. The adjacency lists are then re-ordered based on the position of their corresponding nodes in the chopped duplicate-free Euler tour: all adjacency lists for nodes in the same chunks form a cluster and the distance in G between any two vertices whose adjacency-lists belong to the same cluster is bounded by $\mathcal{O}(\sqrt{B})$.

The randomized *parallel cluster growing method* first chooses $|V|/\mu$ random source vertices. Then it iteratively constructs the local BFS levels around these sources concurrently. In this process each non-source vertex is assigned to a source within closest distance. Since the parallel BFS fronts may touch each other but not overlap, the process ends once all vertices have been assigned to a source, which happens after $\mathcal{O}(\mu \cdot \log |V|)$ iterations with high probability (whp).

The new fringes of the local BFS searches in iteration i are computed by sorting and scanning the fringes of iterations $i-1$ and $i-2$ similarly to the BFS algorithm by Munagala and Ranade [51] with the difference that all required adjacency lists, i.e. for all sources, are concurrently retrieved in a single scan of the graph representation. Once all vertices have been assigned to sources, this information is used to actually form the clusters of adjacency lists and store them consecutively on disk.

The randomized parallel clustering growing approach is actually worse than the Euler tour based approaches since the vertices within a cluster may have larger maximum distances: $\bar{d} = \mathcal{O}(\mu \cdot \log |V|)$ whp. versus $\bar{d} = \mathcal{O}(\mu)$. Also, its I/O complexity is linearly dependent on the choice of μ whereas the Euler tour clustering always takes $\mathcal{O}(\text{sort}(|V|+|E|)+ST(|V|,|E|))$ I/Os, where $ST(|V|,|E|)$ is the I/O-cost of computing a minimum spanning tree (e.g, $\text{sort}(|V|+|E|)$ I/Os using a randomized approach [1]). However, the parallel cluster growing method is conceptually easier.

The BFS Phase. The actual BFS computation is carried out using an appropriately modified version of the $\mathcal{O}(|V| + \text{sort}(|V| + |E|))$-I/O BFS algorithm by Munagala and Ranade [51]. The key difference is to load whole preprocessed clusters into some external sorted list data structure (called hot pool) at the expense of few I/Os, since the clusters are stored contiguously on disk and contain vertices in neighboring BFS levels. This way, the neighboring nodes $N(l)$ of some BFS level l can be computed by scanning only the hot pool. Similar to the parallel cluster growing, removing the nodes visited in levels $l-1$ and l from $N(l)$ by concurrent scanning yields the nodes in the next BFS level. However, the nodes in l and consequently their neighbors in $N(l)$ may belong to different clusters and unstructured I/Os are required to import them once into the hot pool, from where they are evicted again as soon as they have been used to create the respective BFS levels. Maintaining the hot pool itself requires $\mathcal{O}(\text{scan}(|V| + |E|) \cdot \bar{d})$ I/Os, whereas importing the clusters into it accounts for $\mathcal{O}(|V|/\mu + \text{sort}(|V| + |E|))$ I/Os.

Putting everything together, for undirected graphs with $|E| \geq |V|$ and a proper choice of μ, **EM BFS** with the Euler tour based clustering can be computed using $\mathcal{O}(\sqrt{|V| \cdot |E|/B} + \text{sort}(|E|) + ST(|V|, |E|))$ I/Os. Applying the randomized spanning tree approach from [1], the I/O-bound for sparse graphs becomes $\mathcal{O}(|V|/\sqrt{B} + \text{sort}(|V|))$.

Implementations. In [6], we presented our implementation of MR_BFS and the randomized variant of MM_BFS and gave a comparative study of the two algorithms on various graph classes. We demonstrated that the usage of these algorithms along with disk parallelism and pipelining can alleviate the I/O bottleneck of BFS on many small diameter graph classes, thereby making the BFS viable for these graphs. As a real world example, the BFS level decomposition of an external web-crawl based graph of around 130 million nodes and 1.4 billion edges was computed in less than 4 hours using a single disk and 2.3 hours using four disks.

However, both MR_BFS and the parallel cluster growing variant of MM_BFS took *days* on large diameter graphs. In [7], we show that the Euler tour clustering variant of MM_BFS coupled with a heuristic can be used for computing the BFS level decomposition of even large diameter graphs in a few *hours*. MM_BFS decomposes the graph into low diameter clusters and maintains an efficiently accessible pool of adjacency lists required in the next few levels. For many large diameter graphs, the pool fits into the internal memory most of the time. By keeping the portion of the pool that fits into the internal memory as a multi-map hash table and by using caching of adjacency lists with two level hierarchy of clusters, we significantly improve the performance of the external BFS algorithm while keeping the worst case I/O bounds of MM_BFS. The details involve careful trade-offs between internal-computation and I/O.

Our graph implementations are based on the software library STXXL [19] that is an implementation of the C++ standard template library STL for processing huge data sets that can fit only on hard disks. It supports parallel disks, overlapping between disk I/O and computation and it is the first I/O-efficient algorithm library that supports the pipelining technique which can save a considerable amount of I/O.

6   External-Memory Dynamic BFS

The objective of a dynamic graph algorithm is to efficiently process an online sequence of update and query operations. In very recent work and based on our previous results in Section 5 we gave the first non-trivial result on dynamic BFS in external-memory [42]. We consider sequences of either $\Theta(|V|)$ edge insertions, but no deletions (*incremental* version) or $\Theta(n)$ edge deletions, but no insertions (*decremental* version). After each edge insertion/deletion the updated BFS level decomposition has to be output. We prove an amortized bound of $O(|V|/B^{2/3} + \text{sort}(|V|) \cdot \log B)$ I/Os per update. In contrast, the currently best bound for static BFS on sparse undirected graphs as seen before is $\Omega(|V|/B^{1/2} + \text{sort}(|V|))$ I/Os. In the following we review the high-level ideas to computing BFS on general undirected sparse graphs in an incremental or decremental setting.

For the BFS phase, let us consider the insertion of the ith edge (u, v) in an incremental setting and refer to the graph (and the shortest path distances from the source in the graph) before and after the insertion of this edge as $G_{i-1}(d_{i-1})$ and $G_i(d_i)$. We first run an external memory connected component algorithm in order to check if the insertion of (u, v) enlarges the connected component C_s of the source node s. If so, we run the Munagala/Ranade BFS algorithm on the nodes in the new component starting from node v (assuming w.l.o.g. that $u \in C_s$) and add $d_i(u) + 1$ ($d_i(u) = d_{i-1}(u)$ in this case) to all the distances obtained.

Otherwise, we run the BFS phase of MM_BFS, with the difference that the adjacency list for v is added to the hot pool $\mathcal{H}$ when creating BFS level $\max\{0, d_{i-1}(v) - \alpha\}$ of G_i, for a certain advance $\alpha > 1$. By keeping the adjacency lists sorted according to node distances in G_{i-1} this can be done I/O-efficiently for all nodes v featuring $d_{i-1}(v) - d_i(v) \leq \alpha$. For nodes with $d_{i-1}(v) - d_i(v) > \alpha$, we import the whole clusters containing their adjacency lists into $\mathcal{H}$ using unstructured I/Os. If it would require more than $\alpha \cdot |V|/B$ random cluster accesses, we increase α by a factor of two, compute a new clustering for G_{i-1} with larger chunk size $\mu = \Theta(\alpha)$ and start a new attempt by repeating the whole approach with the increased parameters.

We would like to note that (the amortized analysis of) our **EM** dynamic BFS approach requires a special kind of clustering that also insures a minimum amount of $\Omega(\alpha)$ vertices per cluster. In [42] this is realized by combining the Euler tour based clustering with a randomized duplicate elimination. However, it is also possible to use the time forward processing technique [13] on the spanning tree itself in order to cut out subtrees of size $\Omega(\alpha)$ and diameter $\mathcal{O}(\alpha)$.

7   Single Source Shortest-Paths

A relatively simple extension of the BFS approach to SSSP with weights in $[w, W]$ as already sketched in [37] increases the I/O bound by a factor of W/w; each edge may be scanned $\sqrt{B} \cdot W/w$ times. Obviously, W/w must be significantly smaller than $\sqrt{B}$ for this algorithm to be efficient. In [46] we show how to reduce the loss factor from W/w to $\sqrt{\log(W/w)}$, thus exponentially increasing the range of efficiently usable edge weights. We exploit that the relaxation of edges of large weights can be delayed because if such an edge is on a shortest path, it takes some time before its other endpoint is settled. Therefore, we maintain long edges in pools that are touched much less frequently than the pools for short edges. This idea alone already works well on graphs with random edge weights, where we can solve SSSP essentially as fast as BFS. For non-random edge weights we even keep short edges in pools that are touched infrequently; we shift these edges to pools that are touched more and more frequently the closer the time of their relaxation draws. This requires a hierarchical clustering of the graph using $\mathcal{O}(\log(W/w))$ levels.

Eventually in [47] we show how to compute single-source shortest paths in undirected graphs with non-negative edge lengths in $O(\sqrt{|V| \cdot |E|/B} \log |V| + MST(|V|, |E|))$ I/Os, where $MST(|V|, |E|)$ is the I/O-cost of computing a minimum spanning tree (e.g, sort($|V| + |E|$) I/Os using a randomized approach [1]). For sparse graphs, the new algorithm performs $O((|V|/\sqrt{B}) \log |V|)$ I/Os. This result removes our previous algorithm's [46] dependence on the edge lengths in the graph. The new bound is obtained by a number of new ideas to implement a recursive shortest-path algorithm that uses a specific partition into "well-separated" subgraphs, allowing the computation of shortest paths in the whole graph using nearly independent computations on these subgraphs.

Worst-case Analysis vs. Actual Behavior. Based on our previous theoretical results mentioned above and our experience in implementations of **EM** BFS [6, 7] the task was to come up with a *practical* **EM** SSSP approach: In [44] we report on initial experimental results for an algorithm on general undirected sparse graphs where the ratio between the largest and the smallest edge weight is reasonably bounded (for example integer weights in $\{1, \ldots, 2^{32}\}$) and the realistic assumption holds that main memory is big enough to keep one bit per vertex (**SEM** setting). While our implementation only guarantees average-case efficiency, i.e., assuming randomly chosen edge-weights, it turns out that its performance on real-world instances with non-random edge weights is actually even better than on the respective inputs with random weights. Furthermore, compared to our **EM** BFS implementation (the unweighted version of SSSP), the running time of our approach always stayed within a factor of five, for the most difficult graph classes the difference was even less than a factor of two.

We are not aware of any previous I/O-efficient implementation for the classic general SSSP in a (semi) external setting: in two recent projects [12, 52], Kumar-Schwabe-like SSSP approaches on graphs of at most 6 million vertices have been tested, forcing the authors to artificially restrict the main memory size, M, to rather unrealistic 4 to 16 MBytes in order not to leave the semi-external setting or produce huge running times for larger graphs: for random graphs of 2^{20} vertices, the best previous approach needed over six hours. In contrast, for a similar ratio of input size vs. M, but on a 128 times larger and even sparser random graph, our approach was less than seven times slower, a relative gain of nearly 20. On a real-world 24 million node street graph, our implementation was over 40 times faster. Even larger gains of over 500 can be estimated for random line graphs based on previous experimental results for Munagala/Ranade-BFS [51].

8   All-Pairs Problems and Diameter Approximation

Computing diameters of huge graphs is a key challenge in complex network analysis. In principle, the diameter $\mathcal{D}$ can be easily computed using $|V|$ SSSP runs (one for each graph vertex as a root) and keeping track of the largest distance found in this process. However, the underlying *All-Pairs Shortest-Paths (APSP)* problem can be solved even more efficiently: two independent publications [9, 15] provide an optimal $\Theta(|V| \cdot \text{sort}(|V|))$ I/O bound for APSP on unweighted undirected sparse graphs. Unfortunately, taking into account that current machines easily feature several gigabytes of RAM, in the external-memory setting where $|V| > M \gg B$, an algorithm spending $\Omega(|V|^2/B)$ I/Os is practically useless.

Chowdhury and Ramachandran [15] also gave an algorithm for computing approximate all-pairs shortest-paths with additive error. However, their approach only takes less I/O than exact EM APSP when $|E| \geq |V| \cdot \log|V|$, which is not the sparse graph case we are typically interested in. Of course, a constant multiplicative factor approximation of the diameter can be obtained using **EM** BFS. Still, this takes $\Omega(|V|/\sqrt{B})$ I/Os in the worst-case [37]. Hence, the question arose which I/O savings could be obtained if we are willing to accept even larger worst-case approximation errors, for example a multiplicative error of $\mathcal{O}(B)$.

While **EM** graph algorithms are usually hard on general sparse graphs they tend to be easy on trees. Therefore, it would be tempting to extract some kind of spanning tree T from the connected input graph G using just $O(\text{sort}(|V|))$ I/Os [1] and then derive the diameter $\mathcal{D}_G$ of G from $\mathcal{D}_T$, the diameter of T, spending another $O(\text{sort}(|V|))$ I/Os. In fact, a recent experimental paper for internal-memory diameter estimation [35] proposes a heuristic along these lines. Unfortunately, since T is not necessarily a BFS tree the ratio $\mathcal{D}_T/\mathcal{D}_G$ may be as high as $\Omega(|V|)$ in the worst case. Hence, drawing conclusions from $\mathcal{D}_T$ is potentially dangerous.

In recent work [43] we provide the first non-trivial results on approximate diameter computation for sparse graphs in external-memory with I/O complexity better than that of BFS. Our first approach works as follows: instead of trying to guess $\mathcal{D}_G$ directly from $\mathcal{D}_T$ we only use T to contract G by Euler-Tour techniques in some controlled way resulting in a graph G' with $|V|/B$ vertices and $O(|V|)$ edges. Then a subsequent BFS run on an arbitrary vertex of G' only takes $\mathcal{O}(\text{sort}(|V|))$ I/Os. If it identifies $\mathcal{L} \geq 1$ BFS levels then $\mathcal{L} \leq \mathcal{D}_{G'} \leq 2 \cdot \mathcal{L}$ and by our controlled reduction we can conclude that $\mathcal{L} \leq \mathcal{D}_G \leq 2 \cdot B \cdot \mathcal{L}$. If we choose to contract G to only $|V|/k$ vertices with $1 < k \leq B$ then the approximation error is reduced to a multiplicative factor of $\mathcal{O}(k)$. However, the respective I/O bound becomes $\mathcal{O}(|V|/\sqrt{k \cdot B} + \text{sort}(|V|))$ I/Os. For $k \ll B$ the first term usually dominates. For example, if only about $\mathcal{O}(|V|/B^{2/3})$ I/Os can be tolerated, multiplicative errors of $\mathcal{O}(B^{1/3})$ may occur.

Using a different randomized contraction scheme with subsequent **EM** SSSP computation (instead of BFS) we achieve another interesting trade-off:

$\mathcal{O}(|V|/\sqrt{k \cdot B/\log k} + k \cdot \text{sort}(|V|))$ I/Os and expected multiplicative errors of only $\mathcal{O}(\sqrt{k})$ instead of $\mathcal{O}(k)$. This time, if we strive for about $\mathcal{O}(|V|/B^{2/3})$ I/Os, we expect multiplicative errors of $\mathcal{O}(B^{1/6} \cdot \sqrt{\log B})$. With block sizes in **EM** implementations steadily increasing over the last years (currently $B \simeq 10^6$ for fast hard disks), $B^{1/6} \cdot \sqrt{\log_2 B}$ has already dropped below $B^{1/3}$, thus making the second approach not only theoretically interesting.

9 Other Memory Hierarchies Models

The cache-oblivious model introduced by Frigo et al. [24] also assumes a two level memory hierarchy with an internal memory of size M and block transfers of B elements in one I/O. Again, the performance measure is the number of I/Os incurred by the algorithm. However, in contrast to the classical I/O model, the cache-oblivious algorithm does not have any knowledge of the values of M and B. Consequently, the guarantees on I/O-efficient algorithms in the cache-oblivious model do not only hold on any machine with multi-level memory hierarchy but also on all levels of the memory hierarchy at the same time.

Jeff Vitter's recent overview [56] shows that for more and more I/O-efficient algorithms, cache-oblivious counterparts have been developed. For example concerning the **EM** BFS/SSSP algorithms for general graphs mentioned in the sections above, cache-oblivious versions of the MM_BFS algorithm are given in [11], for the Kumar/Schwabe SSSP algorithm in [11, 14], and for the bounded-weights SSSP algorithm in [8].

Recently we also started the investigation of models, algorithms, and data structures for flash memory [5]. Originally used in small portable devices, this block based solid-state storage technology is predicted to become a new standard level in the PC memory hierarchy, partially even replacing hard disks. Unfortunately, the read/write/erase performance of flash memory is quite different from that of hard disks. Therefore, even cache-efficient implementations of most classic algorithms may not exploit the benefits of flash. Flash-efficient BFS is discussed in [4], experimental results for **SEM** SSSP using a flash device appear in [44].

10 Conclusions

In this paper we have sketched the recent development of I/O-efficient BFS/SSSP algorithms for general input graphs. Despite all improvements for undirected graphs, significant gaps remain compared to the respective $\mathcal{O}(\text{sort}(|V| + |E|))$-I/O results for special graph classes like planar graphs (see [56] for an overview). The differences are even larger for directed graphs. For the future, one of the most challenging questions will be to either identify appropriate lower bounds or to try and narrow these gaps. A first result in the latter direction was provided by Haverkort and Toma [30]. They characterized a rather wide class of *near-planar* directed graphs and gave I/O-efficient algorithms for them.

References

1. Abello, J., Buchsbaum, A., Westbrook, J.: A functional approach to external graph algorithms. Algorithmica 32(3), 437–458 (2002)
2. Aggarwal, A., Vitter, J.S.: The input/output complexity of sorting and related problems. Communications of the ACM 31(9), 1116–1127 (1988)
3. Ahuja, R.K., Magnanti, T.L., Orlin, J.B.: Network flows: Theory, Algorithms and Applications. Prentice Hall, Englewood Cliffs (1993)
4. Ajwani, D., Beckmann, A., Jacob, R., Meyer, U., Moruz, G.: On computational models for flash memory devices. In: Proc. 8th Int. Symposium on Experimental Algorithms (SEA). LNCS. Springer, Heidelberg (2009)
5. Ajwani, D., Malinger, I., Meyer, U., Toledo, S.: Characterizing the performance of flash memory storage devices and its impact on algorithm design. In: McGeoch, C.C. (ed.) WEA 2008. LNCS, vol. 5038, pp. 208–219. Springer, Heidelberg (2008)
6. Ajwani, D., Dementiev, R., Meyer, U.: A computational study of external-memory BFS algorithms. In: SODA, pp. 601–610 (2006)
7. Ajwani, D., Meyer, U., Osipov, V.: Improved external memory BFS implementation. In: Proc. Workshop on Algorithm Engineering and Experiments, ALENEX. SIAM, Philadelphia (2007)
8. Allulli, L., Lichodzijewski, P., Zeh, N.: A faster cache-oblivious shortest-path algorithm for undirected graphs with bounded edge lengths. In: Proceedings of the 18th annual ACM-SIAM Symposium on Discrete Algorithms (SODA), pp. 910–919 (2007)
9. Arge, L., Meyer, U., Toma, L.: External memory algorithms for diameter and all-pairs shortest-paths on sparse graphs. In: Díaz, J., Karhumäki, J., Lepistö, A., Sannella, D. (eds.) ICALP 2004. LNCS, vol. 3142, pp. 146–157. Springer, Heidelberg (2004)
10. Arge, L., Meyer, U., Toma, L., Zeh, N.: On external-memory planar depth first search. J. Graph Algorithms Appl. 7(2), 105–129 (2003)
11. Brodal, G., Fagerberg, R., Meyer, U., Zeh, N.: Cache-oblivious data structures and algorithms for undirected breadth-first search and shortest paths. In: Hagerup, T., Katajainen, J. (eds.) SWAT 2004. LNCS, vol. 3111, pp. 480–492. Springer, Heidelberg (2004)
12. Chen, M., Chowdhury, R.A., Ramachandran, V., Roche, D.L., Tong, L.: Priority queues and Dijkstra's algorithm. Technical Report TR-07-54, The University of Texas at Austin, Department of Computer Sciences (October 2007)
13. Chiang, Y.-J., Goodrich, M.T., Grove, E.F., Tamasia, R., Vengroff, D.E., Vitter, J.S.: External memory graph algorithms. In: Proc. 6th Ann. Symposium on Discrete Algorithms, pp. 139–149. ACM-SIAM, New York (1995)
14. Chowdhury, R.A., Ramachandran, V.: External-memory exact and approximate all-pairs shortest-paths in undirected graphs. In: Proc. 16th Annual ACM-SIAM Symposium on Discrete Algorithms (SODA), pp. 735–744 (2005)
15. Chowdury, R., Ramachandran, V.: External-memory exact and approximate all-pairs shortest-paths in undirected graphs. In: Proc. 16th Ann. Symposium on Discrete Algorithms (SODA), pp. 735–744. ACM-SIAM, New York (2005)
16. Cormen, T.H., Leiserson, C.E., Rivest, R.L.: Introduction to Algorithms. McGraw-Hill, New York (1990)
17. Crauser, A., Mehlhorn, K., Meyer, U., Sanders, P.: A parallelization of dijkstra's shortest path algorithm. In: Brim, L., Gruska, J., Zlatuška, J. (eds.) MFCS 1998. LNCS, vol. 1450, pp. 722–731. Springer, Heidelberg (1998)

18. de Kunder, M.: World wide web size (2009), http://www.worldwidewebsize.com/
19. Dementiev, R., Kettner, L., Sanders, P.: STXXL: Standard Template library for XXL data sets. Software: Practice and Experience 38(6), 589–637 (2008)
20. Dijkstra, E.W.: A note on two problems in connexion with graphs. Numerische Mathematik 1, 269–271 (1959)
21. Donato, D., Laura, L., Leonardi, S., Meyer, U., Millozzi, S., Sibeyn, J.F.: Algorithms and experiments for the webgraph. J. Graph Algorithms Appl. 10(2), 219–236 (2006)
22. Edelkamp, S., Meyer, U.: Theory and practice of time-space trade-offs in memory limited search. In: Baader, F., Brewka, G., Eiter, T. (eds.) KI 2001. LNCS, vol. 2174, pp. 169–184. Springer, Heidelberg (2001)
23. Fredman, M.L., Tarjan, R.E.: Fibonacci heaps and their uses in improved network optimization algorithms. Journal of the ACM 34, 596–615 (1987)
24. Frigo, M., Leiserson, C.E., Prokop, H., Ramachandran, S.: Cache-oblivious algorithms. In: 40th Annual Symposium on Foundations of Computer Science, pp. 285–297. IEEE Computer Society Press, Los Alamitos (1999)
25. Goldberg, A.V.: A simple shortest path algorithm with linear average time. In: Meyer auf der Heide, F. (ed.) ESA 2001. LNCS, vol. 2161, pp. 230–241. Springer, Heidelberg (2001)
26. Goldberg, A.V.: Shortest path algorithms: Engineering aspects. In: Eades, P., Takaoka, T. (eds.) ISAAC 2001. LNCS, vol. 2223, pp. 502–513. Springer, Heidelberg (2001)
27. Goldberg, A.V.: A practical shortest path algorithm with linear expected time. SIAM J. Comput. 37(5), 1637–1655 (2008)
28. Golden, B., Magnanti, T.: Transportation planning: Network models and their implementation. Studies in Operations Management, 365–518 (1978)
29. Hagerup, T.: Simpler computation of single-source shortest paths in linear average time. Theory Comput. Syst. 39(1), 113–120 (2006)
30. Haverkort, H., Toma, L.: I/O-efficient algorithms on near-planar graphs. In: Correa, J.R., Hevia, A., Kiwi, M. (eds.) LATIN 2006. LNCS, vol. 3887, pp. 580–591. Springer, Heidelberg (2006)
31. Jájá, J.: An Introduction to Parallel Algorithms. Addison-Wesley, Reading (1992)
32. Kumar, V., Schwabe, E.J.: Improved algorithms and data structures for solving graph problems in external memory. In: Proc. 8th Symp. on Parallel and Distrib. Processing, pp. 169–177. IEEE, Los Alamitos (1996)
33. Lipton, R.J., Tarjan, R.E.: A separator theorem for planar graphs. SIAM Journal on Applied Mathematics 36(2), 177–189 (1979)
34. Madduri, K., Bader, D.A., Berry, J.W., Crobak, J.R.: An experimental study of a parallel shortest path algorithm for solving large-scale graph instances. In: Proc. 9th Workshop on Algorithm Engineering and Experiments (ALENEX). SIAM, Philadelphia (2007)
35. Magnien, C., Latapy, M., Habib, M.: Fast computation of empirically tight bounds for the diameter of massive graphs (submitted, 2007), http://www-rp.lip6.fr/~latapy/Diameter/
36. Maheshwari, A., Zeh, N.: I/O-efficient planar separators. SIAM J. Comput. 38(3), 767–801 (2008)
37. Mehlhorn, K., Meyer, U.: External-memory breadth-first search with sublinear I/O. In: Möhring, R.H., Raman, R. (eds.) ESA 2002. LNCS, vol. 2461, pp. 723–735. Springer, Heidelberg (2002)

38. Meyer, U.: External memory BFS on undirected graphs with bounded degree. In: Proc. 12th Ann. Symp. on Discrete Algorithms, pp. 87–88. ACM–SIAM, New York (2001)

39. Meyer, U.: Heaps are better than buckets: Parallel shortest paths on unbalanced graphs. In: Sakellariou, R., Keane, J.A., Gurd, J.R., Freeman, L. (eds.) Euro-Par 2001. LNCS, vol. 2150, pp. 343–351. Springer, Heidelberg (2001)

40. Meyer, U.: Buckets strike back: Improved parallel shortest paths. In: Proc. 16th Intern. Parallel and Distributed Processing Symposium (IPDPS 2002). IEEE, Los Alamitos (2002)

41. Meyer, U.: Design and Analysis of Sequential and Parallel Single–Source Shortest–Paths Algorithms. PhD thesis, Universität des Saarlandes (2002)

42. Meyer, U.: On dynamic Breadth-First Search in external-memory. In: Proc. 25th Annual Symposium on Theoretical Aspects of Computer Science (STACS), pp. 551–560. IBFI Dagstuhl (2008)

43. Meyer, U.: On trade-offs in external-memory diameter-approximation. In: Gudmundsson, J. (ed.) SWAT 2008. LNCS, vol. 5124, pp. 426–436. Springer, Heidelberg (2008)

44. Meyer, U., Osipov, V.: Improved external memory BFS implementation. In: Proc. 11th Workshop on Algorithm Engineering and Experiments (ALENEX), pp. 85–96. ACM-SIAM, New York (2009)

45. Meyer, U., Sanders, P.: Parallel shortest path for arbitrary graphs. In: Bode, A., Ludwig, T., Karl, W.C., Wismüller, R. (eds.) Euro-Par 2000. LNCS, vol. 1900, pp. 461–470. Springer, Heidelberg (2000)

46. Meyer, U., Zeh, N.: I/O-efficient undirected shortest paths. In: Di Battista, G., Zwick, U. (eds.) ESA 2003. LNCS, vol. 2832, pp. 434–445. Springer, Heidelberg (2003)

47. Meyer, U., Zeh, N.: I/O-efficient undirected shortest paths with unbounded edge lengths. In: Azar, Y., Erlebach, T. (eds.) ESA 2006. LNCS, vol. 4168, pp. 540–551. Springer, Heidelberg (2006)

48. Meyer, U.: Average-case complexity of single-source shortest-paths algorithms: lower and upper bounds. J. Algorithms 48(1), 91–134 (2003)

49. Meyer, U., Sanders, P.: Δ-stepping: a parallelizable shortest path algorithm. J. Algorithms 49(1), 114–152 (2003)

50. Moore, E.F.: The shortest path through a maze. In: Proc. Intern. Symp. on the Theory of Switching, pp. 285–292. Harvard University Press (1959)

51. Munagala, K., Ranade, A.: I/O-complexity of graph algorithms. In: Proc. 10th Ann. Symposium on Discrete Algorithms (SODA), pp. 687–694. ACM-SIAM, New York (1999)

52. Sach, B., Clifford, R.: An empirical study of cache-oblivious priority queues and their application to the shortest path problem (February 2008), http://www.cs.bris.ac.uk/~sach/COSP/

53. Schwartz, M., Stern, T.E.: Routing techniques used in computer communication networks. IEEE Transactions on Communications, 539–552 (1980)

54. Sibeyn, J.F., Abello, J., Meyer, U.: Heuristics for semi-external depth first search on directed graphs. In: Proc. Symposium on Parallel Algorithms and Architectures (SPAA), pp. 282–292. ACM, New York (2002)

55. Smith, J.R.: Parallel algorithms for depth-first searches. I. planar graphs. SIAM Journal on Computing 15(3), 814–830 (1986)

56. Vitter, J.S.: Algorithms and Data Structures for External Memory. Foundations and Trends in Theoretical Computer Science. Now Publishers (2008)

Part IV

Computational Geometry and Geometric Graphs

The Computational Geometry of Comparing Shapes

Helmut Alt

Institut für Informatik, Freie Universität Berlin
`alt@mi.fu-berlin.de`

Abstract. This article is a survey on methods from computational geometry for comparing shapes that we developed within our work group at Freie Universität Berlin. In particular, we will present the ideas and complexity considerations for the computation of two distance measures, the *Hausdorff distance* and the *Fréchet distance*. Whereas the former is easier to compute, the latter better captures the similarity of shapes as perceived by human observers. We will consider shapes modelled by curves in the plane as well as surfaces in three-dimensional space. Especially, the Fréchet distance of surfaces seems computationally intractable and is of yet not even known to be computable. At least the decision problem is shown to be recursively enumerable.

Keywords: computational geometry, shapes, distance measures, computability.

1   Introduction

The recognition and comparison of patterns and shapes are subject of the field of computer vision and a very essential part of modern information technology. Most algorithms for this purpose are based on the representation of images by pixels as they are usually produced by cameras and scanners.

Alternatively, in applications where not color or grey values but only shape and geometric configurations of the objects displayed matter, images can be represented and processed much more efficiently by vector graphics. Here, data are represented by their coordinates in two- or three-dimensional space and it makes sense to develop methods in computational geometry for processing these data.

Consequently, in this article we will consider shapes and patterns in two- or three-dimensional space that consist of sets of points, line segments, or triangles, see Figure 1. It is a survey about research done in our workgroup at Freie Universität Berlin. The methods presented in most cases cannot be applied to practical problems directly, but need to be enhanced with some heuristics which we did successfully for the recognition of similarities in trademarks [1]. The purpose of this paper, however, is to show that the various geometric problems related to shape matching and recognition lead to interesting questions and challenges within the area of algorithms and complexity.

S. Albers, H. Alt, and S. Näher (Eds.): Festschrift Mehlhorn, LNCS 5760, pp. 235–248, 2009.

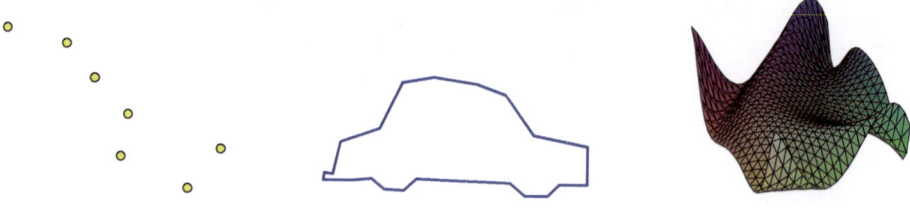

Fig. 1. Geometric patterns and shapes

A natural problem for shapes represented as described above is how much two shapes A and B resemble each other. In order to model resemblance rigorously, it is necessary to define a metric or distance measure $\delta(A, B)$ on the set of shapes which, hopefully, comes close to the intuitive human notion of resemblance. In this article, essentially, we will discuss two distance measures and the intriguing algorithmic problems involved in their computation.

2   The Hausdorff Distance

Suppose that shapes are considered as compact subsets of $\mathbb{R}^2$ or $\mathbb{R}^3$ and let us denote by $\|.\|$ the Euclidean metric. The distance measure for shapes that comes to mind immediately is to consider for any point on shape A the distance to the closest point on shape B and to maximize over all these values:

$$\vec{\delta}_H(A, B) = \max_{a \in A} min_{b \in B} \|a - b\|$$

This distance measure is called the *directed Hausdorff distance* and it measures, how close A is to some part of B, see Figure 2. Obviously, if A and B are compact

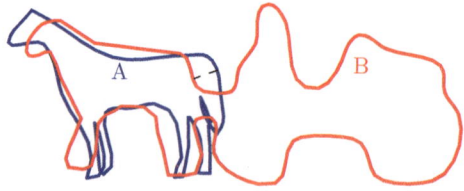

Fig. 2. The directed Hausdorff distance from A to B is attained at the dashed line. (Figure from MPEG7 Core Experiment CE-Shape-1 dataset.)

it is zero exactly if A is a subset of B. In order to compare complete shapes this distance measure is made symmetric and called the *Hausdorff distance*:

$$\delta_H(A, B) = \max(\vec{\delta}_H(A, B), \vec{\delta}_H(B, A)).$$

For an example, see Figure 3.

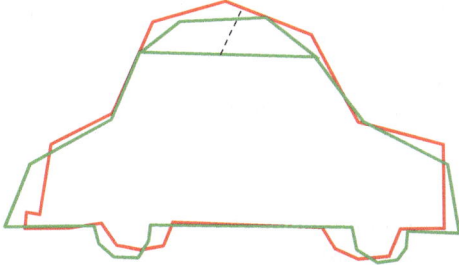

Fig. 3. The Hausdorff distance is attained at the dashed line

Let us consider now the problem of computing the Hausdorff distance for different kinds of shapes A, B. We will explain the ideas developed in [2].

If A and B are "point patterns", i.e., finite sets of n and m points, respectively, there is a straightforward algorithm to compute $\delta_H(A, B)$ in time $O(nm)$.

There is no similar straightforward algorithm if A and B are two sets of line segments, more precisely, the set of points lying on these line segments. In fact, in this case the Hausdorff distance cannot only occur between endpoints of segments, but also between points inside the segments, see Figure 3.

With methods from computational geometry we can find an efficient algorithm for computing the Hausdorff distance for two sets A, B of disjoint line segments which is also the most efficient one for the special case of finite sets of points. We first observe: Suppose that we move in one direction on one of the line segments of A, say, and consider the distance to the closest point in set B. Then, as long as we are in the same Voronoi cell of a segment of B the distance is a unimodal function, i.e., monotone decreasing and then monotone increasing. So the maximum distance on this segment can only be attained at an endpoint or at the boundary to another Voronoi cell of B as in Figure 3. We conclude:

Lemma 1. *The Hausdorff distance between two sets A, B of disjoint line segments can only occur at points that are either endpoints of line segments or intersection points of the Voronoi diagram of one of the sets with a segment of the other.*

This property reduces the set of possible points on the segments where the Hausdorff distance can occur to finitely many candidates, in fact only $O(nm)$. For each of them we can determine the distance to the other figure and return the maximum of these values as the Hausdorff distance.

A yet more efficient algorithm can be found by the following observation: Suppose we traverse a (straight or parabolic) Voronoi edge e of a set of line segments and consider the distance to one of the sites whose Voronoi cell is bounded by e. A simple geometric argument shows that this distance again is a unimodal function. Therefore, if we consider the value of this function at all intersection points of e with segments from the other set, its maximum will be attained at one of the two extreme intersection points, so only these have to be considered for each Voronoi edge. Consequently, we have reduced the number of candidate points where the Hausdorff distance can occur to $O(n + m)$, namely, the endpoints of segments and two intersection points per Voronoi edge. In order to achieve subquadratic runtime we have to make sure to only determine the extreme intersection points. These considerations can be combined into a sweep line algorithm to determine the directed Hausdorff distance between A and B.

Algorithm 1

1. *Compute the Voronoi diagram of B and cut the Voronoi edges at all points where they have vertical tangents. Let V_B be the resulting set of parabolic arcs.*
2. *Initialize the event point schedule of the sweep line algorithm with all endpoints of segments in A and arcs in V_B.*
3. *Perform a line sweep from left to right across the segments in A and arcs in V_B. Whenever a point p is encountered, that is an endpoint of a segment of A or an intersection point between a segment of A and an arc a of V_B, determine the distance of p to the corresponding site in B and update the maximum distance found so far. If p is an intersection point delete arc a from the scene.*
4. *Perform a line sweep from right to left in the same manner and return the maximum of the distances found in both sweeps as the directed Hausdorff distance $\vec{\delta}_H(A, B)$.*

The deletion in steps 3 and 4 makes sure that only the extreme intersection points on each Voronoi edge are determined. Since only a linear number of event points are traversed by the line sweep, the total runtime is $O((n + m) \log(n + m))$. The (undirected) Hausdorff distance between A and B of course can be determined by running the algorithm a second time with the roles of A and B exchanged.

We conclude:

Theorem 1. *The Hausdorff distance of two planar shapes consisting of n and m disjoint line segments can be computed by Algorithm 1 in time $O((n+m) \log(n+m))$.*

Algorithm 1 is asymptotically optimal in the algebraic decision tree model of computation, since even for one-dimensional sets of points A, B their Hausdorff distance is 0 exactly if they are equal, and this set equality problem has a lower bound of $\Omega(n \log n)$.

3   The Fréchet Distance

If shapes are represented by *curves* the intuitive understanding is that two shapes should be compared by traversing both curves and determining how close the courses of the two curves stay together. The Hausdorff distance does not capture this intuition in all cases, in fact, it gives "false positives" in the sense that it considers curves as "close" whose courses are not similar at all, see Figure 4.

A distance measure between curves that takes into account their courses, i.e., continuous parameterizations $\alpha, \beta : [0, 1] \to \mathbb{R}^2$ is the Fréchet distance, given by Maurice Fréchet in 1906 [3], which is defined as follows:

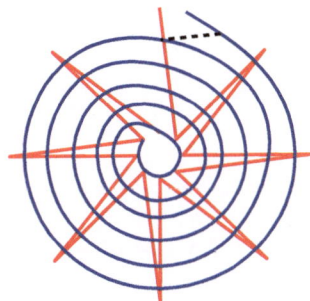

Fig. 4. Example of extremely different shapes that have a small Hausdorff distance

Definition 1. *Let two curves in the plane be defined by (and identified with) their parameterizations $\alpha, \beta : [0,1] \to \mathbb{R}^2$. Then the Fréchet distance between α and β is defined by*

$$\delta_F(\alpha, \beta) = inf_{\sigma, \tau} \max_{t \in [0,1]} \|\alpha(\sigma(t)) - \beta(\tau(t))\|$$

where $\sigma, \tau : [0,1] \to [0,1]$ range over all strictly monotone increasing continuous functions

A popular illustration of the Fréchet distance is as follows: Suppose a man is walking his dog, the man is walking an curve α, the dog on curve β. Both are allowed to control their speed (by functions σ and τ) but not to walk backward on their respective curves. The Fréchet distance then is the minimum length leash that is possible.

Observe, that it suffices to specify one of the functions σ and τ in Definition 1, so an equivalent definition is

$$\delta_F(\alpha, \beta) = inf_{\tau} \max_{t \in [0,1]} \|\alpha(t) - \beta(\tau(t))\|$$

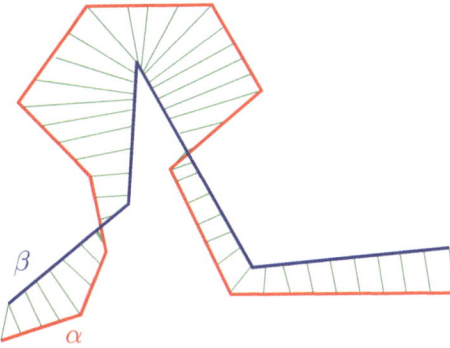

Fig. 5. Realizing the Fréchet distance: the green lines are between points where man and dog are located at the same time

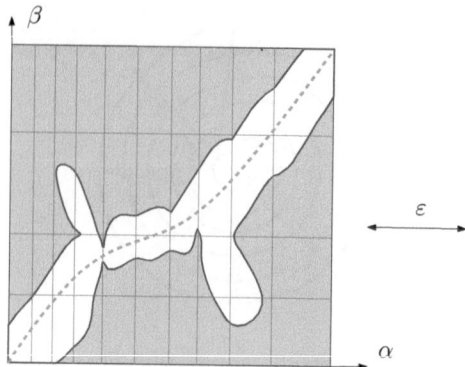

Fig. 6. The free space diagram for α, β of Figure 5 and distance ε

Therefore, if τ is given, we obtain a unique assignment $\alpha(t) \mapsto \beta(\tau(t))$ between the points on both curves, see Figure 5.

The computation of the Fréchet distance looks difficult at first sight since we minimize over a set of functions rather than a set of values. But for polygonal curves, which are given by piecewise linear parameterizations of their edges, we found efficient algorithms [4] whose ideas we will explain in this section.

Let as first consider the *decision problem* whether for two given curves α and β, and a given $\varepsilon \geq 0$, $\delta_F(\alpha, \beta) \leq \varepsilon$. To solve this problem we consider the so-called *free space* which is the subset of $[0, 1]^2$ defined by $\{(s, t)\| \|\alpha(s) - \beta(t)\| \leq \varepsilon\}$. The *free space diagram* is the subdivision of $[0, 1]^2$ in the free space and the rest. For the two curves of Figure 5, the free space diagram is shown in Figure 6.

Now the answer to the decision problem is positive, exactly if there is a continuous path, monotone increasing in both coordinates, from the lower left corner

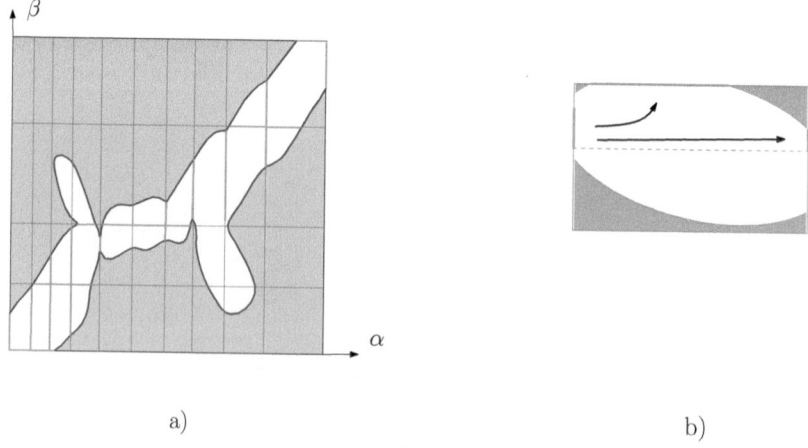

a)   b)

Fig. 7. Determining the existence of a valid path. a) Parts of cell boundaries reachable by a monotone path from the origin are green. b) Iteratively computing the green portions.

(" the origin") to the upper right corner of the diagram, let us call it a *valid path*, see Figure 6. As was mentioned before, in Definition 1 the map σ could be the identity and then the map τ would be given by the curve in the free space diagram.

If the curves α and β are polygonal chains of n and m edges, respectively, the free space diagram consists of nm cells as indicated in Figure 6. Each cell corresponds to the free space diagram for a pair of edges where the free space is the intersection of a rectangle with an ellipse, as easily can be verified. Consequently, we can compute the complete free space diagram in time $O(nm)$.

It remains to determine whether there is a valid path within the diagram. This can be done by considering the cells of the free space diagram row by row from left to right starting with the bottom row. At the upper and right boundaries of each cell we can mark those intervals that can be reached by a monotone path from the origin, if this information is available for the lower and left boundary (see Figure 7 b)). We conclude that the decision problem for the Fréchet distance can be solved for polygonal curves α and β of n and m edges, respectively, in time $O(nm)$.

To solve the problem of *computing* the Fréchet distance, we could do a kind of binary search on an interval $[0, a]$, where a is a value guaranteed to be at least as large as the Fréchet distance (for example the diameter of the bounding box containing both curves) involving the algorithm for the decision problem in each search step. This way we obtain an algorithm of runtime $O(knm)$ to get k correct bits of the result.

Alternatively, we can identify a certain finite set of *critical values* which must contain the real Fréchet distance and apply the technique of *parametric search*, see [4]. Then, we obtain

Theorem 2. *The Fréchet distance between two polygonal curves α and β of n and m edges, respectively, can be computed in time $O(nm \log(nm))$.*

The Fréchet distance can be extended in a natural way to closed curves which is of course also of significant interest for comparing shapes. In Definition 1 we just need to minimize additionally over all possible starting points of the parameterizations. Our algorithm can be modified so that both, the decision and the computation problem can be solved with an additional $\log(nm)$- factor, see [4].

It is an natural open problem whether the quadratic runtime for computing the Fréchet distance or even solving the decision problem can be improved. Since no subquadratic algorithm could be found, one might also conjecture that the decision problem belongs to the class of so-called *3SUM-hard* problems, see [5], but we could find no evidence so far for that hypothesis either.

4 Computing the Distance between Surfaces

In practice, shapes are not restricted to two dimensions and (bounding) curves, but of course surfaces in three dimensions are an important issue in this context.

So the question arises, how the distance measures of the previous sections can be transferred to this higherdimensional setting. Let us assume that we have triangulated surfaces.

4.1 Hausdorff Distance

As far as the *Hausdorff distance* is concerned, there are no essential difficulties, as we showed in detail in [6]. In fact, Lemma 1 can be extended to sets of triangles in higher dimensions:

Lemma 2. *The Hausdorff distance between two sets A, B of disjoint triangles can only occur at points that are*

- *vertices of triangles, or*
- *points inside an edge of a triangle, with the same distance to two different triangles of the other set, i.e., intersecting a facet of its Voronoi diagram, or*
- *points in the interior of a triangle with the same distance to three different triangles of the other set, i.e., intersecting an edge of its Voronoi diagram.*

Lemma 2 reduces the number of possible occurrences of the Hausdorff distance to finitely many, in fact, since the complexity of the Voronoi diagrams is quadratic, their number is $O(n^2m + m^2n)$. So, it gives a straightforward polynomial time algorithm to compute the Hausdorff distance between A and B. The runtime can be further improved with randomized techniques.

The lemma can even be extended from sets of triangles in $\mathbb{R}^3$ to sets of k-dimensional simplices in $\mathbb{R}^d$ for arbitrary k, d with $k \leq d$, see [6].

4.2 Fréchet Distance

The situation is not at all that easy with the Fréchet distance. Even it is not obvious to transfer the definition of the Fréchet distance to surfaces, which was done in a second article by Fréchet, see [7]. In fact, assuming that surfaces are given by continuous parameterizations $\alpha, \beta : [0, 1]^2 \to \mathbb{R}^3$, Definition 1 can be adopted nearly literally:

Definition 2. *Let two surfaces in d-dimensional space, $d \geq 2$ be defined by (and identified with) their parameterizations $\alpha, \beta : [0, 1]^2 \to \mathbb{R}^d$. Then the Fréchet distance between α and β is defined by*

$$\delta_F(\alpha, \beta) = inf_{\sigma, \tau} \max_{t \in [0,1]^2} \|\alpha(\sigma(t)) - \beta(\tau(t))\|$$

where $\sigma, \tau : [0, 1]^2 \to [0, 1]^2$ range over all orientation preserving homeomorphisms.

"Orientation preserving" homeomorphism means that if any Jordan curve C (e.g., the boundary of $[0, 1]^2$) is traversed clockwise, the induced traversal of $\sigma(C)$ (in the example again the boundary) is clockwise, as well. Observe, that also the term "strictly monotone increasing continuous functions" in Definition 1 is in the one-dimensional setting equivalent to "orientation preserving homeomorphisms".

For computational considerations we assume that surfaces are triangulated and given by parameterizations that are affine functions on each triangle. Computing the Fréchet distance of (triangulated) surfaces turned out to be a very challenging computational problem whose complexity is yet unsolved. The first major result in this direction was obtained by Godau in his Ph.D. thesis [8] who showed

Theorem 3. *Computing the Fréchet distance of two given surfaces α, β is NP-hard.*

In fact, it was shown that even the decision problem whether for a given $\varepsilon > 0$ $\delta_F(\alpha, \beta) < \varepsilon$ is NP-hard. This result is independent of the dimension of the space containing the surfaces, in fact, it was shown for (two-dimensional) surfaces in $\mathbb{R}^2$.

The idea of the proof is that input surfaces can be designed where, in order to achieve a distance less than ε, homeomorphisms according to Definition 2 are forced to twist the triangles of one surface in one of two possible directions. This binary information can be propagated along chains of triangles. Therefore, it is possible to model edges and nodes of any graph which is an instance of the *planar 3-satisfiability problem* which is known to be NP-complete.

Observe that Theorem 3 does not claim that the decision problem is NP-*complete* and indeed it is unknown whether it is in NP. Even more surprisingly, it is of yet not known to be *decidable* at all, so the Fréchet distance is not known to be *computable*.

As is well known, most problems in computational geometry can easily be shown to be decidable, since they can be formulated in the *first order theory of the reals* whose theorems are decidable [9]. Here, we have a different situation, since in the definition of the Fréchet distance we not only minimize over real numbers but also over all possible homeomorphisms, i.e., functions. So, with this definition, the statement that $\delta_F(\alpha, \beta) < \varepsilon$ can be expressed as a *second order* but not as a first order formula.

In connection with the Ph.D. thesis of Maike Buchin [10, 11] we were able to obtain more insight into the problem. Since the Fréchet distance $\delta_F(\alpha, \beta)$ is a function whose values are real numbers, it is necessary to define more precisely, what it means that the Fréchet distance is *computable*. We follow the definitions of the research community concerned with the computability and complexity of real functions (see, e.g., [12]). Essentially, a real valued function φ is called computable, if and only if there exists an algorithm (Turing machine) which with an argument x of φ as input produces an infinite sequence of rational numbers which converges to $\varphi(x)$. Together with each approximate function value in the sequence, it is required to output an error bound, and the error bounds form a sequence which converges to zero.

As was said before, the Fréchet distance between triangulated surfaces is not known to be computable. However, we could show that it is computable in a weak sense according to the following definition:

Definition 3. *A function* $\varphi : \mathbb{N} \to \mathbb{R}$ *is called* upper (lower) semi-computable *iff there is a Turing machine which on input x outputs an infinite, monotone decreasing (increasing) sequence of rational numbers converging to $\varphi(x)$.*

The result we could obtain is:

Theorem 4. *The Fréchet distance between two triangulated surfaces in space* $\mathbb{R}^d$, $d \geq 2$, *is upper semi-computable.*

Observe, that semi-computability is in some sense a weak property concerning the computability of functions. In fact, any arbitrary prefix of the infinite output sequence does not reveal much of the real function value, which still could be far away from the last approximation value produced. However, the concept is not unknown in classic computability theory. In fact, Theorem 4 immediately implies the following corollary, where $\langle \alpha, \beta, a \rangle$ denotes some standard encoding of a triple consisting of two triangulated surfaces α and β, and some rational $a > 0$.

Corollary 1. *The set* $\{\langle \alpha, \beta, a \rangle | \ \delta_F(\alpha, \beta) < a\}$, *i.e., the decision problem for the Fréchet distance is* recursively enumerable.

In fact, consider the Turing machine producing a monotone decreasing sequence converging to $\delta_F(\alpha, \beta)$ which exists by Theorem 4. Stop this Turing machine and accept as soon as it produces a value less than a. Thus, the algorithm will eventually halt for all triples $\langle \alpha, \beta, a \rangle$ in the language and it will run forever for the ones not in the language. Curiously, the corollary cannot be derived any more, if we replace the $<$-sign by a $\leq$-sign.

The complete **proof of Theorem 4** is quite technical using results from piecewise linear topology. Here, we will only explain the underlying ideas.

Since we assume that the two input surfaces α and β are triangulated and given by a parameterization that is affine on the single triangles there must be corresponding triangulations K^0 and L^0 of the parameter space $[0,1]^2$. By K^1 and L^1 we denote the refined triangulations obtained from K^0 and L^0 by *barycentric subdivision*, i.e., dividing each triangle into nine smaller triangles by connecting the centers of the edges with the opposite vertex through a line segment. Continuing this process we obtain sequences $K^0, K^1, K^2, ..$ and $L^0, L^1, L^2, ..$ of more and more refined triangulations where the diameters of the triangles converge to 0, see Figure 8.

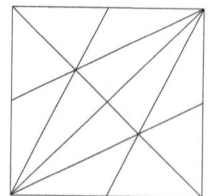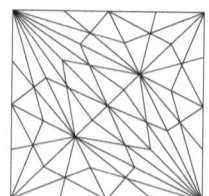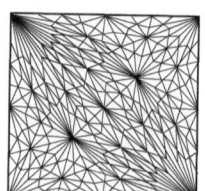

Fig. 8. Barycentric subdivisions K^0, K^1, K^2, K^3

Observe that, as in the case of curves, we can assume that one of the homeo-morphisms in Definition 2 is the identity, we will assume here that it is τ. The main idea is now that any homeomorphism σ on $[0,1]^2$ realizing the Fréchet distance (or coming arbitrarily close to it) can be approximated by a so called *mesh homeomorphism*. By mesh homeomorphism we mean a homeomorphism that maps any edge of some refinement K^n to a polygonal path consisting of edges of some refinement L^m, see Figure 9. This property can be shown in two steps: First, it is well known in topology that any homeomorphism σ, e.g., on $[0,1]^2$, can be approximated arbitrarily closely by a piecewise linear (PL) home-omorphism h', i.e., one which is linear on each triangle in some triangulation of $[0,1]^2$, see [13]. Secondly, we could show that any PL-homeomorphism can be approximated arbitrarily closely by a mesh homeomorphism , see Figure 9.

Observe that there is only a countable number of possible mesh homeomor-phisms. What we have to do now is to make sure that the infinitely running algorithm generates and tests them all. Since the sizes of the triangles in the subdivisions get smaller and smaller, it suffices to check the distances of the ver-tices in a triangle of K^n to all vertices of the assigned region in L^m. Altogether, we obtain the following algorithm:

Algorithm 2

Input
Triangulated surfaces α, β, including triangulations K, L of the parameter spaces, in a finite description.

Output
A monotone decreasing sequence of rational numbers converging to $\delta_F(\alpha, \beta)$

> set $D = \infty$;
> **forall** $(n, m) \in \mathbb{N} \times \mathbb{N}$ **do**
>> *generate the barycentric subdivisions K^m of K and L^n of L;*
>> *let $E = \{e_1, ..., e_k\}$ be the set of edges in K^m;*
>> **forall** *k-tuples $(\pi_1, ..., \pi_k)$ of simple polygonal chains in L^n* **do**
>>> *assign to the edge e_i the polygonal chain π_i for $i = 1, ..., k$;*
>>> **if** *this assignment results in an orientation preserving*
>>>   *homeomorphic image of K^m* **then**
>>>> set $M = 0$;
>>>> **forall** *triangles Δ of K^m* **do**
>>>>> *let $H_\Delta \subset |L^n|$ be the region in L^n assigned to Δ;*
>>>>> **forall** *vertices v of Δ and vertices w of H_Δ* **do**
>>>>>> set $M = \max(M, \|f(v) - g(w)\|)$;
>>>>> end
>>>> end
>>>> set $D = \min(D, M)$;
>>>> *output D;*
>>> **end**
>> **end**
> **end**

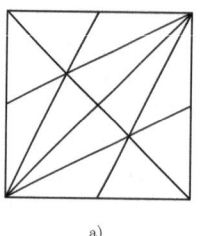  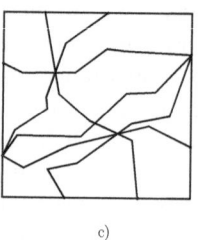 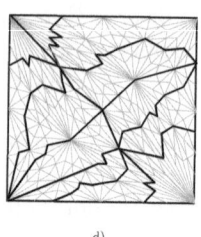

a)   b)   c)   d)

Fig. 9. Approximating a homeomorphism σ by a mesh homeomorphism: a) K^1 b) $\sigma(K^1)$ c) approximation by a PL-homeomorphism d) approximation by a mesh homeomorphism

The first **forall**-loop can be realized by some standard enumeration method for pairs of integers. We assume, that in instead of the Euclidean distance a rational approximation is computed, where the approximation error tends to zero with growing m and n.

The number of k-tuples of polygonal chains of L^n in the second **forall**-loop is finite. In fact, it is bounded by $(l!)^k$ where l is the number of edges in L^n, which itself is exponential in n, whereas k is exponential in m. But efficiency is not the issue here.

Finally, let me remark that a relaxed *variant of the Fréchet distance*, which we called *weak Fréchet distance* does not cause all these computational difficulties. The weak Fréchet distance is defined as in Definitions 1 and 2, where we assume that σ and τ range, instead of over all homeomorphisms, over all surjective maps on $[0, 1]$ or $[0, 1]^2$, respectively. For the man-dog illustration this means that man and dog may run forward, but also backward, as well. It turns out [11], that the weak Fréchet distance can be computed in polynomial time.

5   Concluding Remarks

Of course, the problem of comparing shapes A, B consists not only of determining their distance with respect to some distance measure δ. It cannot be assumed that they are already in a position minimizing $\delta(A, B)$, but that, say, B can undergo some *transformation* in order to be *matched* with A as good as possible. Natural sets of transformations are *translations, rigid motions* (=translation and rotation), *similarities* (scaling and rigid motion), or more general, arbitrary *affine transformations*. So, formally we can define the resemblance of two shapes under distance measure δ and transformation set $\mathcal{T}$ as

$$r_{\delta, \mathcal{T}} = \min_{t \in \mathcal{T}} \delta(A, t(B))$$

and the matching problem is to find a transformation $t \in \mathcal{T}$ where this minimum is attained.

We also worked on the development of shape matching algorithms with respect to the Hausdorff distance and the Fréchet distance. Again, it turned out that matching shapes under the Hausdorff distance is efficiently possible, in principle [2]. We also found polynomial time algorithms with respect to the Fréchet distance of curves [14] but the exponents in the runtimes are forbiddingly high. Finally, we resorted to probabilistic methods for shape matching, which also can be applied for practical tasks [15].

We are aware, that geometric proximity is only one aspect of the similarity of shapes. Further criteria are, e.g., the slope and curvature of curves and surfaces, which should be handled with methods from differential geometry. In this context, the work of David Mumford appears to be particularly interesting, see, e.g., [16].

Beyond the geometric grasp are aspects that are related to *semantic* proximity of shapes, i.e., connections that are established based on our experience. Consider, e.g., a drawing of an elephant from the front and one from the side where every child will recognize the similarity, but it is not possible to be captured by shape matching.

Nevertheless, as our cooperation with a company that recognizes similarities in trademarks shows, geometric aspects of comparing shapes can help to some extent to solve the various extremely challenging problems of computer vision [1]. And it is my feeling, that the more theoretical challenges concerning complexity and computability of even the most simple questions in this context are intriguing enough to deserve a thorough investigation.

Let me conclude with the remark that this line of research that kept us busy for many years and also was carried on by other groups in the computational geometry community was initiated more than twenty years ago by an article investigating algorithmic aspects of comparing point patterns [17]. We did that work together with a researcher, whom I had the privilege to have as a Ph.D. advisor thirty-three years ago and to whose birthday this Festschrift is dedicated, Kurt Mehlhorn.

References

1. Alt, H., Scharf, L., Scholz, S.: Probabilistic matching and resemblance evaluation of shapes in trademark images. In: Proceedings of the ACM International Conference on Image and Video Retrieval (CIVR), Amsterdam, The Netherlands, pp. 533–540 (2007)
2. Alt, H., Behrends, B., Blömer, J.: Approximate matching of polygonal shapes. Ann. Math. Artif. Intell. 13, 251–266 (1995)
3. Fréchet, M.: Sur quelques points du calcul fonctionnel. Rendiconti Circ. Mat. Palermo 22, 1–74 (1906)
4. Alt, H., Godau, M.: Computing the Fréchet distance between two polygonal curves. Internat. J. Comput. Geom. Appl. 5, 75–91 (1995)
5. Gajentaan, A., Overmars, M.H.: On a class of $O(n^2)$ problems in computational geometry. Comput. Geom. 5, 165–185 (1995)
6. Alt, H., Braß, P., Godau, M., Knauer, C., Wenk, C.: Computing the Hausdorff distance of geometric patterns and shapes. In: Aronov, B., Basu, S., Pach, J., Sharir, M. (eds.) Discrete and Computational Geometry. The Goodman–Pollack Festschrift. Algorithms and Combinatorics, vol. 25, pp. 65–76. Springer, Berlin (2003); Special Issue: The Goodman-Pollack-Festschrift (Aronov, B., Basu, S., Pach, J., Sharir, M. (eds.)

7. Fréchet, M.: Sur la distance de deux surfaces. Ann. Soc. Polonaise Math. 3, 4–19 (1924)
8. Godau, M.: On the complexity of measuring the similarity between geometric objects in higher dimensions. PhD thesis, Freie Universität Berlin, Germany (1998), http://www.diss.fu-berlin.de/1999/1/indexe.html
9. Tarski, A.: A decision method for elementary algebra and geometry. Santa Monica CA: RAND Corp. (1948)
10. Buchin, M.: On the Computability of the Frechet Distance Between Triangulated Surfaces. PhD thesis, Freie Universität Berlin, Germany (2007), http://www.diss.fu-berlin.de/
11. Alt, H., Buchin, M.: Can we compute the similarity between surfaces? J. on Disc. and Comput. Geom. (to appear), http://www.citebase.org/abstract?id=oai:arXiv.org:cs/0703011
12. Weihrauch, K., Zheng, X.: Computability on continuous, lower semi-continuous, and upper semi-continuous real functions. Theoretical Computer Science 234, 109–133 (2000)
13. Moise, E.E.: Geometric Topology in Dimensions 2 and 3. Graduate Texts in Mathematics, vol. 47. Springer, Heidelberg (1977)
14. Alt, H., Knauer, C., Wenk, C.: Matching polygonal curves with respect to the fréchet distance. In: Ferreira, A., Reichel, H. (eds.) STACS 2001. LNCS, vol. 2010, pp. 63–74. Springer, Heidelberg (2001)
15. Alt, H., Scharf, L.: Shape matching by random sampling. In: Das, S., Uehara, R. (eds.) WALCOM 2009. LNCS, vol. 5431, pp. 381–393. Springer, Heidelberg (2009)
16. Sharon, E., Mumford, D.: 2d-shape analysis using conformal mapping. Int. J. Comput. Vision 70(1), 55–75 (2006)
17. Alt, H., Mehlhorn, K., Wagener, H., Welzl, E.: Congruence, similarity, and symmetries of geometric objects. Discrete & Computational Geometry 3, 237–256 (1988)

Finding Nearest Larger Neighbors
A Case Study in Algorithm Design and Analysis

Tetsuo Asano[1], Sergey Bereg[2], and David Kirkpatrick[3]

[1] School of Information Science, JAIST, Japan
[2] Department of Computer Science, University of Texas at Dallas, USA
[3] Department of Computer Science, University of British Columbia, Canada

Abstract. Designing and analysing efficient algorithms is important in practical applications, but it is also fun and frequently instructive, even for simple problems with no immediate applications. In this self-contained paper we try to convey some of fun of algorithm design and analysis. Hopefully, the reader will find the discussion instructive as well.

We focus our attention on a single problem that we call the *All Nearest Larger Neighbors Problem*. Part of the fun in designing algorithms for this problem is the rich variety of algorithms that arise under slightly different optimization criteria. We also illustrate several important analytic techniques, including amortization, and correctness arguments using non-trivial loop invariants.

We hope, in this modest way, to reflect our deep admiration for the many contributions of Kurt Mehlhorn to the theory, practice and appreciation of algorithm design and analysis.

1 What Is the ANLN Problem?

Here is a general definition of our *All Nearest Larger Neighbors (ANLN) Problem*: Given a set S of n objects, find, for each object x in S, an object y in S (if one exists) that is (i) larger than x and (ii) at least as close to x as any other object z that is larger than x. We implicitly assume that any two objects are comparable, that is, one of them is larger than the other or they are equal. As we shall see, the problem is simplified considerably if we can assume that all elements are distinct.

Although the ANLN problem seems very natural and worthy of study even without specific applications, it is easy to imagine scenarios in which it could arise. For example, in emergency situations we often rely on protocols for sending messages from all individuals to some specified *leader* (or the reverse). It is easy to see the desirability of a tree-like protocol where (i) individual links are "short" and (ii) nodes closer to the root (leader) have greater *authority* (measured,

S. Albers, H. Alt, and S. Näher (Eds.): Festschrift Mehlhorn, LNCS 5760, pp. 249–260, 2009.

perhaps, by transmission power/capacity). To enact such a protocol, where each node has an assigned authority, it suffices to associate with every node its closest neighbor with higher authority.

2   The ANLN Problem for Linear Arrays

We begin with the simplest possible situation: the objects are n real numbers presented in an array $A[1..n]$. For each element $A[i]$, we want to determine an element $A[j]$ among those with keys larger than $A[i]$ that is closest to $A[i]$, that is for which $|j - i|$ is minimized, with ties broken in favor of the lower indexed neighbor.

Of course, an element with the largest key has no such larger neighbor. For this reason, as well as to simplify the presentation of some of our algorithms, we will assume that the array A has been implicitly extended to $A[-n..3n]$, with $A[-n] = A[3n] = \infty$ and $A[j] = -\infty$ for $j \in [-n+1..0] \cup [n+1..3n-1]$. It should be clear that with this extension the nearest larger neighbors of all non-maximal elements of $A[1..n]$ are unaltered and the nearest larger neighbor of all maximal elements of $A[1..n]$ is $A[-n]$.

[ANLN Problem for a Linear Array]

Input: An array $A[1..n]$ of n real numbers.
Output: An array $NLN[1..n]$ such that $A[NLN[i]]$ is the nearest larger neighbor of $A[i]$. If $A[i]$ is a largest element in $A[1..n]$ then $NLN[i] = -n$.

Figure 1 shows an example of an array A containing 10 numbers together with its associated NLN array.

	1	2	3	4	5	6	7	8	9	10
A	87	32	12	54	28	35	14	61	18	53
NLN	-10	1	2	1	4	4	6	1	8	8

Fig. 1. An example of an array A containing 10 elements together with its associated NLN array

2.1   A Simple Linear-Time Stack-Based Algorithm

The ANLN problem has a very straightforward linear-time solution by computing NLN values to both the left (LNLN) and right (RNLN), using a stack:

Algorithm 1. (Double Stack-Scan): Compute both Left and Right NLN values

---

Input: Array $A[1..n]$ of keys.
Output: Associated array $NLN[1..n]$.

1 **begin**
2   initialize stack S as $\emptyset$
3   **for** $i = 1$ **to** n **do**
4   **while** $S \neq \emptyset$ **and** $A[i] \geq A[\text{top}(S)]$ **do**
5   pop(S)
6   **if** $S = \emptyset$ **then** $LNLN[i] = -n$ **else** $LNLN[i] = \text{top}(S)$
7   push $A[i]$ onto S
8   reinitialize stack S as $\emptyset$
9   **for** $i = n$ **to** 1 **do**
10   **while** $S \neq \emptyset$ **and** $A[i] \geq A[\text{top}(S)]$ **do**
11   pop(S)
12   **if** $S = \emptyset$ **then** $RNLN[i] = 3n$ **else** $RNLN[i] = \text{top}(S)$
13   push $A[i]$ onto S
14   **for** $i = 1$ **to** n **do**
15   **if** $i - LNLN[i] \leq RNLN[i] - i$ **then** $NLN[i] = LNLN[i]$ **else**
        $NLN[i] = RNLN[i]$

16 **end**

The analysis of Algorithm 1 is completely straightforward. While its linear time complexity is clearly (asymptotically) optimal, it comes at the cost of using linear space, both for the stack S and the intermediate NLN results. In recognition of the constraints of software embedded in highly functional devices like digital cameras and scanners, there has been a considerable amount of attention paid recently to the development of algorithms that use only a modest amount of working space. Thus, we are motivated to ask to what extent this is possible for the ANLN problem.

In practice, it is quite reasonable to allow working space of size that is logarithmic in the input size, since otherwise it is quite hard to incorporate recursion in designing algorithms. In this paper we restrict our attention to even more severely space-restricted algorithms. An algorithm is called a *constant-working-space* algorithm if it satisfies the following conditions.

Input: Input data are provided using a read-only array.

Output: Output data must be written to a write-only array.

Recursion: If recursive calls are included in the algorithm, stack area for recursive calls is considered as a part of the working space.

Working space: Arrays may be available in the algorithm, but their sizes must be constant independent of input sizes. Each element (variable or array element) may contain $O(\log n)$ bits where n is the input size.

2.2   A Simple Linear-Time Stack-Based Algorithm

The most obvious way to compute NLN values with limited space is to do so one
at a time. Since we need to identify only the closer of the two (left and right)
nearest larger neighbors, it is natural to adopt a bidirectional search strategy
(extending the neighborhood outward one step at a time) until we find a larger
element.

It is easy to see that the following algorithm satisfies the constant working-
space conditions:

Algorithm 2. (Bidirectional Scan): Scan neighborhoods by increasing
distance.

Input: Array $A[1..n]$ of keys.
Output: Associated array $NLN[1..n]$.

```
1  begin
2      for i = 1 to n do
3          NLN[i] = -n
4          for k = 1 to max{i - 1, n - i} do
5              if i - k ≥ 0 and A[i - k] > A[i] then
6                  NLN[i] = i - k
7                  break
8              if i + k ≤ n and A[i + k] > A[i] then
9                  NLN[i] = i + k
10                 break

11 end
```

Lemma 1. *Algorithm 2 correctly computes the array $NLN[1..n]$ in $O(n^2)$ time.
Furthermore, there is an example which requires $\Omega(n^2)$ time.*

Proof. The algorithm consists of double loops, and clearly runs in $O(n^2)$ time.
It is also easy to see that if n given numbers are all the same then Algorithm 1
takes $\Omega(n^2)$ time.   □

Somewhat surprisingly, Algorithm 2 is considerably more efficient, in the worst
case, if we restrict attention to input arrays with no duplicate keys:

Theorem 2. [Power of Bidirectional Search]
*Assuming that the elements of $A[1..n]$ are all distinct, Algorithm 2 computes the
array $NLN[1..n]$ in $\Theta(n \log n)$ time in the worst case.*

Proof. By the assumption that every input number is distinct, there is a unique
largest element in $A[1..n]$. Let $d_i = |i - NLN[i]|$ and define σ to be a permutation
for which

$$d_{\sigma(1)} \geq d_{\sigma(2)} \geq \cdots \geq d_{\sigma(n-1)}. \tag{1}$$

Let $k \in \{1..n\}$ and consider the distribution of the k elements $A[\sigma(1)], \ldots,$ $A[\sigma(k)]$ in the array (see Figure 2). The smallest distance between two such elements, realized say by $A[\sigma(i)]$ and $A[\sigma(j)]$, must be no more than $n/(k+1)$. By our element distinctness assumption the two elements $A[\sigma(i)]$ and $A[\sigma(j)]$ cannot be equal; without loss of generality assume that $A[\sigma(i)] < A[\sigma(j)]$. Then $d_{\sigma(k)} \leq d_{\sigma(i)} \leq n/(k+1)$.

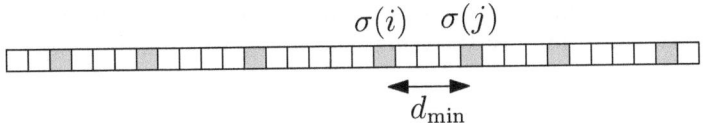

Fig. 2. Distribution of array elements with k largest distances (shaded elements) and two such elements with the minimum gap

Since the running time of the algorithm is $\Theta(\sum_{k=1}^{n} d_k)$, the upper bound on the cost of Algorithm 2 follows from the fact that

$$\sum_{k=1}^{n} d_k = \sum_{k=1}^{n} d_{\sigma(k)} \leq \sum_{k=1}^{n} \frac{n}{k+1} = O(n \log n). \tag{2}$$

It is also easy to show that there is an example for which Algorithm 2 takes $\Omega(n \log n)$ time. Let $n = 2^{t+1} - 1$. We construct a permutation of $(1, 2, \ldots, 2^{t+1} - 1)$ by (i) assigning the keys $(1, 2, \ldots, 2^{t+1} - 1)$ to the nodes of a perfectly balanced binary tree T (of height t) in max-heap order, and (ii) listing the keys by an in-order traversal of T. It is easy to see that all 2^{t-h} keys at height $h < t$ have NLN distances equal to 2^h. Thus the cost of Algorithm 2 is $\Theta(t2^t) = \Theta(n \log n)$ in this case. $\square$

It remains to see if we can improve upon the simple bidirectional search strategy when the input array contains duplicate elements.

2.3   Improvements with Duplicate Keys

Two Distinct Numbers. Suppose that the input sequence contains elements in $\{0, 1\}$.

Lemma 3. *NLN-values for any $\{0, 1\}$-array $A[1..n]$ can be found in $O(n)$ time using constant working space.*

Proof. First, set $NLN[i] = 0$ for all i such that $A[i] = 1$ since all such elements have no larger neighbor. It remains to show how to assign $NLN[i]$ for all i such that $A[i] = 0$.

Suppose that at least one such element exists (otherwise, nothing remains to be done). Let $A[i..j]$ be any maximal interval of 0 elements within $A[1..n]$. If

$i = 1$ then, by maximality, $A[j+1] = 1$ and so $NLN[x] = j+1$, for all $x \in [i,j]$. Similarly, if $j = n$ then $NLN[x] = i - 1$, for all $x \in [i,j]$. Alternatively, it must be the case that both $A[i-1] = 1$ and $A[j+1] = 1$ and hence $NLN[x] = i - 1$, for $x \in [i,m]$, and $NLN[x] = j + 1$, for $x \in [m+1,j]$, where $m = \lceil (i+j)/2 \rceil$.

Since both identifying maximal sequences of 0's and assigning NLN-values, as specified, can be done with constant working space, the result follows. □

Remark. Suppose that the input sequence contains elements in $\{-1, 0, 1\}$. It is straightforward to modify the algorithm above to compute $NLN[x]$ for all x such that $A[x] = 0$.

k Distinct Numbers. Suppose now that the input array A contains k distinct numbers. Assume for simplicity that they are $1, 2, \ldots, k$. By definition $NLN[x] = 0$, for all x such that $A[x] = k$. For each $v = 1, 2, \ldots, k - 1$, we use the above algorithm using the following implicit transformation of A. Replace $A[j]$ by -1 (respectively, 0 or 1) if it is $<$ (respectively,$=$ or $>$) v. By the result of the previous subsection, the running time to compute the NLN-values associated with the 0-elements (of the implicit array) is $O(n)$ and the working space is $O(1)$. Thus, the total running time is $O(kn)$.

General Algorithm–Not Necessarily Distinct Inputs. Suppose that the input contains k distinct numbers, where k is any number from 1 to n. We design an algorithm that achieves the running time of two above algorithms: the time is $O(n \min\{k, \log n\})$.

Our algorithm is presented in pseudo code as Algorithm 3. We assume that the input is given by an array $A[1..n]$ of real numbers. Recall that, to make $NLN[i]$ defined for largest elements in $A[1..n]$ and to avoid cluttering the pseudo code with indexing checks, we have further assumed that the array A has been extended so that $A[-n] = A[3n] = \infty$ and $A[j] = -\infty$, for $j \in [-n+1..0] \cup [n+1..3n]$.

The high level structure of Algorithm 3 is a repetition, for all elements $a \in [1..n]$ of a process that we call ScanFrom(a). If, for some $s > 0$, the elements of the subarrays $A[a-2s..a-s-1]$ and $A[a+1..a+s]$ all have value at most $A[a]$, the elements of the subarray $A[a-s+1..a-1]$ all have value less than $A[a]$, and the element $A[a-s]$ has value equal to $A[a]$, then we say that the index a is *passive*; otherwise, it is *active*. ScanFrom(a) either aborts without reporting any NLN-values (having discovered that index a is passive) or it reports the NLN-value of a and all indices $i > a$ for which $A[i] = A[a]$, up to (but not including) the first (smallest) such index that is active.

The algorithm is presented in a way that emphasizes the fact that the elements $a \in [1..n]$ can be treated in any order (or even in parallel).

Algorithm 3. Finding nearest-larger-neighbours in an array A.

Input: Array $A[1..n]$ of keys.
Output: Associated array $NLN[1..n]$.

```
1  begin
2  │   forall  a ∈ {1,...,n} do
                                                    /* ScanFrom(a) */
3  │   │   x ← A[a];
4  │   │   s ← 1; e ← a;
5  │   │   while  A[a − s] < x and A[a + s] ≤ x do      /* Invariant 1 */
6  │   │   │   if A[a + s] = x then e ← a + s;
7  │   │   └   s ← s + 1;
                /* (A[a − s] ≥ x or A[a + s] > x) and Invariant 1 */
8  │   │   if A[a + s] > x then
9  │   │   │   r ← a + s ;                                /* A[r] > x */
10 │   │   │   if A[a − s] > x then l ← a − s else l ← a − s − 1;
                        /* A[l] ≤ x ⇒ r − a < a − l */
11 │   │   else                   /* A[a − s] ≥ x and Invariant 1 */
12 │   │   │   p ← s;
13 │   │   │   while A[a − p] ≤ x do              /* Invariant 2 */
14 │   │   │   │   if p = 2s then abort ScanFrom(a)      /* Index a is
               │   passive */
15 │   │   │   └   p ← p + 1;
                /* A[a − p] > x,  p ≤ 2s and Invariant 2 */
                                                    /* A[l] > x */
16 │   │   │   l ← a − p ;
17 │   │   │   t ← a + s;
18 │   │   │   while t − e < e − l and A[t] ≤ x do    /* Invariant 3 */
19 │   │   │   │   if A[t] = x then e ← t ;
20 │   │   │   └   t ← t + 1;
                        /* A[t] ≤ x ⇒ t − e ≥ e − l */
21 │   │   └   r ← t;
                                                    /* Invariant 4 */
22 │   │   if r − a ≥ a − l then NLN[a] ← l else NLN[a] ← r;
23 │   │   j ← a; i ← a + 1;
24 │   │   while i ≤ e do                         /* Invariant 5 */
25 │   │   │   if A[i] = x then                    /* j − l > i − j */
26 │   │   │   │   if r − i ≤ i − j then          /* Index i is active */
27 │   │   │   │   │   abort ScanFrom(a);
28 │   │   │   │   else                          /* Index i is passive */
29 │   │   │   │   └   if r − i ≥ i − l then NLN[i] ← l else NLN[i] ← r;
30 │   │   │   └   j ← i;
31 │   │   └   i ← i + 1;

32 end
```

The following invariants describe the evolving knowledge about the structure of the array A, and serve to support our proof of correctness.

Invariant 1

    (i) $A[k] < x$, for $a - s < k < a$;

    (ii) $A[a] = x$;

    (iii) $A[k] \leq x$, for $a < k < e$;

    (iv) $A[e] = x$; and

    (v) $A[k] < x$, for $e < k < a + s$.

Invariant 2

    (i) $A[k] \leq x$, for $a - p < k < a$;

    (ii) $A[a - s] \geq x$;

    (iii) $A[k] < x$, for $a - s < k < a$;

    (iv) $A[a] = x$;

    (v) $A[k] \leq x$, for $a < k < e$;

    (vi) $A[e] = x$; and

    (vii) $A[k] < x$, for $e < k < a + s$.

Invariant 3

    (i) $A[k] \leq x$, for $l < k < a$;

    (ii) $A[a] = x$;

    (iii) $A[k] \leq x$, for $a < k < e$;

    (iv) $A[e] = x$; and

    (v) $A[k] < x$, for $e < k < t$.

Invariant 4

    (i) $A[k] \leq x$, for $l < k < a$;

    (ii) $A[a] = x$;

    (iii) $A[k] \leq x$, for $a < k < e$;

    (iv) $A[e] = x$;

    (v) $A[k] < x$, for $e < k < r$;

    (vi) $A[l] \leq x$

    $\Rightarrow (A[r] > x \ \& \ r - a < a - l)$;

    (vii) $A[r] \leq x$

    $\Rightarrow (A[l] > x \ \& \ r - e \geq e - l)$.

Invariant 5

    (i) $A[k] \leq x$, for $l < k < j$;

    (ii) $A[j] = x$;

    (iii) $A[k] < x$, for $j < k < i$;

    (iv) $A[k] \leq x$, for $i \leq k < e$;

    (v) $A[e] = x$; and

    (vi) $A[k] < x$, for $e < k < r$.

Confirmation of these invariant properties (at the appropriate locations in the algorithm) is a lengthy, but completely straightforward, exercise.

Lemma 4. *All NLN-values assigned by Algorithm 3 are correct*

Proof. *NLN*-values are assigned in lines 22 and 29. The correctness of these assignments follows immediately from Invariant 4.   □

We refer to the interval of array indices encountered during ScanFrom(a) as the *reach* of ScanFrom(a). Index u in the reach of ScanFrom(a) satisfying $A[u] = A[a]$ is said to be a *predecessor* (respectively, *successor*) of index a if $u < a$ (respectively, $u > a$). Note that if a is passive then it must have a predecessor and hence, by transitivity, a nearest active predecessor.

Lemma 5. *For all a, $1 \leq a \leq n$, $NLN[a]$ is assigned exactly once. If a is active then $NLN[a]$ is assigned during ScanFrom(a). If a is passive then $NLN[a]$ is assigned during ScanFrom(u), where u is the nearest active predecessor of a.*

Proof. It is straightforward to confirm that ScanFrom(a) aborts at line 14 if and only if index a is passive. Hence, if index a is active then $NLN[a]$ is reported at line 22. Since ScanFrom(a) never reports NLN-values for indices $i < a$, or for active indices $i > a$, the lemma clearly holds for active indices.

Suppose now that index a is passive. Since ScanFrom(a) aborts at line 14, if $NLN[a]$ is reported it must occur during ScanFrom(u), for one or more active indices $u < a$. But, by line 26, $NLN[a]$ is not reported during ScanFrom(u) if there exists an active index w between u and a. Thus, if $NLN[a]$ is reported it must occur during ScanFrom(u), where u is the nearest active predecessor of a.

Suppose that a is the smallest passive index such that $NLN[a]$ is not assigned during ScanFrom(u), where u is the nearest active predecessor of a. By the minimality of a, we can assume that all passive predecessors of a following u are assigned during ScanFrom(u). It follows, by lines 26-29, that a must not lie within the reach of ScanFrom(u). Hence, by Invariant 3, the the gap separating a from its closest predecessor v, must be at least the distance from v to the left reach of ScanFrom(u). But this contradicts our assumption that index a is passive. $\square$

One can easily convert A into a implicit array of distinct numbers by replacing $A[i]$ by the pair $(A[i], i)$ and using lexicographic order on the pairs. Let d_i be the distance of $A[i]$ to its nearest larger neighbor in this lexicographic order. Let $D = \sum_{i=1}^{n} d_i$.

Lemma 6. *Algorithm 3 runs in $O(D)$ time and uses constant working space.*

Proof. It is clear from inspection of the pseudo code that Algorithm 3 uses constant working space and that the cost associated with ScanFrom(a) is proportional to the size of the reach of ScanFrom(a). Thus, to show that Algorithm 3 uses $O(D)$ time it suffices, in light of Lemma 5, to show that the reach of ScanFrom(a) is (i) $O(d_a)$, if a is passive, and (ii) $O(\sum_{i \in R(a)} d_i)$, where $R(a)$ denotes the set of all indices whose NLN-values are set in ScanFrom(a), otherwise.

Case (i) is clear since, when ScanFrom(a) aborts, the reach of ScanFrom(a) is $[a - 2d_a .. a + d_a]$. For case (ii), we first note that if the nearest larger neighbor of a has an index larger than a, or if a has no successor within the reach of ScanFrom(a), then the reach of ScanFrom(a) is at most $[a - 2d_a .. a + 2d_a]$. So we can assume that l (the left reach of ScanFrom(a)) is the nearest larger neighbor of a and that a has one or more successors within the reach of ScanFrom(a). Let w denote the rightmost such successor.

By lines 18-20, the size of the reach of ScanFrom(a) is at most twice the gap from l to w. Since the gap from l to a is at most $2d_a$, and the gap from any inactive successor u of a to its immediate predecessor is exactly d_u, the result follows directly from Lemma 5 if all of the successors of a are inactive (and hence belong to $R(a)$). On the other hand, if a has an active successor, and v denotes the nearest such successor, then the gap between v and its immediate predecessor u must be less than the gap from l to u (otherwise v would not be in the reach of ScanFrom(a)). Furthermore, since v is active, it also follows that the gap between v and the right reach of ScanFrom(a) is no more than the gap between v and u. But, as we have already already argued, the gap between l and u is $O(\sum_{i \in R(a)} d_i)$. It follows that the full reach of ScanFrom(a) is also $O(\sum_{i \in R(a)} d_i)$. $\square$

Lemma 7. *If array $A[1..n]$ contains k distinct elements then $D = O(n\min (k, \log n))$. Furthermore, in the worst case, $D = \Theta(n\min(k, \log n))$.*

Proof. The result when $k > \log n$ follows immediately from the proof of Theorem 2. So suppose that $k \leq \log n$. To show that $D = O(nk)$ consider any number x in $A[1..n]$. Let $i_1, i_2, \ldots, i_r$ be the indices of x, i.e. $x = A[i_1] = A[i_2] = \ldots A[i_r]$. By definition, $d_{i_p} \leq i_{p+1} - i_p$, for $1 \leq p < r$, and hence $d_{i_1} + d_{i_2} + \ldots d_{i_r} = O(n)$. It follows that $D = O(nk)$.

It remains to show that there exists a sequence of n numbers from the set $\{0, 1, \ldots, k-1\}$ such that $D = \Omega(nk)$. Suppose, for simplicity, that $n = 2^{t+1} - 1$. We construct a sequence by (i) assigning the key $k - s$ to all 2^{s-1} nodes on level s, for $1 \leq s < k$, of a perfectly balanced binary tree T (of height t) in max-heap order, (ii) assigning key 0 to nodes on all other levels and (iii) listing the keys by an in-order traversal of T. It is easy to see that all 2^{k-x-1} keys of value x have NLN distances equal to 2^{t-k+x}, for $1 \leq x < k - 1$. Thus the cost of Algorithm 3 is $\Theta(2^t k) = \Theta(nk)$ in this case. $\square$

3   NLN Problem in Higher Dimensions

It is natural to consider the NLN problem in higher dimensions as well. In two dimensions, suppose we have an array A of size $m \times m$. For each element $A[i, j]$ we want to find a larger element $A[p, q]$ that is nearest to $A[i, j]$. We consider the situation where distance between two such array elements is measured using the L_1 metric ($|p - i| + |q - j|$) or the L_2 (Euclidean) metric ($\sqrt{(p - i)^2 + (q - j)^2}$).

A natural idea (that generalizes the one-dimensional bidirectional search) for solving the nearest larger neighbors problem is to examine neighbors in the increasing order of their distances. We call this algorithm the *Distance Heuristic*.

For L_1 distance the Distance Heuristic has a straightforward constant-space implementation that exploits the fact that the L_1 k-neighborhood of an element $A[i, j]$ (the set of elements $A[p, q]$ whose L_1 distance from $A[i, j]$ is exactly k) forms a rhombus within A centered at $A[i, j]$ and has a simple constant-space enumeration.

If the nearest larger neighbor of $A[i, j]$ has L_1 distance $d_{i,j}$, then the time required to determine $NLN[i, j]$ is $O(d_{i,j}^2)$. Let d_k denotes the k-th largest among the m^2 NLN-distances and consider the distribution of the k elements of A whose NLN-distances are at least d_k. Since d_k is the smallest L_1 distance between two such elements, it must satisfy the inequality $k(d_k/2)^2 \leq m^2$. Thus we have $d_k^2 \leq 4m^2/k$. It follows, by the same analysis used in the proof of Theorem 2, that the Distance Heuristic runs in $O(m^2 \log m)$ time if all elements are distinct.

The problem of implementing the Distance Heuristic using L_2 distances is slightly more involved. The main difficulty is that there seems to be no efficient way of generating neighbors in increasing order of their L_2 distances from a fixed element $A[i, j]$, using only constant working space. Fortunately, we do not really need to generate neighbors in strictly increasing order.

The idea is the following. We first find the L_1 NLN-distance $d_{i,j}$ associated with $A[i,j]$, as above. We then find the element, among all elements larger than $A[i,j]$ with L_1 distance between $d_{i,j}$ and $\sqrt{2}d_{i,j}$, that minimizes the L_2-distance from $A[i,j]$. (See the pseudocode for Algorithm 4 for more detail.)

Algorithm 4. L_2-Distance Heuristic.

Input: Array $A[1..m, 1..m]$ of keys
Output: Associated array $NLN[1..m, 1..m]$

```
 1  begin
 2      for each element A[i, j] in A do
 3          // First Phase: find an L₁ nearest larger neighbor //
 4          k = 0; D = ∞
 5          repeat
 6              for each element A[p, q] in the L₁ k-neighborhood of A[i, j] do
 7                  if A[p, q] > A[i, j] then
 8                      D = (p − i)² + (q − j)²
 9                      NLN[i, j] = (p, q)
10          until k = m or D < ∞
11          if D = ∞ then NLN[i, j] = (−m, −m) else
12              t = k
13              repeat
14                  for each element A[p, q] in the L₁ t-neighborhood of A[i, j]
                    do
15                      if A[p, q] > A[i, j] and (p − i)² + (q − j)² < D then
16                          D = (p − i)² + (q − j)²
17                          NLN[i, j] = (p, q)
18              until t > √2k
19  end
```

Theorem 8. *The L_2-Distance Heuristic solves the Nearest Larger Neighbors Problem in $O(m^2 \log m)$ time for any $m \times m$ array A, under the assumption that all elements are distinct. Furthermore, there is an example of such an array for which $\Theta(m^2 \log m)$ time is required.*

Proof. The correctness of the algorithm using the L_1 distance is obvious since we generate neighbors in the increasing order of the L_1 distances. For the L_2 distance we also generate neighbors using the L_1 distances. We exploit the fact that L_1 distance overestimates L_2 distance by a factor between 1 and $\sqrt{2}$. See Figure 3.

The following example shows that the bound is tight, in the worst case. Suppose that $m = 2^k + 1$. We construct a $m \times m$ array A with keys from 1 to

$m^2 = 2^{2k} + 2^{k+1} + 1$ such that Distance Heuristic takes $\Omega(m^2 \log m)$ time to solve NLN problem for A. Our construction is recursive and assumes that the keys in the first and last row and column are the larger than any key in the interior of a given subarray. We initialize this invariant by assigning the $4m - 4$ largest keys arbitrarily to the first and last rows and columns of A. To continue we first assign the next largest key to the central position $A[2^{k-1} + 1, 2^{k-1} + 1]$ and then the next $4m - 5$ largest elements arbitrarily to the unassigned positions in the middle row and column of A. In effect this partitions A into four subarrays of size $2^{k-1} \times 2^{k-1}$ (with overlapping boundaries) that satisfy the invariant assumption. Thus, we are free to continue the assignment of keys recursively in each of these submatrices.

The specified assignment has the property that that each of the 4^t central elements assigned at the t-th level of recursion has a nearest larger neighbor at distance 2^{k-1-t}, for $0 \leq t < k$. Thus, summing over these central elements only we see that the Distance Heuristic takes time at least $\sum_{t=0}^{k-1} 4^t (2^{k-1-t})^2$ which is $\Theta(k2^k)$ or $\Theta(m^2 \log m)$. $\square$

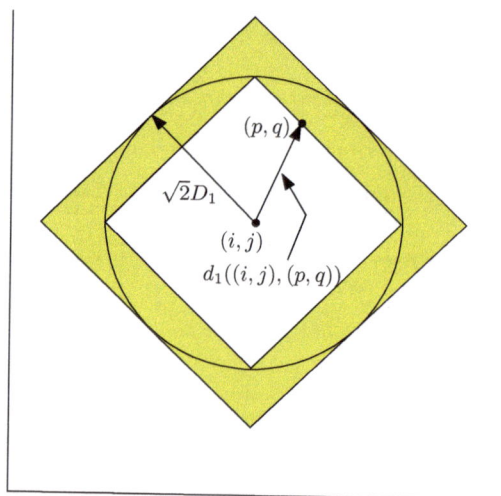

Fig. 3. The region to check whether a closer larger neighbor exists

4   Extensions

It is not hard to see how to extend the Distance Heuristic (for both L_1 and L_2) to work on d-dimensional arrays of total size m^d in time $O(m^d \log m)$, assuming all keys are distinct. However, when $d > 1$ it remains an open problem how to achieve this same bound if the distinctness assumption is dropped.

For the one-dimensional $ANLN$ problem, the reader may find it an interesting exercise to implement Algorithm 3 as a distributed (message passing) algorithm on a ring of processors.

Multi-core Implementations of Geometric Algorithms*

Stefan Näher and Daniel Schmitt

Department of Computer Science, University of Trier, Germany
{naeher,schmittd}@uni-trier.de

Abstract. This paper presents a framework for multi-core implementations of divide and conquer algorithms and shows its efficiency and ease of use by applying it to some fundamental problems in computational geometry. The framework supports automatic parallelization of any D&C algorithm. It is only required that the algorithm is implemented by a C++ class implementing a so-called job-interface. We also report on experimental results and discuss some aspects of the automatic parallelization of randomized incremental algorithms. Some results of this paper have been presented in the 20th Annual Canadian Conference on Computational Geometry ([13]).

1 Introduction

Performance gain in computing is no longer achieved by increasing cpu clock rates but by multiple cpu cores working on shared memory and a common cache. In order to benefit from this development software has to exploit parallelism by multi-threaded programming. In this paper we present a framework for the parallelization of divide and conquer (D&C) algorithms and show its efficiency and ease of use by applying it to a fundamental geometric problem: computing the convex hull of a point set in two dimensions.

In general our framework supports parallelization of divide and conquer algorithms taking as input a linear container of objects (e.g. an array of points). We use the STL iterator interface ([1]), i.e., the input is defined by two iterators left and right pointing to the leftmost and rightmost element of the container. The framework is generic. It can be applied to any D&C-algorithm algorithm that is implemented by a C++ class template implementing the interface defined in Section 2.1.

The paper is structured as follows. In Section 2 we discuss some aspects of the parallelization of D&C-algorithms. Section 2.1 defines the job-interface which has to be used for the algorithms, such that the solvers presented in Section 2.3 can be applied. In Section 3 we discuss a similar approach for the parallelization of incremental algorithms. Section 4 presents experimental results, in particular the speedup achieved for different numbers of cpu cores and different problem instances. Finally, Section 5 gives some conclusions and reports on current and ongoing work.

---

* This work was supported by DFG-Grant Na 303/2-1.

S. Albers, H. Alt, and S. Näher (Eds.): Festschrift Mehlhorn, LNCS 5760, pp. 261–274, 2009.

2   Divide and Conquer

Divide and conquer algorithms solve problems by dividing them into subproblems, solving each subproblem recursively and merging the corresponding results to a complete solution. All subproblems have exactly the same structure as the original problem and can be solved independently from each other, and so can easily be distributed over a number of parallel processes or threads. This is probably the most straightforward parallelization strategy. However, in general it cannot be guaranteed that always enough subproblems exist, which leads to non-optimal speedups. This is in particular true for the first divide step and the final merging step but is also a problem in cases where the recursion tree is unbalanced such that the number of open sub-problems is smaller than the number of available threads.

   Therefore, it is important that the divide and merge steps are solved in parallel when free threads are available, i.e. whenever the current number of sub-problems is smaller than the number of available threads. Our framework basically implements a management system that assigns jobs to threads in such a way that all cpu cores are busy.

2.1   The Job Interface

In the proposed framework a *job* represents a (sub-)problem to be solved by a D&C-algorithm. The first (or root) job represents the entire problem instance. Jobs for smaller sub-problems are created in the divide steps. As soon as the size of a job is smaller than a given constant it is called a *leaf job* which is solved directly without further recursion. As soon as all children of a job have been solved the merge step of the D&C-algorithm is applied and computes the result of the entire problem by combining the results of its children.

   In this way jobs represent sub-problems as well as the corresponding solutions. Note that the result of a job is either contained in the corresponding interval of the input container or has to be represented in a separate data structure, e.g. a separate list of objects. Quicksort is an example for the first case and Quickhull (as presented in Section 2.2) for the second case.

   The algorithm is implemented by member functions of the job class which must have the following interface.

```
class job
{ job(iterator left, iterator right);
  bool  is_leaf();
  void  handle_leaf();
  list<job> divide();
  void  merge(list<job>& L);
};
```

In the constructor a job is normaly created by storing two iterators (e.g. pointers into an array) that define the first and last element of the problem. If the `is_leaf` predicate returns true recursion stops and the problem is solved directly by calling the handle_leaf operation. The `divide` operation breaks a job into smaller jobs and returns them in a list, and the `merge` operation combines the solutions of sub-jobs (given as a list of jobs) to a complete solution. There are no further requirements to a job class.

2.2 Examples

We present job definitions for some well known divide and conquer algorithms. We use Quicksort as an introductory example and then discuss two convex hull algorithms, Gift Wrapping and Quickhull.

Quicksort. Quicksort takes as input an array given by the random access iterators `left` and `right`. Functions `merge` and `handle_leaf` are trivial. The `divide` operation calls a function `partition(1,r)` that performs the partition step with a randomly selected pivot element. It returns the position of the pivot element as an iterator m. Finally, it creates two jobs for the two sub-problems.

```
template<class iterator> class qs_job {
 iterator left, right;

public:

qs_job(iterator l, iterator r): left(l),right(r){}
int  size()    { return right - left + 1; }
bool is_leaf() { return size() <= 1; }
void  handle_leaf() {}
void  merge(list<qh_job>& children){}

list<qh_job> divide()
{ iterator m = partition(left,right);
  list<qh_job> L;
  L.push_back(qs_job(left,m));
  L.push_back(qs_job(m + 1,right));
  return L;
}
};
```

Gift Wrapping. The well-known Gift Wrapping algorithm constructs the convex hull by folding a halfplane around the set of input points such that all points always lie on the same side of the halfplane. In the recursive version of the algorithm two disjoint convex hulls are combined by computing tangents to both hulls. The divide and conquer algorithm is designed as follows:

Partition the input points at some pivot position according to the lexicographical ordering of the cartesian coordinates in two sets L and R, such that the convex hulls of L and R are disjoint. Then compute the convex hull L and R

and the extreme points min and max of both hulls recursively. Finally compute the upper and lower tangents starting with line segment $(max(L), min(R))$.

We assume that the input is unsorted and use the Quicksort partitioning step for creating the two sub-problems. This gives an expected running time of $O(n \log n)$. Note that we sort the input and compute the convex hull at the same time by exploiting the fact that Quicksort has a trivial merge and Gift Wrapping a trivial divide operation.

The corresponding `gw_job` class is derived from `qs_job`. It inherits the input iterators and the operations `size` and `divide`. The convex hull is stored in a doubly-linked list `result`. The class contains in addition iterators `min` and `max` pointing to the extreme points of the hull. Function `handle_leaf()` treats the trivial case of input size one.

The `merge` operation is illustrated in Figure 1. The auxiliary function `compute_tangents()` does the main work by computing the two tangents as described above.

```
template<class iterator> class gw_job : qs_job
{
 list<point> result;
 list_iterator min, max;

public:

qs_job(iterator l, iterator r): qs_job(l,r){}
void  handle_leaf()
{ if (size() == 1) {
  result.push_back(*left);
  max = min = result.begin();}
}
void  merge(list<qh_job>& children)
{ qh_job jleft = children.front();
  qh_job jright = children.back();
  result = compute_tangents(jleft.result,
      jleft.max,jright.min,jright.result);
  min = jleft.min;
  max = jright.max;
}
};
```

Quickhull. We show how to define a job class `qh_job` implementing the well-known Quickhull algorithm ([3]) for computing the convex hull of a point set. For simplicity we consider a version of the algorithm that only computes the upper hull of the given point set and we assume that the input is given by a pair of iterators `left` and `right` into an array of points such that `left` contains the minimal and `right` the maximal point in the lexicographical xy-ordering. The result of a `qh_job` instance is the sequence of points of the upper hull lying

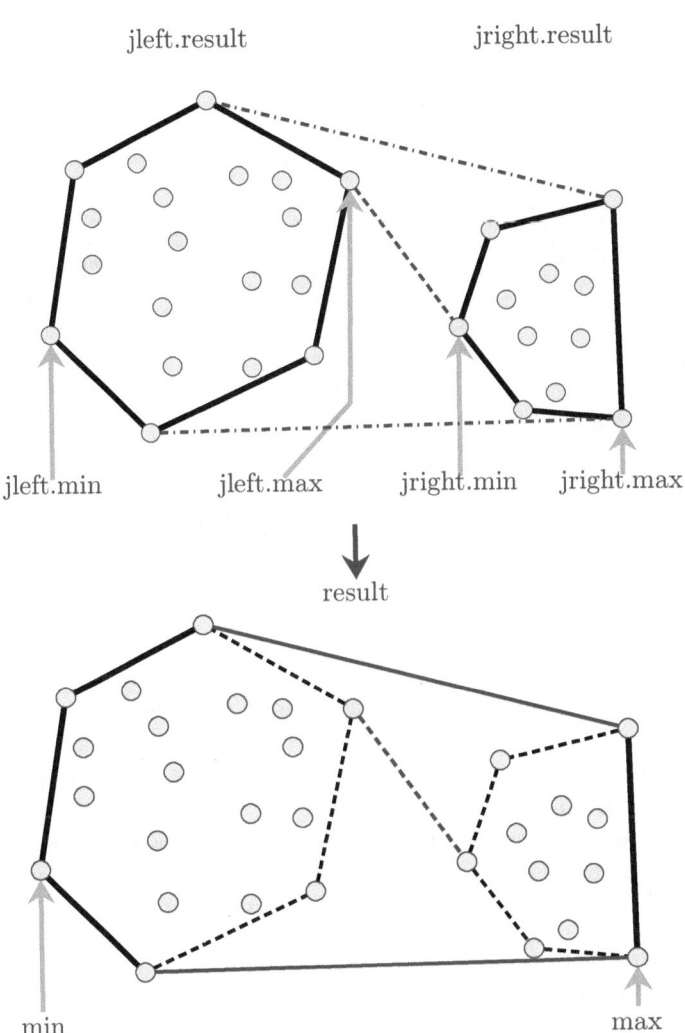

jleft.result   jright.result

jleft.min   jleft.max   jright.min   jright.max

result

min   max

Fig. 1. The merge operation of Gift Wrapping

between left and right. In this scenario any job of size two (only the leftmost and rightmost point) represents a leaf problem and has the empty list as result. Consequently, the handle_leaf operation is trivial (keeping an empty result list).

The divide operation is using two auxiliary functions: farthest_point(l,r) computes a point between l and r with maximal distance to the line segment (l, r) and partition_triangle implements the partition step of Quickhull as shown in Figure 2 and returns the generated sub-problems as a list of jobs. We tried different variants of this partition function. In particular, one using only one thread and one using all available threads. The latter version is similar to the parallel partition strategy proposed in [10] for a multi-core implementation

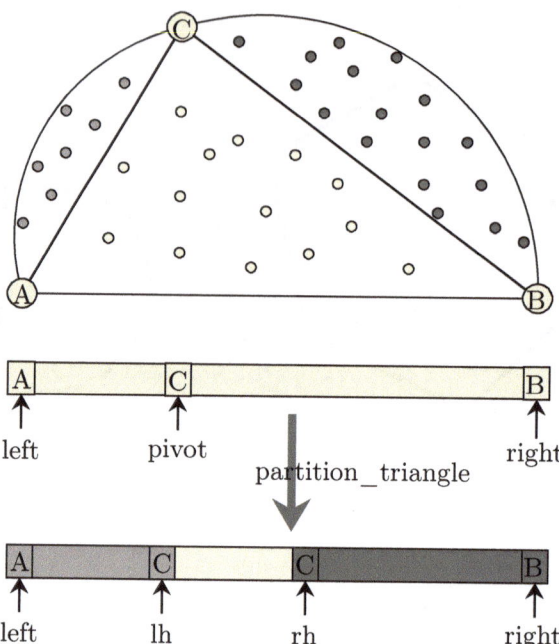

Fig. 2. The partition step of Quickhull

of Quicksort. In the experiments in Section 4) we will see that this can have a dramatic effect on the speedup achieved.

Finally, the **merge** operation takes a list of (two) jobs as input, concatenates their result lists, and inserts the right-most point of the first problem in between. The complete implementation is given by the following piece of C++ code.

```
template<class iterator> class qh_job {
iterator left;
iterator right;
list<point> result;

public:

qh_job(iterator l, iterator r): left(l),right(r) {}
int  size()    { return right - left + 1; }
bool is_leaf() { return size() == 2; }
void  handle_leaf() {}

list<qh_job> divide()
{ iterator pivot = farthest_point(left,right);
iterator lh,rh;
partition_triangle(pivot,left,right,lh,rh);
list<qh_job> L;
```

```
L.push_back(qh_job(left,lh));
L.push_back(qh_job(rh,right));
return L;
}

void  merge(list<qh_job>& children)
{ qh_job j1 = children.front();
qh_job j2 = children.back();
result.conc(j1.result);
result.push_back(*j1.right);
result.conc(j2.result);
}
};
```

2.3   Solvers

Our framework provides different *solvers* which can be used to compute the result of a job. As a very basic and introductory example we give the code for a generic *serial* recursive solver. It can be implemented by a simple C++ function template.

```
template <class job>
void solve_recursive(job& j)
{ if (j.is_leaf()) j.handle_leaf();
else { list<job> Jobs = j.divide();
   job x;
   forall(x,Jobs) solve_recursive(x);
   j.merge(Jobs);
  }
};
```

Note that `solve_recursive` is a generic dc-solver. It accepts any job type *job* that implements the `dc_job` interface. We can now use it easily to implement a serial Quickhull function taking an array of points as input.

```
list<point> QH_SERIAL(array<point>& A)
{ int n = A.size();
qh_job<point*> j(A[0],A[n-1]);
solve_recursive(j);
list<point> hull = j.result;
hull.push_front(A[0]);
hull.push_back(A[n-1]);
return hull;
};
```

It is an easy exercise to write a non-recursive version of this serial solver: simply push all jobs created by divide operations on a stack and use an inner loop processing all jobs on the stack.

Parallel Solvers. Parallel solvers are much more complex. They maintain open jobs, build the recursion tree while the algorithm proceeds and check for the mergeability of sub-jobs. They also have to administrate all threads working in parallel. In our framework all threads use a common job queue which has to be synchronized using a mutex variable.

There are different solver versions according to different requirements. The simplest solver handles problems with a trivial merge step in which case it is not necessary to store the recursion tree explicitly. In our framework solvers distinguish two types of threads. Primary threads work in parallel in different parts of the recursion tree, and secondary threads parallelize basic operations like partitioning and merging. A solver always tries to employ as much primary threads as possible.

There are more parameters that can be changed by corresponding methods of the class. For instance, a limit d for the minimal problem size for any thread. If the size of job gets smaller than d it will not be divided into new jobs but solved by the same thread using a serial algorithm. Using this limit the overhead of starting a huge number of threads on very small problem instances can be avoided. We implement parallel solvers by C++ class templates. In the example one sees the interface of a solver class. The constructor takes as argument the number of threads to be used for solving the problem. The computation starts with calling the run function with a list of root jobs.

```
template <class Job>
class dc_parallel_solver {
public:
  dc_parallel_solver(int thread_num);
  void set_limit(int d);
  void run(list<Job*> j)
};
```

We now can use the parallel solver template to implement a parallel version of the Quickhull function.

```
list<point> QH_PARALLEL(array<point>& A, int thr_n)
{ int n = A.size();
  dc_parallel_solver<job<point*> > solver(thr_n);
  job<point*> j(A[0],A[n-1]);
  solver.run(j);
  list<point> hull = j.result;
  hull.push_front(A[0]);
  hull.push_back(A[n-1]);
  return hull;
};
```

3   Randomized Incremental Construction

Beside the divide and conquer paradigm *randomized incremental construction* is one of the most important and efficient strategies in computational geometry.

In this section we discuss some aspects of a framework for the automatic parallelization of incremental algorithms. The goal is similar as for divide and conquer algorithms. Provided the algorithm is implemented according to a given simple interface it can automatically be parallelized on multi-core platforms. This is work in progress. The final results will be published in [14].

In general an incremental construction algorithm works on a dynamic data structure D representing the solution of the problem for a subset of the input objects such that new objects can be added efficiently.

In the beginning D is either empty or intialized to a simple initial structure e.g. the convex hull of three points of the input set. Adding a new object x consists of two steps. First the *locate* step finds the position of x in the data structure and then the *update* step insert x at this position.

In our approach to automatic parallelization the data structure D has to be implemented as a C++ class D and the locate and insert steps must be member functions of this class with a given predefined interface. Furthermore it is assumed that the data structure supports a position concept similar to the STL iterator or LEDA item concept. We call the corresponding type `elem_pointer`. Then the interface of D can be defined by the following piece of code.

```
template<class T>
class data_structure {
  typedef .... elem_pointer;
  bool locate(T x, elem_pointer& v);
  bool update(T x, elem_pointer v);
```

The locate operation returns the position in a reference parameter of type. The update operation takes this position as its second argument. Both operations have return result of type `bool` indication whether the operation succeeded or not. Since several threads (running on different cpu cores) will be running simultaneously on the data structure the `locate` and `update` methods have to be implemented in a thread save way. This can be achieved by using standard synchronization techniques such as mutex variables.

In the case of incremental algorithms a *solver* initializes the data structure, creates and joins threads, and distributes the input data over all threads. See [14] for details of this approach.

4   Experiments

All experiments were executed on a Linux PC with an Intel quad-core processor running at a speed of 2.6 GHz. As implementation platform we used a thread-safe version of LEDA [7]. In particular, we used the exact geometric primitives of the rational geometry kernel and some of the basic container types such as arrays and lists. All programs were compiled with gcc 4.1.

In the Quicksort experiments, we sort arrays of integers of various size. The implementation make use of a parallel partition function.

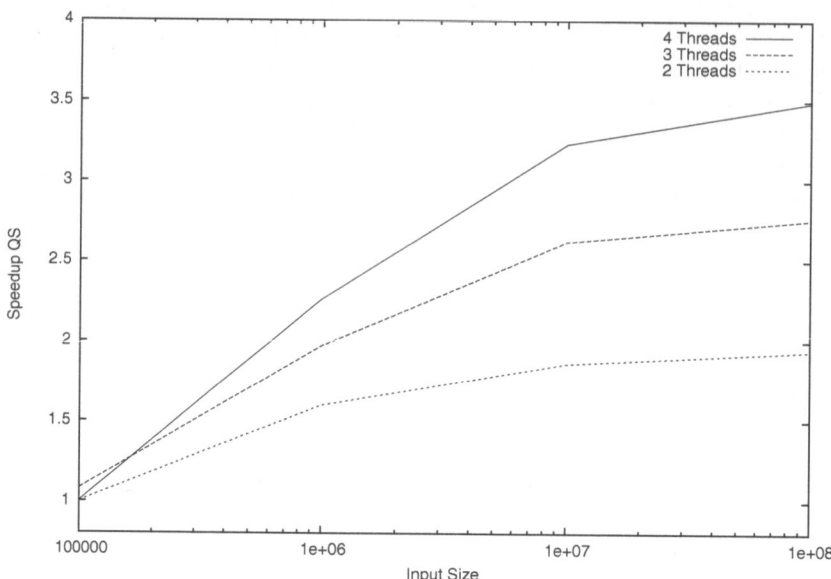

Fig. 3. Quickhull: Speedup of Quicksort

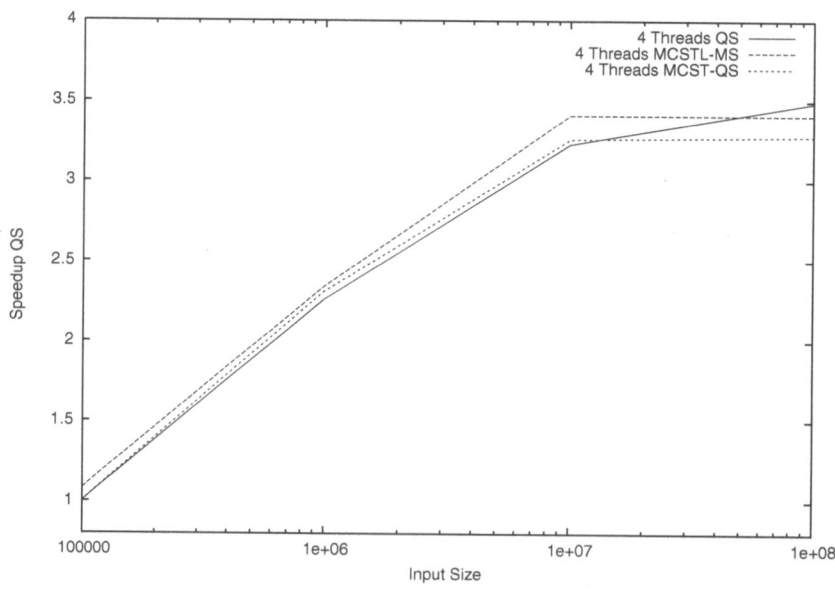

Fig. 4. Quickhull: Comparison Quicksort with MCSTL

In Figure 3 we see the speedup growing near optimal values when the size of the input gets large. To achieve a benefit from parallization we need sufficient problem sizes.

Table 1. Running times in seconds of three sorting algorithms on a set of 10^8 integers with different numbers of threads

	1	2	3	4
QS	10.31	5.32	3.74	2.96
MCSTL-MS	10.47	5.45	3.94	3.07
MCSTL-QS	10.47	5.92	4.02	3.19

Figure 4 compares our Quicksort implementation with two implementations from the MCSTL library ([9]). The MCSTL implementation with the better speedup is based on Mergesort the other on Quicksort. All implementations have a very good speedup progress. However the best speedup need not to correspond with the best absolute running time. Table 1 shows the running times of an input of 10^8 integers and different thread numbers. Our implementation is marginal faster.

For the convex hull experiments we used three different problem generators: random points lying in a square, random points near a circle, and points lying exactly on a circle. Figure 5 shows that our framework achieves a good speedup behavior for points on or near a circle, which is the difficult case for Quickhull because only a few or none of the points can be eliminated in the partitioning step. Note that the 1.0 baseline indicates the performance of a serial version of the algorithm (using only one thread). It turned out that $n/100$ was a good choice for the limit mentioned in Section 2.3.

For random points in a square Quickhull eliminates almost all of the input points in the root job of the algorithm (with high probability), i.e. almost the

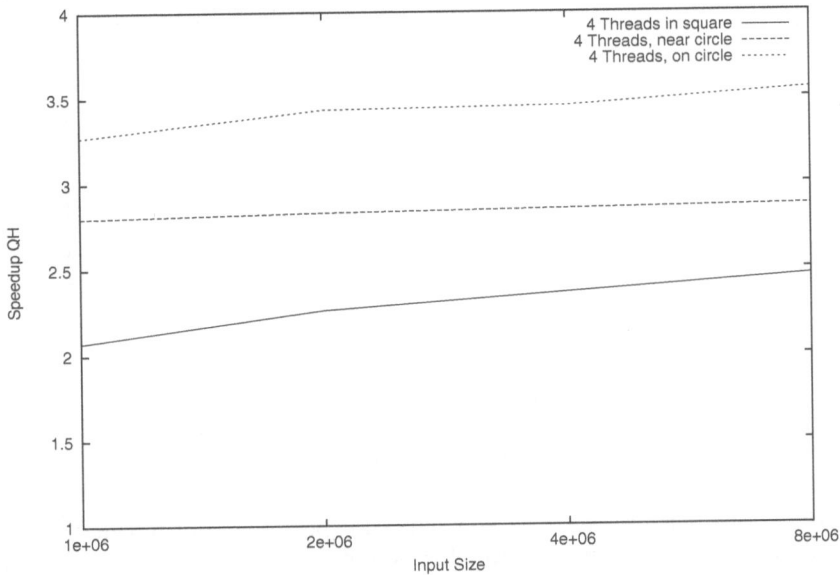

Fig. 5. Quickhull: Speedup with 4 cores

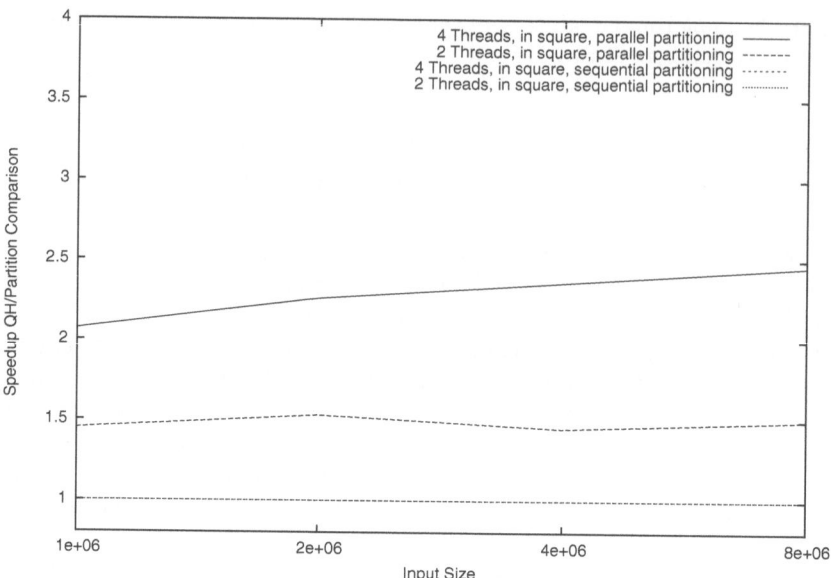

Fig. 6. Quickhull: The effect of parallel partitioning

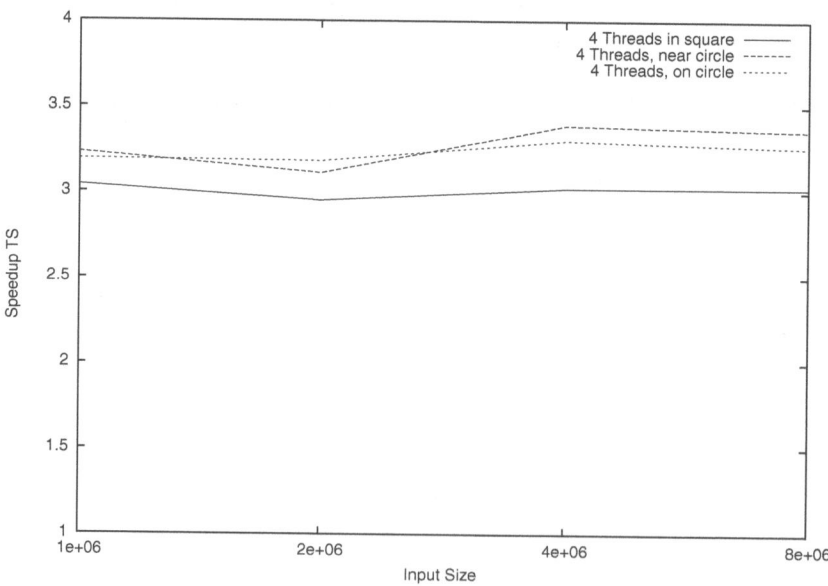

Fig. 7. Tangent Search: Speedup with 4 cores

entire work is done here. In this case the achieved speedup is not optimal. However, Figure 6 shows that without parallelization of the partitioning step we have no speedup at all. We have some ideas to improve the parallel partitioning and hope to improve the results for this kind of problem instances.

We also want to mention here that we ran experiments with different D&C algorithms for convex hulls. In particular, a recursive version of the Gift Wrapping method where the merge step does most of the work by constructing two tangents. Figure 7 shows the speedup behavior of this algorithm for the same set of input instances.

5 Conclusions

We have presented an approach for the automatic parallelization of divide and conquer algorithms and discussed some aspects and ideas for randomized incremental algorithms. The proposed frameworks are generic (by using C++ templates) and can be used very easily. The experiments show that a considerable speedup can be achieved by using two or four threads on a quad core machine. We have some ideas to improve the parallel partitioning of the Quickhull algorithm and hope to be able to improve the efficiency in cases where most of the work is done in the root job. In particular, our framework shows a very good performance also on basic D&C algorithms such as Quicksort. We work on the parallelization of more incremental algorithms for geometric problems and higher dimensional problems, preliminary implementations of incremental construction algorithms for Delaunay triangulations and Voronoi diagrams already exist. Here, one of the major problems is the need of advanced thread-safe dynamic data structures such as graphs or polyhedra.

References

1. Austern, M.H.: Generic programming and the STL. Addison-Wesley, Reading (2001)
2. Bradford Barber, C., Bobkin, D.P., Huhdanpaa, H.: The Quickhull Algorithm for Convex Hulls. ACM Transactions on Mathematical Software 22(4), 469–483 (1996)
3. Bykat, A.: Convex hull of a finite set of points in two dimensions. IPL 7, 296–298 (1978)
4. Eddy, W.F.: A new convex hull algorithm for planar sets. ACM Trans. Math. Softw. 3 (1977)
5. Kallay, M.: The Complexity of Incremental Convex Hull Algorithms. Info. Proc. Letters 19, 197 (1984)
6. Kettner, L., Näher, S.: Two Computational Geometry Libraries: LEDA and CGAL. In: Goodman, J.E., O'Rourke, J. (eds.) Handbook of Discrete and Computational Geometry, pp. 1435–1463. Chapmann & Hall/CRC, Boca Raton (2004)
7. Mehlhorn, K., Näher, S.: The LEDA Platform for Combinatorial and Geometric Computing. Cambridge University Press, Cambridge (1999)
8. Näher, S.: LEDA, a Platform for Cominatorial and Geometric Computing. In: Mehta, D.P., Sahni, S. (eds.) Handbook of Data Structures and Applications, pp. 41.1–41.18. Chapmann & Hall/CRC, Boca Raton (2004)
9. Putze, F., Sanders, P., Singler, J.: MCSTL: The Multi-Core Standard Template Library. Technical Report, MCSTL Version 0.70-beta
10. Tsigas, P., Zhang, Y.: A Simple, Fast Parallel Implementation of Quicksort and its Performance Evaluation on SUN Enterprise 10000

11. Clarkson, K.L.: Applications of Random Sampling in Computational Geometry. Discrete and Computational Geometry 4, 387–421 (1989)
12. Guibas, L.J., Knuth, D.E., Sharir, M.: Randomized Incremental Construction of Delaunay and Voronoi Diagrams. In: Proceedings of the 17th int. colloquium on Automata, languages and programming, pp. 414–431 (1990)
13. Näher, S., Schmitt, D.: A Framework for Multi-Core Implementations of Divide and Conquer Algorithms and its Application to the Convex Hull Problem. In: Proceedings of the 20th Annual Canadian Conference on Computational Geometry (2008)
14. Schmitt, D.: Multi-Core Implementierung geometrischer Algorithmene. Dissertation (in preparation, 2009)

The Weak Gap Property in Metric Spaces of Bounded Doubling Dimension*

Michiel Smid

School of Computer Science, Carleton University, Ottawa, Canada

Abstract. We introduce the weak gap property for directed graphs whose vertex set S is a metric space of size n. We prove that, if the doubling dimension of S is a constant, any directed graph satisfying the weak gap property has $O(n)$ edges and total weight $O(\log n) \cdot wt(MST(S))$, where $wt(MST(S))$ denotes the weight of a minimum spanning tree of S. We show that 2-optimal TSP tours and greedy spanners satisfy the weak gap property.

1 Introduction

Consider a directed graph $G = (S, E)$, where S is a set of n points in $\mathbb{R}^d$, and each edge (p, q) in E has a weight (or length) which is equal to the Euclidean distance $|pq|$ between the points p and q. We consider the problem of estimating the weight $wt(E)$ of the edge set E, which is defined to be the sum of the weights of the edges in E. Clearly, in order to obtain a non-trivial estimate, we need to make some assumptions about the edge set E.

Using geometric properties of $\mathbb{R}^d$, Chandra et al. [5] showed that the directed edge set of the greedy spanner algorithm (to be introduced later) can be partitioned into $O(1)$ subsets, such that each subset satisfies the *gap property*: For any two distinct edges (p, q) and (r, s) that are in the same subset, the distance $|pr|$ is at least proportional to the weight of the shorter of (p, q) and (r, s). They proved that the gap property implies that the weight of the subset is $O(\log n)$ times the weight $wt(MST(S))$ of a minimum spanning tree of the point set S. As a result, the weight of the greedy spanner is $O(\log n) \cdot wt(MST(S))$. (A much more complicated analysis shows that, in fact, the weight of the greedy spanner is $O(wt(MST(S)))$.)

Later, Chandra et al. [6] showed that any tour T that is computed by the 2-opt heuristic (to be introduced later) for the traveling salesperson problem can be analyzed by the same approach: By again using geometric properties of $\mathbb{R}^d$, the directed edge set of T can be partitioned into $O(1)$ subsets, each of which satisfies the gap property. Thus, by the result in [5], the weight of T is $O(\log n) \cdot wt(MST(S))$. Since $wt(MST(S))$ is less than the minimum weight $wt(TSP(S))$ of any tour of S, it follows that $wt(T) = O(\log n) \cdot wt(TSP(S))$. Chandra et al. also showed that, for the case when $d \geq 2$, there exists a 2-opt tour having weight $\Omega(\log n / \log \log n) \cdot wt(TSP(S))$.

---

* This work was supported by NSERC.

S. Albers, H. Alt, and S. Näher (Eds.): Festschrift Mehlhorn, LNCS 5760, pp. 275–289, 2009.

Both results mentioned above use the fact that the input points are in Euclidean space $\mathbb{R}^d$, for some constant $d \geq 1$. This leads to the natural question whether these results hold in an arbitrary *metric space*. Recall that a set S, together with a distance function $|pq|$ for any two points p and q in S, is called a metric space, if for all p, q, and r in S,

1. $|pp| = 0$,
2. $|pq| > 0$ if $p \neq q$,
3. $|pq| = |qp|$, and
4. $|pq| \leq |pr| + |rq|$.

The fourth property is called the *triangle inequality*.

The proof of Chandra *et al.* [5] of the fact that any directed edge set satisfying the gap property has weight $O(\log n) \cdot wt(MST(S))$ holds in any metric space. Narasimhan and Smid [13, Section 6.2] showed that this upper bound is tight, even in the one-dimensional Euclidean metric. On the other hand, in the metric space in which $|pq| = 1$ for all distinct points p and q, the weight of the greedy spanner is $\Theta(n) \cdot wt(MST(S))$. Chandra *et al.* [6] proved that, in any metric space, any tour that is computed by the 2-opt heuristic has weight $O(\sqrt{n}) \cdot wt(TSP(S))$; they also showed that this upper bound is tight. Thus, the weights of the outputs of the greedy spanner algorithm and the 2-opt heuristic behave very differently in Euclidean space $\mathbb{R}^d$ than they do in a general metric space.

The analyses in [5, 6] for the Euclidean metric use the notion of angles. In particular, they use the fact that any set of vectors in $\mathbb{R}^d$, in which any two elements make an angle of at least θ, contains $O(1/\theta^{d-1})$ elements. This is basically a *packing argument*: The number of "large" objects that can be packed in another slightly "larger" object is bounded from above by a constant (which depends on the dimension d). In Euclidean space, the validity of such an argument follows from the fact that a "large" object has a "large" volume. In a general metric space, however, a packing argument cannot be applied.

In this paper, we consider the weights of the greedy spanner and 2-opt tours in metric spaces in which a packing argument is valid. Such metric spaces are called metric spaces of bounded doubling dimension. We will prove that, in such spaces, the weights of the greedy spanner and 2-opt tours are $O(\log n) \cdot wt(MST(S))$. We obtain these results by generalizing the gap property to the so-called weak gap property. We then show that any edge set satisfying the weak gap property has weight $O(\log n) \cdot wt(MST(S))$. Since both the greedy spanner and 2-opt tours satisfy the weak gap property, we obtain the same upper bounds on their weights.

Thus, the contributions of this paper are twofold: First, by introducing the weak gap property, we obtain alternative proofs of known results for the Euclidean metric. Second, our analysis shows that these results in fact hold for any metric space whose doubling dimension is a constant.

2   The Doubling Dimension of Metric Spaces

Let S be a finite metric space. For any two points p and q in S, we denote their distance by $|pq|$. If p is a point in S and $R > 0$ is a real number, then the *ball* with center p and radius R is defined to be the set $\{q \in S : |pq| \leq R\}$.

We now define the notion of doubling dimension, which is due to Assouad [2]; see also Heinonen [11]:

Definition 1. *Let S be a finite metric space and let λ be the smallest integer such that the following is true: For each real number $R > 0$, every ball in S of radius R can be covered by at most λ balls of radius $R/2$.* The doubling dimension *of the metric space S is defined to be $\log \lambda$.*

It is not difficult to show that the doubling dimension of any finite set of points in the Euclidean metric space $\mathbb{R}^d$ is $\Theta(d)$.

2.1 Non-Euclidean Spaces of Doubling Dimension 1

In this section, we give an example of a family of metric spaces having doubling dimension 1. As we will see, this family contains metric spaces whose properties are very different from Euclidean space $\mathbb{R}^d$ for any constant d.

Let $S = \{p_1, p_2, \ldots, p_n\}$ and let $0 < \epsilon < 1$ be a real number. For each i and j with $1 \leq i \leq j \leq n$, we define

$$|p_i p_j| = |p_j p_i| = \begin{cases} 0 & \text{if } i = j, \\ 4^j \text{ or } 4^j + \epsilon & \text{if } i < j. \end{cases}$$

This family of metric spaces occurs in Har-Peled and Mendel [10] (they use powers of 2 instead of powers of 4). For any three pairwise distinct indices i, j, and k with $i < j$, we have

$$|p_i p_j| \leq 4^j + \epsilon \leq 4^{\max(i,k)} + 4^{\max(k,j)} \leq |p_i p_k| + |p_k p_j|,$$

implying that this distance function satisfies the triangle inequality. It follows that S is a metric space.

Lemma 1. *The doubling dimension of the metric space S is equal to one.*

Proof. Let $R > 0$ be a real number and let B be a ball with radius R. We have to show that B can be covered by at most two balls of radius $R/2$.

First assume that $R < 4\epsilon$. Then $R < 4$. Since the minimum distance between any two distinct points of S is at least 16, the ball B contains only its center. Therefore, B can be covered by one ball of radius $R/2$.

Now assume that $R \geq 4\epsilon$. Let j be the largest index such that $p_j \in B$. If $j = 1$, then B contains only one point, so that this ball can be covered by one ball of radius $R/2$. Assume that $j \geq 2$. We define B_1 to be the ball with center p_j and radius $R/2$, and define B_2 to be the ball with center p_{j-1} and radius $R/2$. We claim that $B \subseteq B_1 \cup B_2$.

We may assume that B contains more than one point, because otherwise, $B = B_1$. Let p_k be the center of B. Observe that $k \leq j$. First assume that $k < j$. Since $p_j \in B$, we have $|p_k p_j| \leq R$. On the other hand, we have $|p_k p_j| \geq 4^j$. Thus, $4^j \leq R$. If $k = j$, then let ℓ be any index less than k for which $p_\ell \in B$. In this case, we have $|p_k p_\ell| \leq R$ and $|p_k p_\ell| \geq 4^k = 4^j$. Thus, also in this case, we have $4^j \leq R$.

We are now ready to complete the proof of the claim that $B \subseteq B_1 \cup B_2$. Let p_i be any point in B. If $i \in \{j-1, j\}$, then obviously $p_i \in B_1 \cup B_2$. If $i \leq j-2$, then

$$|p_{j-1}p_i| \leq 4^{j-1} + \epsilon \leq R/4 + \epsilon \leq R/2$$

and, therefore, $p_i \in B_2$.   □

As Har-Peled and Mendel [10] show, this family of metric spaces can be used to show that the time complexity for solving the all-nearest-neighbors problem is $\Theta(n^2)$. Indeed, consider a metric space in the family such that, for each j with $2 \leq j \leq n$, there is exactly one index i_j with $i_j < j$ such that $|p_{i_j}p_j| = 4^j$, whereas for all indices i' with $i' < j$ and $i' \neq i_j$, we have $|p_{i'}p_j| = 4^j + \epsilon$. Then, the (unique) nearest neighbor of p_j is the point p_{i_j}. Thus, any algorithm that solves the all-nearest-neighbors problem must find all indices i_j. By an adversial argument, it is easy to show that this takes $\Omega(n^2)$ time in the worst case.

Recall that the all-nearest-neighbors problem in Euclidean space $\mathbb{R}^d$, for a constant dimension d, can be solved in $O(n \log n)$ time; see Vaidya [16]. Thus, even though the metric space defined above has doubling dimension one, its behavior with respect to the all-nearest-neighbors problem is very different from the Euclidean metric in $\mathbb{R}^d$.

Now consider the case when

$$|p_i p_j| = |p_j p_i| = \begin{cases} 0 & \text{if } i = j, \\ 4^j & \text{if } i = 1 \text{ and } j > 1, \\ 4^j + \epsilon & \text{otherwise.} \end{cases}$$

For each j with $2 \leq j \leq n$, p_1 is the nearest neighbor in S of the point p_j. Therefore, the all-nearest-neighbors graph of S is the star-graph consisting of all edges $\{p_1, p_j\}$, $2 \leq j \leq n$. In fact, the minimum spanning tree is also equal to this star-graph. Thus, in this metric space, both the all-nearest-neighbors graph and the minimum spanning tree have maximum degree $n-1$. It is well known that in Euclidean space $\mathbb{R}^d$, the maximum degree of both these graphs is bounded by a constant that depends only on d.

Since the minimum spanning tree contains the all-nearest-neighbors graph, the time complexity for computing the minimum spanning tree in a metric space of constant doubling dimension is $\Theta(n^2)$. On the other hand, in Euclidean space $\mathbb{R}^d$, the minimum spanning tree can be computed in $O(n \log n)$ time if $d = 2$ (see Preparata and Shamos [14, Section 6.1]) and $o(n^2)$ time if $d > 2$ (see Yao [17]).

2.2 The Packing Lemma

We now show that a packing argument can be applied in any metric space of constant doubling dimension. The following lemma states that a ball of radius R cannot contain many points whose pairwise distances are at least proportional to R.

Lemma 2 (Packing Lemma). *Let S be a finite metric space with doubling dimension d, let $R > 0$ and $\alpha > 0$ be real numbers, let B be a ball in S of radius R, and let X be a subset of S, such that*

1. $X \subseteq B$ and
2. the distance between any two distinct points of X is at least αR.

Then, the number of points in the set X is at most

$$2^{d \cdot \max(0, 2 + \lfloor \log(1/\alpha) \rfloor)}.$$

Proof. If $\alpha > 2$, then the lemma holds because X contains at most one point. Assume that $0 < \alpha \leq 2$. Let $m = 2 + \lfloor \log(1/\alpha) \rfloor$. Then $m \geq 1$ and $m > 1 + \log(1/\alpha)$, implying that $R/2^{m-1} < \alpha R$. By repeatedly applying the definition of doubling dimension, the ball B can be covered by 2^{md} balls B_i ($1 \leq i \leq 2^{md}$) of radius $R/2^m$. Any two points in the same ball B_i have distance at most $2R/2^m < \alpha R$. Therefore, each ball B_i can contain only one point of X. Thus, X contains at most 2^{md} points. $\qquad\square$

3   The Weak Gap Property

We mentioned the gap property in Section 1. Here, we give a formal definition of this property, as well as a weaker version of it. The weak gap property will be the main focus of the rest of this paper.

Definition 2. *Let S be a finite metric space and let E be a set of directed edges whose endpoints are in S.*

1. *For a real constant $w > 0$, we say that E satisfies the w-gap property, if for any two distinct edges (p, q) and (r, s) in E,*

$$|pr| \geq w \cdot \min(|pq|, |rs|).$$

2. *For a real constant $w > 0$, we say that E satisfies the weak w-gap property, if for any two distinct edges (p, q) and (r, s) in E,*

$$|pr| \geq w \cdot \min(|pq|, |rs|)$$

or

$$|qs| \geq w \cdot \min(|pq|, |rs|).$$

Chandra *et al.* [5] proved that, in any metric space S and for any constant $w > 0$, any set of directed edges satisfying the w-gap property has weight $O(\log n) \cdot wt(TSP(S))$, where n is the number of points in S. The example in Narasimhan and Smid [13, Section 6.2] shows that this upper bound is tight, even in one-dimensional Euclidean space.

In general, there is no non-trivial upper bound on the weight of an edge set satisfying the weak gap property. Indeed, consider the metric space S on n points in which $|pq| = 1$ for all $p \neq q$, and take for E the edge set of the complete graph, by giving each edge an arbitrary direction. Then E satisfies the weak 1-gap property, $wt(E) = \Theta(n) \cdot wt(TSP(S))$, and E contains $\Theta(n^2)$ edges. The main result of this paper is a proof of the claims that, if the doubling dimension of the metric space is a constant, (i) the upper bound of $O(\log n) \cdot wt(TSP(S))$ does hold, and (ii) E contains $O(n)$ edges.

Theorem 1 (Gap Theorem). *Let S be a metric space with n points, let the doubling dimension of S be a constant, let $w > 0$ be a real constant, and let E be a set of directed edges whose endpoints are in S. If E satisfies the weak w-gap property, then*

1. *the total weight $wt(E)$ of all edges in E satisfies*

$$wt(E) = O(\log n) \cdot wt(TSP(S)),$$

2. *E contains $O(n)$ edges.*

Recall that $wt(TSP(S))$ and $wt(MST(S))$ differ by a factor of at most 2. Therefore, the Gap Theorem is still valid if we replace $TSP(S)$ by $MST(S)$.

Before we turn to the proof of the Gap Theorem, we consider two algorithms whose outputs satisfy the weak gap property.

3.1   The 2-Opt Heuristic for the Traveling Salesperson Problem

Let S be an arbitrary finite metric space. The *2-opt heuristic* is a well-known approach for heuristically solving the traveling salesperson problem; it was introduced by Lin [12]. The algorithm starts with an arbitrary initial tour along the points of S. Then it repeatedly tries to improve the current tour by making small local changes.

Let T be the current (directed) tour, and assume that T contains two distinct edges (p, q) and (r, s) such that

$$|pr| + |qs| < |pq| + |rs|.$$

Then the algorithm replaces (p, q) and (r, s) by the edges (p, r) and (q, s), and reverses the direction of the edges on the path from q to r. These replacements result in a shorter (directed) tour. The algorithm continues making these replacements until the current tour T has the property that

$$|pq| + |rs| \leq |pr| + |qs| \tag{1}$$

for any two distinct edges (p, q) and (r, s) of T. A tour having this property is called a *2-optimal tour*.

If S is an arbitrary metric space, then the weight of any 2-optimal tour is $O(\sqrt{n}) \cdot wt(TSP(S))$; see Chandra *et al.* [6]. These authors also showed that this upper bound is tight. On the other hand, Chandra *et al.* [6] used the gap property to show that, in the Euclidean metric $\mathbb{R}^d$, any 2-optimal tour has weight $O(\log n) \cdot wt(TSP(S))$.

We will use the Gap Theorem to prove that, if the doubling dimension of the metric space is constant, any 2-optimal tour has weight $O(\log n) \cdot wt(TSP(S))$.

Lemma 3. *Let T be a (directed) 2-optimal tour of the points in S. Then the edge set of T satisfies the weak 1-gap property.*

Proof. Let (p, q) and (r, s) be two distinct edges of T. We may assume without loss of generality that $|pq| \leq |rs|$. Thus, we have to show that $|pr| \geq |pq|$ or $|qs| \geq |pq|$. If $|pr| \geq |pq|$, then we are done. Assume that $|pr| < |pq|$. By (1), we have $|pq|+|rs| \leq |pr|+|qs|$. Combining this with our assumption that $|pr| < |pq|$, we obtain

$$|pq| + |rs| \leq |pr| + |qs| \leq |pq| + |qs|,$$

which implies that $|qs| \geq |rs| \geq |pq|$.   □

Lemma 3 and the Gap Theorem imply the following result.

Theorem 2. *Let S be a metric space with n points and constant doubling dimension. The 2-opt heuristic computes a tour along the points of S, whose weight is $O(\log n) \cdot wt(TSP(S))$.*

3.2   The Greedy Spanner Algorithm

Given a metric space S consisting of n points and a real constant $t > 1$, an undirected graph $G = (S, E)$ is called a *t-spanner* for S, if the following is true: For any two points p and q in S, there exists a path in G between p and q whose weight is at most $t|pq|$. Any such path is called a *t-spanner path* between p and q.

Althöfer *et al.* [1] introduced the following simple greedy algorithm for computing such a spanner (according to them, this algorithm was discovered independently by Bern in 1989):

> **Algorithm.** GREEDYSPANNER(S, t):
> sort the $\binom{n}{2}$ pairs of distinct points in non-decreasing order of their distances and store them in a list L;
> $E = \emptyset$;
> $G = (S, E)$;
> **for each** $\{p, q\} \in L$ ($*$ in sorted order $*$)
> **do** δ = weight of a shortest path in G between p and q;
>   **if** $\delta > t|pq|$
>   **then** $E = E \cup \{\{p, q\}\}$;
> $G = (S, E)$
>   **endif**
> **endfor**;
> output the graph G

It is obvious that this algorithm computes a t-spanner for S. If $t < 2$ and the metric space S has the property that $|pq| = 1$ for all distinct points p and q, then the greedy spanner contains $\binom{n}{2}$ edges and its total edge weight is $\Theta(n) \cdot wt(MST(S))$. On the other hand, Soares [15] proved that, in the Euclidean metric $\mathbb{R}^d$, the greedy spanner has bounded degree and, thus, contains only $O(n)$ edges. Again in $\mathbb{R}^d$, Chandra *et al.* [5] used the gap property to show that the weight of the greedy spanner is $O(\log n) \cdot wt(MST(S))$. In fact, the latter bound was improved to $O(wt(MST(S)))$ by Das *et al.* [7, 8]. (Chapter 14 in [13] contains a complete proof of this claim.)

It is not difficult to show that the greedy spanner contains the all-nearest-neighbors graph. Therefore, by the results in Section 2.1, the maximum degree of the greedy spanner can be as large as $n-1$, even when the doubling dimension is equal to one.

We will use the Gap Theorem to prove that, if the doubling dimension of the metric space is constant, the greedy spanner has weight $O(\log n) \cdot wt(MST(S))$ and contains $O(n)$ edges.

Consider the t-spanner G that is computed by the greedy algorithm. Let G^* be the directed graph obtained by giving each edge of G an arbitrary direction. Thus, each edge $\{p, q\}$ of G appears in G^* either as (p, q) or as (q, p).

Lemma 4. *Let w be a real number with $0 < w < 1 - 1/t$ and let $w' = \min(w, 1 - w - 1/t)$. Then the set of directed edges of G^* satisfies the weak w'-gap property.*

Proof. Let (p, q) and (r, s) be two distinct edges of G^*. We may assume without loss of generality that algorithm GREEDYSPANNER(S, t) considers the pair $\{p, q\}$ before $\{r, s\}$. Thus, we have $|pq| \leq |rs|$. The lemma will follow from the claim that $|pr| \geq w|pq|$ or $|qs| \geq (1 - w - 1/t)|rs|$.

Assume that $|pr| < w|pq|$ and $|qs| < (1 - w - 1/t)|rs|$. Then both $|pr|$ and $|qs|$ are less than $|rs|$. Consider the iteration in which the algorithm adds the edge $\{r, s\}$ to the graph G. Assuming that $p \neq r$ and $q \neq s$, the algorithm has already considered the pairs $\{p, r\}$, $\{p, q\}$, and $\{q, s\}$. Thus, at the start of this iteration, the graph G contains (i) a t-spanner path between p and r, (ii) the edge $\{p, q\}$, and (iii) a t-spanner path between q and s. Obviously, (i) also holds if $p = r$ and (iii) also holds if $q = s$. Since

$$
\begin{aligned}
t|pr| + |pq| + t|qs| &\leq tw|pq| + |pq| + t(1 - w - 1/t)|rs| \\
&\leq (tw + 1 + t(1 - w - 1/t))|rs| \\
&= t|rs|,
\end{aligned}
$$

the graph G contains a t-spanner path between r and s and, therefore, the algorithm does not add the edge $\{r, s\}$ to G. This is a contradiction.   □

If we choose $w = \frac{1}{2}(1 - 1/t)$ in Lemma 4, then the Gap Theorem implies the following result.

Theorem 3. *Let S be a metric space with n points and constant doubling dimension, and let $t > 1$ be a real constant. Algorithm GREEDYSPANNER(S, t) computes a t-spanner for S having $O(n)$ edges and total edge weight $O(\log n) \cdot wt(MST(S))$.*

4   Proof of the Gap Theorem

In this section, we present a proof of the Gap Theorem. This proof will consist of the following four steps:

1. We start by showing that, in any metric space, if a directed edge set E with vertex set S satisfies the gap property and all edges in E have approximately the same weight, then the total weight of E is $O(wt(TSP(S)))$.

2. We then show that, again in any metric space, the total weight of any set E of edges with vertex set S, such that all edges in E are "short", is $O(wt(TSP(S)))$.

3. Next, we show that, in any metric space of constant doubling dimension, and for any directed edge set E with vertex set S, such that E satisfies the weak gap property and all edges in E are "long", the following holds: First, the edge set E can be partitioned into $O(\log n)$ subsets, such that within each subset, edges have approximately the same weight. Then, we show how to further partition each subset into $O(1)$ subsets, each one satisfying the gap property. The analysis in the first step then shows that the total weight of E is $O(\log n) \cdot wt(TSP(S))$.

4. In the final step, we use the well-separated pair decomposition to show that, in any metric space of constant doubling dimension, any directed edge set satisfying the weak gap property contains $O(n)$ edges.

4.1   Edges of Similar Weights Satisfying the Gap Property

We start by considering directed edges having approximately the same weights and that satisfy the gap property (as opposed to the weak gap property). The proof of the following lemma is a simple modification of a proof technique introduced in Chandra et al. [5].

Lemma 5. *Let S be a metric space, let $w > 0$ be a real constant, and let E be a set of directed edges whose endpoints are in S and that satisfy the w-gap property. Assume that $1/2 \leq |pq|/|rs| \leq 2$ for any two edges (p,q) and (r,s) in E. Then, the total weight $wt(E)$ of all edges in E satisfies*

$$wt(E) \leq \frac{2}{w} \cdot wt(TSP(S)).$$

Proof. For any (directed) edge (p,q) of E, we call p the *source* of this edge. First observe that, by the definition of the w-gap property, each point of S can be the source of only one edge in E. Consider the traveling salesperson tour of S. By walking along this tour (starting at some arbitrary point), we visit the sources of the edges in E in some order. We number the edges of E as $e_0, e_1, \ldots, e_{m-1}$, as given by this order. For each i with $0 \leq i < m$, we write the edge e_i as $e_i = (p_i, q_i)$ and define T_i to be the portion of the tour that starts at p_i and ends at p_{i+1} (where indices are read modulo m). By the triangle inequality, we have

$$|p_i p_{i+1}| \leq wt(T_i),$$

whereas, by the w-gap property and our assumption that edges in E differ in weight by a factor of at most two,

$$|p_i p_{i+1}| \geq w \cdot \min(|p_i q_i|, |p_{i+1} q_{i+1}|) \geq (w/2)|p_i q_i|.$$

Thus, we have

$$|p_i q_i| \leq (2/w) \cdot wt(T_i),$$

which implies that

$$wt(E) = \sum_{i=0}^{m-1} |p_i q_i| \leq (2/w) \sum_{i=0}^{m-1} wt(T_i) = (2/w) \cdot wt(TSP(S)). \qquad \square$$

4.2   The Weight of Short Edges

Lemma 6. *Let S be a metric space with n points, let E be a set of directed edges whose endpoints are in S, let D be the weight of a longest edge in E, and let E' be the subset of E consisting of all edges having weight at most D/n^2. Then $wt(E') \leq wt(TSP(S))$.*

Proof. The lemma follows from the observations that E' contains at most $\binom{n}{2} \leq n^2$ edges and $wt(TSP(S)) \geq D$. $\qquad \square$

4.3   Long Edges Satisfying the Weak Gap Property

Let S be a metric space with n points and constant doubling dimension d, let $w > 0$ be a real constant, and let E be a set of directed edges whose endpoints are in S and that satisfy the weak w-gap property.

Let D be the weight of a longest edge in E, let E' be the subset of E consisting of all edges having weight at most D/n^2, and let $E'' = E \setminus E'$. Thus, E'' consists of all edges in E having weight more than D/n^2. For each j with $0 \leq j \leq \lfloor 2 \log n \rfloor$, we define

$$E_j = \{e \in E'' : D/2^{j+1} < wt(e) \leq D/2^j\}.$$

Thus, the edge sets E_j form a partition of E''.

We fix an index j with $0 \leq j \leq \lfloor 2 \log n \rfloor$ and analyze the total weight of all edges in E_j. Even though edges in E_j differ in weight by a factor of at most two, Lemma 5 cannot be applied to E_j, because we only know that E_j satisfies the weak gap property. Our approach will be to further partition the set E_j into $O(1)$ subsets, each of which satisfies the w-gap property. Then, Lemma 5 can be applied to each of these subsets.

Let $L = D/2^{j+1}$, so that

$$E_j = \{e \in E'' : L < wt(e) \leq 2L\}.$$

We define an undirected graph H with vertex set E_j, in which any two distinct vertices (p, q) and (r, s) are connected by an edge if and only if

$$|pr| < w \cdot \min(|pq|, |rs|).$$

Lemma 7. *The maximum degree of the graph H is $O(1)$.*

Proof. Consider any vertex (p, q) of H. We have to prove an upper bound on the number of elements (r, s) in E_j for which $(p, q) \neq (r, s)$ and $|pr| < w \cdot \min(|pq|, |rs|)$.

Let B be the ball with center p and radius $2wL$. For any edge (r, s) in E_j with $|pr| < w \cdot \min(|pq|, |rs|)$, we have $|pr| \leq 2wL$ and, thus, $r \in B$.

By applying the definition of doubling dimension twice, the ball B can be covered by 2^{2d} balls B_i $(1 \leq i \leq 2^{2d})$ of radius $wL/2$. For each i with $1 \leq i \leq 2^{2d}$, we define

$$E_j^i = \{(r, s) \in E_j : (r, s) \neq (p, q), |pr| < w \cdot \min(|pq|, |rs|), r \in B_i\}.$$

Then

$$\{(r, s) \in E_j : (p, q) \neq (r, s), |pr| < w \cdot \min(|pq|, |rs|)\} = \bigcup_{i=1}^{2^{2d}} E_j^i. \qquad (2)$$

Since the degree of (p, q) in H is equal to the size of the set on the left-hand side in (2), we need an upper bound on $\sum_{i=1}^{2^{2d}} |E_j^i|$.

Consider a fixed value of i with $1 \leq i \leq 2^{2d}$. If (r, s) is an edge in E_j^i, then

$$|ps| \leq |pr| + |rs| \leq w \cdot \min(|pq|, |rs|) + |rs| \leq 2wL + 2L = 2(1 + w)L.$$

If (r, s) and (r', s') are two distinct edges in E_j^i, then r and r' are both in B_i and thus

$$|rr'| \leq wL < w \cdot \min(|rs|, |r's'|).$$

Therefore, by the weak w-gap property, we have

$$|ss'| \geq w \cdot \min(|rs|, |r's'|) \geq wL.$$

In particular, $s \neq s'$. Thus, if we define X to be the set of the sinks s of all edges (r, s) in E_j^i, then (i) X contains the same number of elements as E_j^i, (ii) all points of X are contained in the ball with center p and radius $2(1 + w)L$, and (iii) the distance between any two distinct points of X is at least wL. By applying Lemma 2 with $\alpha = w/(2(1 + w))$, it follows that

$$|E_j^i| \leq 2^{d \cdot \max(0, 2 + \log(1/\alpha))} = 2^{d \cdot \max(0, 3 + \log(1 + 1/w))}.$$

Thus, we have shown that the degree of the vertex (p, q) in H is at most

$$\sum_{i=1}^{2^{2d}} |E_j^i| \leq 2^{d(2 + \max(0, 3 + \log(1 + 1/w)))}.$$

Since d and w are constants, the proof is complete. $\qquad \square$

Let m be the maximum degree of any vertex in H. We color the vertices of H (i.e., the elements of E_j) using $m + 1$ colors, such that any two adjacent vertices have distinct colors. For each k with $0 \leq k \leq m$, let

$$E_{jk} = \{e \in E_j : e \text{ has color } k\}.$$

The subsets E_{jk}, $0 \leq k \leq m$, partition the set E_j.

Lemma 8. *For each k with $0 \leq k \leq m$, the set E_{jk} satisfies the w-gap property.*

Proof. Let (p, q) and (r, s) be two distinct edges in E_{jk}. Since these two edges have the same color, they are not connected by an edge in H. Then, the definition of H implies that $|pr| \geq w \cdot \min(|pq|, |rs|)$. $\qquad \square$

We are now able to complete the proof of the first claim in the Gap Theorem: We started by partitioning the edge set E into E' and E''. Lemma 6 gives an upper bound on $wt(E')$. Then, we partitioned E'' into $O(\log n)$ subsets E_j, and further partitioned each E_j into $m + 1 = O(1)$ subsets E_{jk}. By Lemmas 5 and 8, the total weight of all edges in each subset E_{jk} is $O(wt(TSP(S)))$.

4.4   The Number of Edges

In this section, we prove the second claim in the Gap Theorem. Our proof will use the well-separated pair decomposition of Callahan and Kosaraju [4].

Let S be a metric space with n points. For any two non-empty subsets A and B of S, we define their distance $|AB|$ as

$$|AB| = \min\{|pq| : p \in A, q \in B\}$$

and the diameter $diam(A)$ of A as

$$diam(A) = \max\{|pq| : p \in A, q \in A\}.$$

For a real number $c > 1$, called the *separation ratio*, we say that A and B are *well-separated*, if

$$|AB| \geq c \cdot \max(diam(A), diam(B)).$$

Thus, if c is large, then (i) all distances between points in A and points in B are approximately equal and (ii) distances within A (or B) are much smaller than distances between points in A and points in B.

Definition 3. *A* well-separated pair decomposition *for S is a sequence*

$$\{A_1, B_1\}, \{A_2, B_2\}, \ldots, \{A_m, B_m\}$$

of pairs of non-empty subsets of S, for some integer m, such that

1. *for each i with $1 \leq i \leq m$, A_i and B_i are well-separated, and*
2. *for any two distinct points p and q of S, there is exactly one index i such that*
   (a) $p \in A_i$ and $q \in B_i$, or
   (b) $p \in B_i$ and $q \in A_i$.

The integer m is called the size *of the well-separated pair decomposition.*

Callahan and Kosaraju [4] showed that, in the Euclidean metric $\mathbb{R}^d$, a well-separated pair decomposition of size $m = O(n)$ exists, and can in fact be computed in $O(n \log n)$ time. Har-Peled and Mendel [10] generalized this result to metric spaces of constant doubling dimension:

Theorem 4. *Let S be a metric space with n points and constant doubling dimension, and let $c > 1$ be a real constant. There exists a randomized algorithm that constructs, in $O(n \log n)$ expected time, a well-separated pair decomposition for S, consisting of $O(n)$ pairs.*

The following lemma completes the proof of the Gap Theorem:

Lemma 9. *Let S be a metric space with n points and constant doubling dimension, let $w > 0$ be a real constant, and let E be a set of directed edges whose endpoints are in S and that satisfy the weak w-gap property. Then, E contains $O(n)$ edges.*

Proof. Choose the separation ratio c to be larger than $1/w$. Consider a well-separated pair decomposition $\{A_i, B_i\}$, $1 \le i \le m$, where $m = O(n)$.

We will prove below that for each i with $1 \le i \le m$, the set E contains (i) at most one edge (p, q) with $p \in A_i$ and $q \in B_i$, and (ii) at most one edge (p, q) with $p \in B_i$ and $q \in A_i$. Therefore, E contains at most $2m = O(n)$ edges.

Assume that (i) is not true. Then, E contains two distinct edges (p, q) and (r, s) with $p, r \in A_i$ and $q, s \in B_i$. We may assume without loss of generality that $|pq| \le |rs|$. Since A_i and B_i are well-separated, we have

$$|pr| \le diam(A_i) \le \frac{1}{c} \cdot |A_i B_i| \le \frac{1}{c} \cdot |pq| < w|pq|$$

and, by a symmetric argument,

$$|qs| < w|pq|.$$

This contradicts the fact that (p, q) and (r, s) satisfy the weak w-gap property.   □

5   Final Remarks

5.1   Open Problems

We have introduced the weak gap property as an alternative to the gap property of Chandra *et al.* [5]. We have shown that, in any metric space whose doubling dimension is constant, any set of edges satisfying the weak gap property has $O(n)$ elements and total weight $O(\log n) \cdot wt(TSP(S))$ or, equivalently, $O(\log n) \cdot wt(MST(S))$. The example in Narasimhan and Smid [13, Section 6.2] shows that this upper bound is tight, even in the one-dimensional Euclidean metric.

We have shown that both 2-optimal tours and greedy spanners satisfy the weak gap property. Thus, in case the doubling dimension is constant, their total weight is within a factor of $O(\log n)$ of the minimum possible weight. These results lead to the following open problems:

Problem 1. Chandra *et al.* [6] showed that, in the Euclidean plane $\mathbb{R}^2$, there exists a 2-optimal tour whose length is $\Omega(\log n / \log \log n) \cdot wt(TSP(S))$.

- Is it true that, in the Euclidean metric in $\mathbb{R}^d$, the weight of any 2-optimal tour is $O(\log n/\log\log n) \cdot wt(TSP(S))$?
- Does there exist a metric space S of constant doubling dimension, such that S contains a 2-optimal tour whose length is $\Omega(\log n) \cdot wt(TSP(S))$?

Problem 2. Das *et al.* [7, 8] showed that, in the Euclidean metric in $\mathbb{R}^d$, the weight of the greedy spanner is $O(wt(MST(S)))$; see Chapter 14 in [13] for a complete proof.

- Is it true that, in any metric space whose doubling dimension is constant, the weight of the greedy spanner is $O(wt(MST(S)))$?

It follows from the proofs of Lemmas 3 and 4 that the (directed) edges of any 2-optimal tour and the greedy spanner satisfy the following slightly stronger property: For any two distinct edges (p, q) and (r, s),

$$|pr| \geq w \cdot \min(|pq|, |rs|)$$

or

$$|qs| \geq w \cdot \max(|pq|, |rs|).$$

It is not clear, however, whether this is of any help to solve Problems 1 and 2.

Problem 3. As we have seen in this paper, edge sets in metric spaces of bounded doubling dimension can be analyzed using a packing argument.

- Find other classes of metric spaces, in which non-trivial upper bounds on the weight of 2-optimal tours, greedy spanners, or other interesting graphs, can be obtained.

5.2   Further Reading

A detailed analysis of tours produced by the 2-opt heuristic, both for general metric spaces and Euclidean spaces, can be found in Chandra *et al.* [6].

Althöfer *et al.* [1] contains an analysis of the greedy spanner for general metric spaces. Recently, Bose *et al.* [3] have shown that, in case the doubling dimension is constant, the greedy spanner can be computed in $O(n^2 \log n)$ time. Observe that, since the greedy spanner contains the all-nearest-neighbors graph, the results in Section 2.1 imply that the greedy spanner cannot be computed in subquadratic time. We remark, however, that, in the Euclidean metric in $\mathbb{R}^d$, an approximation of the greedy spanner can be computed in $O(n \log n)$ time; see Gudmundsson *et al.* [9]. A detailed description of their algorithm, as well as many other algorithms for constructing spanners, can be found in Narasimhan and Smid [13].

Acknowledgements. The author thanks Hubert Chan, Anupam Gupta, and Giri Narasimhan for helpful discussions during the workshop *Geometric Networks and Metric Space Embeddings*, which was held in Dagstuhl (Germany) in November 2006.

References

1. Althöfer, I., Das, G., Dobkin, D.P., Joseph, D., Soares, J.: On sparse spanners of weighted graphs. Discrete & Computational Geometry 9, 81–100 (1993)
2. Assouad, P.: Plongements lipschitziens dans $\mathbb{R}^N$. Bulletin de la Société Mathématique de France 111, 429–448 (1983)
3. Bose, P., Carmi, P., Farshi, M., Maheshwari, A., Smid, M.: Computing the greedy spanner in near-quadratic time. To appear in Algorithmica
4. Callahan, P.B., Kosaraju, S.R.: A decomposition of multidimensional point sets with applications to k-nearest-neighbors and n-body potential fields. Journal of the ACM 42, 67–90 (1995)
5. Chandra, B., Das, G., Narasimhan, G., Soares, J.: New sparseness results on graph spanners. International Journal of Computational Geometry & Applications 5, 125–144 (1995)
6. Chandra, B., Karloff, H., Tovey, C.: New results on the old k-opt algorithm for the traveling salesman problem. SIAM Journal on Computing 28, 1998–2029 (1999)
7. Das, G., Heffernan, P., Narasimhan, G.: Optimally sparse spanners in 3-dimensional Euclidean space. In: Proceedings of the 9th ACM Symposium on Computational Geometry, pp. 53–62 (1993)
8. Das, G., Narasimhan, G., Salowe, J.: A new way to weigh malnourished Euclidean graphs. In: Proceedings of the 6th ACM-SIAM Symposium on Discrete Algorithms, pp. 215–222 (1995)
9. Gudmundsson, J., Levcopoulos, C., Narasimhan, G.: Fast greedy algorithms for constructing sparse geometric spanners. SIAM Journal on Computing 31, 1479–1500 (2002)
10. Har-Peled, S., Mendel, M.: Fast construction of nets in low-dimensional metrics and their applications. SIAM Journal on Computing 35, 1148–1184 (2006)
11. Heinonen, J.: Lectures on Analysis on Metric Spaces. Springer, Berlin (2001)
12. Lin, S.: Computer solutions of the traveling salesman problem. Bell Systems Technical Journal 44, 2245–2269 (1965)
13. Narasimhan, G., Smid, M.: Geometric Spanner Networks. Cambridge University Press, Cambridge (2007)
14. Preparata, F.P., Shamos, M.I.: Computational Geometry: An Introduction. Springer, Berlin (1988)
15. Soares, J.: Approximating Euclidean distances by small degree graphs. Discrete & Computational Geometry 11, 213–233 (1994)
16. Vaidya, P.M.: An $O(n \log n)$ algorithm for the all-nearest-neighbors problem. Discrete & Computational Geometry 4, 101–115 (1989)
17. Yao, A.C.: On constructing minimum spanning trees in k-dimensional spaces and related problems. SIAM Journal on Computing 11, 721–736 (1982)

On Map Labeling with Leaders

Michael Kaufmann

Wilhelm-Schickard-Institut für Informatik
Universität Tübingen, 72116 Tübingen, Germany

Abstract. We present an overview over recent work on map labeling with emphasis on labeling with leaders. The labels are assumed to be placed on the boundary of the map while they are connected to the corresponding graphical features by polygonal lines, the so-called leaders. On various models and methods we demonstrate an impressive example for an application where a whole range of algorithmic techniques have successfully been applied.[1]

1 Traditional Map Labeling

The visualization of the information on a map, i.e. the appropriate association of text labels with graphical features is one of the main challenges and a key task in the process of producing high-quality maps.

Considering the importance and complexity of ancient and not so ancient hand-drawn maps [21] one could even talk about the art of map labeling. The cartographers have developed certain rules and techniques to place labels effectively [13, 28]. One of them is that the labels should be pairwise disjoint and close to the graphical object they belong to.

There is a little story how the topic has been introduced to the field of algorithmics [19]. Around the year 1990 Rudi Krämer, a former student of Kurt Mehlhorn, now working for the city of Munich, asked his professor for help when facing the problem of labeling a map with a set of landmarks for ground water level of the city. Kurt Mehlhorn shared that problem with his students, including the author, and also with his collegues, including the computational geometry group in Berlin, headed by Emo Welzl and Helmut Alt. The result of this collaboration was a mathematical model for the problem and a series of algorithms published in several papers arised that laid the foundations for a new field in algorithmics. The driving forces behind this development were the works by M. Formann, F. Wagner [11], and later on by A. Wolff [26] as well as M. van Kreveld and T. Strijk [20] from Utrecht. Unfortunately, the majority of map labeling problems are shown to be NP-complete [2, 25]. Due to this fact, graph drawers and computational geometers have suggested labeling approximations [2, 11, 25] and heuristics [27] as well as methods from optimization [23, 10], which often try to maximize either the label size or the number of features with labels. A nice and short overview on map labeling with emphasize on graph drawing has been

---

[1] Large pieces of this review and pictures have been taken from [3, 4, 6, 7, 15].

S. Albers, H. Alt, and S. Näher (Eds.): Festschrift Mehlhorn, LNCS 5760, pp. 290–304, 2009.

presented by G. Neyer [18]. A more detailed bibliography on map labeling mainly from the computer science aspects can be found in [24]. In general, map labeling has found its place in the research landscape not only of cartography, but also geographic information systems and more general computational geometry. The ACM Computational Geometry Task Force [9] has identified label placement as an important area of research.

Recently, a new exciting variant has been considered that is regularly used in practice but that has not considered systematically by algorithmics people: boundary labeling. The motivating example for this project was a map of the school infrastructure of Greece where very large labels have been used and clearly the traditional labeling models were not sufficient. The large labels have no space close to their corresponding point sites and therefore they are placed somewhere on the outside of the map. The point sites are connected to their corresponding labels by polygonal lines, which are called leaders later on. In a first paper, Bekos, Kaufmann, Symvonis and Wolff introduced the formal models and basic algorithms [3], and later on extended in various directions.

In the following, we will review the results related to boundary labeling and will only sketch the used techniques such that an algorithmicist could fill in the details. The aim of the description is to demonstrate the variety of algorithmic techniques and the power of formal models that really help to achieve useful practical results.

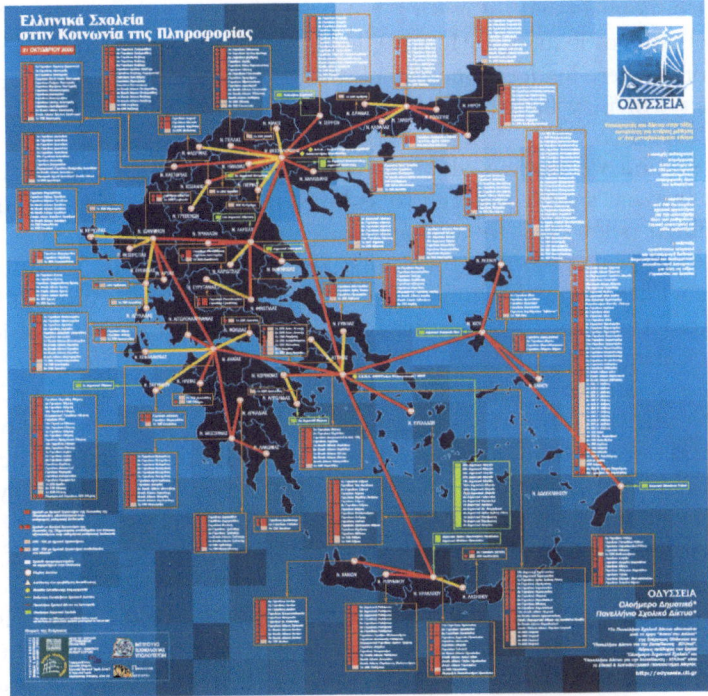

Fig. 1. The blue map: the motivation example

2   Boundary Labeling - Models and Methods

In this section, we introduce the basic models and algorithms for boundary labeling. Abstracting from the so-called blue map of Greece, and neglecting the details like an underlying graph, and many textual labels that are placed in the traditional way close to the corresponding point sites we concentrate on the new feature, namely the labels connected by the leaders. We formulate the following very simple model:

We assume that a set $P = p_1, ..., p_n$ of point sites inside an axis-parallel rectangle R. Each site p_i is associated with an axis-parallel rectangular label l_i. The labels have to be placed and connected to their corresponding sites by polygonal leaders c_i, such that (a) no two labels intersect, (b) no two leaders intersect, and (c) the labels lie outside R but touch R. Concerning the location of the labels and the type of leaders, different variants are considered.

For a given axis-parallel rectangle R of width W and height H, and a set P of n point sites $p_i = (x_i, y_i)$, for $1 \le i \le n$ in R, each associated with an axis-parallel rectangular open label l_i of width w_i and height h_i, we aim to find a legal or an optimal leader-label placement. Our formal criteria for a legal leader-label placement are the following:

1. Labels have to be disjoint.
2. Labels have to lie outside R but touch the boundary of R.
3. Leader c_i connects site p_i with label l_i for $1 \le i \le n$.
4. Intersections of leaders with other leaders, sites or labels are not allowed.
5. The ports where leaders touch labels may be prescribed (the center of a label edge, say) or may be arbitrary (sliding ports).

In this section we present algorithms that compute legal leader-label placements (for brevity, simply referred to as boundary labelings) for various types of leaders defined below, but we also optimize the boundary placements according to the following two objective functions: Short leaders (minimum total length) and simple leader layout (minimum number of bends). These criteria have been adopted from the area of graph drawing as established abstract objectives to approach clearness and readability in drawings.

A rectilinear leader consists of a sequence of axis-parallel segments that connects a site with its label. These segments are either parallel (p) or orthogonal (o) to the side of the bounding rectangle R to which the label is attached. This notation yields a classification scheme for rectilinear leaders: let a type be an alternating string over the alphabet $\{p, o\}$. Then a leader of type $t = t_1, \ldots, t_k$ consists of an $x-$ and y-monotone connected sequence $(e_1, \ldots, e_k)$ of segments from site to label, where each segment e_i has the direction that the letter t_i prescribes. At first we focus on leaders of the types opo and po, see Figure 2. In general, type-opo leaders, or shortly opo-leaders, are routed with one or three segments such that the first segment connects the site to the boundary of the rectangle where the corresponding label is placed. We further insist that the parallel p-segment is immediately outside the bounding rectangle R and is routed

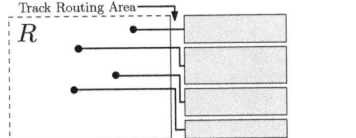 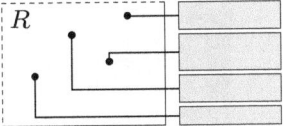

Fig. 2. Illustrating opo- and po-leaders

in the so-called track routing area. We consider the special case of type-o leaders to be of type opo and of type po as well. Generalizations of this concepts are straightline (type-s) leaders that consist of one single segment which is not restricted to be axis-parallel, and leaders that have o- and p-segments but also diagonal segments (type-d) (see subsection 3.3).

Additionally, we assume that the sites are in general position such that rectilinear leaders may not overlap with their first segments.

2.1 Opo-Leaders

We start with the assumption that the labels are placed on one, say the right side of the rectangle (Fig. 3) and that they have uniform and maximal size. Observe that the vertical order of the leaders is the same as the vertical order of the points. When we are just interested in a legal boundary labeling, we can sort the point sites and the labels as well by y-coordinates, assign sites to corresponding label positions and route the opo-leaders.

Theorem 1. *Given a rectangle R of sufficient size, a side s of R, a set P of n point sites in R in general position and a rectangular label for each site, there is an $O(n \log n)$-time algorithm that attaches labels to s and connects them to the corresponding sites with non-intersecting opo-leaders.*

In a second step, we assume sliding ports and aim to minimize the number of bends on the leaders. This means that we maximize the number of leaders without bends. Labels with uniform and maximal size do have a fixed position, so we do not have freedom for optimization. With labels l_i of different height h_i, we are allowed to vary the vertical spacing between the labels. With a dynamic programming approach that incrementally adds the leaders from bottom to top minimizing the number of used bends we can prove

Theorem 2. *Given a rectangle R of sufficient size, a side s of R, a set P of n sites in R in general position and a rectangular label for each site, there is an $O(n^2)$-time algorithm that attaches the labels to s and connects them to the corresponding sites with non-intersecting opo-leaders such that the total number of leader bends is minimized.*

In the next extension, we optimize the total leader length. This is obvious if the labels are only on one side, as the algorithms above already give optimal labelings for this case. Therefore, we assume labels on the left and the right sides, but at first we consider the case of uniform labels. Here we apply a similar

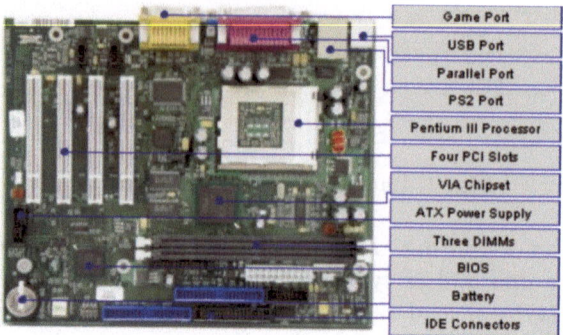

Fig. 3. A simple example for opo-leaders connecting labels to one side

dynamic programming algorithm as above. Sweeping from bottom to top, we compute at each site i values $T(i, k)$ that give the optimal leader length for the case that k of the i lowest leaders have been routed to the left boundary. This gives

Theorem 3. *Given a rectangle R with $n/2$ uniform labels of maximum height on each of its left and right sides, and a set P of n sites in R in general position, there is an $O(n^2)$-time algorithm that attaches each site to a label with non-intersecting opo-leaders such that the total leader length is minimized.*

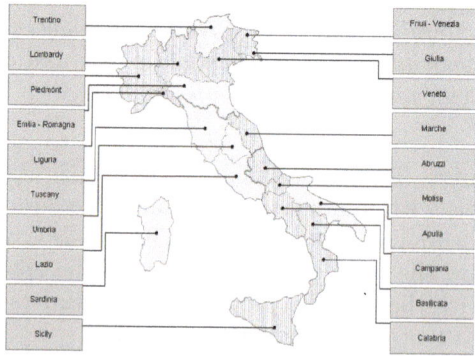

Fig. 4. Minimal leader lengths for labels on two opposite sides of the rectangle

This dynamic program can be refined such that varying heights are allowed:

Theorem 4. *Given a rectangle R of height H, a set P of n sites in R in general position where site p_i is associated with label l_i of height h_i, there is an $O(nH^2)$-time algorithm that places the labels to the sides of the rectangle and attaches the corresponding sites with non-intersecting type-opo leaders such that the total leader length is minimum.*

More general, we allow labels on all four sides of the rectangle boundary. To find a legal solution for opo-leaders, we have to assign the point sites to the side where their label is placed. Hence, we have to partition the point set according to the number of labels on each side into four sets, such that the leaders from different partitions do not intersect. This can be ensured by a partition of the rectangle such that each polygon is convex. By two rotational sweeps around appropriate corners of the rectangle [17], this partition can be constructed.

Theorem 5. *Given a rectangle R of sufficient size, a set P of n sites in R in general position, square uniform labels, one per site, and numbers $n_1, \ldots, n_4$ that express how many labels are to be attached to which side of R, there is an $O(n \log n)$-time algorithm that attaches the labels to R and connects them to the corresponding sites with non-intersecting opo-leaders.*

More interesting is the optimization version of the four-sided problem. We formulate this problem as a bipartite matching problem. On one side, we have the n point sites, on the other side the label positions. For each point and each possible label position, we create an edge having the weight corresponding to the unique length of the potential opo-leader. Solving this weighted matching problem [22] gives an length-minimum assignment of the labels to the points. We still have to argue that although this assignment might have crossings in some rare cases, these crossings can be avoided without increasing the total leader length.

Theorem 6. *Consider four-side opo-labeling of n sites with uniform labels and sliding ports. A crossing-free solution of minimum total length can be computed in $O(n^2 \log^3 n)$ time.*

2.2 Po-Leaders

When turning to po-leaders, first observe that the leaders have at most one bend. Secondly unlike in the opo-case, there are non-unique crossing-free routings of leaders. The main idea to handle po-leaders is to take the opo-solutions and transform them to po-solutions. Note that each nontrivial opo-leader can be transformed into a po-leader by flipping the first two segments. Unfortunately, this might create crossings. But crossings can be resolved by exchanging the corresponding labels between the two crossing leaders. Iterating $O(n^2)$ resolution steps leads to a crossing-free po-solution.

Theorem 7. *Given a rectangle R, a side s of R, a set P of n sites in R in general position and a rectangular uniform label for each site, there is an $O(n^2)$-time algorithm that attaches the labels to s and connects them to the corresponding sites with non-intersecting po-leaders.*

Applying the same technique as above - transforming opo-leaders into po-leaders and resolve crossings - we can achieve length-minimal po-solutions, even for the case that labels are placed on two opposite sides of the rectangles. In the latter case, we again use dynamic programming.

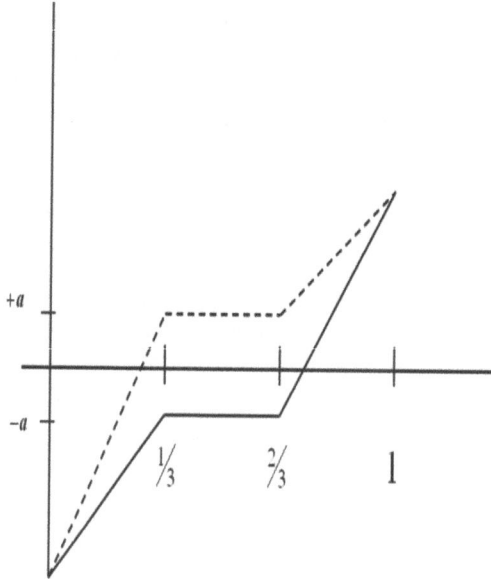

Fig. 4.

Many of the extracts in the axioms are functions, and they are given using lambda notation, so for example, $\lambda(x.\ x)$ is the notation for the identity function. This notation is especially simple for the first axiom where the extract is simply a function that takes two points p_1 and p_2 as inputs and then takes some proof *neq* that the points are not equal, and returns a ray connecting the points and two axiomatic equalities about how to obtain the origin and destination of the ray.

5.2 Euclid-Like Axioms

Axiom 1. To draw a straight line segment from any point to any other point.

$\forall p_1, p_2 : Point.\ p_1 \neq p_2\ \Rightarrow \exists r : Ray.\ origin(r) = p_1\ \&\ dest(r) = p_2.$
extract $\lambda(p_1, p_2.\lambda neq.\ pair(ray(p_1; p_2); pair(axiom; axiom))).$

Axiom 2. To produce a finite line continuously in a straight line.

$\forall r : Ray.\ \exists l : Line.\ \forall p : Point.\ (p\ on\ r \Rightarrow p\ on\ l).$ **extract**
$\lambda(r.\ pair(line(r)); \lambda(p.\ \lambda(hyp.\ axiom))).$

The line l is oriented by the ray r whose origin and destination are two points on l. We define the relations $p\ right\ l$ and $p\ left\ l$ with reference to the direction of r.

We can prove Euclid's axiom by using the segment given by the ray, namely

$\forall s : Seg.\ \exists l : Line.\ \forall p : Point.\ (p\ on\ s \Rightarrow p\ on\ l).$ **extract**
$\lambda(s.\ pair(line(r)); \lambda(p.\ \lambda(hyp.\ axiom))).$

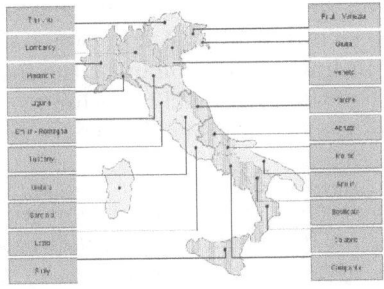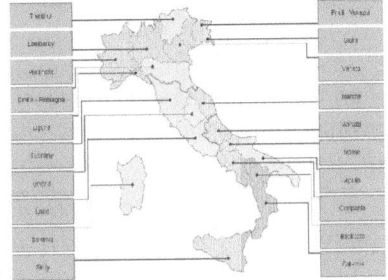

Fig. 7. Example for min-total-length po-leaders to labels on two sides. For comparison, we added the corresponding example for opo-leaders.

Proof idea: Observe that the lowest type-s leader connects the point and labels on the convex hull including points and labels. Using a semi-dynamic convex-hull data structure which only supports deletion of points [12] leads to the theorem above.

If we have labels on all four sides, we again have to partition the point set into four convex polygons related to one side each, such that each polygon section can be routed separately. This can be done by basic geometric means, namely a combination of appropriate linear and rotational sweeps.

Theorem 11. *A legal four-side type-s leader-label placement for fixed uniformly distributed labels with fixed ports can be computed in $O(n \log n)$ time.*

For optimization of the total leader length we can formulate the problem as an Euclidean min-cost bipartite matching problem between points and labels. The solution is crossing-free and can be obtained in time $O(n^{2+\epsilon})$ for arbitrary $\epsilon > 0$ [1]. This clearly holds not just for labels on one side but also for the case of four-sides-labels.

Theorem 12. *A legal one-side type-s leader-label placement of minimum total leader length for fixed labels with sliding ports can be computed in time $O(n^{2+\epsilon})$ time for any $\epsilon > 0$.*

Theorem 13. *A legal four-side type-s leader-label placement of minimum total leader length for fixed labels with sliding ports can be computed in time $O(n2 + \epsilon)$ time for any $\epsilon > 0$.*

3   Extending the Standard Models

3.1   Multi-stack Labeling

In this subsection, we review a variant of boundary labeling problems where we assume that there is not enough space for all labels. Therefore we seek to obtain labelings with labels arranged on more than one stacks placed at the same side of the rectangle R. We refer to problems of this type as multi-stack boundary labeling problems. We aim for maximizing the uniform label size for boundary labeling with two and three stacks of labels. The key component of our algorithms is a dynamic programming technique that combines the merging of lists and the bounding of the search space of the solution. The technique could be refined such that the necessary bends in the opo-routing are always close to the corresponding stack of labels.

Theorem 14. *Given a rectangle R of integer height H and a set P of n points in R in general positions, there exists an $O(n^4 \log H)$ time algorithm that produces a legal multi-stack labeling with two stacks of labels on the same side of R and with opo-leaders such that the uniform integer height of the labels is maximum.*

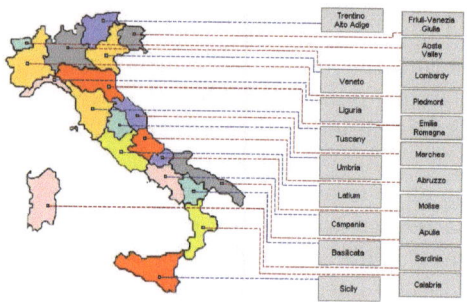

Fig. 8. An example for two-stack-opo boundary labeling

Theorem 15. *Given a rectangle R of integer height H and a set P of n points in R in general positions, there exists an $O(n^4 \log H)$ time algorithm that produces a legal multi-stack labeling with three stacks of labels on the same side of R and with opo-leaders such that the uniform integer height of the labels is maximum and the leaders connected to labels at the i-th stack are restricted to bend in the i-th track routing area.*

On the negative side, we could prove NP-completeness for the case of variable label heights, even if each label can only be connected to two candidate points.

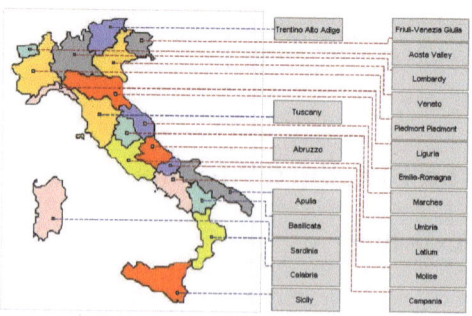

Fig. 9. An example for two-stack-opo with leaders bending at the i-th track routing area

Theorem 16. *Given a rectangle R of height H, a set P of n line segments (sites) in R that are parallel to the y-axis and a label of height h_i for each site $s_i \in P$, it is NP-hard to place all labels at two stacks on one side of R with non-intersecting opo-leaders.*

Theorem 17. *Given a rectangle R of height H, a set P of n sites in R, each associated with two candidate points, and a label of height h_i for each site $s_i \in P$, it is NP-hard to place all labels at two stacks on one side of R with nonintersecting opo-leaders.*

3.2   Labeling Polygons with Leaders

In this subsection, we consider the model where the graphical features are not point sites but area features, e.g. a region of a map. Here we apply again our technique to formulate the assignment problem as a mincost-bipartite matching problem between region and label candidate. The problem here is to determine the related edge weights which correspond to the minimum Manhattan distances between the polygonal region and the labels. We assume that the regions are generalized canonical polygons (gc-polygons) whose edges are vertical, horizontal or diagonal. We allow that the leader may attach to the polygon on any position of the boundary. Using basic computational geometry methods we can compute those distances efficiently.

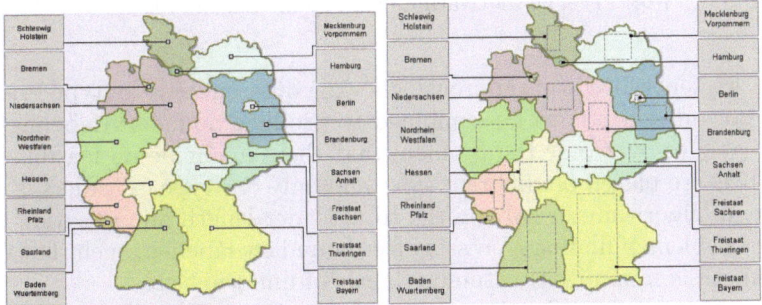

Fig. 10. A motivating example for boundary labeling for polygons. Observe the visual improvement when area features are used instead of points.

Theorem 18. *Consider the case where the labels are placed in fixed positions on all four sides of rectangle R, m is the number of labels, n is the number of gc-polygons, $k_0 = O(k + m)$ and k is the maximum number of corners that a site of type gc-polygon. The minimum distance under the Manhattan metric between any label and any polygon can be computed in time $O(n(k_0 + m) \log k_0)$.*

After applying the min-cost bipartite matching algorithm, eventually crossings have to be resolved in a similar way as described for the opo-leaders in section 2.

3.3   Boundary Labeling with Diagonal Leaders

In 2007, Benkert, Haverkort, Kroll and Nöllenburg introduced a new type of leaders which seems to overcome most of the aesthetical drawbacks of opo-, po and
straightline leaders, they also allowed diagonal segments. In their publication [8],
they presented a dynamic programming approach for length-minimization but
they restricted to one side only, and more seriously, they restricted to do-leaders,
where only the first segment of the leader might be diagonal. Unfortunately, this
restriction is not very natural but leaves many problems unfeasible (e.g., if many
points are relatively close to the center of the boundary.

Fig. 11. The motivating example for diagonal leaders

In [7], we generalized this approach allowing different pairs of type leaders (i.e.
do and pd, od and pd) to be combined to produce boundary labelings. Thus,
we were able to overcome the problem that there might be no feasible solution
when labels are placed on different sides and only one type of leaders is allowed.
We gave an algorithm for solving the total leader length minimization problem
(i.e., the problem of finding a crossing free boundary labeling, such that the total
leader length is minimized) assuming labels of uniform size.

The formal definition for the new type of leaders is the following:

- Type-od leaders: The first line segment of a leader of type od is orthogonal
  (o) to the side of R containing the label it leads to. Its second line segment
  is diagonal (d) to that side.
- Type-pd leaders: The first line segment of a leader of type pd is parallel (p)
  to the side of R containing the label it leads to. Its second line segment is
  diagonal (d) to that side.
- Type-do leaders: The first line segment of a leader of type do is diagonal (d)
  to the side of R containing the label it leads to. Its second line segment is
  orthogonal (o) to that side.

We could prove the following theorem restricting to two types of leaders.
Note that after application of the appropriate bipartite matching technique [16]
a crossing resolution phase is necessary even for the case of one-side-leaders.

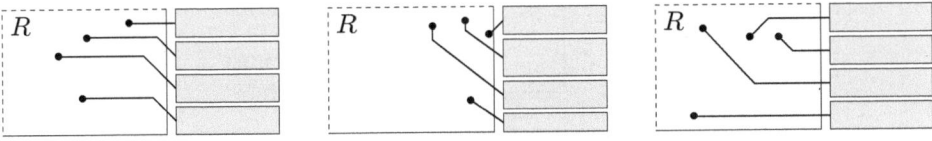

Fig. 12. Three different kind of leaders with diagonals: od, pd, do

Theorem 19. *Given a set P of n sites and a set L of n labels of uniform size placed at fixed positions on one side of the enclosing rectangle R, we can compute in $O(n^3)$ total time a legal boundary labeling of minimum total leader length with type do and pd leaders.*

Allowing labels on more sides, we could apply again the bipartite matching technique but the crossing resolution phase is much more involved.

Theorem 20. *Given a set P of n sites and a set L of n labels of uniform size, placed at fixed positions, on two opposite or even all four sides of the enclosing rectangle R, we can compute in $O(n^3)$ total time a legal boundary labeling of minimum total leader length with either type do and pd leaders or with od and pd leaders.*

For the case, that we only aim for a legal but not length-minimal solution, we could improve the time bound by partitioning the rectangle in four appropriate regions and then using a greedy routing of the leaders followed by a crossing resolution phase.

Theorem 21. *Given a set P of n sites and a set L of n labels of uniform size placed at fixed positions on all four sides of the enclosing rectangle R, we can compute in $O(n^2)$ total time a legal boundary labeling with type od and pd leaders.*

Finally, we could prove NP-completeness for the case of arbitrary label heights by a reduction of an appropriate scheduling problem called total discrepancy problem.

Theorem 22. *Given a set P of n sites, a label l_i of height h_i for each site s_i and an integer k, it is NP-complete to decide whether there exists a legal boundary labeling of total leader length no more than k assuming type do and pd leaders.*

3.4   Many-to-One Labeling

A very interesting alternative generalization of the basic model has been considered in [15]. Namely, in certain applications, more than one site may be required to be connected to a common label. In this case, the presence of crossings among leaders often becomes inevitable. Minimizing the total number of crossings in boundary labeling becomes a critical design issue as crossing is often regarded as the main source of confusion in visualization. Lin, Kao, and Yen introduce the model for multi-site-to-one-label boundary labeling, or many-to-one labeling,

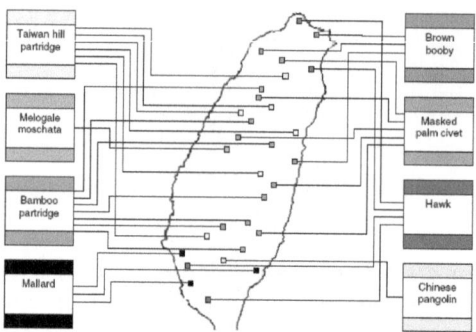

Fig. 13. An example for many-to-one boundary labeling

and concentrated on the crossing minimization problem. They give nice reductions to the well-known two-layer-crossing minimization problem of bipartite graphs which is NP-complete and show NP-completeness-results for the crossing minimization problem for various one-side and two-side labeling schemes.

Theorem 23. *Given an integer k, a set P of N sites and a set L of n labels, and a many-to-one function mapping the points from P to the labels in L, it is NP-complete to decide if there is a set of opo/po-leaders with at most k crossings for the case that the labels are placed on one side of the rectangle, and for the case that the labels are placed on two opposite sides.*

Even more interesting, the authors present some approximation methods inspired by other approximation algorithms for crossing minimization from the graph drawing literature. For example, they use an approximation for max-bisection to achieve an approach for the two-sided problem for labels of maximal size.

Theorem 24. *Given a set P of N sites and a set L of n labels, and a many-to-one function mapping the points in P to the labels in L that are placed on two opposite sides of the rectangle. For the case of opo-leaders the crossing minimization problem can be solved within a factor of C from the optimum.*

Note that the constant C is a somewhat complicated expression depending on the weights of the potential leaders.

4   Discussion and Open Problems

We have demonstrated a nice example of how algorithmics can effectively work. Starting from a challenging problem ('the blue map') we derived a very simple model that abstracts from many practical parameters and applied various algorithmic techniques to solve the problem. Gradually, we generalized the basic models allowing more flexibility in the model constraints and found out which of the methods is robust enough to carry over. Clearly, the aim to provide effective algorithms for the labeling problem in full generality is not reached yet, therefore we formulate the following open problems as possible extensions:

1. Reformulate Vaidya's geometric matching algorithm such that it can be applied to the od/do - metric.
2. Develop algorithms for a mixed model, where internal labels and boundary labels are allowed.
3. Allow the placement of leader-connected labels not only at the boundary but anywhere in the map where there is empty space.
4. Develop approximations for the NP-complete problems for boundary labeling.
5. Evaluate which is the most appropriate model for boundary labeling (opo, po, do, mixed, ...).

Acknowledgments

The author would like to thank Michalis Bekos and Antonis Symvonis for the wonderful collaboration on this topic.

References

1. Agarwal, P.K., Efrat, A., Sharir, M.: Vertical decomposition of shallow levels in 3-dimensional arrangements and its applications. In: Proc. 11th Annu. ACM Sympos. Comput. Geom., pp. 39–50 (1995)
2. Agarwal, P., van Kreveld, M., Suri, S.: Label placement by maximum independent set in rectangles. Computational Geometry: Theory and Applications 11, 209–218 (1998)
3. Bekos, M., Kaufmann, M., Symvonis, A., Wolff, A.: Boundary labeling: Models and efficient algorithms for rectangular maps. In: Pach, J. (ed.) GD 2004. LNCS, vol. 3383, pp. 49–59. Springer, Heidelberg (2005); An extended version appeared in Computational Geometry: Theory and Applications 36(3), 215–236 (2006)
4. Bekos, M.A., Kaufmann, M., Potika, K., Symvonis, A.: Multi-stack Boundary Labeling Problems. In: Arun-Kumar, S., Garg, N. (eds.) FSTTCS 2006. LNCS, vol. 4337, pp. 81–92. Springer, Heidelberg (2006)
5. Bekos, M., Kaufmann, M., Potika, K., Symvonis, A.: Polygons Labelling of Minimum Leader Length. In: Proc. Asia Pacific Symposium on Information Visualisation (APVIS 2006). CRPIT 60, pp. 15–21 (2006)
6. Bekos, M., Kaufmann, M., Potika, K., Symvonis, A.: Area-Feature Boundary Labeling. The Computer Journal (to appear)
7. Bekos, M., Kaufmann, M., Nöllenburg, M., Symvonis, A.: Boundary labeling with octilinear leaders. In: Gudmundsson, J. (ed.) SWAT 2008. LNCS, vol. 5124, pp. 234–245. Springer, Heidelberg (2008); An extended version is to appear in Algorithmica
8. Benkert, M., Haverkort, H., Kroll, M., Nöllenburg, M.: Algorithms for multi-criteria one-sided boundary labeling. In: Hong, S.-H., Nishizeki, T., Quan, W. (eds.) GD 2007. LNCS, vol. 4875, pp. 243–254. Springer, Heidelberg (2008)
9. Chazelle, B., 36 co-authors.: The computational geometry impact task force report. In: Chazelle, B., Goodman, J.E., Pollack, R. (eds.) Advances in Discrete and Computational Geometry, vol. 223, pp. 407–463. AMS (1999)

10. Christensen, J., Marks, J., Shieber, S.: Algorithms for cartographic label placement. In: Proceedings of the American Congress on Surveying and Mapping, vol. 1, pp. 75–89 (1993)

11. Formann, M., Wagner, F.: A packing problem with applications to lettering of maps. In: Proc. 7th ACM Symp. Comp. Geom (SoCG 1991), pp. 281–288 (1991)

12. Hershberger, J., Suri, S.: Applications of a semi-dynamic convex hull algorithm. BIT 32, 249–267 (1992)

13. Imhof, E.: Positioning Names on Maps. The American Cartographer 2, 128–144 (1975)

14. Iturriaga, C., Lubiw, A.: NP-hardness of some map labeling problems. Technical Report CS-97-18, University of Waterloo (1997)

15. Lin, C.-C., Kao, H.-J., Lin, C.-C.: Many-to-one boundary labeling. Journal of Graph Algorithms and Applications (JGAA) 12(3), 319–356 (2008)

16. Kuhn, H.W.: The Hungarian method for the assignment problem. Naval Research Logistic Quarterly 2, 83–97 (1955)

17. Mehlhorn, K.: Data Structures and Algorithms 3: Multi-dimensional Searching and Computational Geometry. EATCS Monographs on Theoretical Computer Science, vol. 3. Springer, Heidelberg (1984)

18. Neyer, G.: Map Labeling with Application to Graph Drawing. In: Kaufmann, M., Wagner, D. (eds.) Drawing Graphs. LNCS, vol. 2025, pp. 247–273. Springer, Heidelberg (2001)

19. Schmidt, V.A.: Reine Forschung, praktische Resultate. Newspaper article in Die Zeit (April 28, 1995)

20. Strijk, T., van Kreveld, M.: Labeling a rectilinear map more efficiently. Information Processing Letters 69(1), 25–30 (1999)

21. Tufte, E.R.: The Visual Display of Quantitative Information. Graphics Press (1983)

22. Vaidya, P.M.: Geometry helps in matching. SIAM J. Comput. 18, 1201–1225 (1989)

23. Verweij, B., Aardal, K.: An optimisation algorithm for maximum independent set with applications in map labelling. In: Nešetřil, J. (ed.) ESA 1999. LNCS, vol. 1643, pp. 426–437. Springer, Heidelberg (1999)

24. Wolff, A.: The Map-Labeling Bibliography (1996), http://i11www.ira.uka.de/map-labeling/bibliography/

25. Wagner, F.: Approximate map labeling is in Omega (n log n). Technical Report B 93-18, Fachbereich Mathematik und Informatik, Freie Universitat Berlin (1993)

26. Wagner, F., Wolff, A.: Map labeling heuristics: provably good and practically useful. In: Proceedings of the eleventh annual symposium on Computational geometry, pp. 109–118 (1995)

27. Wagner, F., Wolff, A.: A combinatorial framework for map labeling. In: Whitesides, S.H. (ed.) GD 1998. LNCS, vol. 1547, pp. 316–331. Springer, Heidelberg (1998)

28. Yoeli, P.: The Logic of Automated Map Lettering. The Cartographic Journal 9, 99–108 (1972)

The Crossing Number of Graphs: Theory and Computation

Petra Mutzel

Technische Universität Dortmund, Informatik LS11, Algorithm Engineering
Otto-Hahn-Str. 14, 44227 Dortmund, Germany
petra.mutzel@tu-dortmund.de
http://ls11-www.cs.tu-dortmund.de

Abstract. This survey concentrates on selected theoretical and computational aspects of the crossing number of graphs. Starting with its introduction by Turán, we will discuss known results for complete and complete bipartite graphs. Then we will focus on some historical confusion on the crossing number that has been brought up by Pach and Tóth as well as Székely. A connection to computational geometry is made in the section on the geometric version, namely the rectilinear crossing number. We will also mention some applications of the crossing number to geometrical problems. This review ends with recent results on approximation and exact computations.

1 Introduction

The crossing number $cr(G)$ of a graph is the smallest number of edge crossings achievable when laying out G in the 2-dimensional plane. The problem originated from Turán in 1944 when he worked in a labor camp [31]:

"There were some kilns where the bricks were made and some open storage yards where the bricks were stored. All the kilns were connected by rail with all the storage yards. The bricks were carried on small wheeled trucks to the storage yards. All we had to do was to put the bricks on the trucks at the kilns, push the trucks to the storage yards, and unload them there ... the trouble was only at the crossings. The trucks generally jumped the rails there, and the bricks fell out of them; in short, this caused a lot of trouble and loss of time ... the idea occured to me that this loss of time could have been minimized if the number of crossings of the rails had been minimized."

In 1952, Turán mentioned this problem to Zarankiewicz, who presented a solution for the crossing number of complete bipartite graphs in 1954 [34]. Unfortunately, Ringel found a gap in the published proof that has not been closed yet.

S. Albers, H. Alt, and S. Näher (Eds.): Festschrift Mehlhorn, LNCS 5760, pp. 305–317, 2009.

1.1   The Crossing Number for Complete Bipartite Graphs

However, the formula presented by Zarankiewicz is conjectured to be correct. His construction provides drawings with exactly

$$Z(m,n) := \left\lfloor \frac{m}{2} \right\rfloor \left\lfloor \frac{m-1}{2} \right\rfloor \left\lfloor \frac{n}{2} \right\rfloor \left\lfloor \frac{n-1}{2} \right\rfloor$$

crossings. Figure 1 shows the construction for $K_{6,6}$.

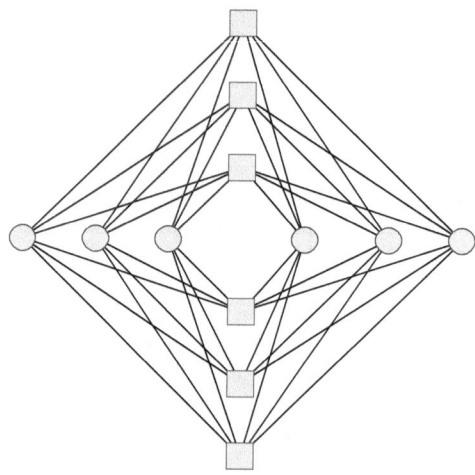

Fig. 1. Zarankiewicz's construction for $K_{6,6}$

The conjecture is known to be true for $K_{m,n}$ with $\min(m,n) \leq 6$ and for $m, n \leq 7$. The smallest unsolved cases are $K_{7,11}$ and $K_{9,9}$ with conjectured values 225 and 256, respectively. New bounding techniques using semi-definite programming [12] have shown that

$$0.8594 Z(m,n) \leq cr(K_{m,n}) \leq Z(m,n).$$

1.2   The Crossing Number for Complete Graphs

The crossing number for the complete graph K_n is not known either. It is generally believed to be given by the formula provided by Guy [18]:

$$Z(n) := \frac{1}{4} \left\lfloor \frac{n}{2} \right\rfloor \left\lfloor \frac{n-1}{2} \right\rfloor \left\lfloor \frac{n-2}{2} \right\rfloor \left\lfloor \frac{n-3}{2} \right\rfloor.$$

For odd n the formula can be written as $Z(n) = \frac{1}{64}(n-1)^2(n-3)^2$. Also Guy presented a general drawing scheme for K_n that produces drawings with exactly $Z(n)$ crossings. Guy's construction for K_8 is shown in Figure 2.

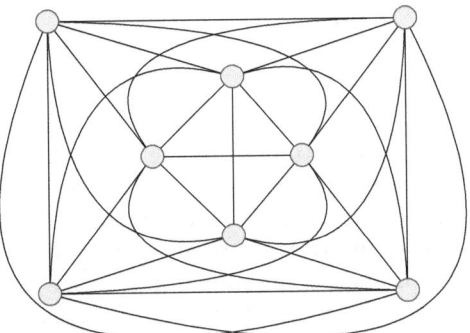

Fig. 2. Guy's construction for K_8

Combinatorial arguments show the following fact: If Guy's conjecture is true for K_{2k-1}, then it is also true for K_{2k}. This is the reason why the proofs concentrate on K_n for odd n. Guy [18] was able to prove his conjecture for K_n with $n \leq 10$. For 35 years, this result could not be improved. Recently, Pan and Richter [24] showed that $cr(K_{11}) = 100$ thus getting $cr(K_{12}) = 150$ for free. The smallest unsolved case is K_{13} with conjectured crossing number 225.

For many graph classes the situation is similar: $cr(G)$ is known for small instances only, while for general n not much is known (e.g., hypercube graphs, toroidal graphs, or generalized Peterson graphs). For a bibliography on the crossing number, see [33].

2   Confusion on the Crossing Number

In their interesting article [23] "What crossing number is it anyway?" Pach and Toth stated that "... some authors might have thought of ..." different crossing numbers. They pointed out that the definitions for the crossing number provided in the literature were not always the same. In order to investigate this further, we will use the formal definitions of Szégely [30].

A *drawing* D of a graph G into the plane is an injection from the vertex set $V(G)$ into the plane, and a mapping Φ of the edge set $E(G)$ into the set of simple planar curves, such that the curve corresponding to the edge has end points $\Phi(u)$ and $\Phi(v)$, and contains no other vertices. The *number of crossings* $cr(D)$ in D is the number of intersection points of all unordered pairs of interiors of edges. The *crossing number* $cr(G)$ of a graph G is the minimum $cr(D)$ over all drawings D of G. A drawing D is *optimal* if it realizes $cr(D) = cr(G)$.

A drawing D is called *normal* if it satisfies

i any two of the curves have finitely many points in common
ii no two curves have a point in common in a tangential way
iii no three curves cross each other in the same point

A drawing is *nice*, if it is normal, and satisfies

iv no two adjacent edges cross
 v any two edges cross at most once

It can be observed that an optimal drawing must satisfy i, ii, iv, and v, and can be transformed to satisfy iii. Therefore, we can restrict ourselves to consider normal or even nice drawings when interested in $cr(G)$.

Pach and Tóth [23] introduced two variants of crossing numbers: The *pairwise crossing number* $cr\text{-}pair(G)$ is the minimum number of edge pairs that cross each other at least once, over all normal drawings of G. The *odd crossing number* $cr\text{-}odd(G)$ is equal to the minimum number of edge pairs that cross each other an odd number of times, over all normal drawings of G.

In [32], Tutte introduces yet another version that Szégely calls the *independent odd crossing number* $cr\text{-}iodd(G)$. It is equal to the minimum number of non-adjacent edge pairs that cross each other odd times, over all normal drawings of G. The reason why Tutte introduced this crossing number was his "... view that crossings of adjacent edges are trivial, and easily get rid of." But so far nobody has shown that this can be done in this setting.

The following relation between these variants is obvious:

$$cr\text{-}iodd(G) \leq cr\text{-}odd(G) \leq cr\text{-}pair(G) \leq cr(G)$$

Pach and Tóth mention that "... perhaps the most exciting open problem in the area ..." is the question: "Are they all equal?"

One of the reasons why researchers thought that these numbers might be equal is an old theorem by Hanani [19], rediscovered by Tutte [32]. It states that every graph that can be drawn such that every pair of non-adjacent edges intersects an even number of times, is planar. Pach and Tóth [23] have generalized this result by showing that one can redraw even edges without crossings even in the presence of odd edges. Pelsmajer, Schaefer, and Štefankovič [25] have shown that this redrawing can be performed without adding pairs of edges that intersect an odd number of times; in particular, the odd crossing number does not increase by the redrawing. It is known that $cr(G) \leq 2(cr\text{-}odd(G))^2$ [23] and that for graphs with $cr\text{-}odd(G) \leq 3$ indeed $cr\text{-}odd(G) = cr(G)$ [25].

Some authors have stated the conjecture that $cr\text{-}odd(G) = cr(G)$. A surprising result by Pelsmajer, Schaefer, and Štefankovič [26] showed that equality of both crossing number variants does not hold. The authors have presented a quite simple infinite family of graphs with $cr\text{-}odd(G) < cr\text{-}pair(G) = cr(G)$.

Figure 3 shows an example of such a graph G. The four distinguished edges a, b, c and d have weights $w_a = 1$, $w_b = w_c = 3$, and $w_d = 4$. We assume that the weights of the edges e along the two main cycles are heavy so that they are not crossed in an optimal drawing (e.g., $w_e = 15$). We can think of replacing an edge with weight w by w parallel edges. It is also possible to get rid of parallel edges by subdividing these edges. The left drawing shows an optimal drawing for $cr(G)$ and $cr\text{-}pair(G)$. The crossing number and the pairwise crossing number is $cr(G) = cr\text{-}pair(G) = 15$. In the right drawing, the edges a and c cross

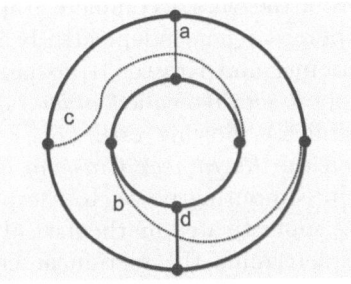 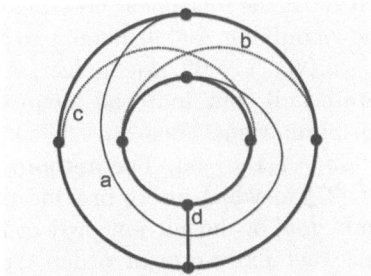

Fig. 3. A graph with $cr\text{-}odd(G) < cr(G)$. The left drawing shows an optimal drawing for $cr(G)$ and $cr\text{-}pair(g)$ with value 15. The right drawing shows an optimal drawing for $cr\text{-}odd(G)$ with value 13.

each other exactly twice providing the amount of 0 to $cr\text{-}odd(G)$ thus giving $cr\text{-}odd(G) = 13 < cr(G)$.

However, the question if $cr(G) = cr\text{-}pair(G)$ is still open. The bound of $cr(G) = O(cr\text{-}pair(G)^2/\log(cr\text{-}pair(G)))$ has been provided by Valtr (mentioned in [22]).

Now that we know that the four crossing numbers are not equal, a question already stated by Szégely is catching our interest [30]: "How is it possible that decades in research of crossing numbers passed by and no major confusion resulted from these foundational problems?" — Perhaps the graph theory community was just lucky that the bounds they provided in all these years apply for all kinds of crossing numbers.

3   The Rectilinear Crossing Number

The geometric version of the crossing number is called the *rectilinear crossing number*, denoted by $cr\text{-}lin$, and requires a drawing in the plane with straight line segments. It is well known that a planar graph always has a planar drawing with straight line segments. Therefore, one may think that $cr\text{-}lin(G)$ and $cr(G)$ are equal or close together. However, already Guy [17] has shown that $cr(K_8) = 18$ while $cr\text{-}lin(K_8) = 19$. Later, Bienstock and Dean [5] have shown that the two numbers are equal ($cr\text{-}lin(G) = cr(G)$) for graphs with small crossing number $cr(G) \le 3$. On the other hand, the authors have also shown that there are graphs with crossing number 4 and arbitrarily large rectilinear crossing number. However, for graphs with bounded degree, the crossing number and the rectilinear crossing number are bounded as functions of one another [4]. In detail, if a graph has maximum degree d and crossing number k, its rectilinear crossing number is at most $O(dk^2)$.

It is conjectured that the construction by Zarankiewicz for the crossing number of complete bipartite graphs provides the correct numbers for $cr\text{-}lin(K_{m,n})$. Due to the nature of the construction, a proof for $cr(K_{m,n}) = Z(m,n)$ would directly lead to $cr\text{-}lin(K_{m,n}) = Z(m,n)$.

Until 2001, the rectilinear crossing number for the class of complete graphs K_n was known only for $n \leq 9$. Then, two groups of researchers independently showed that $cr\text{-}lin(K_{10}) = 62$. Aichholzer, Aurenhammer and Krasser [1] exhaustively enumerated all combinatorial inequivalent point sets (so-called *order types*) of size 10. Similar methods have been successful for showing $cr\text{-}lin(K_{11}) = 102$ and $cr\text{-}lin(K_{12}) = 153$. The authors initiated the *Rectilinear Crossing Number Project* [27] in which users provide their own computing power to the project. The main goal of the current project is to use sophisticated mathematical methods (abstract extension of order types) to determine the rectilinear crossing number for small values of n, and to compute all existing combinatorial non-isomorphic minimal drawings. Currently, the rectilinear crossing number is known for all K_n with $n \leq 21$ ($cr\text{-}lin(K_{21}) = 2055$). In contrast to $cr(K_n)$ there is no conjecture following some formula for arbitrarily large n for $cr\text{-}lin(K_n)$. However, the gaps between the lower and upper bounds for n up to 100 are quite small, e.g., $1.459.912 \leq cr\text{-}lin(K_{100}) \leq 1.463.970$.

4   Applications of the Crossing Number

Székely has used bounds for the crossing number $cr(G)$ for providing a simple proof of the Szemerédi-Trotter theorem, that is an important result in combinatorial geometry. It asks for the maximum number of incidences of n points and m curves in the plane such that each pair of curves intersects at maximal $O(1)$ points, and there are no more than $O(1)$ curves passing through each pair of points. The answer in this case is $O(m + n + (mn)^{2/3})$. The idea of Szégely's proof was to build a graph in which the vertex set is associated with the point set and the edges with the curve segments. Using bounds for the crossing number for complete graphs essentially provided the solution.

Szemerédi-Trotter like theorems can be used for proving hard Erdős problems in combinatorics, in number theory, in analysis or geometric measure theory. E.g., Elekes [14] used it to show that any n distinct real numbers have $\Omega(n^{1.25})$ distinct sums or products. And a famous problem of Erdős in geometry asks for the maximum number of unit distances that are possible among n points in the plane. Application of the Szemerédi-Trotter theorem provides $O(n^{4/3})$, and this is the best known estimate so far [29].

Dey [13] used the bounds known for the rectilinear crossing number for proving upper bounds on geometric k-sets. This led to considerable improvement on this bound after its early solution about 27 years ago.

The rectilinear crossing number of a complete graph is essentially the same as the minimum number of convex quadrilaterals determined by a set of n points in general position. It is known that the number of quadrilaterals is proportional to the fourth power of n, but the precise constant is unknown [28].

An important application of crossing numbers is in graph drawing and VLSI-design. This is why the author started research in this area in 1995. Figure 4 shows a graph with 120 objects and 161 edges that originated at an insurance company. The original drawing had 122 crossings, while the crossing minimal drawing has only 6 crossings.

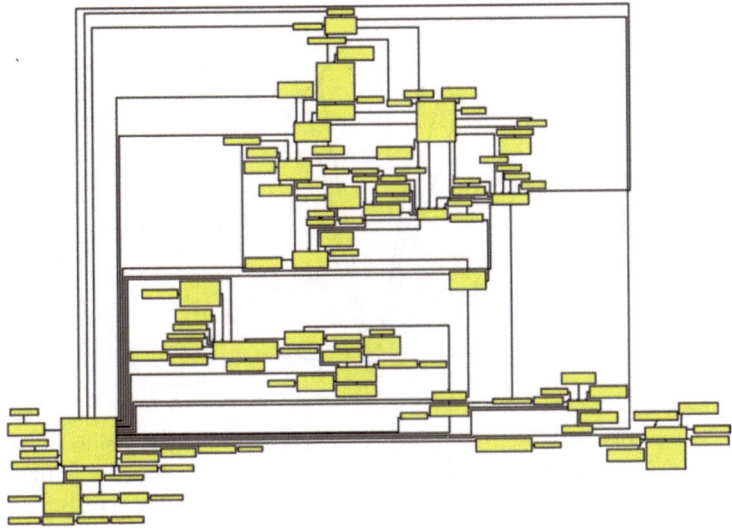

Fig. 4. A drawing of the insurance graph with 6 crossings

5 Approximation

Garey and Johnson have shown NP-completeness of the decision variant of the crossing number problem [21]. Pach and Tóth have shown the same for the crossing number variants *cr-odd* and *cr-pair*. Bienstock has shown that also the computation of *cr-lin*(G) is NP-hard [3].

No polynomial time algorithm is known for approximating $cr(G)$ for general graphs within some non-trivial factor. Bhatt and Leighton [2] suggested the first algorithm which approximates $|V| + cr(G)$ for a bounded degree graph $G = (V, E)$ with a factor of $O(\log^4 |V|)$. This approximation factor has been improved by Even, Guha and Schieber [15] in 2002 to $O(\log^3 |V|)$. For sparse graphs, when $cr(G) = o(|V|)$, this approximation does not guarantee good results. Until recently, polynomial time algorithm approximating $cr(G)$ was known, not even for special graph classes.

Recently, the first approximation results have been achieved that do depend on the maximum degree and $cr(G)$ only. The approximation results concern the graph classes of *almost planar graphs* and *apex graphs*. A graph $G = (V, E)$ is called *almost planar* if G is non-planar, but there does exist an edge $e \in E$ so that $G - \{e\}$ is planar. Given a planar embedding Π of the remaining graph $G - \{e\}$, the edge e can be re-inserted with the minimal number of crossings via a shortest path in the extended geometric dual graph of Π. Gutwenger, Mutzel, and Weiskircher [16] have presented a linear time algorithm (based on the data structure of SPQR-trees) which is able to find the optimal embedding Π_0 of $G - \{e\}$, so that inserting e into Π_0 leads to a crossing minimum drawing over the set of all possible planar embeddings Π. The natural question arises, if this approach does approximate the crossing number $cr(G)$ by some small factor.

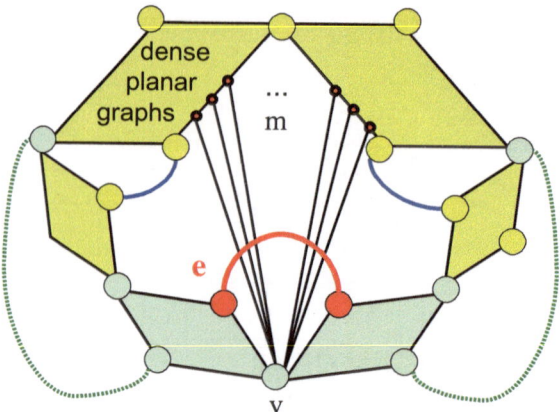

Fig. 5. Inserting edge e back into the planar graph $G - \{e\}$ yields m crossings if $deg(v) \geq 2m$. Note that the shaded blocks are dense triconnected subgraphs. The crossing minimal drawing of G with 2 crossings can be found when flipping the two blocks adjacent to v.

Figure 5 shows an example for which the edge e will be inserted into the planar graph $G - \{e\}$ with m crossings. However, the minimum crossing number of G is 2. This can be achieved by flipping the two lower components adjacent to v. Hliněný and Salazar [20] observed that the approximation factor in this case depends on the degree of vertex v. If the degree is bounded, then this example does not hurt the approximation anymore. Hliněný and Salazar have shown that the above algorithm provides crossing numbers of at most $\Delta(G - \{e\})cr(G)$, where $\Delta(G)$ denotes the maximum degree of G. This number has later been improved to $(\Delta(G - \{e\})/2)cr(G)$ by Cabello and Mohar [7]. This provides the first constant approximation algorithm for almost planar graphs with bounded degree graphs.

Very recently, these results could be generalized to apex-graphs. A graph $G = (V, E)$ is called an *apex graph* if G is non-planar, but there does exist a vertex $v \in V$ so that $G - \{v\}$ is planar. Chimani, Gutwenger, Mutzel, and Wolf [9] have shown that v and all its incident edges can be re-inserted into an optimal embedding Π_0 of $G - \{v\}$ (which the algorithm will identify) with the minimum number of crossings in polynomial time. Chimani, Hliněný, and Mutzel [10] have shown that this algorithm will find solutions which are at most a factor of $deg(v)\Delta(G - \{v\})/2$ away from the optimum solution $cr(G)$.

Both approximation results are (almost) tight: for almost planar graphs, there is an example showing that the approximation factor can be reached, while for apex graphs the example is still a factor of 2 away.

Some open questions arise:

– Is there a polynomial time algorithm for computing the crossing number $cr(G)$ for almost planar graphs? Cabello and Mohar [7] have shown that the weighted version is NP-hard.
– Can the above results be generalized, e.g., for graphs that are planar after deleting a fixed number of edges?

6 Exact Computation

Very few publications exist for computing the crossing number of general graphs exactly. Grohe has shown in 2001 that the crossing number problem is fixed-parameter tractable. However, the used concepts are based on the theoretical results of Robertson and Seymour. Recently, Kawarabayashi and Reed 2007 have improved the quadratic running time to a linear time algorithm for fixed k. Both approaches reduce the graph to one with bounded tree-width and the same crossing number and then test if the graph has crossing number at most k. There is common agreement that this approach is purely theoretical. Related to this, the following open problem is among the most important ones in the area: Can $cr(G)$ be computed in polynomial time for graphs with bounded tree-width?

Until 2005, no practically efficient algorithm for computing the crossing number was known. Today, there exist two approaches for computing the exact crossing number for general graphs. The approaches are based on two integer linear programming (ILP) formulations of the crossing number problem that can be solved by branch-and-cut algorithms.

The ILP formulation by Buchheim et al. [6] is called the *subdivision crossing minimization approach* (SOCM) and optimizes over the set of all simple drawings. A drawing is called *simple* if every edge is only crossed at most once. In order to provide an optimal solution for $cr(G)$, we need to subdivide all edges in G into a path of length $|E|$. The variables $x_{e,f}$ are associated with all non-adjacent edge pairs $(e, f) \in E^2$. The constraints come essentially from Kuratowski's theorem stating that a graph G is planar if and only if it does not contain a subdivision of $K_{3,3}$ or K_5. Besides the Kuratowski-constraints and the $0/1$-constraints, the ILP also contains constraints that guarantee to get simple drawings of the subdivided graph.

The second approach by Bomze, Chimani and Mutzel [11] is called the ordering-based ILP model (OOCM). This is not restricted to simple drawings. Instead, additional linear ordering variables y_{efg} are introduced for each edge $e \in E$ that may be crossed more than once. The variables y_{efg} for edges $e, f, g \in E$ provide the information in which order an edge e is crossed by f and g. In this ILP we need Kuratowski constraints on the x variables, linear ordering constraints on the y variables, $0/1$-constraints for all variables, and additional linking constraints between the x and y variables.

We solve both ILP models with branch-and-cut algorithms. In order to get these algorithms to work in practice, we needed to come up with new preprocessing techniques as well as new combinatorial column generation methods. Our computational experiments on a benchmark set of about 11,000 graphs show that we can compute the exact crossing numbers for general sparse graphs with up to 100 vertices and crossing number up to 37 within 30 minutes.

Figure 6 shows the percentage of instances that have been solved within 30 minutes of computation time for about 11.000 graphs of the Rome library. The x-asis shows the number of vertices $|V|$. The number of edges of the Rome graphs is below $1.5|V|$ on average. The crossing number of almost all graphs with up to 60 vertices could be solved to provable optimality within 30 minutes CPU-time.

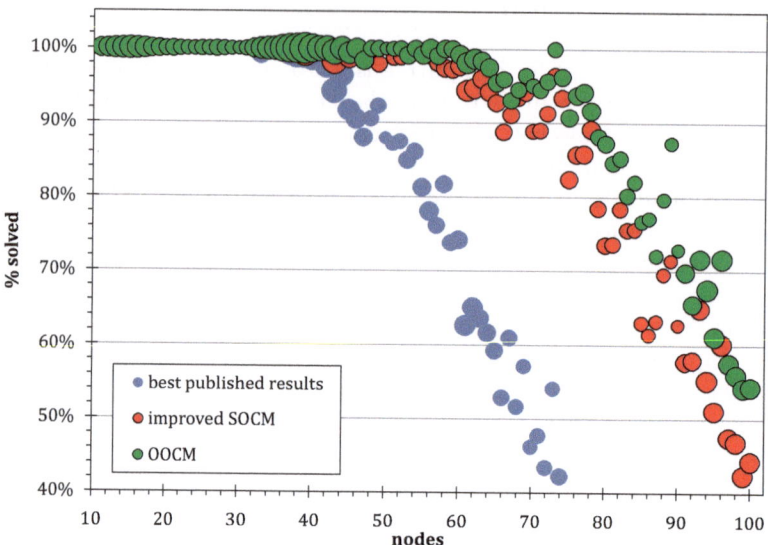

Fig. 6. The average percentage of instances solved within 30 minutes of computation time of the ILP models SOCM and OOCM

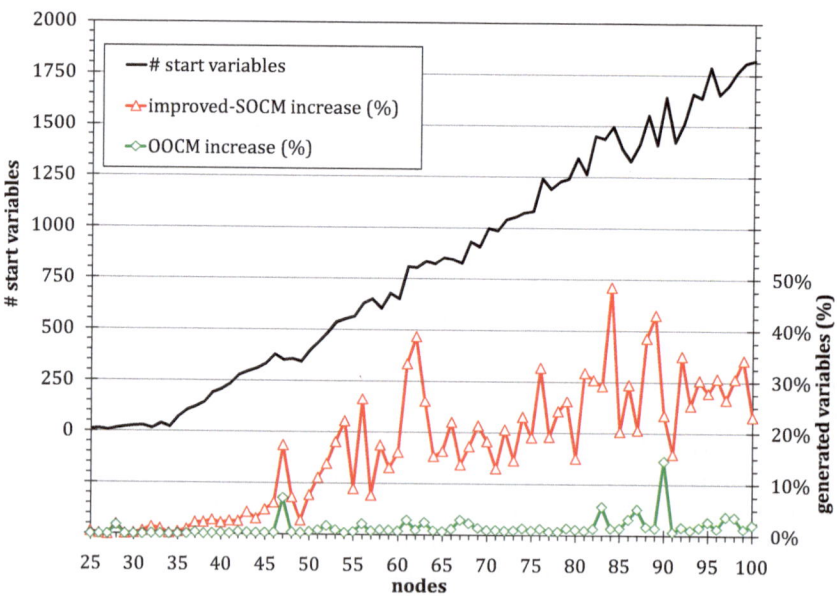

Fig. 7. The dark line shows the average number of start variables in our branch-and-cut approach (left axis). The right axis shows the percentage of additional variables generated by the two approaches.

The graphs with 100 vertices are much harder to solve. But still, more than 50% of the instances with 100 vertices could be solved to optimality.

It seems that the second formulation based on linear ordering dominates our first ILP model. For most instances, we need far more variables in our SOCM model than in the OOCM model. The dark line in Figure 7 shows the average number of start variables (left axis) in our branch-and-cut approach. For graphs with 100 vertices the average number was about 1800 variables. During the run of the algorithm, column generation adds in additional variables. These numbers are much higher for the SOCM approach. While SOCM added about 50% new variables with respect to the start variables, OOCM only had to add 18% additional variables for the 100-vertex graphs.

We find it surprising that in the OOCM model only very few y-variables are needed in order to find the optimum solution. Detailed experiments and results can be found in [8, 11].

7   Solved Open Problems

We close our survey with selected experimental results for special graph classes. While we could verify the crossing number of the complete graph on 11 (and 12) vertices (with an alternative optimum solution), one more vertex is still a challenge.

On the other hand, we are able to compute the crossing number of generalized Petersen graphs $P_{n,4}$ up to $n = 44$ which was unknown before. Based on our computed results, we came up with a conjecture of the crossing number of this graph class.

Moreover, we are confident that we are about to be able to compute the crossing number of the smallest toroidal grid graph $T_{8,8}$ whose crossing number is still unknown (and conjectured to be 48).

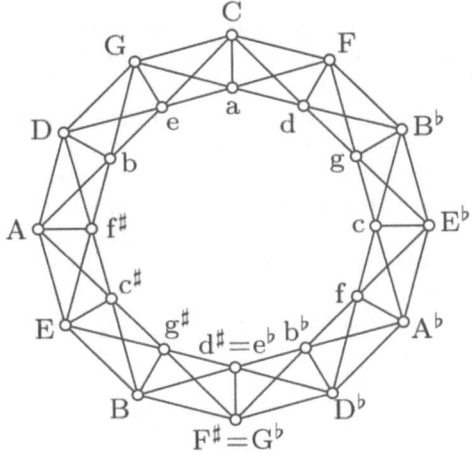

Fig. 8. Knuth's musical graph

In Exercise 133 of The Art of Computer Programming, Volume 4, Draft of Section 7, Donald E. Knuth uses the "musical graph" on page 73 of *Graphs* by R. J. Wilson and J. J. Watkins (1990) (see Fig. 8).

It represents simple modulations between key signatures. While all kinds of properties of this graph are easily analyzed, the question "Can it be drawn with fewer than 12 crossings?" remained open. After 9.71 seconds of computation time, our program proved that the crossing number is indeed 12 and produced an alternative embedding that is not as nice as the original, though.

Acknowledgments. I am grateful to Michael Jünger who pointed out the open problem of Knuth's forthcoming book. Many thanks to Markus Chimani who was co-author of most cited papers on the crossing number, and who did all computations with his code.

References

1. Aichholzer, O., Aurenhammer, F., Krasser, H.: On the crossing number of complete graphs. In: Symposium on Computational Geometry, pp. 19–24 (2002)
2. Bhatt, S.N., Leighton, F.T.: A framework for solving vlsi layout problems. Journal of Computer and System Sciences 28, 300–343 (1984)
3. Bienstock, D.: Some provably hard crossing number problems. Discrete & Computational Geometry 6, 443–459 (1991)
4. Bienstock, D., Dean, N.: New results on rectilinear crossing numbers and plane embeddings. J. Graph Theory 16(5), 389–398 (1992)
5. Bienstock, D., Dean, N.: Bounds for rectilinear crossing numbers. J. Graph Theory 17(3), 333–348 (1993)
6. Buchheim, C., Chimani, M., Ebner, D., Gutwenger, C., Jünger, M., Klau, G.W., Mutzel, P., Weiskircher, R.: A branch-and-cut approach to the crossing number problem. Discrete Optimization, Special Issue in memory of George B. Dantzig 5(2), 373–388 (2008)
7. Cabello, S., Mohar, B.: Crossing and weighted crossing number of near-planar graphs. In: Tollis, I.G., Patrignani, M. (eds.) GD 2008. LNCS, vol. 5417, pp. 38–49. Springer, Heidelberg (2009)
8. Chimani, M.: Exact Crossing Minimization. PhD thesis, Fakultät für Informatik, Technische Universität Dortmund (2008)
9. Chimani, M., Gutwenger, C., Mutzel, P., Wolf, C.: Inserting a vertex into a planar graph. In: SODA, pp. 375–383. SIAM, Philadelphia (2009)
10. Chimani, M., Hliněný, P., Mutzel, P.: Vertex insertion approximates the crossing number for apex graphs (submitted, 2009)
11. Chimani, M., Mutzel, P., Bomze, I.: A new approach to exact crossing minimization. In: Halperin, D., Mehlhorn, K. (eds.) ESA 2008. LNCS, vol. 5193, pp. 284–296. Springer, Heidelberg (2008)
12. de Klerk, E., Pasechnik, D.V., Schrijver, A.: Reduction of symmetric semidefinite programs using the regular *-representation. Math. Program. 109(2-3), 613–624 (2007)
13. Dey, T.K.: Improved bounds for planar k -sets and related problems. Discrete & Computational Geometry 19(3), 373–382 (1998)

14. Elekes, G.: On the number of sums and products. Acta Arithm. 81(4), 365–367 (1997)
15. Even, G., Guha, S., Schieber, B.: Improved approximations of crossings in graph drawings and VLSI layout areas. SIAM Journal on Computing 32(1), 231–252 (2002)
16. Gutwenger, C., Mutzel, P., Weiskircher, R.: Inserting an edge into a planar graph. Algorithmica 41(4), 289–308 (2005)
17. Guy, R.K.: The decline and fall of Zarankiewicz's theorem. In: Proof techniques in Graph Theory, pp. 63–69. Academic Press, London (1969)
18. Guy, R.K.: Crossing numbers of graphs. In: Proc. Graph Theory and Applications, pp. 111–124. LNM (1972)
19. Chojnacki, C., Hanani, H.: Über wesentlich unplättbare Kurven im drei-dimensionalen Raume. Fundam. Math. (1934)
20. Hliněný, P., Salazar, G.: Crossing and weighted crossing number of near-planar graphs. In: Kaufmann, M., Wagner, D. (eds.) GD 2006. LNCS, vol. 4372, pp. 162–173. Springer, Heidelberg (2007)
21. Johnson, M.R., Johnson, D.S.: Crossing number is NP-complete. SIAM J. Algebraic Discrete Methods 4(3), 312–316 (1983)
22. Kolman, P., Matousek, J.: Crossing number, pair-crossing number, and expansion. J. Combin. Theory Ser. B 92, 99–113 (2003)
23. Pach, J., Tóth, G.: Which crossing number is it anyway? Journal of Combinatorial Theory, Series B 80, 225–246 (2000)
24. Pan, S., Richter, R.B.: The crossing number of K_{11} is 100. Journal of Graph Theory 56(2), 128–134 (2007)
25. Pelsmajer, M.J., Schaefer, M., Štefankovič, D.: Removing even crossings on surfaces. Electronic Notes in Discrete Mathematics 29, 85–90 (2007)
26. Pelsmajer, M.J., Schaefer, M., Štefankovič, D.: Odd crossing number and crossing number are not the same. Discrete & Computational Geometry 39(1-3), 442–454 (2008)
27. The rectilinear crossing number project, http://dist.ist.tugraz.at/cape5/index.php
28. Scheinerman, E.R., Wilf, H.: The rectilinear crossing number of a complete graph and sylvester's "four point problem" of geometric probability. American Mathematical Monthly 101, 939–943 (1994)
29. Spencer, J., Szemerédi, E., Trotter, W.T.: Unit distances in the euclidean plane. Graph Theory and Combinatorics, 293–308 (1984)
30. Székely, L.A.: A successful concept for measuring non-planarity of graphs: the crossing number. Discrete Mathematics 276(1-3), 331–352 (2004)
31. Turán, P.: A note of welcome. Journal of Graph Theory 1, 7–9 (1977)
32. Tutte, W.T.: Toward a theory of crossing numbers. Journal of Combinatorial Theory (1970)
33. Vrt'o, I.: Bibliography on crossing numbers, ftp://ftp.ifi.savba.sk/pub/imrich/crobib.pdf
34. Zarankiewicz, K.: On a problem of P. Turan concerning graphs. Fund. Math. (1954)

Part V
Algorithm Engineering, Exactness, and Robustness

Algorithm Engineering –
An Attempt at a Definition

Peter Sanders*

Universität Karlsruhe, 76128 Karlsruhe, Germany
sanders@ira.uka.de

Abstract. This paper defines algorithm engineering as a general methodology for algorithmic research. The main process in this methodology is a cycle consisting of algorithm design, analysis, implementation and experimental evaluation that resembles Popper's scientific method. Important additional issues are realistic models, algorithm libraries, benchmarks with real-world problem instances, and a strong coupling to applications. Algorithm theory with its process of subsequent modelling, design, and analysis is not a competing approach to algorithmics but an important ingredient of algorithm engineering.

1 Introduction

Algorithms and data structures are at the heart of every computer application and thus of decisive importance for permanently growing areas of engineering, economy, science, and daily life. The subject of *Algorithmics* is the systematic development of efficient algorithms and therefore has pivotal influence on the effective development of reliable and resource-conserving technology. We only mention a few spectacular examples.

Fast search in the huge data space of the internet (e.g. using Google) has changed the way we handle knowledge. This was made possible with full-text search algorithms that are harvesting matching information out of petabytes of data within fractions of a second and by ranking algorithms that process graphs with billions of nodes in order to filter *relevant information* out of heaps of result data. Less visible yet similarly important are algorithms for the efficient distribution and caching of frequently accessed data under massive load fluctuations or even distributed denial of service attacks.

One of the most far-reaching results of the last years was the ability to read the human genome. Algorithmics was decisive for the early success of this project [1]. Rather than just processing the data coming our of the lab, algorithmic considerations shaped the implementation of the applied shotgun sequencing process.

The list of areas where sophisticated algorithms play a key role could be arbitrarily continued: computer graphics, image processing, geographic information systems, cryptography, planning in production, logistics and transportation,...

---

* Partially supported by DFG grant SA 933/4-1.

S. Albers, H. Alt, and S. Näher (Eds.): Festschrift Mehlhorn, LNCS 5760, pp. 321–340, 2009.

How is algorithmic innovation transferred to applications? Traditionally, algorithmics used the methodology of *algorithm theory* which stems from mathematics: algorithms are designed using simple models of problem and machine. Main results are provable performance guarantees for all possible inputs. This approach often leads to elegant, timeless solutions that can be adapted to many applications. The hard performance guarantees lead to reliably high efficiency even for types of inputs that were unknown at implementation time. From the point of view of algorithm theory, taking up and implementing an algorithmic idea is part of application development. Unfortunately, it can be universally observed that this mode of transferring results is a slow process. With growing requirements for innovative algorithms, this causes growing gaps between theory and practice: Realistic hardware with its parallelism, memory hierarchies etc. is diverging from traditional machine models. Applications grow more and more complex. At the same time, algorithm theory develops more and more elaborate algorithms that may contain important ideas but are usually not directly implementable. Furthermore, real-world inputs are often far away from the worst case scenarios of the theoretical analysis. In extreme cases, promising algorithmic approaches are neglected because a mathematical analysis would be difficult.

Since the early 1990s it therefore became more and more apparent that algorithmics cannot restrict itself to theory. So, what else should algorithmicists do? *Experiments* play a pivotal here. Algorithm engineering (AE) is therefore sometimes equated with *experimental algorithmics*. However, in this paper we argue that this view is too limited. First of all, to do experiments, you also have to *implement* algorithms. This is often equally interesting and revealing as the experiments themselves, needs its own set of techniques, and is an important interface to software engineering. Furthermore, it makes little sense to view design and analysis on the one hand and implementation and experimentation on the other hand as separate activities. Rather, a feedback loop of design, analysis, implementation, and experimentation that leads to new design ideas materializes as the central process of algorithmics.

This cycle is quite similar to the cycle of theory building and experimental validation in Popper's scientific method [2]. We can learn several things from this comparison. First, this cycle is driven by *falsifiable hypotheses* validated by experiments – an experiment cannot prove a hypothesis but it can support it. However, such support is only meaningful if there are conceivable outcomes of experiments that prove the hypothesis wrong. Hypotheses can come from creative ideas or result from *inductive reasoning* stemming from previous experiments. Thus we see a fundamental difference to the *deductive reasoning* predominant in algorithm theory. Experiments have to be *reproducible*, i.e., other researchers have to be able to repeat an experiment to the extent that they draw the same conclusions or uncover mistakes in the previous experimental setup.

There are further aspects of AE as a methodology for algorithmics, outside the main cycle. Design, analysis and evaluation of algorithms are based on some *model* of the problem and the underlying machine. Since gaps between theory and practice often relate to these models, they are an important aspect of AE.

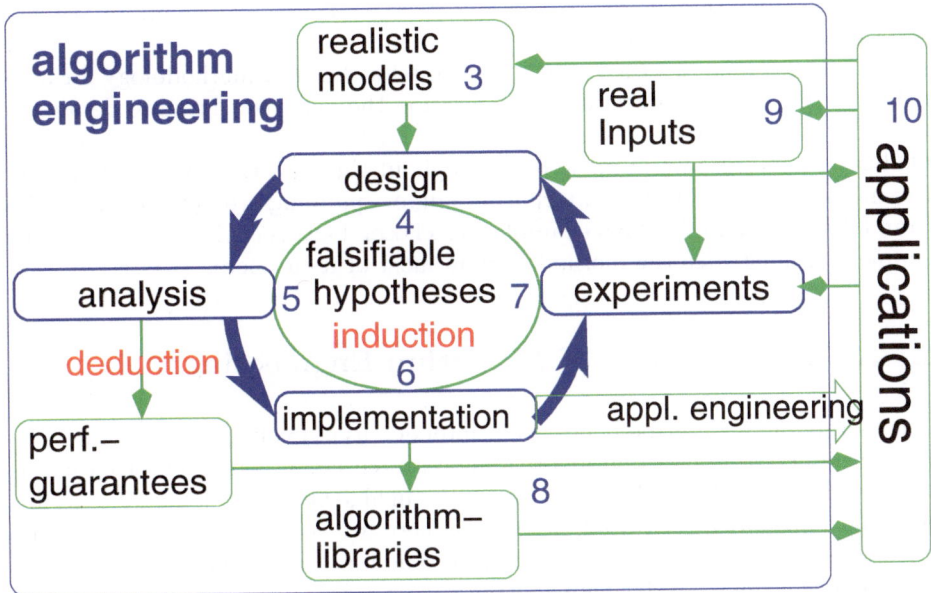

Fig. 1. Algorithm engineering as a cycle of design, analysis, implementation, and experimental evaluation driven by falsifiable hypotheses. The numbers refer to sections.

Since we aim at practicality, *applications* are an important aspect. However we choose to view applications as being outside the methodology of AE since it would otherwise become too open ended and because often one algorithm can be used for quite diverse applications. Also, every new application will have its own requirements and techniques some of which may be abstracted away for algorithmic treatment. Still, in order to reduce gaps between theory and practice, as many interactions as poissible between the application and the activities of AE should be taken into account: Applications are the basis for *realistic* models, they influence the kind of analysis we do, they put constraints on useful implementations, and they supply *realistic inputs* and other design parameters for experiments. On the other hand, the results of analysis and experiments influence the way an algorithm is used (fast enough for real time or interactive use?,...) and implementations may be the basis for software used in applications. Indeed, we may view *application engineering* as a separate process living in both AE and a concrete application domain where methods from both areas are used to adapt an algorithm to a particular application. Applications engineering bridges remaining unavoidable gaps between experimental implementations and production quality code. Note that there are important differences between these two kinds of code: fast development, efficiency, and instrumentation for experiments are very important for AE, while thorough testing, maintainability, simplicity, and tuning for particular classes of inputs are more important for the applications. Furthermore, the algorithm engineers may not even know all the applications for which their algorithms will be used. Hence, *algorithm libraries* of highly tested codes with clear simple user interfaces are an important link between AE and applications.

Figure 1 summarizes the resulting schema for AE as a methodology for algorithmics. The following sections will describe the activities in more detail. We give examples of challenges and results that are a more or less random sample biased to results we know well. Throughout this paper, we will demonstrate the methodology using the external minimum spanning tree (MST) algorithm from [3] as an example. This example was chosen because it is at the same time simple and illustrates the methodology in most of its aspects.

2   A Brief "History" of Algorithm Engineering

The methodology described here is not intended as a revolution but as a description of observed practices in algorithmic research being compiled into a consistent methodology. Basically, all the activities in algorithm development described here have probably been used as long as there are computers. However, in the 1970s and 1980s algorithm theory had become a subdiscipline of computer science that was almost exclusively devoted to "paper and pencil" work. Except for a few papers around D. Johnson, the other activities were mostly visible in application papers, in operations research, or J. Bentley's programming pearls column in *Communications of the ACM*. In the late 1980s, people within algorithm theory began to notice increasing gaps between theory and practice leading to important activities such as the Library of Efficient Data Types and Algorithms (LEDA, since 1988) by K. Mehlhorn and S. Näher and the DIMACS implementation challenges (`http://dimacs.rutgers.edu/Challenges/`). It was not before the end of the 1990s that several workshops series on experimental algorithmics and algorithm engineering were started.[1] There was a Dagstuhl workshop in 2000 [4], and several overview papers on the subject were published [5, 6, 7, 8, 9].

The term "algorithm engineering" already appears 1986 in the Foreword of [10] and 1989 in the title of [11]. No discussion of the term is given. At the same time T. Beth started an initiative to move the CS department of the University of Karlsruhe more into the direction of an engineering discipline. For example, a new compulsory graduate-level course on algorithms was called "Algorithmentechnik" which can be translated as "algorithm engineering". Note that the term "engineering" like in "mechanical engineering" means the *application* oriented use of science whereas our current interpretation of algorithm engineering has applications not as its sole objective but equally strives for general scientific insight as in the natural sciences. However, in daily work the difference will not matter much.

P. Italiano organized the "Workshop on Algorithm Engineering" in 1997 and also uses "algorithm engineering" as the title for the algorithms column of EATCS

---

[1] The Workshop on Algorithm Engineering (WAE) is not the engineering track of ESA. The Alex workshop first held in Italy in 1998 is now the ALENEX workshop held in conjuction with SODA. WEA, now SEA was first organized in 2002.

in 2003 [12] with the following short abstract: "Algorithm Engineering is concerned with the design, analysis, implementation, tuning, debugging and experimental evaluation of computer programs for solving algorithmic problems. It provides methodologies and tools for developing and engineering efficient algorithmic codes and aims at integrating and reinforcing traditional theoretical approaches for the design and analysis of algorithms and data structures." Independently but with the same basic meaning, the term was used in the influential policy paper [5]. The present paper basically follows the same line of argumentation attempting to work out the methodology in more detail and providing a number of hopefully interesting examples.

3   Models

A big difficulty for defining models for problems and machines is that (apparently) only complex models are adequate images of reality whereas only simple models lead to simple, widely usable, portable, and analyzable algorithms. Therefore, AE must simultaneously and carefully abstract from application problems and refine theoretical models.

A successful example for a machine model is the external memory model (or I/O model) [13, 14, 15] which is a careful refinement of the von Neumann model [16]. Instead of a uniform memory, there are two levels of memory. A fast memory of limited size M and and a slow memory that is accessed in blocks of size B. While only counting I/O steps in this model can become a highly theoretical game, we get an abstraction useful for AE if we additionally take internal work into account and if we are careful to use the right values for the parameters M and B^2. Algorithms good in the I/O model are often good in practice although the model oversimplifies aspects like rotational delays, seek time, disk data density depending on the track use, cache replacement strategies [17], flexible block sizes, etc. Sometimes it would even be counterproductive to be too clever. For example, a program carefully tuned to minimize rotational delays and seek time might experience severe performance degradation as soon as another application accesses the disk.

An practical modelling issue that will be important for our MST example is the maximal reasonable size for the external memory. In the last decades, the cost ratio between disk memory and RAM has remained at around 200. This ratio is not likely to increase dramatically as long as RAM and hard disk capacities improve at a similar pace. Hence, in a *balanced* system with similar investments for both levels of memory, the ratio between input size and internal memory size is not huge. In particular, the *logarithm* of this ratio is bounded by a fairly small constant.

---

[2] A common pitfall when applying the I/O model to disks is to look for a natural *physical* block size. This can lead to values (e.g. the size of 512 byte for a decoding unit) that are four orders of magnitude from the value that should be chosen – a value where data transfer takes about as long as the average latency for a small block.

The I/O model has been successfully generalized by adding parameters for the number of disks D, number of processors P, or by looking at the cache-oblivious case [18] where the parameters M and B are not known to the program.

An example for application modelling is the simulation of traffic flows. While microscopic simulations that take the actual behavior of every car into account are currently limited to fairly small subnetworks, it may soon become possible to simulate an entire country by only looking at the paths taken by each car.

4   Design

As in algorithm theory, we are interested in efficient algorithms. However, in AE, it is equally important to look for simplicity, implementability, and possibilities for code reuse. Furthermore, efficiency means not just asymptotic worst case efficiency, but we also have to look at the constant factors involved and at the performance for real-world inputs. In particular, some theoretically efficient algorithms have similar best case and worse case behavior whereas the algorithms used in practice perform much better on all but contrived examples. An interesting example are maximum flow algorithms where the asymptotically best algorithm [19] is much worse than theoretically inferior algorithms [20, 21].

We now present a similar, yet simpler example. Consider an undirected connected graph G with n nodes and m edges. Edges have nonnegative weights. An MST of G is a subset of edges with minimum total weight that forms a spanning tree of G. The MST problem can be solved in $\mathcal{O}(\text{sort}(m))$ expected I/O steps [22] where $\text{sort}(N) = \mathcal{O}(N/B \log_{M/B} N/B)$ denotes the number of I/O steps required for external sorting [13]. There is also a deterministic algorithm that requires $\mathcal{O}(\text{sort}(m) \lceil \log \log(nB/m) \rceil)$ I/Os [23].

However, before [3] there was no actual implementation of an external MST algorithm (or for any other nontrivial external graph problem). The reason was that previous algorithms were complicated to implement and have large constant factors that have never been exposed in the analysis. We therefore designed a new algorithm.

The base case of our algorithm is a simple *semiexternal* variant of Kruskal's algorithm [22] (A *semiexternal* graph algorithm is allowed $\mathcal{O}(n)$ words of fast memory): The edges are sorted (externally) by weight and scanned in sorted order. An edge is accepted into the MST if it connects two components of the forest defined by the previously found MST edges. This decision is supported by an internal memory union-find data structure. Even this simple algorithm is a good example for algorithm reuse since it can call highly tuned external sorting codes such as the routine in the external implementation of the STL, STXXL [24]. The *pipelining* facility of the STXXL saves up to 2/5 of the I/Os by directly feeding the sorted output into the final scan. For semiexternal algorithms, constant factors are particularly important for the space consumption in fast memory. Therefore, we developed a variant of the union-find data structure with path compression and union-by-rank [25] that needs only $\lceil \log(n + 1 + \log n) \rceil \approx \log n$

bits for each node. The trick is that root nodes of the data structure need rank information while only non-root nodes need parent information. Since ranks are at most $\log n$, the values $n..n + \log n$ can be reserved for rank information.

If $n > M$, all known external MST algorithms rely on a method for reducing the number of nodes. Our algorithmically most interesting contribution is *Sibeyn's* algorithm for node reduction based on the technique of *time forward processing*. The most abstract form of Sibeyn's algorithm is very simple. In each iteration, we remove a random node u from the graph. We find the lightest edge $\{u, v\}$ incident to u. By the well known cut-property that underlies most MST algorithms, $\{u, v\}$ must be an MST edge. So, we output $\{u, v\}$, remove it from E, and *contract* it, i.e., all other edges $\{u, w\}$ incident to u are replaced by edges $\{v, w\}$. If we store the original identity of each edge, we can reconstruct the MST from the edges that are output.

We transform the algorithm into a *sweeping algorithm* by renumbering the nodes using a random permutation π and then removing the nodes in the order $n..M$. When this is finished, the remaining problem can be solved using the semiexternal algorithm.

There is a very simple external realization of Sibeyn's algorithm based on priority queues of edges. Edges are stored in the form $((u, v), c, e_{old})$ where (u, v) is the edge in the current graph, c is the edge weight, and e_{old} identifies the edge in the original graph. The queue normalizes edges (u, v) in such a way that $u \geq v$. We define a priority order $((u, v), c, e_{old}) < ((u', v'), c', e'_{old})$ iff $u > u'$ or $u = u'$ and $c < c'$. With these conventions in place, the algorithm can be described using the simple pseudocode in Figure 2. This algorithm is not only conceptually simple but also easy to implement because it can again reuse software by relying on the external priority queue in STXXL [24]. Note that while a sophisticated external priority queue needs thousands of lines of code, the actual implementation of Figure 2 is not much longer than the pseudo code.

```
ExternalPriorityQueue: Q
foreach (e = (u, v), c) ∈ E do Q.insert(((π(u), π(v)), c, e))     -- rename
currentNode := -1                          -- node currently being removed
i := n                                     -- number of remaining nodes
while i > n' do
    ((u, v), c, e_old) := Q.deleteMin()
    if u ≠ currentNode then                -- lightest edge out of a new node
        currentNode := u                   -- node u is removed
        i--
        relinkTo := v
        output e_old                                            -- MST edge
    elsif v ≠ relinkTo then Q.insert((v, relinkTo), c, e_old)   -- relink non-self-loops
```

Fig. 2. An external implementation of Sibeyn's algorithm using a priority queue

5   Analysis

Even simple and proven practical algorithms are often difficult to analyze and this is one of the main reasons for gaps between theory and practice. Thus, the analysis of such algorithms is an important aspect of AE. For example, randomized algorithms are often simpler and faster than their best deterministic competitors but even simple randomized algorithms are often difficult to analyze.

Many complex optimization problems are attacked using *meta heuristics* like (randomized) local search or evolutionary algorithms. Algorithms of this type are simple and easily adaptable to the problem at hand. However, only very few such algorithms have been successfully analyzed (e.g. [26]) although performance guarantees would be of great theoretical and practical value.

An important open problem is partitioning of graphs into approximately equal sized blocks such that few edges are cut. This problem has many applications, e.g., in scientific computing. Currently available algorithms with performance guarantees are too slow for practical use. Practical methods first contract the graph while preserving its basic structure until only few nodes are left, compute an initial solution on this coarse representation, and then improve by local search. These algorithms, e.g., [27] are very fast and yield good solutions in many situations yet no performance guarantees are known.

An even more famous example for local search is the simplex algorithm for linear programming. Simple variants of the simplex algorithm need exponential time for specially constructed inputs. However, in practice, a linear number of iterations suffices. So far, only subexponential expected runtime bounds are known – for inpracticable variants. However, Spielmann and Teng were able to show that even small random perturbations of the coefficients of a linear program suffice to make the expected run time of the simplex algorithm polynomial [28]. This concept of *smoothed analysis* is a generalization of *average case analysis* and an interesting tool of AE also outside the simplex algorithm. Beier and Vöcking were able to show polynomial smoothed complexity for an important family of NP-hard problems [29]. For example, this result explains why the knapsack problem can be efficiently solved in practice and has also helped to improve the best knapsack solvers. There are interesting interrelations between smoothed complexity, approximation algorithms, and pseudopolynomial algorithms that is also an interesting approach to practical solutions of NP-hard problems.

Our randomized MST edge reduction algorithm is actually quite easy to analyze.

Theorem 1. *The expected number of edges inspected by the abstract algorithm until the number of nodes is reduced to n' is bounded by $2m \ln \frac{n}{n'}$.*

Proof. In the iteration when i nodes are left (note that $i = n$ in the first iteration), the expected degree of a random node is at most $2m/i$. Hence, the expected number of edges, X_i, inspected in iteration i is at most $2m/i$. By the linearity of expectation, the total expected number of edges processed is

$$\sum_{n'<i\le n} \mathbb{E}\left[X_i\right] \le \sum_{n'<i\le n} \frac{2m}{i} = 2m \sum_{n'<i\le n} \frac{1}{i} = 2m \left(\sum_{1\le i\le n} \frac{1}{i} - \sum_{1\le i\le n'} \frac{1}{i} \right)$$

$$= 2m(H_n - H_{n'}) \le 2m(\ln n - \ln n') = 2m \ln \frac{n}{n'}$$

where $H_n = \ln n + 0.577\cdots + \mathcal{O}(1/n)$ is the n-th harmonic number.   ∎

Plugging in the complexity of the priority queue used [30] we obtain an I/O complexity of $\mathcal{O}(\text{sort}(m) \lceil \log(n/M) \rceil)$. This is actually asymptotically *worse* than the previous theoretical algorithms by a factor up to $\log(n/M)$. However, recall from Section 4 that $\log(n/M)$ is a *constant* in balanced machines. A close analysis of the constant factors involved [3] in the theoretical algorithms reveals that all things considered, Sibeyn's algorithm needs a factor at least *four less* I/Os in realistic situations. Hence, our MST example exemplifies that closer looks at constant factors are an important aspect of algorithm analysis in AE and that constant factors can beat asymptotic behavior.

Sibeyn's algorithm is also a good example for the importance of looking at non-worst case instances. It turns out that for planar graphs the factor $\log(n/M)$ is not needed since planar graphs remain planar under edge contraction and thus we always have constant average degree if we a careful enough to collapse parallel edges.

An MST algorithm can also be used to find connected components [31]. Since edges have no weights now, we are free to choose any edge. Choosing the edge leading to the node with smallest index actually looks like a good idea since this measure delays reconsidering the relinked edges. It looks like this should reduce the "suboptimality" of the algorithm to $\log \log(n/M)$. However, a full analysis remains an open problem[3]. This is an example for a simple randomized algorithm that is difficult to analyze because there are subtle dependencies to be taken into account.

6   Implementation

Implementation only appears to be the most clearly prescribed and boring activity in the cycle of AE. One reason is that there are huge semantic gaps between abstractly formulated algorithms, imperative programming languages, and real hardware. A typical example for this semantic gap is the implementation of an $\mathcal{O}(nm \log n)$ matching algorithm in [32]. Its abstract description requires a sophisticated data structure whose efficient implementation only succeeded in [32].

An extreme example for the semantic gap are geometric algorithms which are often designed assuming exact arithmetics with real numbers and without considering degenerate cases. The robustness of geometric algorithms has therefore become an important branch of AE [33, 34, 35].

Even the implementation of relatively simple basic algorithms can be challenging. You often have to compare several candidates based on small constant

---

[3] In [31] there is no proof of the stated bounds.

factors in their execution time. Since even small implementation details can make a big difference, the only reliable way is to highly tune all competitors and run them on several architectures. It can even be advisable to compare the generated machine code (e.g. [30, 36], [37]).

Often only implementations give convincing evidence of the correctness and result quality of an algorithm. For example, an algorithm for planar embedding [38] was the standard reference for 20 years although this paper only contains a vague description how an algorithm for planarity testing can be generalized. Several attempts at a more detailed description contained errors (e.g. [39]). This was only noticed during the first correct implementation [40]. Similarly, for a long time nobody suceeded in implementing the famous algorithm for computing three-connected components from [41]. Only an implementation in 2000 [42] uncovered and corrected an error. For the related problem of computing a maximal planar subgraph there was a series of publications in prominent conferences uncovering errors in the previous paper and introducing new ones – until it turned out that the proposed underlying data structure is inadequate for the problem [43].

An important consequence for planning AE projects is that important implementations cannot usually be done as bachelor or master theses but require the very best students or long term attendance by full time researchers or scientific programmers.

Our MST code was implemented as a Bachelor thesis [44], however by one of the best programmers I have seen and reusing tens of thousands of lines of code from the STXXL that were the basis for a PhD thesis [45]. A particular challenge was the exploitation of parallel disks since it turned out that the code was compute bound. We obtained a considerable speedup by implementing a special purpose bucket priority queue that exploits the properties the problem: We only sort by the node ID of the larger endpoint. The minimum weight incident edge is found by extracting all incident edges. This is no actual overhead since those edges will later be relinked anyway.

7   Experiments

Meaningful experiments are the key to closing the cycle of the AE process. For example, experiments on crossing minimization in [46] showed that previous theoretical results were too optimistic so that new algorithms became interesting.

Experiments can also have a direct influence on the analysis. For example, reconstructing a curve from a set of measured points is a fundamental variant of an important family of image processing problems. In [47] an apparently expensive method based on the travelling salesman problem is investigated. Experiments indicated that "reasonable" inputs lead to easy instances of the travelling salesman problem. This observation was later formalized and proven. A quite different example of the same effect is the astonishing observation that arbitrary access patterns to data blocks on disk arrays can be almost perfectly balanced when two redundant copies of each block are placed on random disks [48].

Compared to the natural sciences, AE is in the privileged situation that it can perform many experiments with relatively little effort. However, the other side of the coin is highly nontrivial planning, evaluation, archiving, postprocessing, and interpretation of results. The starting point should always be falsifiable hypotheses on the behavior of the investigated algorithms which stem from the design, analysis, implementation, or from previous experiments. The result is a confirmation, falsification, or refinement of the hypothesis. The results complement the analytic performance guarantees, lead to a better understanding of the algorithms, and provide ideas for improved algorithms, more accurate analysis, or more efficient implementation.

Successful experimentation involves a lot of software engineering. Modular implementations allow flexible experiments. Clever use of tools simplifies the evaluation. Careful documentation and version management help with reproducibility – a central requirement of scientific experiments, that is challenging due to the frequent new versions of software and hardware.

Experiments with external memory algorithms are challenging because they require huge inputs and execution times measuring in hours. In particular, when you compare against a *bad* algorithm, running times can easily reach months. Perhaps this is the reason why [3] was the first actual implementations of an external graph algorithm. Many previous implementations of external algorithms relied on artificially restricted main memory sizes to achieve small running times. We believed that this is inacceptable for results intended to convince practitioners to use external algorithms. Our solution was to use carefully configured yet relatively cheap machines that can be dedicated to the experiments for weeks, high performance implementations, and careful planning of experiments.

Our starting point for designing experiments was the study by Moret and Shapiro [49]. We have adopted the instance families for *random* graphs with random edge weights and random *geometric* graphs where random points in the unit square are connected to their d closest neighbors. In order to obtain a simple family of planar graphs, we have added *grid* graphs with random edge weights where the nodes are arranged in a grid and are connected to their (up to) four direct neighbors. We have not considered the remaining instance families in [49] because they define rather dense graphs that would be easy to handle semiexternally or they are specifically designed to fool particular algorithms or heuristics. We have chosen the parameters of the graphs so that m is between $2n$ and $8n$. Considerably denser graphs would be either solvable semiexternally or too big for our machine.

The experiments have been performed on a low cost PC-server (around 3 000 Euro in July 2002) with two 2 GHz Intel Xeon processors, 1 GByte RAM and 4×80 GByte disks (IBM 120GXP) that are connected to the machine in a bottleneck-free way (see [50] for more details on the hardware). This machine runs Linux 2.4.20 using the XFS file system. Swapping was disabled. All programs were compiled with g++ version 3.2 and optimization level -O6. The total computer time spent on the experiments was about 25 days producing a total I/O volume of several dozen Terabytes.

Figure 3 summarizes the results using bucket priority queues. The internal implementations were provided by Irit Katriel [51]. The curves only show the

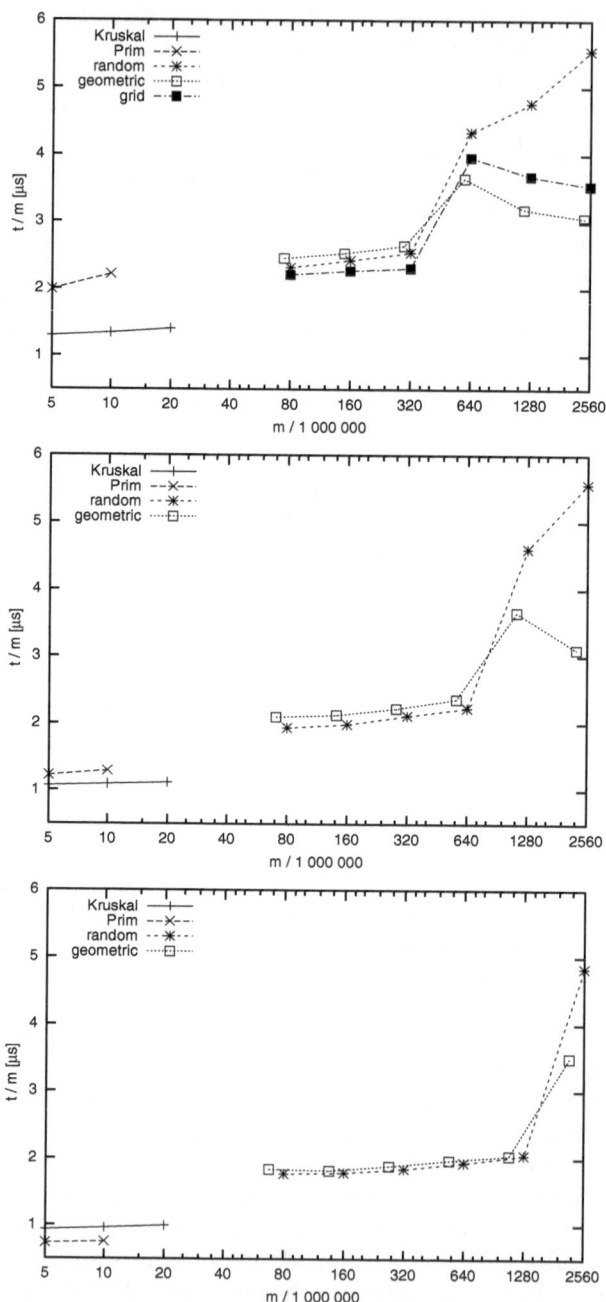

Fig. 3. Execution time per edge for $m \approx 2 \cdot n$ (top), $m \approx 4 \cdot n$ (center), $m \approx 8 \cdot n$ (bottom)

internal results for random graphs — at least Kruskal's algorithm shows very similar behavior for the other graph classes. Our implementation can handle up to 20 million edges. Kruskal's algorithm is best for very sparse graphs ($m \leq 4n$) whereas the Jarník-Prim algorithm (with a fast implementation of pairing heaps) is fastest for denser graphs but requires more memory. For $n \leq 160\,000\,000$, we can run the semiexternal algorithm and get execution times within a factor of two of the internal algorithm. The curves are almost flat and very similar for all three graph families. This is not astonishing since Kruskal's algorithm is not very dependent on the structure of the graph. Beyond $160\,000\,000$ nodes, the full external algorithm is needed. This immediately costs us another factor of two in execution time: We have additional costs for random renaming, node reduction, and increasing the size of an edge from 12 bytes to 20 bytes (for renamed nodes). For random graphs, the execution time keeps growing with n/M as predicted by the upper bound from Theorem 1.

The behavior for grid graphs is much better than predicted by Theorem 1 because planar graphs remain sparse under edge contraction. It is interesting that similar effects can be observed for geometric graphs. This is an indication that it is worth removing parallel edges for many nonplanar graphs. Interestingly, the time per edge *decreases* with m for grid graphs and geometric graphs. The reason is that the time for the semiexternal base case does not increase proportionally to the number of input edges. For example, $5.6 \cdot 10^8$ edges of a grid graph with $640 \cdot 10^6$ nodes survive the node reduction, vs. $6.3 \cdot 10^8$ edges of a grid graph with twice the number of edges.

Another observation is that for $m = 2560 \cdot 10^6$ and random or geometric graphs we get the worst time per edge for $m \approx 4n$. For $m \approx 8n$, we do not need to run the node reduction very long. For $m \approx 2n$ we process less edges than predicted by Theorem 1 even for random graphs simply because one MST edge is removed for each node.

8 Algorithm Libraries

Algorithm libraries are made by assembling implementations of a number of algorithms using the methods of software engineering. The result should be efficient, easy to use, well documented, and portable. Algorithm libraries accelerate the transfer of know-how into applications. Within algorithmics, libraries simplify comparisons of algorithms and the construction of software that builds on them. The software engineering involved is particularly challenging, since the applications to be supported are *unknown* at library implementation time and because the separation of interface and (often highly complicated) implementation is very important. Compared to applications-specific reimplementation, using a library should save development time without leading to inferior performance. Compared to simple, easy to implement algorithms, libraries should improve performance. In particular for basic data structures with their fine-grained coupling between applications and library this can be very difficult. To summarize, the triangle between generality, efficiency, and ease of use leads to challenging tradeoffs because often optimizing one of these aspects will deteriorate the others. It is also worth mentioning that *correctness* of algorithm libraries is even more important than for other software because it is extremely difficult for a user to debug library code that has not been written by his team. Sometimes it is not even sufficient for a library to be correct as long as the user does not *trust* it sufficiently to first look for bugs outside the library. This is one reason why result checking, certifying algorithms, or even formal verification are an important aspect of algorithm libraries. All these difficulties imply that implementing algorithms for use in a library is several times more difficult / expensive / time consuming / frustrating /⋯ than implementations for experimental evaluation. On the other hand, a good library implementation might be *used* orders of magnitude more frequently. Thus, in AE there is a natural mechanism leading to many exploratory implementations and a few selected library codes that build on previous experimental experience.

Let us now look at a few successful examples of algorithm libraries. The Library of Efficient Data Types and Algorithms LEDA [21] has played an important part in the development of AE. LEDA has an easy to use object-oriented C++ interfaces. Besides basic algorithms and data structures, LEDA offers a variety of graph algorithms and geometric algorithms.

Programming languages come with a run-time library that usually offers a few algorithmic ingredients like sorting and various collection data structures (lists, queues, sets, ...). For example, the C++ standard template library (STL) has a very flexible interface based on templates. Since so many things are resolved at compile time, programs that use the STL are often equally efficient as hand-written C-style code even with the very fine-grained interfaces of collection classes. This is one of the reasons why our group is looking at implementations of the STL for advanced models of computation like external computing (STXXL [24]) or multicore parallelism (MCSTL, GNU C++ standard library [52]). We should also mention disadvantages of template based libraries: The more flexible their offered functionality, the more cumbersome it is to use (the upcoming new C++ standard might slightly improve the situation). Perhaps the worst aspect is coping with extra long error messages and debugging code with thousands of tiny inlined functions. Writing the library can be frustrating for an algorithmicist since the code tends to consist mostly of trivial but lengthy declarations while the algorithm itself is shredded into many isolated fragments.

The Boost C++ libraries (www.boost.org) are an interesting concept since they offer a forum for library designers that ensures certain quality standards and offers the possibility of a library to become part of the C++ standard.

The Computational Geometry Algorithms Library (CGAL) www.cgal.org that is a joined effort of several AE groups is perhaps one of the most sophisticated examples of C++ template programming. In particular, it offers many *robust* implementations of geometric algorithms that are also efficient. This is achieved for example by using floating point interval arithmetics most of the time and switching to exact arithmetics only when a (near)-degenerate situation is detected. The mechanisms of template programming make it possible to hide much of these complicated mechanisms behind special number types that can be used in a similar way as floating point numbers.

As already mentioned, our external MST algorithm successfully uses the STXXL thus importing much of its code from a library. This is particularly true for the priority queue based implementation that has acceptable performance when using a single disk.

9 Instances

Collections of realistic problem instances for benchmarking have proven crucial for improving algorithms. There are interesting collections for a number of NP-hard problems like the travelling salesman problem [4], the Steiner tree problem, satisfiability, set covering, or graph partitioning. In particular for the first three problems the benchmarks have helped enable astonishing breakthroughs. Using deep mathematical insights into the structure of the problems one can now compute optimal solutions even for large, realistic instances of the travelling salesman problem [53] and of the Steiner tree problem [54]. It is a bit odd that similar benchmarks for problems that are polynomially solvable are sometimes more difficult to obtain. For route planning in road networks, realistic inputs have become available in 2005 [55] enabling a revolution with speedups of up to six orders of magnitude over Dijkstra's algorithm and a perspective for many applications [56]. In string algorithms and data compression, real-world data is also no problem. But for many typical graph problems like flows, random inputs are still common practice. We suspect that this often leads to unrealistic results in experimental studies. Naively generated random instances are likely to be either much easier or more difficult than realistic inputs. With more care and competition, such as for the DIMACS implementation challenges, generators emerge that drive naive algorithms into bad performance. While this process can lead to robust solutions, it may overemphasize difficult inputs. Another area with lack of realistic input collections are data structures. Apart from some specialized scenarios like IP address lookup, few inputs are available for hash tables, search trees, or priority queues.

Our MST example lives on the dark side of the world of problem instance collections. This is a (moderate) risk for evaluating Sibeyn's algorithm since it is not clear how the density of the graph behaves in practice and whether nodes with very high degree might emerge during the computation. But even simple internal algorithms have such input dependencies: Kruksal's algorithm is much faster if we can use bucket sorting and the Jarník–Prim algorithm suffers when a lot of decrease-key operations are needed. Some literature/web search revealed that there is no lack of actual applications of the MST problem. Interestingly, clustering by removing MST edges seems to be more important than the classical network design motivation. However, it was difficult to find applications where a) huge inputs look important, b) the inputs are sparse (otherwise semiexternal algorithms are fine), and, c) *generating* the input is faster than finding the MST.

---

[4] http://www.iwr.uni-heidelberg.de/groups/comopt/software/TSPLIB95/

Later, our code turned out to be a useful tool for implementing external breadth-first-search and shortest path computations [57].

10   Applications

We could discuss many important applications where algorithms play a major role and a lot of interesting work remains to be done. Since this would go beyond the scope of this paper, we only want to mention a few: Bioinformatics (e.g. sequencing, folding, docking, phylogenetic trees, DNA chip evaluations, reaction networks); information retrieval (indexing, ranking); algorithmic game theory; traffic information, simulation and planning for cars, busses, trains, and air traffic; geographic information systems; communication networks; machine learning; real time scheduling.

An example for application engineering is recently started work on MSTs for image segmentation where satellite images define huge grid graphs. Here, aggressive exploitation of the special structure of the problems leads away from Sibeyn's algorithm. We exploit the simple 2D structure of the inputs and their small integer edge weights and also use parallelism. Furthermore, processing the edges in sorted order allows identifying the segments in a single pass. All this led a to a highly specialized parallel variant of Kruskal's algorithm.

The effort for implementing algorithms for a particular application usually lies somewhere between the effort for experimental evaluation and for algorithm libraries depending on the context.

An important goal for AE should be to help shaping the applications (as in the example for genome sequencing mentioned in the introduction) rather than act as an ancillary science for other disciplines like physics, biology, mechanical engineering,...

11   Conclusions

We hope to have demonstrated that AE is a "round" methodology for the development of efficient algorithms which simplifies their practical use. We want to stress, however, that it is not our intention to abolish algorithm theory. The saying that there is nothing as practical as good theory remains true for algorithmics because an algorithm with proven performance guarantees has a degree of generality, reliability, and predictability that cannot be obtained with any number of experiments. However, this does not contradict the proposed methodology since it views algorithm theory as a subset of AE, making it even more rich by asking additional interesting kinds of questions (e.g. simplicity of algorithms, care for constant factors, smoothed analysis,...). We also have no intention of criticizing some highly interesting research in algorithm theory that is less motivated from applications than by fundamental questions of theoretical computer science such as computability or complexity theory. However, we do want to criticize those papers that begin with a vague claim of relevance to some fashionable application area before diving deep into theoretical constructions that look completely irrelevant for the claimed application. Often this is not intentionally misleading but more like a game of "Chinese whispers" where a research area starts as a sensible abstraction of an application area but then develops a life of itself, mutating into a mathematical game with its own rules. Even this can be interesting but researchers should constantly ask themselves why they are working on an area, whether there are perhaps other areas where they can have larger impact on the world, and how false claims for practicality can damage the reputation of algorithmics in practical computer science.

Acknowledgements

I would like to thank the coinitiators of the DFG SPP 1307, Kurt Mehlhorn, Rolf Möhring, Burkhard Monien, and Petra Mutzel for their advice and fruitful discussions that led to the definition presented here. I would like to thank Jop Sibeyn for his elegant MST algorithm and many other insights, Roman Dementiev for his excellent work on STXXL, and Dominik Schultes for the very good implementation of Sibeyn's algorithm. Discussions with many other colleagues, in particular with Rudolf Fleischer have helped to shape my view of algorithm engineering.

References

1. Venter, J.C., Adams, M.D., Mayers, E.W., et al.: The sequence of the human genome. Science 291(5507), 1304–1351 (2001)
2. Popper, K.R.: Logik der Forschung. Springer (1934) English Translation: The Logic of Scientific Discovery, Hutchinson (1959)
3. Dementiev, R., Sanders, P., Schultes, D., Sibeyn, J.: Engineering an external memory minimum spanning tree algorithm. In: IFIP TCS, Toulouse (2004)
4. Fleischer, R., Moret, B., Schmidt, E.M.: Experimental Algorithmics. LNCS, vol. 2547. Springer, Heidelberg (2002)
5. Aho, A.V., Johnson, D.S., Karp, R.M., Kosaraju, S.R., McGeoch, C.C., Papadimitriou, C.H., Pevzner, P.: Emerging opportunities for theoretical computer science. SIGACT News 28(3), 65–74 (1997)
6. Moret, B.M.E.: Towards a discipline of experimental algorithmics. In: 5th DIMACS Challenge. DIMACS Monograph Series (2000) (to appear)
7. McGeoch, C.C., Precup, D., Cohen, P.R.: How to find big-oh in your data set (and how not to). In: Liu, X., Cohen, P.R., Berthold, M.R. (eds.) IDA 1997. LNCS, vol. 1280, pp. 41–52. Springer, Heidelberg (1997)
8. McGeoch, C., Moret, B.M.E.: How to present a paper on experimental work with algorithms. SIGACT News 30(4), 85–90 (1999)
9. Johnson, D.S.: A theoretician's guide to the experimental analysis of algorithms. In: Goldwasser, M., Johnson, D.S., McGeoch, C.C. (eds.) Proceedings of the 5th and 6th DIMACS Implementation Challenges. American Mathematical Society (2002)
10. Beth, T., Clausen, M. (eds.): AAECC 1986. LNCS, vol. 307. Springer, Heidelberg (1988)
11. Beth, T., Gollman, D.: Algorithm engineering for public key algorithms. IEEE Journal on Selected Areas in Communications 7(4), 458–466 (1989)

12. Demetrescu, C., Finocchi, I., Italiano, G.F.: Algorithm engineering, algorithmics column. Bulletin of the EATCS 79, 48–63 (2003)
13. Aggarwal, A., Vitter, J.S.: The input/output complexity of sorting and related problems. Communications of the ACM 31(9), 1116–1127 (1988)
14. Vitter, J.S., Shriver, E.A.M.: Algorithms for parallel memory, I: Two level memories. Algorithmica 12(2/3), 110–147 (1994)
15. Meyer, U., Sanders, P., Sibeyn, J. (eds.): Algorithms for Memory Hierarchies. LNCS, vol. 2625. Springer, Heidelberg (2003)
16. von Neumann, J.: First draft of a report on the EDVAC. Technical report, University of Pennsylvania (1945)
17. Mehlhorn, K., Sanders, P.: Scanning multiple sequences via cache memory. Algorithmica 35(1), 75–93 (2003)
18. Frigo, M., Leiserson, C.E., Prokop, H., Ramachandran, S.: Cache-oblivious algorithms. In: 40th Symposium on Foundations of Computer Science, pp. 285–298 (1999)
19. Goldberg, A.V., Rao, S.: Beyond the flow decomposition barrier. Journal of the ACM 45(5), 1–15 (1998)
20. Cherkassky, B.V., Goldberg, A.V.: On implementing push-relabel method for the maximum flow problem. In: Balas, E., Clausen, J. (eds.) IPCO 1995. LNCS, vol. 920. Springer, Heidelberg (1995)
21. Mehlhorn, K., Näher, S.: The LEDA Platform of Combinatorial and Geometric Computing. Cambridge University Press, Cambridge (1999)
22. Abello, J., Buchsbaum, A., Westbrook, J.: A functional approach to external graph algorithms. Algorithmica 32(3), 437–458 (2002)
23. Arge, L., Brodal, G., Toma, L.: On external memory MST, SSSP and multi-way planar graph separation. In: Halldórsson, M.M. (ed.) SWAT 2000. LNCS, vol. 1851, pp. 433–447. Springer, Heidelberg (2000)
24. Dementiev, R., Kettner, L., Sanders, P.: STXXL: Standard Template Library for XXL data sets. Software Practice & Experience 38(6), 589–637 (2008)
25. Tarjan, R.E.: Efficiency of a good but not linear set merging algorithm. Journal of the ACM 22, 215–225 (1975)
26. Wegener, I.: Simulated annealing beats metropolis in combinatorial optimization. In: Caires, L., Italiano, G.F., Monteiro, L., Palamidessi, C., Yung, M. (eds.) ICALP 2005. LNCS, vol. 3580, pp. 589–601. Springer, Heidelberg (2005)
27. Karypis, G., Kumar, V.: Multilevel k-way partitioning scheme for irregular graph. J. Parallel Distrib. Comput. 48(1) (1998)
28. Spielman, D., Teng, S.H.: Smoothed analysis of algorithms: why the simplex algorithm usually takes polynomial time. In: 33rd ACM Symposium on Theory of Computing, pp. 296–305 (2001)
29. Beier, R., Vöcking, B.: Typical properties of winners and losers in discrete optimization. In: 36th ACM Symposium on the Theory of Computing, pp. 343–352 (2004)
30. Sanders, P.: Fast priority queues for cached memory. ACM Journal of Experimental Algorithmics 5 (2000)
31. Sibeyn, J.F.: External connected components. In: Hagerup, T., Katajainen, J. (eds.) SWAT 2004. LNCS, vol. 3111, pp. 468–479. Springer, Heidelberg (2004)
32. Mehlhorn, K., Schäfer, G.: Implementation of weighted matchings in general graphs: The power of data structure. ACM Journal of Experimental Algorithmics 7 (2002)

33. Burnikel, C., Könemann, J., Mehlhorn, K., Näher, S., Schirra, S., Uhrig, C.: Exact geometric computation in leda. In: SCG 1995: 11th annual symposium on Computational geometry, pp. 418–419. ACM, New York (1995)

34. Berberich, E., Eigenwillig, A., Hemmer, M., Hert, S., Kettner, L., Mehlhorn, K., Reichel, J., Schmitt, S., Schömer, E., Wolpert, N.: EXACUS: Efficient and exact algorithms for curves and surfaces. In: Brodal, G.S., Leonardi, S. (eds.) ESA 2005. LNCS, vol. 3669, pp. 155–166. Springer, Heidelberg (2005)

35. Bast, H., Funke, S., Sanders, P., Schultes, D.: Fast routing in road networks with transit nodes. Science 316(5824), 566 (2007)

36. Sanders, P., Winkel, S.: Super scalar sample sort. In: Albers, S., Radzik, T. (eds.) ESA 2004. LNCS, vol. 3221, pp. 784–796. Springer, Heidelberg (2004)

37. Brodal, G.S., Fagerberg, R., Vinther, K.: Engineering a cache-oblivious sorting algorithm. In: 6th Workshop on Algorithm Engineering and Experiments (2004)

38. Hopcroft, J., Tarjan, R.E.: Efficient planarity testing. J. of the ACM 21(4), 549–568 (1974)

39. Mehlhorn, K.: Data Structures and Algorithms. EATCS Monographs on Theoretical CS, vol. I — Sorting and Searching. Springer, Heidelberg (1984)

40. Mehlhorn, K., Mutzel, P.: On the embedding phase of the Hopcroft and Tarjan planarity testing algorithm. Algorithmica 16(2), 233–242 (1996)

41. Hopcroft, J.E., Tarjan, R.E.: Dividing a graph into triconnected components. SIAM J. Comput. 2(3), 135–158 (1973)

42. Gutwenger, C., Mutzel, P.: A linear time implementation of SPQR-trees. In: Marks, J. (ed.) GD 2000. LNCS, vol. 1984, pp. 77–90. Springer, Heidelberg (2001)

43. Jünger, M., Leipert, S., Mutzel, P.: A note on computing a maximal planar subgraph using PQ-trees. IEEE Transactions on Computer-Aided Design 17(7), 609–612 (1998)

44. Schultes, D.: External memory minimum spanning trees. Bachelor thesis, Max-Planck-Institut f. Informatik and Saarland University (2003), http://algo2.iti.uni-karlsruhe.de/schultes/emmst/

45. Dementiev, R.: Algorithm Engineering for Large Data Sets. PhD thesis, Saarland University (2006)

46. Jünger, M., Mutzel, P.: 2-layer straightline crossing minimization: Performance of exact and heuristic algorithms. Journal of Graph Algorithms and Applications (JGAA) 1(1), 1–25 (1997)

47. Althaus, E., Mehlhorn, K.: Traveling salesman-based curve reconstruction in polynomial time. SIAM Journal on Computing 31(1), 27–66 (2002)

48. Sanders, P., Egner, S., Korst, J.: Fast concurrent access to parallel disks. In: 11th ACM-SIAM Symposium on Discrete Algorithms, pp. 849–858 (2000)

49. Moret, B.M.E., Shapiro, H.D.: An empirical assessment of algorithms for constructing a minimum spanning tree. DIMACS Series in Discrete Mathematics and Theoretical Computer Science 15, 99–117 (1994)

50. Dementiev, R., Sanders, P.: Asynchronous parallel disk sorting. In: 15th ACM Symposium on Parallelism in Algorithms and Architectures, San Diego, pp. 138–148 (2003)

51. Katriel, I., Sanders, P., Träff, J.L.: A practical minimum spanning tree algorithm using the cycle property. In: Di Battista, G., Zwick, U. (eds.) ESA 2003. LNCS, vol. 2832, pp. 679–690. Springer, Heidelberg (2003)

52. Singler, J., Sanders, P., Putze, F.: MCSTL: The multi-core standard template library. In: Kermarrec, A.-M., Bougé, L., Priol, T. (eds.) Euro-Par 2007. LNCS, vol. 4641, pp. 682–694. Springer, Heidelberg (2007)

53. Applegate, D., Bixby, R., Chvátal, V., Cook, W.: Implementing the Dantzig-Fulkerson-Johnson algorithm for large traveling salesman problems. Math. Programming 97(1-2), 91–153 (2003)
54. Polzin, T., Daneshmand, S.V.: Extending reduction techniques for the Steiner tree problem. In: Möhring, R.H., Raman, R. (eds.) ESA 2002. LNCS, vol. 2461, pp. 795–807. Springer, Heidelberg (2002)
55. Sanders, P., Schultes, D.: Highway hierarchies hasten exact shortest path queries. In: Brodal, G.S., Leonardi, S. (eds.) ESA 2005. LNCS, vol. 3669, pp. 568–579. Springer, Heidelberg (2005)
56. Delling, D., Sanders, P., Schultes, D., Wagner, D.: Engineering route planning algorithms (2008) (submitted for publication),
   http://i11www.ira.uka.de/extra/publications/dssw-erpa-09.pdf
57. Ajwani, D., Dementiev, R., Meyer, U.: A computational study of external-memory BFS algorithms. In: ACM-SIAM Symposium on Discrete Algorithms, pp. 601–610 (2007)

Of What Use Is Floating-Point Arithmetic in Computational Geometry?

Stefan Funke

Institut für Mathematik und Informatik,
Ernst-Moritz-Arndt-Universität Greifswald

Abstract. We give a sketchy and informal overview of the use of floating-point arithmetic in the implementation of geometric algorithms. First we point out the pitfalls of a too naive use of floating-point arithmetic and then talk about less naive ways which do not compromise the sensibility of the outcome. Accidentally, Kurt Mehlhorn and his collaborators had a finger in the pie all the time.

1 Introduction

Geometric algorithms are usually designed and proven to be correct in a computational model that assumes exact computation over the real numbers. Since no computer provides exact arithmetic on real numbers in hardware, programmers must find some substitution when implementing these algorithms. Quite commonly, they resort to fast finite precision arithmetic due to its support by hard- and software as well as its convenient use. For some problems and restricted sets of input data, this approach works well, but in many implementations the effects of squeezing the infinite set of real numbers into the finite set of floating-point numbers can cause catastrophic errors in practice.

There are several ways geometric algorithms may misbehave when exact arithmetic is replaced by floating-point arithmetic. In the best case, they produce quite usable results in spite of some incorrect decisions, but most algorithms do not; they either produce completely inconsistent results, crash or loop.

To give you an idea how easily a simple predicate can be decided incorrectly when replacing exact arithmetic by finite precision computation, look at the example in Figure 1:

Consider a line f given by the equation $y = f(x) = 1.4 \cdot x / 2.7$. What we are interested in, is the position of the point $P(0.76/0.40)$ with respect to f, i.e. does P lie above or below the line f. This test occurs in almost any geometric algorithm and is called the *sidedness* or *orientation predicate*. Using exact arithmetic, it is not hard to see that P actually lies *above* f as

$$f(P_x) = f(0, 76) = 1.4 \cdot 0.76 / 2.7 < 0.40 = P_y$$

Now assume we are restricted to a floating-point system with base 10, mantissa length 2, and rounding to nearest, i.e. after an arithmetic operation the result

S. Albers, H. Alt, and S. Näher (Eds.): Festschrift Mehlhorn, LNCS 5760, pp. 341–354, 2009.

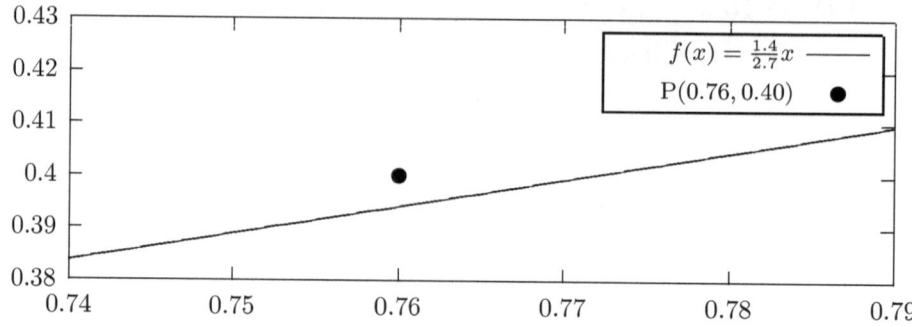

Fig. 1. Point P is clearly above line f

is always rounded to two significant digits. Let's do the calculation with this restricted precision. $\odot, \oslash$ denote the floating-point counterparts of multiplication and division.

$$f(P_x) = f(0.76) = 1.4 \odot 0.76 \oslash 2.7 = 1.1 \oslash 2.7 = 0.41 > 0.40 = P_y$$

And hence we conclude that P is *below* f which is clearly wrong !

Conditionals like this are the most critical parts in a program because they determine the flow of control. If in every test the same decision is made as if all computations would have been done over the reals, the algorithm is always in a state equivalent to its theoretical counterpart. But still, if some predicates are decided incorrectly, why is this such a big problem ? It might only produce a slightly perturbed output. The problem is, that conditionals are usually not independent. So if due to roundoff errors a conditional is decided incorrectly, this might contradict some other conditionals already decided or going to be decided in the future. Algorithms are usually not designed to cope with such inconsistencies, so they crash, loop or produce garbage output. See [1] for some instructive examples[1].

The rest of this paper provides a brief and rather informal overview of what use floating-point arithmetic nevertheless is for the implementation of geometric algorithms. Most of the mentioned results were obtained by Kurt himself or members of his research group.

2   Ideally – Robust Floating-Point Algorithms

When reality does not match the computational model under which algorithms are developed, it seems natural to adapt the model to reflect what happens in practice. People have tried to develop algorithms that right from the beginning take into account the possible inaccuracies induced by floating-point arithmetic. Unfortunately, designing such *robust* algorithms, is a quite challenging task.

---

[1] See also "finger in the pie".

We know of only few such attempts, for example in a series of papers Milenkovic and Fortune [2, 3] consider several (rather basic) geometric problems in the context of floating-point arithmetic. Each of these problems – for example in [3] the computation of a line arrangement – requires individual consideration and no generic methodology for deriving robust algorithms is known.

Another example in this category is the work by Sugihara et al. in [4]. Here the authors describe a robust algorithm to compute the Voronoi diagram of point sets in $\mathbb{R}^2$. The algorithm maintains certain *combinatorial* invariants throughout the algorithm and uses geometric predicates only to steer the computation in certain directions. So even if the outcome of all predicates was purely random, it would still compute something. With increasing accuracy of the predicate decisions, the outcome converges towards the correct result. But again, the reasoning and derivation of this algorithm is very specific to the problem of computing the Voronoi diagram and seems difficult to generalize to a generic method for deriving robust algorithms.

3   Realistically – Speeding Up Exact Arithmetic

Since changing the model of computation and designing *robust* algorithms seems very challenging, people have looked for other approaches, of which the most successful one is the *exact computation paradigm* ([5]); it advocates to guarantee correctness of the implementation by ensuring that every single predicate is evaluated correctly.

As we have seen in the introductory example, the evaluation of a geometric predicate amounts to the computation of the sign of an arithmetic expression. So the naive way to compute the sign of an expression is to compute the value of the expression exactly and to read off the sign from the value.

There are several exact arithmetic schemes designed specifically for computational geometry; most of them are methods for exactly evaluating the sign of a determinant using IEEE double precision floating-point arithmetic, and hence can be used to perform e.g. the orientation and incircle tests or even the insphere test. Difficulties arise, if the tests to be performed involve previously computed geometric objects which require extended precision to be exactly represented.

A more general approach, which is not specific to determinants or even predicates, are multiprecision packages like GMP, CORE or LEDA [6, 7, 8]. They allow arbitrary precision arithmetic on integers, fixed-point numbers or floating-point numbers. Of course, exact arithmetic with these packages has its cost, which is considerably higher than floating-point arithmetic. Depending on the input bit-length the arbitrary precision primitives are at least about 10-100 times slower than their floating-point counterparts.

3.1   Floating-Point Filtering

Floating-point filters [9, 10] partly provide remedy for the problem of excessive running times with exact arithmetic packages. A floating-point filter computes an approximate value of an expression (using floating-point arithmetic) and a bound for the maximal deviation from the true value. If the error bound is smaller than the absolute value of the approximation, approximation and exact value have the same sign and hence the sign of the approximation may be returned. In this way the true sign can be obtained quickly. The advocates of floating-point filters claim that filters at the predicate level realize the exact computation paradigm at little cost; the running time is claimed to be no more than twice the running time of a pure floating-point implementation.

Of course, this statement is only true if the floating-point filter always succeeds in deciding the predicate, and the floating-point filter mechanism can be applied for the whole computation.

In the following we will give a brief account about the derivation of error bounds for floating-point computations and how to instrument those in a floating-point filter.

3.2   Deriving Error Bounds for Floating-Point Computations

Let us have a closer look at floating-point arithmetic. If the floating-point arithmetic on a machine complies to the IEEE standard, one can guarantee an error bound for the error occurring in *one single* operation: Let $x = x_1 \mathrm{op} x_2$ be the *exact* outcome of an arithmetic operation on two floating-point numbers x_1, x_2, $\widetilde{x} = x_1 \mathrm{fop} x_2$ the result under floating-point arithmetic. Then the IEEE standard guarantees that $|\widetilde{x} - x| \leq 2^{-(p+1)}$ where p is the mantissa length of the floating-point representation ($p = 52$ for the C/C++ type double).

But as we evaluate complex expressions involving more than one operator, errors are propagated in some way from the "earlier stages" of computation to the final result. Assuming we have evaluated a complex expression e with floating-point arithmetic to $\widetilde{e}$, what we then want is an upper bound err for the error of this value, i.e. something like:

$$|e - \widetilde{e}| \leq err$$

We briefly present several techniques that can be used to compute this error bound. For a more detailed description of these techniques and their proofs see [11, 12, 13].

Fully-Dynamic Error Analysis. Fully-dynamic error analysis means that the error bound is computed completely at run-time, and hence can make use of the actual values of the expressions. We present two schemes for fully-dynamic error analysis:

Relative Error Bounds. In [11], relative error bounds are derived for floating-point computations, i.e. for every expression e and its floating-point evaluation $\widetilde{e}$, an ϵ_e is calculated such that the following is true at all times:

$$|\widetilde{e} - e| \leq \epsilon_e \cdot |\widetilde{e}|$$

The following shows a list of the formulas for inductively computing the ϵ_e-values always assuming that the for the operands x, y the invariant is fulfilled:

$$\epsilon_{x\pm y} = 2^{-p-1} + |\tfrac{\tilde{x}}{\tilde{e}}| \cdot \epsilon_x + |\tfrac{\tilde{y}}{\tilde{e}}| \cdot \epsilon_y$$
$$\epsilon_{x \cdot y} = (\epsilon_x + \epsilon_y + \epsilon_x \cdot \epsilon_y) \cdot (1 + 2^{-p-1}) + 2^{-p-1}$$
$$\epsilon_{x/y} = 2^{-p-1} + \tfrac{\epsilon_x + \epsilon_y}{1 - \epsilon_y} \cdot (1 + 2^{-p-1})$$
$$\epsilon_{\sqrt{x}} = \epsilon + \epsilon_x \cdot (1 + 2^{-p-1})$$

For more detailed information and the proofs see [11]. Note though that

- The relative errors must be computed dynamically at runtime, since computing the relative error of an addition requires the actual values of the floating-point approximations.
- The overhead compared to simple floating-point evaluation is quite high. For example, the addition $x + y$ requires – apart from the computation $x + y$ itself – 2 additions, 2 multiplications, 2 divisions and 2 absolute values for computing the new relative error.

These error bounds are used as built-in floating-point filter of the LEDA type real to filter out easy tests which do not require the arbitrary precision calculation.

Interval arithmetic. Burnikel, Brönnimann, and Pion in [12] presented another fully-dynamic scheme for computing error bounds, which was previously known in the numerical analysis community but not used in the context of computational geometry before. Their approach is based on the possibility to switch rounding modes if the floating-point arithmetic on a machine complies with the IEEE standard. A value x is represented by an interval $[\tilde{x}] = [\tilde{x}_l, \tilde{x}_u]$. Assuming x and y are represented by intervals $[\tilde{x}]$ and $[\tilde{x}]$, the following rules are used to compute the interval resulting from an arithmetic operation:

$$[\tilde{x}] + [\tilde{y}] = [\tilde{x}_l + \tilde{y}_l, \tilde{x}_u + \tilde{y}_u]$$
$$[\tilde{x}] - [\tilde{y}] = [\tilde{x}_l - \tilde{y}_u, \tilde{x}_u - \tilde{y}_l]$$
$$[\tilde{x}] \cdot [\tilde{y}] = [\min\{\widetilde{x_l y_l}, \widetilde{x_l y_u}, \widetilde{x_u y_l}, \widetilde{x_u y_u}\}, \max\{\widetilde{x_l y_l}, \widetilde{x_l y_u}, \widetilde{x_u y_l}, \widetilde{x_u y_u}\}]$$
$$[\tilde{x}]/[\tilde{y}] = \begin{cases} [\tilde{x}] \cdot [1/\widetilde{y_u}, 1/\widetilde{y_l}] , & 0 \notin [\tilde{y}] \\ \mathbf{R} & , \text{ otherwise} \end{cases}$$
$$[\tilde{x}]^{1/2} = \begin{cases} [\widetilde{x_l}^{1/2}, \widetilde{x_u}^{1/2}] , & 0 \notin [\tilde{y}] \\ \mathbf{R} & , \text{ otherwise} \end{cases}$$

If one of the intervals is infinite, we set the resulting interval to $\mathbf{R} = [-\infty, \infty]$. Since the computed intervals $[\tilde{x}]$ in general have bounds x_l, x_u which are not exactly representable by a floating-point number, we always round downwards (resp. upwards) to obtain an interval $[\tilde{x}_l, \tilde{x}_u]$ that encloses the 'real' interval and is exactly representable by floating-point upper and lower bounds. This rounding can be cheaply implemented by switching the rounding mode of the IEEE floating-point unit. Unfortunately, not all currently used platforms adhere to the IEEE floating-point standard.

Again we note, though, that the overhead compared to pure floating-point arithmetic is rather high. In case of the multiplication, 4 times more operations have to be performed, not even counting the comparisons to determine the maximum and minimum for lower and upper bounds. Furthermore, switching the rounding-modes on most architectures is a very costly operations. Still, at the cost of this higher overhead, interval arithmetic gives the best possible error bounds.

Semi-Static Error Analysis. As we have seen, fully-dynamic error analysis has the drawback of implying a rather large overhead during runtime.

This suggests dividing the computation of the error bounds in a static part and a dynamic part. The static part can be precomputed before runtime without any knowledge of the actual values of the expressions, whereas the dynamic part is computed during runtime, but with hopefully much less operations than the error calculations we have seen in the last paragraph. We will review the semi-static scheme which we have presented in [14]. For sake of simplicity we neglect the problems of underflow and refer the reader to [13] or [15] for a complete discussion.

For every expression e we not only compute the floating-point approximation $\tilde{e}$ but also an upper bound $\widetilde{e_{sup}}$ for $|\tilde{e}|$, called the supremum of $\tilde{e}$, and an integer ind_e – the index of e–, such that the following bound for the absolute error of the floating-point approximation is true at all times:

$$|\tilde{e} - e| \leq \widetilde{e_{sup}} \cdot ind_e \cdot 2^{-p} \tag{1}$$

An input value x exactly representable by a double has the floating-point approximation $\tilde{x} = x$, the supremum $\widetilde{x_{sup}} = |\tilde{x}|$ and the index 0. An input value not exactly representable by a double has the floating-point approximation $\tilde{x} = round(x)$, the supremum $\widetilde{x_{sup}} = |\tilde{x}| = |round(x)|$ and the index 1.

The index ind_e may be computed statically whereas $\tilde{e}$ and $\widetilde{e_{sup}}$ must be computed at runtime using the inductively given rules in table 1. $+, -, \cdot, /, .^{\frac{1}{2}}$ denote exact addition, subtraction, multiplication, division, and square root, whereas $\oplus, \ominus, \odot, /, \sqrt{}$ denote their floating-point counterparts.

We see that the computation of the supremum is quite similar to the computation of the floating-point approximation itself and therefore the implied overhead is reasonably small. In case of $+, -, \cdot$, only twice as many operations

Table 1. Rules for computing approximations, suprema and indices

expr. e	approx. $\tilde{e}$	supremum $\widetilde{e_{sup}}$	index ind_e
$x + y$	$\tilde{x} \oplus \tilde{y}$	$\widetilde{x_{sup}} \oplus \widetilde{y_{sup}}$	$1 + MAX(ind_x, ind_y)$
$x - y$	$\tilde{x} \ominus \tilde{y}$	$\widetilde{x_{sup}} \oplus \widetilde{y_{sup}}$	$1 + MAX(ind_x, ind_y)$
$x \cdot y$	$\tilde{x} \odot \tilde{y}$	$\widetilde{x_{sup}} \odot \widetilde{y_{sup}}$	$1 + ind_x + ind_y$
x/y	$\tilde{x} \oslash \tilde{y}$	$\dfrac{(|\tilde{x} \oslash \tilde{y}|) \oplus (\widetilde{x_{sup}} \oslash \widetilde{y_{sup}})}{(\widetilde{y_{sup}}) \ominus (ind_y + 1) \cdot 2^{-p}}$	$1 + MAX(ind_x, ind_y + 1)$
$x^{\frac{1}{2}}$	$\sqrt{\tilde{x}}$	$\begin{cases} (\widetilde{x_{sup}} \oslash \tilde{x}) \odot \sqrt{\tilde{x}} & if\ \tilde{x} > 0 \\ \sqrt{\widetilde{x_{sup}}} \odot 2^{\frac{p}{2}} & if\ \tilde{x} = 0 \end{cases}$	$1 + ind_x$

are needed, three times as many in case of $\sqrt{}$. The division has a considerably higher overhead but its use often can be avoided in practice.

Fully Static Error Analysis. If the input data consists of integer values of a bound bit-length and $+, -, \cdot$ are the only operators used, an upper bound for $\widetilde{e_{sup}}$ can be determined statically such that no overhead for the computation of the supremum occurs at runtime . (This is even possible for non-integer input values where the bit-length of $\lceil |e| \rceil$ is bounded.) An upper bound for the bit-length of the supremum can be inductively computed using the following formulas:

$$bitlen_{\pm} = 1 + MAX(bitlen_{op_1} + bitlen_{op_2})$$

$$bitlen_{\cdot} = bitlen_{op_1} + bitlen_{op_2}$$

So there is no overhead at runtime for computing the error bound $ind_e \cdot 2^{bitlen_{e_{sup}}} \cdot 2^{-p}$.

3.3   Filtering of Geometric Predicates

As we have seen with these schemes for computing error bounds of floating-point calculations, there is always a tradeoff between tightness of the error bound and the run-time overhead implied by the computation of the error bound. This suggests the following evaluation strategy for the sign evaluation of an arithmetic expression e:

1. evaluate e using double arithmetic to $\widetilde{e}$
2. check the sign of $\widetilde{e}$ with the fully-static determined error bound (if available); only if this fails, continue
3. compute the semi-static error bound and check the sign of $\widetilde{e}$ with that; only if this fails, continue
4. compute the fully-dynamic error bound and check the sign of $\widetilde{e}$ with that; only if this fails continue
5. evaluate e using exact arithmetic to obtain the sign of e

In this way, easy instances are always decided in early stages of this cascaded evaluation and the implied overhead is reasonably small. The generation of such cascaded evaluation schemes can even be automated by tools like EXPCOMP [13, 14] which we have developed.

The overhead observed in practical applications when comparing such an exact but filtered implementation with a (not necessarily reliable) floating-point implementation very often is around a factor of 2 (only measuring the time to evaluate the predicates and not taking into account the time spent on the combinatorial part of the algorithm). It gets worse, though, if

– the input data exhibits a lot of (near-)degeneracy, i.e. many of the critical expressions end up with a value of zero or close to zero, which makes it much harder for the filter stages to decide the predicate, of course.

– the predicates do not only operate on input data but on geometric objects constructed during the course of the algorithm; remember that if all filter stages fail, the last stage of the predicate evaluation falls back to exact arithmetic assuming that the input data is available in an exact representation. So in these cases the filtering only takes place in the predicate evaluation but not in the constructions.

3.4   Structural Filtering

In the remainder of this section we want to discuss filtering strategies in general. We view the execution of an algorithm as a sequence of steps. A step may be anything from the execution of a single instruction over the execution of a large subprogram to the execution of the entire program. If every step of an algorithm produces the correct result, the entire computation will produce the correct result.

The execution of a step consists of the evaluation of conditionals (predicates) and the execution of the straight-line code between the conditionals. The simplest way to ensure the correct execution of a step is to guarantee that all conditionals in the step are evaluated correctly.

An alternative way to ensure the correct execution of a step is to allow errors in the evaluation of the conditionals, to check at the end of the step whether the step performed correctly, and, if not, to repair the errors made. Of course, this approach is only viable if the "unsafe" execution of a step is faster than its "safe" execution, if the correctness check is simple, if errors occur rarely, and if the repair is simple. Observe that there are four "ifs" in the preceding sentence. We will show that there are many situations where the answer to all four ifs is yes.

We start by refining our view of the execution of an algorithm. We view algorithms as manipulating an underlying data structure and distinguish between search and update steps. Update steps are pieces of code that may change the underlying data structure and search steps are pieces of code that do not change the underlying data structure but are otherwise arbitrary. Structural filtering applies to search steps. It does not modify update steps. Thus the underlying data structure stays correct. We give three examples to illustrate the concepts.

i Any algorithm falls under the paradigm if we call the value of all program variables the underlying data structure, the evaluation of each predicate[2] in a conditional a search step (the step "searches" for the value of the expression), and call the straight-line pieces of code between conditionals update steps.
ii Consider a dictionary implementation based on a balanced tree. The tree constitutes the data structure manipulated by the algorithm. An insert operation consists of a search step, which determines the position in the tree at which the new key is to be added, followed by an update step, which adds the key to the tree.

---

[2] We assume that predicates in conditionals have no side-effects, a minor restriction. In geometric programs the predicates in conditionals are typically the evaluation of the sign of an arithmetic expression.

iii Consider an incremental algorithm for constructing Delaunay diagrams. The data structure is the current Delaunay triangulation and a search structure for locating points in the triangulation. An insertion of a new point consists of a search step, which locates the triangle of the current triangulation containing the new point, and an update step which inserts the point, performs flips to construct the new Delaunay triangulation, and modifies the search structure.

We postulated that a search step does not change the underlying data structure. A search step computes information (= the value of a predicate, a position in a tree, a triangle in a triangulation) which the subsequent update step uses to perform changes on the data structure. A search step evaluates some number of predicates. We assume that a predicate can be evaluated in two ways; the expensive way guarantees the correct value and the cheap way will usually give the correct result, but may err. In this general discussion we make no assumption about when a cheap comparison errs. In the context of geometric programs a cheap evaluation of a predicate is the evaluation with floating-point arithmetic, and an expensive evaluation is the evaluation with exact arithmetic (maybe with a floating-point filter).

The safe way to perform a step is to use only expensive predicate evaluations. Assume now that we use cheap predicate evaluations instead. The following observation is trivial but powerful. *If a search step amounts to a walk in an acyclic graph where predicate evaluations are used to determine the edges to be followed, then a search step will always terminate.* In our three examples above the search is a walk in an acyclic graph[3].

The search step, if executed with cheap predicates, may not end in the right sink of the acyclic graph. We postulate that it is easy to check whether the correct sink is reached. In our first example, the check amounts to the error-bound computation in the floating point evaluation of the underlying arithmetic expression, in our second example, the check amounts to the (exact) comparison with the two neighboring elements, and in the third example, the check amounts to orientation tests with three sides of a triangle.

If the search step ends in the correct sink of the search graph, we are done at this point. If the check reveals an error, we still have to find the correct sink. There is a generic way of reaching the correct sink. Repeat the search with expensive predicate evaluations. Observe that this is possible because we postulated that a search step does not change the underlying data structure. In our first example, the generic strategy amounts to an evaluation with exact arithmetic. In the two other examples, there are better ways to correct the error. In the second example, we may walk along the leaves of the tree and in the third example, we may use a walk through the triangulation.

Let us summarize. Structural filtering applies to search steps. If the search step amounts to the walk in an acyclic graph then it can be performed with cheap comparisons without the danger of looping. An error in the search step can

---

[3] In the first example the graph is a tree with three nodes. In the root the boolean expression is evaluated and the two children correspond to true and false.

always be corrected by redoing the search with expensive comparisons. Better strategies may exist and we gave two examples. The verification of the search step is problem dependent. With the generic solution to error correction, *only* the verification requires additional programming.

What can we hope to gain by structural filtering? The cost of an update step is unchanged. The cost of a search step is its cost when executed with cheap comparisons, plus the cost of the check, plus the cost of the repair. Structural filtering is particularly useful if the search steps dominate the running time of the algorithm. This is the case for our second and third example and, more generally, for many incremental constructions in geometry. In an insertion into a tree, the search step has cost $O(\log n)$ and the update step has cost $O(1)$. The same holds true for randomized incremental algorithms for convex hulls, Delaunay triangulations, Voronoi diagrams, and many other problems.

There is a second phenomenon which is exploited by structural filtering. Predicate evaluations may be redundant. There may be several paths to the correct sink and hence errors in predicates may be corrected by later predicates. In [16] we apply the idea of structural filtering to several problems, ranging from simple sorting problems to the randomized incremental construction of Delaunay triangulations. With relatively simple modifications we could obtain considerable gains in terms of the running time.

We will next compare structural filtering with filtering on the predicate level and filtering on the algorithm level.

Filtering on Predicate Level. Filtering at the predicate level was discussed in detail earlier in this section. Let us consider the extreme cases. If the floating-point computation always computes the correct sign, the cheap evaluation never errs and saves the computation of the error bound. The computation of the error bound has typically about the same cost as the computation of the sign and hence a cheap comparison has about half the cost of an expensive comparison. Thus we may expect that structural filtering can make significant savings; we should not expect to see a factor of two since the search step has to do some work outside the predicate evaluations and since structural filtering has to verify the result of the search.

If the floating-point computation never computes the correct sign, predicate filtering always has to resort to exact arithmetic. Since the cost of exact arithmetic is significantly larger than the cost of floating-point arithmetic (around 10-100 times the cost; see [17], for example), stage three will dominate the cost of an expensive predicate evaluation and a cheap comparison is much cheaper than an expensive comparison. Thus, even with the generic repair technique, the cost of structural filtering is not much larger than the cost of predicate filtering; observe that the cost of the search step with cheap predicates will be much smaller than the cost of the search with expensive predicates.

The advantage of predicate filtering is its genericity. Once "filtered" versions of the predicates are available, all algorithms using them benefit. There is no change required in an algorithm to switch from unfiltered predicates to filtered predicates. Moreover, the techniques for writing filtered predicates are well developed and even software supported [14].

The disadvantage of predicate filtering is the fact that the error-bound computation is always made. Structural filtering avoids it at the cost of the verification of the search step.

Filtering on Algorithm Level. While the filters on predicate level work on the level of the most basic (low-level) operations of an algorithm, filters on algorithm level work on the highest level possible. Here the idea is: compute with floating-point arithmetic, check the result, and repair, if necessary, to get the exact result.

There are two problems with filtering at the algorithm level. First, as we have heard before, the design of robust algorithms using only floating-point arithmetic is a difficult task even if robustness only means that the program should always run to completion. Second, the repair step is non-trivial if the floating-point algorithm does not come with a strong guarantee of what it computes. The purpose of restricting filtering to the search steps is precisely to guarantee that errors in predicate evaluations do not corrupt the data structure. Only the paper [18] discusses filtering at the algorithm level and the repair step. The main disadvantage of filtering at the algorithm level is that there are no widely applicable techniques for obtaining robust floating-point implementations.

Of course, filtering at the algorithm level approach also has its advantages. If no cheap evaluation errs, the result will be correct, and the only additional cost is the cost of checking.

4   Strangely – Not Computing What You Want, But at Least Exactly

Of course, when asking practicioners about how they cope with rounding errors of floating-point arithmetic they will tell you quite a few different strategies. One very interesting strategy is the following: before executing the (unsafe floating-point) implementation of an algorithm $\mathcal{A}$ on some input x, we perturb x randomly by some amount δ to obtain $\widetilde{x}$ on which we actually execute $\mathcal{A}$.

For some reason it turns out that running $\mathcal{A}$ on $\widetilde{x}$ is much more reliable than running $\mathcal{A}$ on the original input x. So in particular when the input x on which $\mathcal{A}$ is run has been obtained by some measurement process, it seems highly attractive to perturb x to $\widetilde{x}$ since the validity of the outcome of running $\mathcal{A}$ is not compromised by a sufficiently small perturbation.

Starting with a paper by Halperin et al [19] people have started to look at this strategy from a more theoretical point of view, trying to understand why it is so successful in practice and turning this into a theoretically sound framework.

4.1   Controlled Perturbation

As we have seen before, geometric algorithms branch on geometric predicates. A basic predicate for two-dimensional geometry is orientation. Given three points decide whether they lie on a common line or form a left turn or form a right turn. The *orientation predicate* for $d+1$ points $(p_0, \ldots, p_d)$ in $\mathbb{R}^d$ is given by the sign of a $(d+1) \times (d+1)$ determinant:

$$orient(p_0, \ldots, p_d) := \text{sign} \begin{vmatrix} p_{01} & \cdots & p_{0d} & 1 \\ \vdots & \cdots & \vdots & \vdots \\ p_{d1} & \cdots & p_{dd} & 1 \end{vmatrix}. \tag{2}$$

The determinant evaluates to zero if and only if the $d+1$ points lie in a common hyperplane. In many algorithms this is considered a degeneracy.

Again, when evaluating an arithmetic formula E using floating-point arithmetic, round-off errors occur which might result in the wrong sign being reported. In order to guard against round-off errors, we postulate the availability of a predicate $\mathcal{G}_E$ with the following *guard property: If $\mathcal{G}_E$ evaluates to true when evaluated with floating point arithmetic, the evaluation of E with floating point arithmetic yields the correct sign.* In an idealistic algorithm A we now guard every sign test by first testing whether the corresponding guard evaluates to true. If not, we abort. We call the resulting algorithm a guarded algorithm and use A_g to denote it. On an input x, A_g will either follow the same execution path as A or abort after an initial segment of it. In the former case, we will say that A_g succeeds on x. When A_g succeeds on x, the combinatorial part of the output will be correct and the numerical part will be a floating point approximation of the exact result. In all applications in this paper, the numerical part of the output will be identical to the input. Also the running time of A_g on x will be at most the running time of A on input x; this assumes that the cost of evaluating a guard is bounded by the cost of evaluating the corresponding expression and ignores constant factors.

The controlled perturbation version of idealistic algorithm A is as follows: Let δ be a positive real. On input x, we first choose a δ-perturbation $\tilde{x}$ of x and then run the guarded algorithm A_g on $\tilde{x}$. If it succeeds, fine. If not, repeat. What is a δ-perturbation? If the input is a set of points, the following definition is natural. A δ-perturbation of a point is a random point in the δ-ball (or δ-cube) centered at the point and for a set of points a δ-perturbation is simply a δ-perturbation of each point in the set. For more complex objects, alternative definitions come to mind, e.g., for a a circle one may want to perturb the center or the center and the radius.

The goal is now to show experimentally and/or theoretically that A_g has a good chance of working on a δ-perturbation of each input and a small value of δ. More generally, one wants to derive a relation between the precision p of the floating point system (= length of the mantissa), a characteristic of the input set, e.g., the number of points in the set and an upper bound on the maximal coordinate of any point in the input, and δ. Halperin et al. have done so for arrangements of polyhedral surfaces, arrangements of spheres, and arrangements of circles.

We want to stress that a guarded algorithm can be used without any analysis. Suppose we want to use it with a certain δ. We execute it with a certain precision p. If it does not succeed, we double p and repeat.

Guard predicates must be safe and should be effective, i.e., if a guard does not fire, the approximate sign computation must be correct, and guards should not fire too often unnecessarily. It is usually difficult to analyze the floating point evaluation of $\mathcal{G}_E$ directly. For the purpose of the analysis, we therefore postulate the existence of a *bound predicate* $\mathcal{B}_E$ with the property: *If $\mathcal{B}_E$ holds, $\mathcal{G}_E$ evaluates to true when evaluated with floating point arithmetic.* When E is evaluated by a straight-line program, it is easy to come up with suitable predicates $\mathcal{G}_E$ and $\mathcal{B}_E$ using the error bounds described already in Section 3.2.

In [20], the controlled perturbation scheme was examined for the randomized incremental construction of Delaunay triangulations, [21] generalizes the scheme to algorithms that can be viewed as decision trees.

5   What Else Is to Expect?

While the computation with simple geometric objects like points and lines is believed to be understood well enough that efficient implementations of most algorithms are feasible, computation with more complex objects like curves and surfaces is still a big challenge. Due to the involved arithmetic expressions being of much higher degree the error bounds that are obtained for floating-point calculations are often too pessimistic to be useful – be it for the filtering or a perturbation approach. There have been attempts, to analyze and design algorithms with the explicit goal of using only low-degree predicates. This might be a fruitful strategy to bring floating-point arithmetic again into play. Also, there has been considerable progress on the hardware front; current graphics cards use processors (GPUs) whose floating-point performance exceeds that of normal general purpose CPUs by orders of magnitudes. It might be interesting to instrument those to allow for exact evaluation of geometric predicates with littel runtime overhead.

References

1. Kettner, L., Mehlhorn, K., Pion, S., Schirra, S., Yap, C.: Classroom examples of robustness problems in geometric computations. Comput. Geom. Theory Appl. 40(1), 61–78 (2008)
2. Milenkovic, V.: Verifiable Implementations of Geometric Algorithms using Finite Precision Arithmetic. Phd thesis, Carnegie Mellon University (1988)
3. Fortune, S.: Vertex-rounding a three-dimensional polyhedral subdivision. Discrete Comput. Geom. 22(4), 593–618 (1999)
4. Sugihara, K., Ooishi, Y., Imai, T.: Topology-oriented approach to robustness and its applications to several Voronoi-diagram algorithms. In: Proc. 2nd Canad. Conf. Comput. Geom., pp. 36–39 (1990)
5. Yap, C.K., Dubé, T.: The exact computation paradigm. In: Du, D.Z., Hwang, F.K. (eds.) Computing in Euclidean Geometry, 2nd edn. Lecture Notes Series on Computing, vol. 4, pp. 452–492. World Scientific, Singapore (1995)
6. Granlund, T.: GMP, The GNU Multiple Precision Arithmetic Library. 2.0.2 edn. (1996), http://www.swox.com/gmp/

7. Mehlhorn, K., Näher, S.: LEDA: A Platform for Combinatorial and Geometric Computing. Cambridge University Press, Cambridge (2000)
8. Karamcheti, V., Li, C., Pechtchanski, I., Yap, C.: The CORE Library Project. 1.2 edn. (1999), http://www.cs.nyu.edu/exact/core/
9. Fortune, S., Van Wyk, C.J.: Static analysis yields efficient exact integer arithmetic for computational geometry. ACM Trans. Graph. 15(3), 223–248 (1996)
10. Karasick, M., Lieber, D., Nackman, L.R.: Efficient Delaunay triangulations using rational arithmetic. ACM Trans. Graph. 10(1), 71–91 (1991)
11. Burnikel, C., Mehlhorn, K., Schirra, S.: The LEDA class "real" number. Technical Report MPI-I-96-1-001, Max-Planck-Institut für Informatik, Saarbrücken (1996)
12. Brönnimann, H., Burnikel, C., Pion, S.: Interval arithmetic yields efficient dynamic filters for computational geometry. In: Proc. 14th Annu. ACM Sympos. Comput. Geom., pp. 165–174 (1998)
13. Funke, S.: Exact arithmetic using cascaded computation. Master's thesis, Universität des Saarlandes (1997)
14. Burnikel, C., Funke, S., Seel, M.: Exact geometric predicates using cascaded computation. In: Proc. 14th Annu. ACM Sympos. Comput. Geom., pp. 175–183 (1998)
15. Burnikel, C.: Exact Computation of Voronoi Diagrams and Line Segment Intersections. PhD thesis, Universitaet des Saarlandes (1996)
16. Funke, S., Mehlhorn, K., Naeher, S.: Structural filtering – a paradigm for efficient and exact geometric programs. In: Proc. 11th Canad. Conf. on Comput. Geom. (1999)
17. Schirra, S.: A case study on the cost of geometric computing. In: Goodrich, M.T., McGeoch, C.C. (eds.) ALENEX 1999. LNCS, vol. 1619, pp. 156–176. Springer, Heidelberg (1999)
18. Kettner, L., Welzl, E.: One sided error predicates in geometric computing. In: Mehlhorn, K. (ed.) Proc. 15th IFIP World Computer Congress, Fundamentals - Foundations of Computer Science, pp. 13–26 (1998)
19. Halperin, D., Shelton, C.R.: A perturbation scheme for spherical arrangements with application to molecular modeling. Computational Geometry: Theory and Applications 10(4), 273–288 (1998)
20. Funke, S., Klein, C., Mehlhorn, K., Schmitt, S.: Controlled perturbation for delaunay triangulations. In: SODA 2005: Proceedings of the sixteenth annual ACM-SIAM symposium on Discrete algorithms, Philadelphia, PA, USA, pp. 1047–1056. Society for Industrial and Applied Mathematics (2005)
21. Mehlhorn, K., Osbild, R., Sagraloff, M.: Reliable and efficient computational geometry via controlled perturbation. In: Bugliesi, M., Preneel, B., Sassone, V., Wegener, I. (eds.) ICALP 2006. LNCS, vol. 4051, pp. 299–310. Springer, Heidelberg (2006)

Car or Public Transport—Two Worlds

Hannah Bast

Max-Planck-Institute for Informatics, Saarbrücken, Germany

Abstract. There are two kinds of people: those who travel by car, and those who use public transport.[1] The topic of this article is to show that the algorithmic problem of computing the fastest way to get from A to B is also surprisingly different on road networks than on public transportation networks.

On road networks, even very large ones like that of the whole of Western Europe, the shortest path from a given source to a given target can be computed in just a few microseconds. Lots of interesting speed-up techniques have been developed to this end, and we will give an overview over the most important ones.

Public transportation networks can be modeled as graphs just like road networks, and most algorithms designed for road networks can be applied for public transportation networks as well. They just happen to perform not nearly as well, and to date we do not know how to route similarly fast on large public transportation networks as we can on large road networks.

The reasons for this are interesting and non-obvious, and it took us a long time to fully comprehend them. Once understood, they are relatively easy to explain, however, and that is what we want to do in this article. Oh, and by the way, happy birthday, Kurt!

1 Introduction

The last five years have seen an exciting surge of research on routing algorithms for large transportation networks. Most of this work has been done on road networks, but some of it was also considering public transportation networks.

Both road networks and public transportation networks can be very naturally modeled as directed graphs. For a road network, each node corresponds to a junction, where two or more road segments meet, and the arcs of the graph correspond to road segments. The cost of an arc is simply the time it takes to travel across the respective road segment. A shortest path in this graph then corresponds to the fastest way to get from a point A to a point B.

Public transportation networks are modeled in a similar way, except that besides the spatial information we also have to deal with time schedules. In the simplest and most natural model, each node corresponds to a departure or arrival

---

[1] Admittedly, there are a few people using both modes of transportation from time to time, but not that many.

S. Albers, H. Alt, and S. Näher (Eds.): Festschrift Mehlhorn, LNCS 5760, pp. 355–367, 2009.

event at a particular station. For example, a node might stand for the event of ICE 500 arriving at Mannheim Hauptbahnhof at 21:24.[2] Arcs between nodes then either correspond to waiting from one event to the next at a particular station, or to taking a particular train (or bus or ...) from one station at a particular time to another station. The cost of an arc is the respective waiting or travel time, so that a shortest path in this graph corresponds to the fastest way to get from a particular station A at a particular time t_A to a particular station B at a particular time t_B.

We will come back to these two models in Section 2, giving slightly more detail and commenting on possible refinements there.

1.1   Dijkstra's Algorithm

The method of choice for computing the shortest path from a give source node to a given target node in a given graph is Dijkstra's algorithm, which dates back to the 1950s [1]. In a nutshell, Dijkstra's algorithm works as follows. Each node is assigned a tentative cost, which is initially 0 for the source node and ∞ for all other nodes in the graph. The algorithm then starts from the source node, and visits all outgoing arcs from there. For each such outgoing arc, it checks whether via this arc it can reach the node at the other side (the so-called *tail*) of the arc at a lower cost than assigned to that node so far. If yes (which is true, in particular, if we reach the node for the first time), its tentative cost is updated to the new, lower cost. This procedure is called *relaxing an arc*. Once all outgoing arcs of a node have been relaxed, that node is called *settled*. In the next round, we pick the node with the smallest tentative cost, which has not been settled so far, and relax its outgoing arcs. We iterate this until the target node is settled.

It is a simple, but non-trivial, elegant three-line proof to show that if the arc costs are non-negative, then once a node is settled, the tentative cost assigned to it at that time is actually the cost of the shortest path from the source to that node. Therefore each node is settled at most once. It is also important to observe that before Dijkstra's algorithm settles the target node, it will have settled (and thus computed the shortest path cost for) all nodes which can be reached from the source at smaller cost. If we color all nodes settled by Dijkstra's algorithm before it reaches the target on a drawing of the road network in the plane, we therefore see a disk-like area around the source node; see Figure 1.

What is the complexity of Dijkstra's algorithm? Each settling of a node requires to find, among the unsettled nodes at that point, that node with the smallest tentative distance. This operation is supported by a data structure called a *priority queue*, and it can be implemented to work in time $O(\log n)$, where n is the number of items in the queue. Relaxing an arc potentially requires to update the tentative cost of a node, and we know that if it is updated it is actually decreased. This operation is therefore called *decrease-key*, and can be supported in amortized constant time, that is, a sequence of m such operations takes $O(m)$

---

[2] An event the author had the pleasure to witness personally many times over the course of the last year.

Fig. 1. The search space of Dijkstra's algorithm on a road network, for a given source and target. Left: the original algorithm. Right: its bidirectional version.

time. The complexity of Dijkstra's algorithm is thus $O(n \cdot \log n + m)$, where n is the number of nodes settled before the target node, and m is the sum of the out-degrees of these nodes. In any case, n is bounded by the total number of nodes in the graph, and m by the total number of arcs, and if source and target are far apart, these bounds are actually tight within a small constant factor.

Interestingly, even 50 years after its invention, it is still not known whether Dijkstra's algorithm is theoretically optimal, or whether an algorithm exists that solves the shortest path problem in linear time $O(n + m)$. For our application, this question is academic, however, since even in the best case, each node and each arc would have to be visited at least once, and that alone is very expensive when the network is very large.

For example, consider the road network of the whole of Western Europe. This can be modeled by a graph with about 20 million nodes and about 50 million arcs. It is hard to make the operations involved in settling a node faster than 100 nanoseconds on a standard PC (that is about the time it takes to read a single cache line, or the time for a single cache miss). But even for such a highly tuned implementation of Dijkstra's algorithm, settling all nodes would take on the order of seconds.

In public transportation networks we have yet more nodes. The local public transportation network of Berlin-Brandenburg alone has around 4 million departure and arrival events. Extrapolating this to the whole of Europe (we don't have the actual data yet, so we can only guess) would give a graph with hundreds of millions of nodes.

2   Models Again

In our introduction above, we already gave a brief description of how to model both road networks and public transportation networks as directed graphs. We here recall these descriptions, and talk about a few more relevant details and possible refinements.

2.1   Road Networks

There isn't much to add to the description for road networks; it really is that simple. We have an arc for each road segment[3], a node for each junction of two or more segments, the cost of the arc is the time it takes to travel along that arc, and the goal is to compute the shortest path from a given source node to a given target node.

A number of recent works have addressed a *time-dependent* variant of this problem, where an arc cost is not just a scalar value, but a piece-wise linear function that maps each possible arrival time at the head of the arc to a travel cost [2] [3]. A simple variant of Dijkstra's algorithm can solve this problem as well. As we will see in the next section, public transportation networks can also be modeled by time-dependent graphs.

2.2   Public Transportation Networks

Let us recall the simple model from the introduction. We have a node for each departure and arrival event, nodes are grouped by stations, and arcs are either waiting arcs (between two nodes of the same station) or transit arcs (between two nodes from different stations).

This modeling leaves out the important issue of *transfer safety buffers and costs*: a change of vehicle takes time, and we want to penalize paths with many changes of vehicle—two issues that do not arise in road networks. A simple and natural way to model this, is by having *two* nodes for each arrival or departure event, which represent the state of being *on board* a vehicle and *at the station*, respectively, at the respective station and time.

In its simplest form, a query is given by a source station, an earliest departure time at that station, and a target station. More realistically, however, source and target are not stations, but *geographic locations*, from which we first have to walk to nearby stations. This is important especially in municipal areas, where it is not at all clear which station is the best to walk to first, and it really is (and hence should be made) part of the routing problem to identify the best such station. We then effectively have *sets* of source and target stations.

Note that this is not an issue in road networks, because these are typically so dense that without significant loss of quality in the results, we can simply *snap* to the nearest road segment or junction when source and target are given as geographic locations.

The model we described so far is known as the *time-expanded* model. As an alternative, we can also represent public transportation networks in the *time-dependent* model described in the previous subsection. Simply have one node per station, and the arc cost of getting from the station at the head of the arc at time x to the tail of the arc is $d - x + t$, where $d \geq x$ is the next departure of

---

[3] A long, curvy piece of road is typically approximated by a sequence of straight-line segments. However, this is done for the purpose of realistic rendering of the network, and is irrelevant for solving the shortest path problem. In fact, the first thing an efficient algorithm would do is contract such sequences to one arc again.

a vehicle to the tail station, and t its travel time. This indeed yields piece-wise linear arc costs.

Asymptotically, the two models do not differ, since a time-aware Dijkstra computation on the time-dependent graph essentially performs the same sequence of operations as an ordinary Dijkstra computation on the time-expanded graph. In practice, a carefully tuned implementation of the time-dependent model can give improvements of a factor of 10 and more over the time-expanded model, but this difference vanishes as soon as realistic features like transfer costs are taken into account. For details on the comparison between the two models, see [4].

2.3   Multi-criteria Cost and Traffic Days

A non-trivial model extension that makes sense for road networks, but is almost mandatory for public transportation networks is to consider *multi-criteria cost functions*. For example, users are typically interested in both travel time and the number of transfers but often not both of them can be minimized at the same time: there may be a connection that takes two hours and does not require any transfers, and there may be a connection which takes only one and a half hours but requires two transfers. Some users will prefer the faster one, and some will prefer the no-transfer one, and so we should (compute and) present both.

Another practical issue that significantly complicates routing on public transportation networks are *traffic days*: certain connections operate only on certain days and not on others.

With respect to their algorithmic solution, both issues are closely related in that they mean that each node in the graph is no longer labeled by only a single cost but by a whole *set of incomparable costs* instead. But again, Dijkstra's algorithm can be easily extended to also deal with this situation. The items in the queue are now individual cost labels (of which a single node can have several), and when settling a cost label, we relax each arc of the node to which the label belongs as before except that we now have to consider the new cost together with *all* the costs of the tail node of the arc, and discard those costs which are no longer optimal.

Obviously, the complexity of relaxing an arc now depends on the number of incomparable costs at the tail of the arc, and, in principle, this number could grow very large. However, we and others have found that with a cost function modeling traveling time and transfer costs, and considering traffic days over periods of a few weeks, the average number of incomparable costs per node is a small constant, and the running time of Dijkstra's algorithm adapted to deal with multiple costs per node lies about a factor of 10 over that of an ordinary Dijkstra computation [5].

2.4   Computing Costs Versus Computing Actual Paths

In the next section we will often tacitly assume that all we want to compute is the *cost* of a shortest path. It indeed typically holds that once we can compute costs fast, we can also compute paths reasonably fast. A very simple, generic way goes as follows. Start at the source node. For each adjacent node v, compute the cost of a shortest path from v to the target, and add the cost of the arc from the source to v. The adjacent node with the smallest such sum lies on the (or rather: a) shortest path. Pick that node, and do the same thing from there. Iterate.

This generic way requires $d \cdot l$ cost computations, where d is the average degree of the nodes on the shortest path, and l is the length of the path. For any of the algorithms mentioned in the following, there are approach-specific ways to do much better this, but we will not get into the details in this paper. The bottom line to remember is that once we can compute costs fast, we can also compute the paths itself fast.

3   Tricks of the Trade ... and Why and When They Work

With the models all set, the rest of the paper is now essentially a list of the most relevant and effective "tricks of the trade" that have been developed for the speeding up of shortest path queries on transportation networks, in particular from the last decade. Most of these tricks have been invented and applied for road networks first, and were only later transferred to public transportation networks (with, as we will see, limited success so far).

The structure of each of the following subsections is as follows: give a short description of the "trick", explain why it works well for road networks, and then say what the problems are when applying it to public transportation networks. Wherever possible, we will roughly quantify the performance gain in terms of asymptotic complexity and / or actual running times, and refer to the respective papers for the detailed experiments.

As a side effect, this section will also be giving an overview of all the fascinating recent work on routing in transportation networks. This overview is by no means complete, however, since we focus on those tricks which turned out to be most successful, and in each case mention only the one or two most representative works using that trick. For a more complete survey of recent techniques on routing in road networks, see, for example, [6]. For an account of routing algorithms for public transportation networks, see, for example, [7].

3.1   Bidirectional Search

A very simple idea to improve over the plain Dijkstra algorithm is to simultaneously search from the source and target node at the same time, until "the two search frontiers meet". More precisely, we maintain two priority queues, one for the search from the source, as for the ordinary Dijkstra, and one for the backward search from the target, which is just a forward search in the reversed graph, that is, the graph where each arc (u, v) is replaced by (v, u). In each round, we settle the node with the smallest *overall* tentative cost, that is, from the source or to the target; for this, a simple comparison of the minima of the two priority queues suffices. Once we settle a node in one queue that is already settled in the other queue, we get the first tentative cost of a shortest path. To guarantee optimality, we have to continue until the sum of the tentative costs of the current minima of the queues is above the current tentative shortest path cost (which then is indeed the cost of the shortest path).

Speaking in terms of Figure 1, bidirectional Dijkstra reduces the search from a single disk with radius r, to two discs with radius $r/2$. That is, the search space (and hence the query processing time), halves. This by itself is not a big improvement, but as we will see in the following subsections, bidirectional search turns out to be a key ingredient in other, more sophisticated speed-up techniques.

In public transportation networks bidirectional search is more complicated, since we know the target *station*, but not the particular *node* at that station at which we are going to arrive. In fact, finding that node is a significant part of the problem we want to solve in the first place. What we can so, however, is to search backwards from the *set of all nodes* at that station. The backward search would then compute, for each node that it settles, the cost of the path to the earliest node of the station which it can reach. Combined with other techniques this becomes yet more complicated, but by itself is not one of the main obstacles.

Summary: Bidirectional search by itself is not very effective, but is an important ingredient in more sophisticated techniques. In public transportation networks, we need to search backwards from a whole set of potential target nodes, which makes things more complicated.

3.2   Hierarchy

Most navigation devices in public use nowadays implement a variant of the following simple routing heuristic. Roads have different *levels of importance*: for example, in the road map of Manhattan in Figure 2 (left), we see white (small) roads, yellow (national) roads, and orange roads (motorways). A simple heuristic is then to do a bidirectional search, that takes into account *all* the roads in close proximity to the source and target, but once a certain distance from the source or target is reached, considers only yellow and orange roads, and at a certain even larger distance from the source or target considers only the orange roads. For an appropriate definition of "close proximity" and "certain distance" most shortest paths indeed have that property, like the path in the Figure 2.

This heuristic very significantly reduces the number of nodes that have to be settled and arcs that have to be relaxed, however, at the price of a certain loss of exactness. In the seminal works of [8] and [9] this heuristic has been turned into an exact algorithm, by actually *computing* a level of importance for each arc (which intuitively correspond to the road colors in Figure 2, but algorithmically have nothing to do with them). On road networks both precomputation and query times are very fast. With the latest version of their algorithm, the importance levels can be computed in about 15 minutes for the complete road network of Western Europe, with subsequent query times on the order of 1 millisecond. The method was, quite appropriately, named *highway hierarchies*.

On public transportation networks, even if we leave the complications of bidirectional search described in Section 3.1 aside, experimental studies ([10] and also our own) have shown that the speed-ups obtained are much less dramatic

 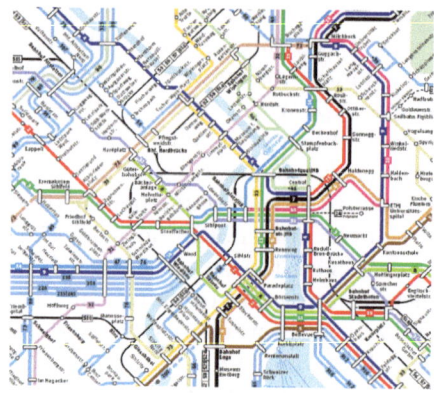

Fig. 2. Left: a shortest path in the road network of Manhattan. Right: a section of the tram + bus network of Zurich.

than for road networks. On large municipal areas, query processing times can even be *worse* than for a well-tuned implementation of Dijkstra's algorithm.

The main reason for this disappointing performance is actually easy to understand. Look at the tram + bus network of Zurich in Figure 2 (right), and think of a few random queries and their solutions. You will find that there is *hardly any hierarchy*. Intuitively, all the trams and buses are equally important, and exactly which tram or bus is chosen for a given query depends more on how well the schedules of the various lines match, than on some connection being more important than others. Once we travel long-distance between cities, a first level of hierarchy does appear (intuitively, the long-distance trains as opposed to the local trams and buses), but not on the intra-city level.

While this may be fine for a relatively small area like that of Zurich (about one thousand stations), this is a major performance problem for large municipal areas like, for example, New York (several tens of thousands of stations, with tens of millions of arrival / departure events). A Dijkstra computation even on this local network takes on the order of seconds, and hierarchical methods are of no use to speed things up there.

Worse than that, also the precomputation time suffers on such networks. In order to identify the first level of hierarchy, a method like highway hierarchy does a local search from each node, until all paths have reached the next level of the hierarchy. But for all nodes within a municipal area, this local search will have to cover the whole municipal area, which can encompass *millions of nodes*. In contrast, we know that for road networks local searches of only *a few hundred nodes* are enough to discover the next level of the hierarchy [9].

Summary: The efficiency of hierarchical approaches in terms of both precomputation and query time is proportional to the extent of the local searches necessary to find the next level of hierarchy. For road networks, a few hundred nodes per local search are typically enough. For public transportation networks, frequently whole municipal areas with millions of nodes need to be explored.

3.3   Shortcuts / Contraction

Recall the footnote in Section 2, where we talked about modeling a long, curvy piece of road as a sequence of short straight-line road segments. As mentioned there, this is done for the purpose of nice rendering. For the purpose of the shortest-path computation, we may as well replace that sequence by a single arc again, thus significantly reducing the number of nodes and arcs in the graph. This replacement is an instance of so-called *contraction*, and the new arc is called a *shortcut*. Some methods also insert shortcuts without actually removing arcs, but instead have a mechanism to consider only selected arcs at query time [11].

For a method like highway hierarchies, contraction is of good use not only on the original graph, but also on the iteratively computed subnetworks. Just think of the subnetwork of all motorways. Most junctions there are of the kind that we either enter or leave the motorway to or from a less important road. With all non-motorways removed from the graph, we will have only very few nodes with degree larger than 2, namely the actual motorway junctions.

In fact, contraction can be taken one step further by also contracting nodes of a degree larger than 2. To contract a node x, we simply look at all pairs u, v, where u is adjacent to an incoming arc and v is adjacent to an outgoing arc, and check whether there is a shortest path containing u, x, v. If yes, we insert the shortcut (u, v). This pays off, provided that we do not insert (many) more shortcuts than we remove arcs by removing x; see [9] for details.

Note that contraction and shortcuts are not so much a stand-alone method, but have instead acted as a catalyzer for a variety of multi-level methods, in particular: [9] [11] [12].

As far as public transportation networks are concerned, consider again the tram and bus network of Zurich from Figure 2. Most stations are "junctions", where more than one line meets, and if one takes the possibility of walking between stations into account (see Section 2.2), the average number of lines to which one can transfer at a given station increases further. This is especially true in cities with many different transportation agencies and therefore many stations in the vicinity of each other. But contraction and / or the introduction of shortcuts is only effective for nodes of low degree.

We have already found in Section 3.2, that the difficult searches are the local ones, where local can mean a whole municipal area. Unfortunately, it is exactly in these area, on the lowest level of the network, that the node degree is too high for contraction to be effective.

Summary: Contraction / Shortcuts don't help us speeding up local searches on the lowest level of the hierarchy, due to the high node degree there.

3.4   Goal Direction

The simplest form of goal direction is to augment Dijkstra's algorithm by a heuristic that for each node in the graph estimates the cost to the given target. Nodes are then retrieved from the priority queue by the *sum* of their tentative cost and the value of the heuristic function. This variant of Dijkstra's algorithm is known under the name A* *(A-star) algorithm*, and was first described in 1968 [13].

The performance gain of A* depends on the quality of the heuristic. It is a three-line proof (very similar to the correctness proof for Dijkstra's algorithm) that A* is correct, whenever the heuristic function *underestimates* the actual cost of the respective node to the target. If the heuristic cost is always zero, we are back to Dijkstra's algorithm. If the heuristic function magically knows the exact cost to the target, A* will be perfect in that it settles only the nodes on the (or rather: a) shortest path.

One simple, non-magical heuristic is to underestimate the cost to the target by the geographic straight-line distance to the target divided by the maximum speed of a vehicle anywhere in the network. This heuristic always underestimates the true cost, sometimes by not much (when the shortest path to the target is geographically relatively straight and uses mainly motorways), sometimes a lot (when the shortest path to the target is long-winding and uses mainly slow roads). Overall, this heuristic gives a notable but not very dramatic improvement in query processing time by a factor of about 2 to 3, for both road and public transportation networks. A more powerful heuristic, based on precomputed distances to so-called *landmarks*, has been presented in [14].

The most powerful form of goal direction is provided by so-called *arc flags* [15] [16]. Here the graph is partitioned into k regions, and for each arc k bits are precomputed, where the ith bit is 1 if and only if that arc is on a shortest path to a node within region i. At query time we can then simply ignore all arcs outside the region containing the target where the bit for that region is set to 0. In an extreme case, where each node forms a region on its own, the arc flags for the target node would then show us the shortest path without any detour.

These arc flags / bits can be computed by running Dijkstra's algorithm separately from each node, in the reversed graph. This, however, is equivalent to a quadratic-cost all-pairs shortest path computation. It is easy to see, that it is enough to consider *only nodes on the boundary* of each region. In a perfect grid graph with n nodes, partitioned into k parts (by $\sqrt{k} - 1$ horizontal and $\sqrt{k} - 1$ vertical cuts), the number of boundary nodes would be on the order of $\sqrt{n \cdot k}$, which still gives an order $n^{3/2}$ cost for the precomputation, even for small k. In real graphs, the cost tends more towards n^2.

A conceptually simple trick to reduce the precomputation cost to almost linear is to work with a multi-level partitioning of the graph. In the precomputation, the backwards Dijkstra computation from a boundary node of a cell in the partitioning can then stop, as soon as all nodes in the containing cell from the next level are settled.

It is here that we meet another fundamental difference between road networks and public transportation networks. Namely, for road networks we can indeed settle all nodes in a geographically bounded region with cost roughly proportional to the number of nodes in that region; see, for example, [6].

In public transportation networks, however, we have a fundamental and very annoying problem, which we will explain by an example. Consider a node in

Zurich and assume that we want to settle all nodes in Zurich and the surrounding villages. Even though the geographic extent of that region is relatively small, there will be several nodes in that region which can be reached only at a very high cost. The reason is simply *bad connectivity*: we might be just too late for the last bus of the day and have to wait overnight for the first bus of the next day, thus getting an optimal connection taking 15 hours. But in 15 hours, we can get to the airport, take a plane to New York and explore half of the city there ... and Dijkstra's algorithm will just do that.

As extreme as it may sound, this phenomenon is actually the rule and not the exception. We consider it a major open problem to come up with an algorithm for local searches in public transportation networks with cost proportional to the number of nodes to be settled.

Summary: Goal direction is potentially very effective but has very high precomputation costs. For public transportation networks, this cost is quadratic due to the lack of efficient algorithms for local searches on such networks.[4]

3.5   Distance Tables

An extreme precomputation would be to compute a table with distances between all pairs of nodes in the given graph. Query times would then be instantaneous (recall the bottom line of Section 2.4 that once we can compute the cost fast, we can also compute the actual paths fast), but the precomputation complexity would be quadratic in both time and space.

Distance tables for a *subset* of the nodes have been used as a "turbo" in various approaches in the past. We here briefly describe *transit node routing*, which works solely with distance tables and is the fastest method for routing in road networks (with at the same time reasonable preprocessing) to date [17] [18].

The *transit nodes* are a subset of nodes with the following "magical" properties: (1) the set is small, on the order of $\sqrt{n}$, where n is the total number of nodes in the network; (2) all shortest paths that cover a certain minimal geographic distance D have at least one transit node on them; (3) the number of transit nodes hit first on shortest paths that start from a fixed node is small; we call these few transit nodes the *access nodes* of a node.

Given such a set of transit nodes, we precompute for each node, the distances to all of its access nodes, and the distance between each pair of transit nodes. For a given query, let x and y be the number of access nodes of the source and target, respectively. To answer the query, we then need to look up a mere $x + y + x \cdot y$ of the precomputed distances and try out all $x \cdot y$ combinations of access node near the source and access node near the target. On road networks the astonishingly low number of 5 access nodes, on average, can be achieved, leading to extremely fast query times on the order of a few *microseconds*. The precomputation can be done in a number of ways, one of which is similar to the precomputation for highway hierarchies and with a comparable complexity [18].

---

[4] The precomputation from [2], although extremely well-tuned, is quadratic, too.

It is important to understand that due to property (2) above, this only works when the source and target are geographically at least a distance of D apart. (Obviously the *short* shortest paths cannot all be hit by a small set of common transit nodes, too.) But when D is small, this is not a problem, since for queries below this threshold any conventional method is good enough; see [18] for details.

Experiments show that also in public transportation networks we can find a good set of transit nodes with properties (1) - (3) above. The number of access nodes per node is by a factor of 5 - 10 higher than in road networks, but still small enough to yield query times on the order of milliseconds.

The problem are (a) the local searches required to precompute, for each node, the distances to its access nodes, and (b) the local searches required at query time when source and target are less than the distance D apart. Both of these can, and often will, involve computing shortest paths of very large cost, and we have no efficient solution for that case for exactly the reasons described at the end of Section 3.4 (the "it can take 15 hours to the nearby village" problem).

Summary: A good set of transit nodes can be found for both road and public transportation networks. However, in public transportation networks, we do not have efficient algorithms for the local searches required to precompute the distances to the access nodes or at query time when source and target are close together.

4   Conclusions

We gave an overview of the main techniques to speed up shortest path computation on transportation networks compared to Dijkstra's algorithm. We specifically looked at: bidirectional search, hierarchies of subnetworks, goal direction, contraction and shortcuts, and distance tables. We found that all of these approaches work well (and some extremely well) for road networks, but none of them gave convincing results for public transportation networks so far. We identified two key open problems which so far have obviated fast routing on very large public transportation networks:

Open Problem 1: (Speed-up despite lack of hierarchy) How to achieve, with reasonable precomputation cost, a significant speed-up over Dijkstra's algorithm in large municipal areas with hardly any hierarchy, for example, in large bus-only networks?

Open Problem 2: (Efficient local searches) How to compute shortest paths to all nodes in a local (for example, geographic) neighborhood efficiently, in the face of (albeit few) shortest paths within that neighborhood of large cost?

Acknowledgments

Thanks to Peter Sanders for a brief, but very fruitful exchange on the right selection and categorization of the tricks of the trade in Section 3.

References

1. Dijkstra, E.: A note on two problems in connexion with graphs. Numerische Mathematik 1, 269–271 (1959)
2. Delling, D.: Time-dependent SHARC-routing. In: Halperin, D., Mehlhorn, K. (eds.) ESA 2008. LNCS, vol. 5193, pp. 332–343. Springer, Heidelberg (2008)
3. Batz, G.V., Delling, D., Sanders, P., Vetter, C.: Time-dependent contraction hierarchies. In: 11th Workshop on Algorithm Engineering and Experiments (ALENEX 2009), pp. 97–105 (2009)
4. Pyrga, E., Schulz, F., Wagner, D., Zaroliagis, C.D.: Efficient models for timetable information in public transportation systems. ACM Journal of Experimental Algorithmics 12 (2007)
5. Müller-Hannemann, M., Schnee, M.: Finding all attractive train connections by multi-criteria pareto search. In: 4th Workshop on Algorithmic Methods for Railway Optimization (ATMOS 2004), pp. 246–263 (2004)
6. Schultes, D., Sanders, P.: Dynamic highway-node routing. In: Demetrescu, C. (ed.) WEA 2007. LNCS, vol. 4525, pp. 66–79. Springer, Heidelberg (2007)
7. Müller-Hannemann, M., Schulz, F., Wagner, D., Zaroliagis, C.D.: Timetable information: Models and algorithms. In: 4th Workshop on Algorithmic Methods for Railway Optimization (ATMOS 2004), pp. 67–90 (2004)
8. Sanders, P., Schultes, D.: Highway hierarchies hasten exact shortest path queries. In: Brodal, G.S., Leonardi, S. (eds.) ESA 2005. LNCS, vol. 3669, pp. 568–579. Springer, Heidelberg (2005)
9. Sanders, P., Schultes, D.: Engineering highway hierarchies. In: Azar, Y., Erlebach, T. (eds.) ESA 2006. LNCS, vol. 4168, pp. 804–816. Springer, Heidelberg (2006)
10. Bauer, R., Delling, D., Wagner, D.: Experimental study on speed-up techniques for timetable information systems. In: 7th Workshop on Algorithmic Methods for Railway Optimization (ATMOS 2007) (2007)
11. Geisberger, R., Sanders, P., Schultes, D., Delling, D.: Contraction hierarchies: Faster and simpler hierarchical routing in road networks. In: McGeoch, C.C. (ed.) WEA 2008. LNCS, vol. 5038, pp. 319–333. Springer, Heidelberg (2008)
12. Bauer, R., Delling, D.: SHARC: Fast and robust unidirectional routing. In: 10th Workshop on Algorithm Engineering and Experiments (ALENEX 2008), pp. 13–26 (2008)
13. Hart, P., Nilsson, N., Raphael, B.: A formal basis for the heuristic determination of minimum cost paths. IEEE Transactions on Systems Science and Cybernetics 4(2), 100–107 (1968)
14. Goldberg, A., Harrelson, C.: Computing the shortest path: A* search meets graph theory. In: 16th Symposium on Discrete Algorithms (SODA 2005), pp. 156–165 (2005)
15. Lauther, U.: An extremely fast, exact algorithm for finding shortest paths in static networks with geographical background. Münster GI-Tage (2004)
16. Köhler, E., Möhring, R.H., Schilling, H.: Acceleration of shortest path and constrained shortest path computation. In: Nikoletseas, S.E. (ed.) WEA 2005. LNCS, vol. 3503, pp. 126–138. Springer, Heidelberg (2005)
17. Bast, H., Funke, S., Matijevic, D.: Ultrafast shortest-path queries via transit nodes. In: DIMACS Implementation Challenge Shortest Paths (2006); An updated version of the paper appears in the upcoming book
18. Bast, H., Funke, S., Matijevic, D., Sanders, P., Schultes, D.: In transit to constant time shortest-path queries in road networks. In: 9th Workshop on Algorithm Engineering and Experiments (ALENEX 2007) (2007)

Is the World Linear?

Rudolf Fleischer[*]

Fudan University
Shanghai Key Laboratory of Intelligent Information Processing
Department of Computer Science and Engineering
Shanghai 200433, China
rudolf@fudan.edu.cn

Dedicated to Kurt Mehlhorn's 60th Birthday

Abstract. Super-resolution is the art of creating nice high-resolution raster images from given low-resolution raster images. Since "nice" is not a well-defined term in mathematics and computer science, we propose a linear model of the world that allows us, under certain conditions, to achieve perfect super-resolution for arbitrarily high resolution. For example, we may now create a larger-than-life picture of Kurt.

1 Reminiscences

I had the great pleasure to spend 15 years in the group of Kurt Mehlhorn. As a sophomore at the Universität des Saarlandes in Saarbrücken, I joined his group in 1983 as a programming slave in the Hill project [16]; several years later, some mysterious system crashes could be traced back to my first (and last) implementation of a red-black tree. After a year in Beijing I returned in 1989 just in time to witness the foundation of the Max-Planck-Institut für Informatik, or short, the MPI, where I continued my academic career as a PhD student. I remember these years as a time when we students could pursue our own independent research endeavors with little interference from our supervisor.[1] Enjoying the academic freedom and superb working conditions at the MPI, I stayed in Saarbrücken until my habilitation in 1999, and then moved to Canada to see a bit more from the world.

Kurt is a researcher with a very braod range of interests, and I consider myself lucky that I had the opportunity to spend so many years of my early career under his guidance. After a short period of work in abstract complexity theory [19] he saw the light and shifted his research focus to more practical problems in the design and analysis of efficient algorithms. Eventually, his efforts to make algorithm

[*] This work was partially supported by a grant from the National Natural Science Foundation of China (No. 60573025), the National High Technology Research and Development Program of China (863 Program) (No. 2007AA01Z176), and the Shanghai Leading Academic Discipline Project (project number B114).

[1] We did meet him regularly at lunch time, though, and sometimes on the tennis court.

S. Albers, H. Alt, and S. Näher (Eds.): Festschrift Mehlhorn, LNCS 5760, pp. 368–379, 2009.

research more practicable and usable culminated his development of the LEDA library for C++ [20]. Besides being an incredibly efficient administrator, Kurt's great strengths have always been his ability to properly model a problem and his emphasis on rigorous proofs (preferably elegant ones). In this note, dedicated to his 60th birthday, I follow this path and propose a new mathematical model for super-resolution that makes it for the first time (as far as I know) possible to prove under which conditions perfect super-resolution can be achieved.

2 Introduction

Super-resolution is the art (or magic) of generating a high-resolution raster image from a given low-resolution raster image[2]. Changing the resolution becomes necessary whenever we change the size of an image, for example when we want to display an image on different devices (e.g., computer screen, TV screen, mobile phone, printer, etc.). For example, we may want to enhance the quality of a photograph taken by a low-resolution mobile phone, or we may want to enhance the resolution to improve the success rate in automatic image recognition applications (e.g., automatic person identification in a camera surveillance system), or we may want to upgrade DVD movies to HDTV. Super-resolution is not restricted to the enhancement of still images, it is also used for video enhancement [26], in acoustics (speech recognition) [4], etc. Unfortunately, super-resolution is an ill-posed optimization problem because there can be infinitely many images yielding the same given low-resolution image using a fixed technology, not to mention the problem of trying to do super-resolution in situations where the imaging technology is unknown. Still, because of its importance in daily life, many algorithms have been proposed for super-resolution in general and for special applications.

Background. The idea of super-resolution, or *upscaling*, was first proposed by Tsai and Huang [31], although the term itself only appeared much later [13]. Initially, the goal was to compute a high resolution image from several slightly different frames of a moving object (e.g., see [10]), for example from a video sequence [14, 28]. Later, single image super-resolution was proposed, which is more demanding since fewer information is available. For surveys on super-resolution techniques see, for example, [34, 35].

The simplest super-resolution algorithm, *pixel replication (PR)*, just replaces each pixel by a square of equally colored pixels, which usually results in a poor image.

Most super-resolution algorithms are based on spatial interpolation [1, 5, 15, 17, 21]. The simplest form is *linear interpolation (LI)* which colors new intermediate pixels by the distance-weighted average of the values of the four closest original pixels. More sophisticated variants are bilinear and bicubic interpolation, and methods based on wavelet transforms [23]. LI is simple and works reasonably well if the objects are not too finely structured. Bilinear and cubic

---

[2] In this paper, we only consider grayscale images, although it is clear that the results generalize to color images.

interpolation are often used in video players to display on HDTV screens, but edges tend to become fuzzy, which requires employing sharpening filters after the interpolation. One of the best general purpose upscaling algorithms, producing the sharpest images and used in many software systems (e.g., see [30]), seems to be the `Lanczos3` algorithm [32]. Unser gave a nice survey on sampling and interpolation [33].

Interpolation is a *reconstruction-based* algorithm. According to Baker and Kanade [3], reconstruction-based algorithms are based on the assumption that there exists some model of the world, i.e., the structure of images and the picture taking device, that can be exploited to guide the upscaling process. For the picture taking device, the usual assumption is that we have a simple digital camera whose pixels are assigned values according to the average light intensity over the pixel. We call this value the *pixel color*. Since a raster image is then simply the result of a downsampling process, super-resolution becomes the problem of finding the most suitable original high-resolution image that might have produced the given low-resolution image [24].

Simple math shows that reconstruction-based algorithms cannot be very successful in general. Baker and Kanade [3] observed that the space of original images mapped to the same target image is growing very fast with increasing magnification factor, which means that even for magnification factors as small as 4 the performance of reconstruction-based super-resolution algorithms can be far from satisfactory. This was later verified by Lin and Shum [18] who proved that the practical limit for the magnification factor is 1.6.

To break this natural barrier, Baker and Kanade [3] suggested to study *recognition-based* algorithms. By adding a learning step, they transformed the reconstruction problem into a recognition problem which has much higher chance of success for super-resolution. Later, many super-resolution schemes were based on the idea of first learning the "best" pixel enlargement on a certain set of "typical" examples [3, 6, 7, 11]. Recently, Sun *et al.* [29] proposed a super-resolution scheme based on the gradient field of an image.

Another way to improve the performance of reconstruction-based algorithms is to add stronger assumptions to the world model. Sajjad *et al.* [27] suggested to color pixels based on local geometric shapes, which works well for images of objects whose boundaries match the proposed shapes. Akins *et al.* [2] suggested to color pixels dependent on their classification in a small neighborhood of pixels, where the classification parameters are learned in a training phase.

Our contribution. We propose a new reconstruction-based algorithm. The algorithm itself is probably not really new, some people might call it a vectorization technique. What is new is that it is based on a clean and simple world model that allows us to formulate super-resolution as a well-defined optimization problem that can be solved to perfection if certain conditions on the structure of the image and the resolution are satisfied. In traditional super-resolution research the quality of a new algorithm can only be determined visually in comparison with various old algorithms, or sometimes by means of some mathematically provable error reduction properties.

We have implemented a rough protoype of our new algorithm so that we can show a few upscaled images. Since our goal was to propose a theoretically sound framework for super-resolution and not to show visual superiority to exisiting algorithms, we have not implemented the best known super-resolution algorithms to compare our images against. Instead, we have implemented two of the trivial algorithms, PR and LI, to demonstrate that our method achieves some non-trivial image enhancement.

Structure of the paper. We define our new model in Section 3. In Section 4, we explain how to achieve good super-resolution in this model. We end with conclusions in Section 5 and a rather large picture of Kurt (Fig. 5).

3   A Linear World

We all know the world is not flat (with a notable exception [12]), but what if it was linear? A typical image consists of non-intersecting plane shapes whose grayscale values are either constant or changing gradually and smoothly. Shape boundaries are characterized by a sharp change in the grayscale value. In reality, object boundaries may be curved or even fractal, but any curve can be locally approximated by a polygonal chain. Since an image is not more than a local approximation of reality, we may as well assume that boundaries are locally piecewise linear. Similarly, we may assume that the changes of light intensity on a surface can be locally described by a linear gradient. This means, there is a direction f such that any line with slope f has constant light intensity, and the intensity changes linearly if we move the line perpendicular to itself. See Fig. 1(a) for an example of an object with linear boundaries (the pattern, in its true white and blue colors, is easily recognized anywhere south of the Weisswurst equator). Figs. 1(b)–1(d) show the same image with magnification factor 32 as produced by our new algorithm, Exact, and the two trivial algorithms PR and LI. Note that the Exact picture looks much sharper than the other two upscaled images, even sharper than the original picture.

The linearity assumption has some interesting consequences. The following two lemmas (which are probably folklore) justify that LI is a good super-resolution algorithm, except at object boundaries which tend to become fuzzy. The following lemma can be proved purely geometrically, without using integrals (the pixel color is the integral of the light intensity function over the pixel).

Lemma 1. *On a surface whose light intensity changes according to a linear gradient, the color of a pixel is exactly the value of the light intensity in the center of the pixel.*

Proof. Let c be the center of the pixel, and let v be the pixel color (see Fig. 2). Let f be the light gradient, and let g be the line of slope f through c, which by definition has constant light intensity, which we denote by w. By definition, v is the average light intensity over the pixel (more formally, we would have to

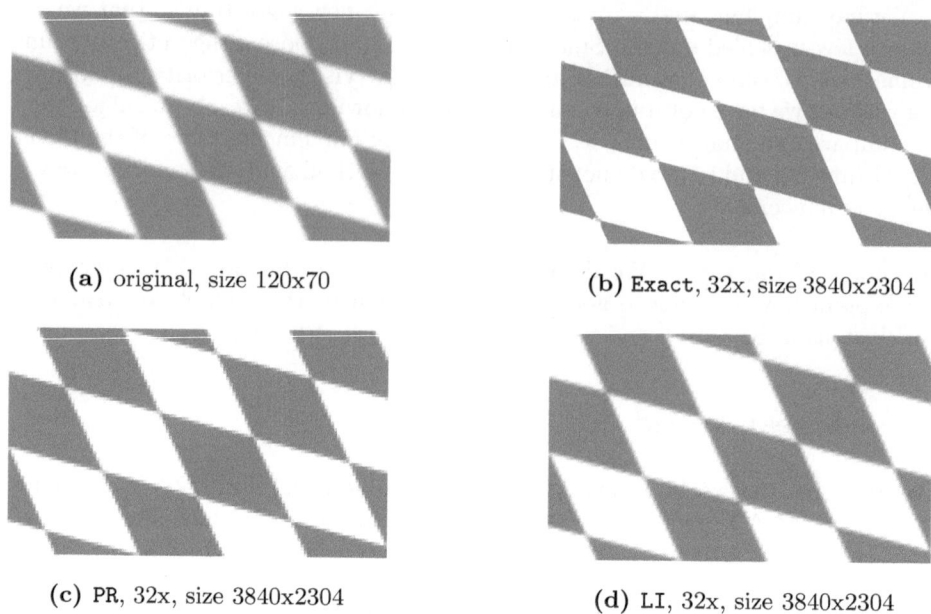

(a) original, size 120x70

(b) Exact, 32x, size 3840x2304

(c) PR, 32x, size 3840x2304

(d) LI, 32x, size 3840x2304

Fig. 1. A white-blue pattern, of restricted geographic importance; magnification 32x

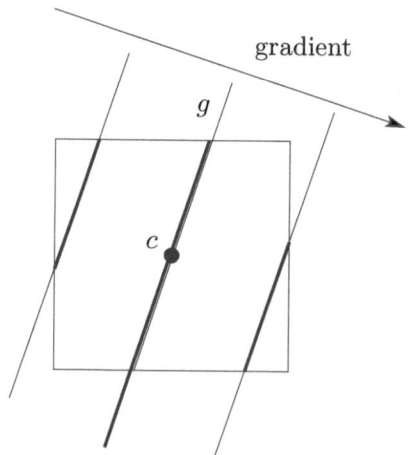

Fig. 2. If the light intensity is given by a linear gradient, the pixel color is exactly the light intensity at its center c

compute an integral). By the linearity of the gradient, two lines of slope f with c in the middle between the two lines contribute the same value w to v as the line g (though with a different weight as g, which would be the length of the cross-section between the line and the pixel). Thus, $v = w$.   □

Corollary 1. *On a surface whose light intensity changes according to a linear gradient,* LI *colors new intermediate pixels with the correct value. That is,* LI *achieves perfect super-resolution on such a surface.*   □

4   Exact Super-Resolution

To achieve perfect super-resolution, all we have to do is to detect the linear object boundaries and the linear gradients for the surfaces between boundaries. There are many edge-detection algorithms for raster images [8, 22, 25], but in our situation it is actually quite easy to detect them locally. Assume a linear boundary line between regions of constant colors c_l and c_r, respectively, crosses a pixel vertically, see Fig. 3. Then the pixel color can be computed as

$$c_b = \frac{a+b}{2} \cdot c_l + (1 - \frac{a+b}{2}) \cdot c_r .$$

Similarly, the combined color value of the two pixels in the row above can be computed as

$$c_t = \frac{b+c}{2} \cdot c_l + (1 - \frac{b+c}{2}) \cdot c_r .$$

Note that c_t is not just the color of the single pixel above our pixel. Since we also know that

$$b = \frac{a+c}{2} ,$$

we can compute a and b from the observed pixel colors c_b and c_t as

$$a = \frac{3c_b - c_t - 2c_r}{2(c_l - c_r)}$$

and

$$b = \frac{c_b + c_t - 2c_r}{2(c_l - c_r)} .$$

Note that this is only one of several cases, where the pixel is crossed vertically. Other cases are that the boundary line crosses three pixels on the top row, or that the boundary line crosses the pixel horizontally (where we need to rotate the two vertical cases by 90 degrees). It is actually possible to compute a and b directly just from the colors of our pixel and the pixel above (or a right neighbor for horizontal boundary lines). However, the equations for b and c become complicated, involving two nested square roots in the worst situation.

This leads to a simple algorithm for super-resolution, which we call **Exact**. For a given raster image, we first determine the *constant pixels* which are those pixels having a neighbor of the same color. Then we find all pixels that correspond to one of the four cases of boundary lines as explained above and compute the equations of the lines crossing the pixels (in a final step we also adjust the lines

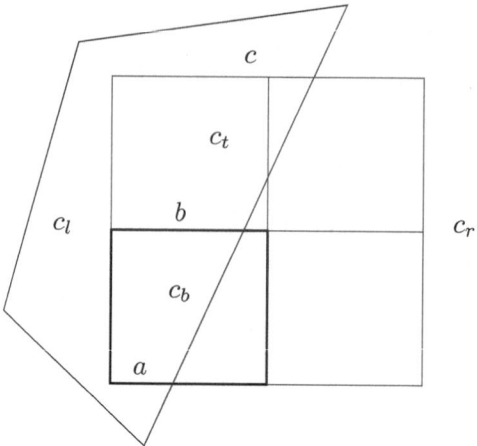

Fig. 3. Two regions of constant color meet in a pixel (thick boundary)

slightly at pixel boundaries so that they form polygonal chains). For these pixels, it is then easy to correctly color the smaller pixels that replace the pixel in the upscaled image. All remaining pixels are labeled as gradient pixels where we use LI to color the smaller upscaled pixels.

Note that our boundary line detection requires that the pixels on both sides of the boundary line can be classified as constant pixels. This is not always possible, for example when several boundary lines are very close together. This happens in particular around pixels where several boundary lines meet. In the current implementation, these areas are not properly handled, see for example Fig. 1(b), where the points where two polygons meet in a vertex are a bit fuzzy in the enlarged image. However we do not consider this a great disadvantage of our method. If boundaries lie very close together, then it probably means that our camera did not record the pixel colors correctly anyway, and we cannot expect our (or any other) super-resolution algorithm to correct such errors. This applies in particular to cases where several boundaries cross a single pixel (without meeting there). In general, if two boundaries are very close together, it seems not wrong to treat them as a gradient (which is what happens in our current implementation).

In all other cases, our algorithm can guarantee a perfect reconstruction of the original image. In that sense, our algorithm behaves similar as, for example, the algorithms that try to reconstruct a smooth curve from a sample of points on the curve [9]. These algorithms only succeed if the points satisfy a certain density condition.

Theorem 1. *Our new algorithm* Exact *can perfectly reconstruct all parts of a downsampled image (in the linear world) where any two object boundaries are separated by at least two pixels of the same color.*   □

We have implemented a prototype of Exact in C++. It is so simple that we did not even need LEDA [20]; the most complicated data structure is a two-dimensional

(a) original, size 324x216

(b) `Exact`, 32x, size 10368x6912

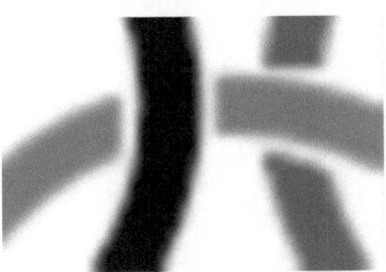

(c) original, detail, size 40x27

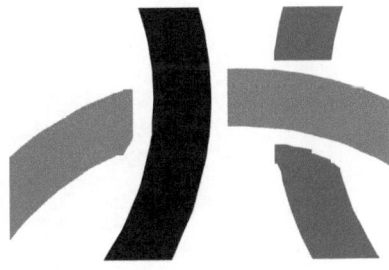

(d) `Exact`, 32x, detail, size 1296x864

(e) `PR`, 32x, detail, size 1296x864

(f) LI, 32x, detail, size 1296x864

Fig. 4. Olympic rings, magnification 32x. Since we cannot show the enlarged images in full size, we show a small detail in Figs. (c)–(f).

(a) original, size 298x414

(b) Exact, 16x, size 4768x6624

Fig. 5. Kurt growing up, magnification 16x. The original picture is taken from http://www.mpi-inf.mpg.de/~mehlhorn/fotos/Kurt1980.jpg.

array. The prototype still has some minor bugs, as can be observed in some minor artefacts in the computed high-resolution images (for example, Fig. 4(d)).

5   Conclusions

Note that Sun *et al.* [29] also proposed to use gradients to detect boundaries. However, their algorithm requires some parameters that must be guessed (or empirically determined), while our algorithm does not require any parameters. Also, our goal is to achieve *provably* good super-resolution (versus super-resolution that produces visually "nice" images).

Of course, at this point a question arises naturally: Is the world linear? And what, if it is not? Fig. 4 shows a famous image of interleaved circles that our algorithm could enlarge properly with magnification 32x. Thus, our simplified linear world model seems to be good enough to allow for high-quality super-resolution of non-linear images. It should also be possible to extend the theory of our local boundary detection to boundaries of higher degree, i.e., instead of computing linear boundaries we could compute quadratic or cubic boundary lines. Alternatively, we could try to smoothen the polygonal chains that we compute, pixel by pixel, by some spline function.

We are currently working on a better implementation of Exact. We hope the new implementation will have an improved boundary detection that allows us to strengthen Thm. 1. For example, it should be possible to correctly identify those pixels where two boundaries meet by interpolating the shape of polygonal chains into areas where several of them are too close together to be properly recognized locally.

Another problem is robustness. Obviously, our algorithm is quite sensitive to noise in the image data. One solution might be to introduce a threshold ϵ such that two pixels are considered to have the same color if their pixel colors differ by at most ϵ. Though the upscaling of the Olympic rings (Fig. 4) looks very nice, boundary detection is much less useful for upscaling of photographs, mainly because photographs do not satisfy our assumptions on the world model: there are usually no clear-cut gradients and boundaries in a photo, even on surfaces of constant color pixel colors may actually vary slightly. For example, in Kurt's photo (Fig. 5), only 7% of all pixels are classified as boundary line pixels, versus 70% gradient pixels which are upscaled using LI. In comparison, the Olympic rings have 3% boundary line pixels and 2% gradient pixels (which are actually wrong classifications due to minor imperfections in the original picture and, maybe, some minor bugs in my program code).

Acknowledgments

My students Nate Xu Xiaoming and Wang Yihui worked out the above-mentioned complicated formulas for determining line equations from pixel colors. Erik Cheng is currently trying to transform my prototype into a useful piece of software. My colleague Zhang Junping is funding this research from his 863 project.

References

1. Acharya, T., Tsai, P.-S.: Computational foundations of image interpolation algorithms. ACM Ubiquity 8 (2007)
2. Atkins, C.B., Bouman, C.A., Allebach, J.P.: Optimal image scaling using pixel classification. In: Proceedings of the 2001 International Conference on Image Processing (ICIP 2001), vol. 3, pp. 864–867 (2001)
3. Baker, S., Kanade, T.: Limits on super-resolution and how to break them. In: Proceedings of the 2000 IEEE Conference on Computer Vision and Pattern Recognition (CPVR 2000), vol. 2, pp. 372–379 (2000)
4. Blomgren, P., Papanicolaou, G., Zhao, H.: Super-resolution in time reversal acoustics. Journal of the Acoustical Society of America 111, 230–248 (2002)
5. Blu, T., Thévenaz, P., Unser, M.: Generalized interpolation: higher quality at no additional cost. In: Proceedings of the 1999 International Conference on Image Processing (ICIP 1999), vol. 3, pp. 667–671 (1999)
6. Candocia, F.M., Principe, J.C.: Superresolution of images with learned multiple reconstruction kernels. In: Guan, L., Kung, S.Y., Larsen, J. (eds.) Multimedia Image and Video Processing, ch. 4, pp. 219–243. CRC Press, New York (2000)
7. Corduneanu, A., Platt, J.C.: Learning spatially-variable filters for super-resolution of text. In: Proceedings of the 2010 International Conference on Image Processing (ICIP 2010), vol. 1, pp. 849–852 (2005)
8. Davis, L.: A survey of edge detection techniques. Computer Graphics and Image Processing 4, 248–270 (1975)
9. Dey, T.K., Mehlhorn, K., Ramos, E.A.: Curve reconstruction: connecting dots with good reason. Computational Geometry: Theory and Applications 10, 289–303 (2000)
10. Farsiu, S., Elad, M., Milanfar, P.: A practical approach to super-resolution. In: Proceedings of the 40th Asilomar Conference on Signals, Systems and Computers (2006) (invited paper)
11. Freeman, W.T., Jones, T.R., Pasztor, E.C.: Example-based super-resolution. IEEE Computer Graphics and Applications, 56–65 (2002)
12. Friedman, T.L.: The World is Flat: A Brief History of the Twenty-First Century. Farrar, Strauss and Giroux (2005)
13. Irani, M., Peleg, S.: Super resolution from image sequences. In: Proceedings of the 10th International Conference on Pattern Recognition (ICPR 1990), vol. 2, pp. 115–120 (1990)
14. Jiang, Z., Wong, T.-T., Bao, H.: Practical super-resolution from dynamic video sequences. In: Proceedings of the 2003 IEEE Conference on Computation Vision and Pattern Recognition (CVPR 2003), vol. 2, pp. 549–554 (2003)
15. Lehmann, T.M., Gönner, C., Spitzer, K.: Survey: interpolation methods in medical image processing. IEEE Transactions on Medical Imaging 18(11), 1049–1075 (1999)
16. Lengauer, T., Mehlhorn, K.: The HILL system: a design environment for the hierarchical specification, compaction, and simulation of integrated circuit layouts. In: Penfield Jr., P. (ed.) Proceedings of the MIT VLSI Conference, Artech House, Inc. (1984)
17. Li, X., Orchard, M.T.: New edge-directed interpolation. IEEE Transactions on Image Processing 10(10), 1521–1527 (2001)
18. Lin, Z., Shum, H.-Y.: Fundamental limits of reconstruction-based superresolution algorithms under local translation. IEEE Transactions on Pattern Analysis and Machine Intelligence 26(1), 83–97 (2004)

19. Mehlhorn, K.: On the size of sets of computable functions. In: Proceedings of the 14th IEEE Symposium on Automata and Switching Theory, pp. 190–196 (1973)
20. Mehlhorn, K., Näher, S.: The LEDA Platform for Combinatorial and Geometric Computing. Cambridge University Press, Cambridge (1999)
21. Meijering, E.: A chronology of interpolation: from ancient astronomy to modern signal and image processing. Proceedings of the IEEE 90(3), 319–342 (2002)
22. Mitra, B.: Gaussian-based edge-detection methods: a survey. IEEE Transactions on Systems, Man, and Cybernetics — Part C: Applications and Reviews 32(3) (2002)
23. Mueller, N., Lu, Y., Do, M.N.: Image interpolation using multiscale geometric representations. In: Proceedings of the 2007 SPIE Conference on Electronic Imaging (2007)
24. Price, J.R., Hayes III, M.H.: Optimal prefiltering for improved image interpolation. In: Proceedings of the 32nd Asilomar Conference on Signals, Systems and Computers, vol. 2, pp. 959–963 (1998)
25. Price, K.: Keith Price Bibliography: evaluation of edge detection algorithms (2009), http://www.visionbib.com/bibliography/edge235.html
26. Raghupathy, A., Chandrachoodan, N., Liu, K.J.R.: Algorithm and VLSI architecture for high performance adaptive video scaling. IEEE MultiMedia 5(4), 489–502 (2003)
27. Sajjad, M., Khattak, N., Jafri, N.: Image magnification using adaptive interpolation by pixel level data-dependent geometrical shapes. Proceedings of the World Academy of Science, Engineering and Technology 25, 88–97 (2007)
28. Shechtman, E.: Space-time super-resolution. Master's thesis, Faculty of MAthematics and Computer Science, The Weizmann Institute of Science (2003)
29. Sun, J., Sun, J., Xu, Z., Shum, H.-Y.: Image super-resolution using gradient profile prior. In: Proceedings of the 2007 IEEE Conference on Computation Vision and Pattern Recognition (CVPR 2007) (2007); Poster
30. Easy Thumbnails User Manual. Fookes Software, Switzerland (2001)
31. Tsai, R.Y., Huang, T.S.: Multiframe image restoration and registration. In: Advances in Computer Vision and Image Processing, ch. 7, vol. 1, pp. 317–339. JAI Press, Greenwich (1984)
32. Turkowski, K.: Filters for common resampling tasks. In: Glassner, A.S. (ed.) Graphics Gems I, pp. 147–165. Academic Press, London (1990)
33. Unser, M.: Sampling — 50 years after Shannon. Proceedings of the IEEE 88(4), 569–587 (2000)
34. van Ouwerkerk, J.D.: A modular approach to image super-resolution algorithms. Ph.D. thesis, Dept. of Media and Knowledge Eng., Delft Univ. of Technology (2006)
35. Wittman, T.: Mathematical techniques for image interpolation, Oral exam paper (2005), http://www.math.umn.edu/~wittman/Poral2.pdf

In Praise of Numerical Computation

Chee K. Yap⋆

Dedicated to Kurt Mehlhorn on His 60th Birthday

Abstract. Theoretical Computer Science has developed an almost exclusively discrete/algebraic persona. We have effectively shut ourselves off from half of the world of computing: a host of problems in Computational Science & Engineering (CS&E) are defined on the continuum, and, for them, the discrete viewpoint is inadequate. The computational techniques in such problems are well-known to numerical analysis and applied mathematics, but are rarely discussed in theoretical algorithms: iteration, subdivision and approximation. By various case studies, I will indicate how our discrete/algebraic view of computing has many shortcomings in CS&E. We want embrace the continuous/analytic view, but in a new synthesis with the discrete/algebraic view. I will suggest a pathway, by way of an exact numerical model of computation, that allows us to incorporate iteration and approximation into our algorithms' design. Some recent results give a peek into how this view of algorithmic development might look like, and its distinctive form suggests the name "numerical computational geometry" for such activities.

*You might object that it would be reasonable enough for me to try to
expound the differential calculus, or the theory of numbers, to you,
because the view that I might find something of interest to say to you
about such subjects is not* prima facie *absurd; but that geometry is, after
all, the business of geometers, and that I know, and you know, and I
know that you know, that I am not one; and that it is useless for me to
try to tell you what geometry is,
because I simply do not know.*

— G.H.Hardy, in "What is Geometry?"

1925 Presidential Address to the Mathematical Association

1 Introduction

This article celebrates the scientific work of Professor Kurt Mehlhorn, a special
friend and colleague. Few computer scientists can match the impact that Kurt
has had in computer science. Even to summarize the scope of his work would be

⋆ The work is supported in part by NSF Grants CCF-043086 and CCF-0728977.

S. Albers, H. Alt, and S. Näher (Eds.): Festschrift Mehlhorn, LNCS 5760, pp. 380–407, 2009.

a daunting task. Since this essay is about numerics, I may let the numbers speak for themselves: his current webpage lists 207 papers, 9 books and 6 software systems. I propose to only highlight one aspect of Kurt's experimental work, as it is a special tribute to say that any theoretician had significant experimental contributions. Over twenty years ago, Kurt began a quest to put the corpus of data structures and algorithms produced by the theoretical Computer Science community into code. That was the birth of the software library known as LEDA [31, 32]. Indeed, around this time, computational geometry witnessed a spurt of experimental geometric software development. But major software development requires sustained effort over a long period of time which, as theoreticians, we may not have the constitution for. Yet today LEDA is the basis of a successful commercial company. Like most large software, LEDA is the work of many hands: Kurt's collaborators include Stefan Näher with whom he wrote the LEDA book [30], Stefan Schirra, Christian Uhrig, Christoph Burnikel and others.

¶1. *What* LEDA *has Achieved.* LEDA has implemented the best practical data structures and discrete algorithms that have been developed in the last 40 years. But the unique part of LEDA lies in its collection of geometric algorithms. Since the late 1980's, computational geometers have become acutely aware of numerical nonrobustness issues in geometric computation. Of all the areas of algorithms, we are especially afflicted. Some have declared the problem intractable, even for problems as simple as the robust intersection of two line segments. In retrospect, what is remarkable about Kurt's foresight was his insistence, from the very first, that LEDA must be fully reliable and practical, *even for geometric algorithms.* Twenty years ago, that was a big wish for a geometric library. A few "robust geometric algorithms" were beginning to appear in the literature, but nothing with which to stock an entire library. Each problem required special treatment, and many approaches were contending to solve nonrobustness issues (see my survey in [54]).

I will classify these approaches into two camps: those wishing to make fast machine arithmetic reliable and those wishing to compute exactly in order to achieve reliable software. Kurt's approach falls under the latter "exact" camp. Many researchers in our community did not think the exact camp could be practical or could compete with machine floating point computation. To place yourself in context, by the late 1980's, machine floating point had become the dominant mode of numerical computation (and has remained so today). Floating point arithmetic has become standardized, enjoys full industry support, and has moved from software into standard hardware in the form of co-processors. This view is summed up by Steve Fortune's foreword in an Algorithmica special issue on implementation issues [22]: "*Floating point arithmetic has numerous engineering advantages: it is well-supported ... the Challenge is to demonstrate that a reliable implementation can result from the use of floating point arithmetic.*"

What about exact computation? It was (and still is) regarded as the domain of specialists and specialized applications. Yu [59] wrote a thesis under Chris Hoffmann that concluded that exact computation will not be practical for Boolean operations on polyhedral objects in the foreseeable future. But LEDA did find a general and systematic solution to nonrobust geometry — not by implementing specialized "robust algorithms" for each problem — but by introducing a general number type called `Leda Real` that has the remarkable property that comparisons are error-free. According to the principles of **exact geometric computation**, this implied that the geometry would be exact and hence free from nonrobustness issues. Now, if a computation involves only rational operations, then this property might not appear impressive (just use a BigRational number package, although you would still run into the efficiency bottleneck described by Yu). But `Leda Real` included square-roots and later, arbitrary real algebraic numbers. Despite this, it remains practical for all the common geometric problems. Today, such algorithms are reasonably competitive with nonrobust machine-precision algorithms. Any programmer can implement a fully robust geometric algorithm (provided the primitives are algebraic) using software such as LEDA. Superficially, it appears that the exact camp has won in a healthy contest of ideas. But lurking behind this triumph, we see some ideas of the other camp are also firmly embedded.

¶2. *Exact Numerical Computation.* How does `Leda Real` do this? There are five key ingredients, the first two well-known and next three novel:

(1) You must use arbitrary precision — but use **BigFloats** (for efficiency, do not use BigRationals).
(2) Track errors automatically — use **interval arithmetic**. Interval arithmetic tells us when a comparison between approximate values is valid.
(3) All numbers must have an exact representation — use **expressions**. This representation supports the the ability to approximate each number to any desired absolute precision. Such approximations must be available on demand.
(4) You must solve the **zero problem**, described later. In practice, we use some **constructive zero bounds** which tell us when a numerical approximation is small enough that we may declare the exact value to be zero. The BFMSS bound [11] from the LEDA group is one of the best zero bounds in this area.
(5) You should exploit adaptivity of numerical computations. A highly effective technique here is **numerical filters** which can decide most comparisons quickly. Thus, through filters, the "engineering advantages of floating point arithmetic" of Fortune is restored. Work from LEDA is in the vanguard of trying to extend such techniques, from cascading filters to filtering of general algorithms [12, 23].

These ideas also appear in my earlier work on the `Real/Expr` [57], the precursor to `Core Library`. Another major library founded on similar principles of exact numerical computation is the CGAL library [21]. The computing principle that urges us to such a distinctive mode of computation is exact geometric computation. But in this paper, I want to look at the broader implications; for this purpose, I call this mode of computation **exact numerical computation** (ENC). Note that "numerical" often has the connotation of inexactness, but no such inference is[1] intended

---

[1] There is an important and related issue of inexact data, which I do not address in this essay.

here. Of course, we will use numerical approximations, but they are used to derive exact conclusions with the help of zero bounds. Actually, exactness in ENC cannot be taken for granted: very little is known about the zero problem in transcendental cases [42]. In such cases, as applied mathematicians know very well, we need carefully circumscribed conditions (smoothness, Morseness, non-singularity, Lipschitz, etc) that allow exact solution. Another way to restore exactness is to modify correctness in the sense of backwards error analysis. All these are within the parameters of ENC.

Exact computation is traditionally the domain of symbolic computation and computer algebra. Nevertheless, ENC has no parallel in the computer algebra literature (e.g., [10] or [17, Chap. 4]). Algebraic computation in computer algebra is greatly influenced by the great subject of algebraic number theory, focusing on algebraic and arithmetical properties of number fields $\mathbb{Q}(\alpha)$ (e.g., [37]). But such approaches do not have the flexibility and adaptivity necessary to be deployed in practical geometric computations. My favorite illustration is the following: to compute a number of the form $\alpha = \sum_{i=1}^{100} \sqrt{n_i}$ (for positive integers n_i), standard computer algebra methods require the computation of a defining polynomial of α which generally has degree 2^{100}, a daunting task. Yet, in a geometric application like computing Euclidean shortest paths, we may have to handle thousands of such α's. Using our ENC approach, most of these computations can be dispatched quite routinely since we only need to construct an expression for each α. The comparison of such α's could be time consuming, but in practice we are saved by ENC's adaptive complexity.

To sum up, I believe that LEDA represents an important achievement in computing history: through the work of LEDA and related work in the computational geometry community, we now understand the fundamental barriers to robust geometric computation and have identified key elements for solving this problem in a systematic way. The existence of commercial libraries such as LEDA and CGAL prove that robust geometric computation is a practical reality today. Kurt's broad insights and leadership in this area have played a major role in this achievement. To read some of Kurt's thoughts on this area, I recommend[2] his article [29]. The societal benefits of robust geometric computation are potentially immense: nonrobust numerical computation has negative impact on programmer/researcher productivity (many of us experience this), represents a huge economic cost [41], stands in the way of full automation in industry, and often plays a role in dramatic disasters.

¶3. *An Apology.* The above quote from Hardy [25] expresses my own ambivalence about writing on numerical computation, for I know that you know that I do not do much numerical computing. What little I know is the combination of numerical computation with algebraic computation. It is this synthesis that I will talk about. My praise of numerical computation represents a slow personal conversion that has grown over time. When I told a colleague what I will write about, the reaction was — *but surely computation is discrete?* I hope to show that there is a deeper issue at stake.

---

[2] It was written for another similar occasion, the festschrift of Thomas Ottmann.

2 Return to the Continuum

What I have discovered over the years, as an unintended consequence of the pursuit of robust geometric computation, is that numerical computation has many virtues that theoretical algorithms fail to recognize. We have been enamored with discrete computation (which is good in itself) but to the exclusion of the continuous. We feel that if a problem or algorithm is numerical, it is the domain of numerical analysts and applied mathematicians. It is true that we should not be amateurs in what others can do better. But the study of ENC has convinced me that some of these numerical concerns should be our concern.

¶4. *Problems in Computational Science & Engineering* Let me briefly clarify what I mean by the "continuum". In some literature, it refers to the real numbers $\mathbb{R}$. But we may expand its reference to any locally compact topological space such as $\mathbb{R}^n$ or $\mathbb{C}$. By **continuum problems**[3] we mean the problems of computing functions whose domains and/or ranges are continua. Unfortunately, the theory of continuum computing, often pronounced as "real computation", is in its relative infancy because its foundations are still very much in dispute [56]. We have no consensus similar to Church's thesis in discrete computation. This is an exciting opportunity for the computability and complexity theorist, but this is not my focus in this essay.

First, I point out what we are missing out on by our totally discrete view of computing. I am especially interested in problems arising in a constellation of subdisciplines, collectively known as **Computational Science & Engineering** (CS&E). For any discipline X of science, mathematics, or engineering, it is possible to identify a subdiscipline called "Computational X". Thus we have computational biology, computational physics, computer algebra, etc. In the earth, atmospheric and ocean sciences, the computational aspect is so central that it is redundant to attach the "computational" prefix. There has been an explosive growth in computational activities in CS&E. Keen observers of the scientific enterprise have identified the CS&E phenomenon as representing a third pathway to scientific discovery. Alongside the two traditional pathways based on theory (deduction) and experimentation (induction), we now have computation (simulation). In many disciplines X, computer simulation is increasingly seen as an alternative to physical experimentation. Computational labs vie with traditional wet labs to provide insights for X.

Where is the Computer Science in computational X? Taking a highly Computer Science-centric view, imagine computational X as a collection of computational problems, and so CS&E is the union of these collections. Let us also regard Computer Science as a collection of computational techniques. Then the relationship between Computer Science and CS&E can be pictured as a matrix where each problem is represented by a column, and each technique represented by a row. Each matrix entry has a numerical score between 0 and 1, indicating the relevance of a technique to a problem. Of course, this is only a cartoon

---

[3] Sometimes called "continuous problems", but this terminology is confusing, especially for geometric computation which is inherently discontinuous.

Table 1. The CS&E Matrix

	Atmospheric Sciences	Comput. Biology	Material Science	Comput. Physics	$\cdots$	Electrical Engineering
Huge Datasets	1.0	0.8	0.2	0.6	$\cdots$	0.1
Optimization	0.2	0.2	0.5	0.2	$\cdots$	0.3
Symbolic Computation	0.1	0.0	0.2	0.7	$\cdots$	0.5
String Algorithms	0.0	0.8	0.0	0.2	$\cdots$	0.0
Parallel Algorithms	0.5	0.2	0.2	0.4	$\cdots$	0.1
$\vdots$		$\vdots$			$\ddots$	

view to make a point. In Table 1, I have further simplified the column space by identifying each computational X with only one column.

Practitioners of Computational X tend to identify themselves with a particular column (as "column scientists"), while computer scientists might view themselves doing row science. The glue that makes CS&E coherent is Computer Science. When we develop algorithms in a particular row, we are often oblivious to the applications. But by aligning our row activities to particular columns we may gain new insights for the science of computing, and we would share in the advancement of X. This would be the "best practice".

But the record of involvement of Computer Science in the Computational X's shows an uneven record. In areas such as computational biology, there is a great synergy, while in many others, the Computer Science component is[4] basically non-existent. This dichotomy is highly correlated with the division between discrete and continuous computation. Computational biology can be reduced to discrete algorithms (strings and trees), and it is easy for computer scientists to make contributions at this level of abstraction. But Computer Science involvement falls off rapidly as the need for numerical computation increases. My general thesis is that there is a large role for theoretical algorithms in such computation, and this ought to be most clearly understood by those of us who work in the field of computational geometry.

¶5. *Two Worlds of Computing.* When I suggest that Computer Science should engage in the continuum problems of CS&E, it invites a clash of two world views on computing. One view is motivated by computing ideal mathematical objects and the other, by physical modeling. Most of us live exclusively in one of these worlds and are oblivious to the other. General claims about computing from one perspective can be quite wrong in the other. We cannot afford to fall into this trap, as we intend to work at their interface.

In the **mathematical world view**, the continuous makes perfect sense and its use in mathematical modeling has been highly successful. Exact computation is also meaningful here, whether it is computing arbitrarily accurate values of

---

[4] Just because Computational X uses computers does not mean that there is Computer Science development in it, any more than there is carpentry in Computer Science although computer scientists use tables and wooden cabinets. David Bindel reminds me that numerical analysis is present in these fields, and so it is clear that in this essay, I use "Computer Science" in a narrow sense without including numerical analysis.

π or in automated proof of geometric theorems. The ideas of exact geometric computation sit comfortably in this world. The algorithmic problems studied in theoretical computer science are also exact ones.

The **physical world view** disdains exact computing, however. It is argued that the physical world is discrete and nondeterministic. In other words, the continuous and deterministic world is a myth. When such arguments are used to dismiss the reality of the other world of computing, they miss the point of myths, whether in folklore or in science. We know that a "point mass" in Physics is a fiction, but try abolishing it from Physics text books. The point (no pun intended) is that continuous models satisfy Occam's Razor in their description of many physical phenomena. Thus, the term "continuum mechanics" is no oxymoron even when applied to the study of particle systems like fluids and gases.

Besides discreteness, the physical world view advances a related argument about finiteness. It is noted that physical constants have limited accuracy and that current 64-bit machine precision seems more than adequate for physical modeling, from the subatomic to astrophysical scales. There is a simple counter to this thought: a well-known phenomenon in ENC is that in order to compute a value up to (say) 1-bit accuracy, the intermediate computed values might require arbitrarily many bits of accuracy. In fact, it is remarkably easy to run out into very high precision.

Many applied fields are ostensibly uninterested in exactness. It may be hard to see why such fields might have any interest in exact computation, so let me provide some examples. In computer graphics, it is arguably unnecessary to compute beyond the accuracy of screen resolution (this is analogous to the limited accuracy of physical constants). Over the years, on quizzing experts in this field, I have been repeatedly surprised by acknowledged[5] nonrobustness issues. Or consider protein folding, an inherently approximate process. Nevertheless, the folded protein may have several distinct minimal energy states: how could we ensure that our numerical simulation has sufficient accuracy to be[6] *qualitatively* correct? Or consider the fact that approximate computation may be best modeled by an idealized mathematical model. Then, the best policy might be to compute exactly, or to try to emulate exact computation. As another interesting example, floating point computer arithmetic must ultimately rely on exact computation to solve the exact rounding problem [58]. Therefore I believe that, like the myths of point masses and continuum mechanics, exact computation has a role to play even in the approximate computations of CS&E.

¶6. *Is Geometry Continuous or Discrete?* Many continuum problems are geometric in nature. So it is useful to understand the general character of geometric computation. Computational geometers have much insight to offer in this regard, as they have had to grapple with this question as they confronted nonrobustness in geometric computation.

---

[5] You are unlikely to see these issues discussed in print.

[6] i.e., closer to the correct minimal energy state than to any others.

Geometry comes in two main forms: analytic geometry and synthetic geometry. The former uses equations and coordinates to define geometry while the latter (e.g., Euclidean geometry) proceeds from axioms. Interestingly, Hardy [25] regards analytic geometry as mundane and considers synthetic geometry as the "higher geometry". But computationally, we see that analytic geometry is by far the more important (cf. [7]). In automatic geometric theorem proving, for instance, the synthetic approach has had limited success, while the analytic approach, especially influenced by Wu Wen-Tsun's insights, has flowered today. Henceforth, I focus exclusively on analytic geometry.

Analytic geometry is the interplay of the continuous and discrete. The continuum enters in two ways. To make this concrete, allow me to introduce a little framework. In the first place, geometric objects are **parametric objects**. Geometric prototypes are points and lines in the plane. A point p is given by $Point(x, y)$ and a line ℓ is given as $Line(a, b, c) : aX + bY + c = 0$ where $x, y, a, b, c \in \mathbb{R}$ are numerical parameters. These parameters might be constrained (e.g., $a^2 + b^2 > 0$). The **parameter space** of geometric objects of each type is therefore a continuum. The space of points may be identified with the Euclidean plane $\mathbb{R}^2$, and the space of lines is a subset of the projective space $\mathcal{P}^2(\mathbb{R})$.

In general, we can treat more complex geometric objects, such as a convex polytope in $\mathbb{R}^n$, as a cell complex in the sense of algebraic topology. The **(combinatorial) type** of a geometric object can be represented by a directed graph G with parametric variables $X_1, \ldots, X_m$ associated with its vertices and edges, together a constraint predicate $C(X_1, \ldots, X_m)$. We will write $G(X_1, \ldots, X_m)$ for this type, with $C(X_1, \ldots, X_m)$ implicit. An assignment of values $a_i \in \mathbb{R}$ to each X_i is valid if the predicate $C(a_1, \ldots, a_m)$ holds. E.g., $C(a, b, c)$ might say that $a^2 + b^2 > 0$. A valid assignment $(a_1, \ldots, a_m)$ to $G(X_1, \ldots, X_m)$ is called an instance, and we write "$G(a_1, \ldots, a_m)$" for the instance. All geometric objects with which we compute can be put in this form (see [54]). For each type $G(X_1, \ldots, X_m)$, we obtain a parametric space comprising all of its instances.

Suppose we have a surface S, viewed as a parametric object $S = \text{Surface}(\mathbf{x})$ with parameters $\mathbf{x} \in \mathbb{R}^m$. These parameters can be approximated by some $\widetilde{\mathbf{x}}$. If $\|\mathbf{x} - \widetilde{\mathbf{x}}\| \leq \varepsilon$, we call $\text{Surface}(\widetilde{\mathbf{x}})$ a **parametric ε-approximation** of S. But we will see a (for us) more important kind of approximation.

The continuum enters geometry in a second way. Geometric objects such as points, hypersurfaces, cell complexes, etc, must live in a common **ambient space** in order to interact. Each geometric object $G = G(a_1, \ldots, a_m)$ is associated with a subset $\lambda(G(a_1, \ldots, a_m))$ of its ambient space, say $\mathbb{R}^n$ (for some n). Call $\lambda(G)$ the **locus** of G; in practice, we often identify G with its locus. For instance, if G is a curve, its locus is a 1-dimensional subset of $\mathbb{R}^n$.

Two geometric objects S and T, not necessarily of the same type, living in a common ambient space, are said to be **ε-close** if the Hausdorff distance between their loci is $\leq \varepsilon$. Thus, we can approximate continua, such as surfaces S, by discrete finite objects such as triangulations T, and we call T an **explicit ε-approximation** of S. The **explicitization problem** is to compute an explicit ε-approximation T from (the parameters of) S. For instance, a real function f :

$\mathbb{R}^n \to \mathbb{R}$ can be explicitly approximated by a triangulation T that approximates its graph $gr(f) := \{(\mathbf{x}, f(\mathbf{x})) : \mathbf{x} \in \mathbb{R}^n\}$. Such approximations are central to our concerns.

Interaction among geometric objects is captured by **geometric predicates** which define relationships via loci. Thus, we have the $OnLine(p, \ell)$ predicate which holds if p lies on ℓ, or the $LeftTurn(p, p', p'')$ predicate which holds if we make a left turn at p' as we move from p to p' and then to p''. Let R be a set of geometric predicates, and let $O_1, \ldots, O_m$ be m sets of geometric objects, each set O_i having a fixed type. For instance, let $m = 2$, O_1 be a set of points, and O_2 a set of lines. The sum total of the geometric relationships defined by R on $(O_1, \ldots, O_m)$ constitute the "geometry" of $(O_1, \ldots, O_m)$ induced by R. *The field of computational geometry is concerned with computing, representing, and querying such geometries.*

Geometric objects can also be constructed from other data: besides constructing the objects directly from their numerical parameters (e.g., $p \leftarrow Point(x, y)$ or $\ell \leftarrow Line(a, b, c)$), we may construct them from other geometric objects (e.g., $p \leftarrow Intersect(\ell, \ell')$ or $\ell \leftarrow Line(p, p')$). Such predicates and constructors become the primitives for geometric computation as we shall see in the next section.

3   Abstract Computational Models

Computational Geometry is primarily concerned with discrete and combinatorial algorithms. These algorithms are largely non-numerical. This last characterization must strike the casual observer as an anomaly. Since geometric data arises from the continuum, surely numerical computations must be central to geometric computation? By nature, computation is discrete: each computational step (sequential or parallel) is a discrete event that transforms the computational data in a well-defined (not necessarily deterministic) way. Despite this discrete nature, we can develop computational models for continuum problems and geometric applications. This paradox also appears in mathematical logic: we build theories of the continuum using a logical language that is countable.

Computation and mathematical logic have much more in common. In discussing computational models, especially for continuum computation, we can take another page from logic. The language of any first order theory is comprised of two parts: (A) a "logical part" that has the standard logical symbols such as Boolean operators, equality, quantifiers and a countable set of variables; (B) an "extra-logical part" that[7] has predicate and operator symbols which are unique to the particular theory (e.g., [46]). Standard rules of logical deduction are supplemented by special axioms for the extra-logical parts of the theory.

We apply a similar approach to computation: each computational model can be divided into a logical part, called the **base model**, and an extra-logical part. Because of the extra-logical part, such models are called **abstract (computational) models**. The base model may be any standard computational model; Turing machines and random access machines (RAM) [1] are commonly used.

---

[7] Also called the "non-logical" part.

Typically, the base models are at least equivalent to Turing machines. The choice of a base model determines the kind of data structures and type of control structures of our algorithms. With Turing machines as the base model, we are making the choice to have strings as our basic data structure. The extra-logical powers typically come from oracles (perhaps countably many). To compute with real numbers, these oracles can represent real functions. Through these oracles, we can access countably many real constants (like π, e) via special string encodings that the oracle understands. Ko [27] uses such oracle Turing machines extensively in his work. Alternatively, in using RAMs as the base model, we are choosing the ability to have random access to storage locations containing integers. To compute with extra-logical objects such as real numbers, we allow the storage locations to store reals, and provide corresponding real predicates and operators.

¶7. *Abstract Pointer Machines.* Thus we may speak of "abstract Turing machines" or "abstract RAMs". But I am especially partial to **abstract pointer machines** that use Schönhage's pointer machines[8] [43] as the base model. Pointer machines directly encode and manipulate structures called **tagged graphs**, i.e., directed graphs whose edges (called **pointers**) have labels (called **tags**) taken from a finite set Δ of symbols. Its operations are instantly recognizable by computer scientists. In brief, its two main operations are

$$\text{Assignment} : w \leftarrow v$$

$$\text{Test} : \text{if } (w \equiv v) \text{ goto } L$$

where $w, v \in \Delta^*$ and L is a natural number (a label of an instruction). A pointer machine, in bare form, is a finite sequence of such instructions. If G is a tagged graph with a designated node called the origin, then a string $w \in \Delta^*$ yields a path that begins at the origin and ends at a node denoted $[w]_G$. The last pointer (edge) in the path w has a special role: execution of the assignment instruction "$w \leftarrow v$" will modify G by re-directing the last pointer of w to point at $[v]_G$. The test instruction "if $(w \equiv v)$ goto L" is also easily understood: the indicated goto is executed if $[w]_G = [v]_G$.

Such machines are naturally extended to operate on algebraic entities such as real numbers, which are stored in the nodes of the graphs (see [56]). Abstract pointer machines are natural for geometric computation which calls for the juxtaposition of combinatorial structures with real numbers. Imagine trying to encode geometric structures into strings in an abstract Turing machine — a most unnatural thought.

In fact, the analogy to logic can be carried even further: just as modern logic does not require logical languages to be associated with any particular model, we can also view our computational models to be pure syntax, with certain syntactic rules. It is up to the application to provide models and "abstract interpretations". But this would take us beyond our immediate interest. *So in the following, I assume that each abstract computational model comes with a standard interpretation.*

---

[8] Also known as storage modification machines.

¶8. Classification of Abstract Models. Once we have an abstract computational model, we can discuss computability and complexity. The simplest complexity model is to charge a unit cost for each operation. I will categorize important abstract computational models from the literature into three classes:

- **Analytic Models.** In the field of computable analysis, models based on Type-2 Theory of Effectivity (TTE) [53] or oracle TM's [27] have been studied. I will shortly discuss numerical models that fall under this classification.
- **Algebraic Models.** The real RAM [1] is an example. The BSS model of Blum-Shub-Smale [5] can be seen as a real RAM with limited random access. Alternatively, it is a Turing machine whose tape cells can store arbitrary real numbers. The extra-logical powers here are the ring operations $(+, -, \times, 0, 1)$ and real comparison $(<, =)$. More generally, we call an abstract model "algebraic" if the extra-logical objects belong to algebraic structures like rings or fields, and the extra-logical functions or predicates take only these objects as arguments. Note that the extra-logical operation $\exp(x)$ or $\sin(x)$ count as "algebraic" in our sense. The Information-Based Complexity approach of Traub and Woźniakowski [50] focus on algorithms in such algebraic models.
- **Geometric Models.** It is tedious to design geometric algorithms directly in the algebraic or analytic models. So most geometric models use the real RAM as the base model, introduce higher level objects such as points or surfaces, and assume geometrically meaningful predicates and constructors such as those discussed earlier. For example, below we discuss a geometric constructor that shoots a ray to obtain a sample point on a surface.

Abstract models are important and useful, regardless of whether they are realistic or not. I stress this point because some have criticized algebraic models (e.g., BSS model) on account of unrealism. But it would be untenable to develop most of the algorithms of computational geometry in an analytic or algebraic model. The unique place[9] of the Turing model is never challenged by any of these abstract models. The abstract models serve other useful purposes, including providing a modular description of algorithms at various levels of abstraction [56]. Thus, the algebraic model, not the standard Turing machines, is most appropriate for describing Strassen's matrix multiplication algorithm.

4   Case Studies in Abstract Models

Computational models greatly influence the kinds of algorithms we design. They can hide or accentuate different computational issues. Of course, we know this. Nevertheless, we might gain some insights into potential pitfalls by looking at four case studies.

---

[9] Although the standard base models are equivalent by Church's thesis, the Turing model captures complexity-theoretic concepts such as space, time, nondeterminism, etc. much better than most. The pointer model is close to the Turing model in this respect.

¶9. *CASE 1: Meshing or, Watch your Implementation Gap.* Most physical simulations require some kind of mesh. For our purposes, we may identify a mesh as a triangulation. I will consider a basic problem in meshing: generating topologically correct ε-close meshes for implicit surfaces (see the survey [6]). In case the surface is algebraic, a well-known algebraic approach to this problem uses some form of cylindrical algebraic decomposition. As these algebraic techniques are expensive and non-adaptive, I will focus on two adaptive approaches based on sampling and subdivision. They differ in their choice of abstract models. Let S be a non-singular surface given by $f(x, y, z) = 0$.

In the **sampling approach**, we construct a mesh T from a finite set P of sample points on S. Typically, $T = T(P, S)$ is a subset of the Delaunay triangles of P. We incrementally add sample points to P until the required ε-closeness criterion is achieved. Such algorithms are based on a geometric model that supports the classical primitive of "ray shooting". The primitive returns the first point p (if any) on S intersected by a given ray. We then add p to the set P. If S is an algebraic surface, the sample points would have algebraic number coordinates. Although it is possible to implement such a ray shooting model exactly (it reduces to computing the first positive root of a polynomial), this is expensive and nontrivial to implement. Should we implement *exact* ray shooting in order to *approximate* a surface? Probably not. Yet there is no known analysis of sampling algorithms based on approximate sample points. This leaves an **implementation gap** in what is otherwise a beautiful exact approach. The next section will expand on this example.

We turn to the **subdivision approach**. Typically, we wish to construct a mesh for the part of the surface lying within some given box B_0. We construct a quadtree rooted at B_0 by repeated subdivision until each leaf box satisfies some criterion. The well-known marching cube algorithm falls under this approach. Subdivision methods are easy to implement and widely used in practice. I want to highlight an algorithm of Plantinga and Vegter (PV) [36] which represents the first complete purely numerical algorithm for the meshing of non-singular surfaces in $\mathbb{R}^3$. This is no mean achievement, considering that there were been several prior attempts (e.g., [49, 47, 40]) that fall short in one way or another. Unlike the sampling approach, this numerical algorithm suffers no implementation gap: it is easy to implement using just a bigFloat number package.

¶10. *CASE 2: Transcendental Comparisons or, Can we really do this?* The previous case study points out the hidden cost of implementing abstract real RAM operations. In our second case, we see an extreme example of this phenomenon. In 2003, while I was visiting the laboratory of Professor Doeksoo Kim in Hanyang University, Korea, he demonstrated his geometric software for computing shortest paths between any two points while avoiding a collection of n discs. In the real RAM model, it is an exercise to reduce this problem to Dijkstra's algorithm on a suitable graph. Of course, an effective implementation needs additional techniques such as the ability to cull away most of the irrelevant discs for any particular query, but that is another story. The implementation uses machine precision arithmetic and I casually suggested that to produce guaranteed results,

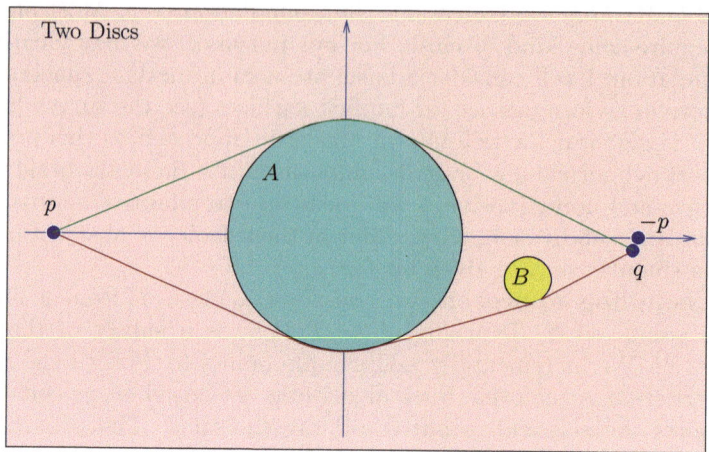

Fig. 1. Shortest path from p to q

they might look into tools like LEDA or Core Library. But upon reflection, I was greatly surprised to discover that I did not know how to solve it. That is because the shortest disc-avoiding path γ between two points consists of an alternating sequence of straightline segments (σ_i) and circular arcs (α_j):

$$\gamma = (\sigma_0, \alpha_1, \sigma_1, \alpha_2, \ldots, \alpha_m, \sigma_m).$$

This is illustrated by Figure 1 which shows two disc obstacles A, B, and two possible shortest paths from p to q. Note that q is close to $-p$, so it is not obvious which is shorter. The length of σ_i is algebraic but the length of α_j is non-algebraic. So the length of γ is an algebraic number plus a transcendental value. Dijkstra's algorithm requires the comparison of two such lengths, and there were no known decision methods here. Eventually, we were able to show the decidability of such comparisons [15] by appealing to Lindemann's theorem in transcendental number theory. Obtaining complexity bounds requires more work, depending on Baker's theory of linear form in logarithms. Although our story ended well, the initial fear that we might have unwittingly invoked an uncomputable form of the real RAM is a lesson not easily forgotten. For our next case, we turn to a problem where the computability remains open.

¶11. *CASE 3: Discrete Morse Theory or, How to take the first step.* A powerful research methodology in algorithms is to develop discrete analogues of continuous theories. In recent years, discrete forms of differential geometry, minimal surfaces, Ricci flows, etc. have been developed. Edelsbrunner, Harer, and Zomorodian [20] developed a discrete Morse theory for triangulated surfaces. Given a triangulated surface with a Morse function, we can compute its discrete Morse complex (which is a quadrangulation) in a purely combinatorial way. They further used discrete Morse theory to compute a simplification hierarchy that has many useful applications.

In some applications, we do not begin with a triangulation, but with a smooth surface S with associated height function. Suppose that we wish to compute its Morse complex. One approach is to first compute a triangulation T of S, then compute a discrete Morse complex using the algorithm of Edelsbrunner et al. Let us write $T \simeq S$ if the discrete Morse complex of T is combinatorially equivalent to the usual Morse complex of S. Unfortunately, we do not know how to compute a T such that $T \simeq S$. Another way to see this difficulty is to ask the simpler problem: given a non-degenerate saddle point, how do we connect it to its two maximas? The issue is to compute the integral lines correctly. No current (numerical) methods can guarantee this. Assuming S is algebraic, we see that the critical points are algebraic and can be located exactly. But the integral lines are probably nonalgebraic, and we have no a priori bounds on how close they can get to other critical points.

In general, the problem is to compute a discrete analogue T of a continuum S such that "$T \simeq S$", meaning that the topological invariants of T are equal to the corresponding invariants of S. Once we have T, the computability of its topological invariants is usually not in question. Computer scientists have gravitated naturally to this discrete computation, but I suggest that we also look at the more fundamental question of computing the transformation $S \mapsto T$ which is largely open.

This *first step*, the transition from continuous-to-discrete, amounts to solving an **explicitization problem** in the sense of ¶6. Surface meshing and computing the Morse complex are two examples. Such examples are easily multiplied: computing discrete representations of vector fields, the numerical solution of partial differential equations, etc. An interesting problem investigated by Nishida and Sugihara is the Voronoi diagram of points in a flow field [34].

¶12. *CASE 4: Numerical Halting Problem or, How to be Adaptive.* Behind each explicitization problem, you will find the zero problem, which I will now explain. For any set E of real expressions, we define a corresponding **zero problem**, $Zero(E)$: given $e \in E$, is the value of e equal to 0? Here, each $e \in E$ is an expression defined over some set Ω of partial functions on $\mathbb{R}$, and e either denotes a unique value $val(e) \in \mathbb{R}$, or $val(e)$ is undefined. Except in the case of algebraic expressions, the decidability of these zero problems is generally open.

Consider the "sum of square-roots" problem. This is the zero problem for the set E_0 comprising the expressions $e = \sum_{i=1}^{m} a_i \sqrt{b_i}$ where $a_i \in \mathbb{Z}$ and $b_i \in \mathbb{N}$. Its complexity is a famous problem in computational geometry (see Blömer [2, 3]). In this case study, I will illustrate the influence of abstract models in addressing the zero problem. If your abstract model is the real RAM with the ability to extract square roots, then $Zero(E_0)$ is trivial: explicitly evaluate the expression $e = \sum_{i=1}^{m} a_i \sqrt{b_i}$ in $3m - 1$ steps and perform the needed comparison to 0 in one more step. But the real RAM is unrealistic when discussing square-roots. So I turn to two other approaches, an algebraic one and an numerical one.

(1) Suppose you use the standard RAM that allows ring operations on arbitrary integers. To solve $Zero(E_0)$, you can use a well-known algebraic method

known as "repeated squaring": in each stage of this process, if you arrange so that one side has exactly those terms involving a given square root, then you can eliminate this square root by squaring both sides of the two-sided equation. Unless you are extremely lucky, you will need to perform m such stages. After the first $\log_2 m$ stages, we expect to see terms which are products of any subset of the original square roots (there are 2^m such terms). So the complexity is at least single exponential in m.

(2) Suppose you use some numerical model (the next section will provide one such) that supports approximations of square roots to any desired precision. For simplicity, assume that approximations are given by enclosing intervals. We can compute a potentially infinite sequence $I_i = [u_i, v_i]$ $(i = 0, 1, 2, \ldots)$ of improving approximations to $val(e)$, where $v_i - u_i \leq 2^{-i}$. If $u_i > 0$ or $v_i < 0$ for any i, we can stop and conclude that $val(e) \neq 0$. Thus, if $|val(e)| \neq 0$, this will stop within $1 - \log_2 \min\{1, |val(e)|\}$ steps. What if $val(e) = 0$? In general, we have no method of stopping. But for $e \in E_0$, we can compute an a priori **zero bound** $\beta(e) \geq 0$ with the property that if $val(e) \neq 0$ then $|val(e)| > \beta(e)$. In this case, if you have not concluded that $val(e) \neq 0$ after $-\log_2 \min\{1, \beta(e)\}$ steps, you can declare $val(e) = 0$. Known bounds for $\beta(e)$ imply that after at most an exponential number of steps, we can declare that $val(e) = 0$. It is not known if this is necessary.

This numerical approach gives rise to the **numerical halting problem**, the problem deciding when to stop computing a potentially infinite sequence I_i $(i = 0, 1, 2, \ldots)$ of approximations. Like the classic halting problem for Turing machines, this decision problem is asymmetrical: one case is easy, and the other is hard. If $val(e) \neq 0$, it is trivial to halt. If $val(e) = 0$, then it is highly non-trivial to halt.

Which method should we prefer? The algebraic method is non-adaptive (all-or-nothing) because, informally, with a measure-zero exception, it requires the worst case complexity bound. The numerical method is adaptive because, again with a measure-zero exception, its complexity depends on $|val(e)|$. Some years ago, I noted two other advantages of the numerical method:

(a) Typically, the zero problem is only a subproblem of the more general **sign problem**: given $e \in E$, we want to know the sign of $val(e)$, assuming $val(e)$ is defined. The algebraic method of repeated squaring requires nontrivial modifications in order to decide sign (there are numerous cases to consider) but signs come for free with the numerical method.

(b) Suppose you need to perform $n \log n$ comparisons of the form $e_i : e_j$ where $1 \leq i < j \leq n$. This problem arises in Fortune's sweepline algorithm. This reduces to the sign problem for the expression $e_i - e_j$. Using the algebraic approach, one must do repeated squaring for each comparison. Using the numerical approach, we have a better option. Assume we have a bound B such that $B \leq \beta(e_i - e_j)$ for all $i < j$. Then you just approximate each expression e_i by some numerical value $\tilde{e}_i$ with error less than $B/2$. It turns out that B is not too large so that the difficulty of this approximation is comparable to performing a repeated squaring comparison. Now the comparison

$e_i : e_j$ is easily decided by comparing the approximations $\widetilde{e}_i : \widetilde{e}_j$ (declare $val(e_i) = val(e_j)$ if $|\widetilde{e}_i - \widetilde{e}_j| < B$). So the algebraic approach requires $n \log n$ difficult computations, but the numerical method only requires n difficult computations (to approximate each $\widetilde{e}_i$).

5   The Exact Numerical Model

Numbers are the fountainhead of analytic geometry. But we are trained to design algorithms exclusively in abstract models that are devoid of numbers. This is the source of the implementation gaps we saw in our case studies. The goal in this section is to develop a numerical computational model that avoids such pitfalls while remaining useful for geometric algorithms.

Smale has observed that numerical analysis has no abstract computational models to investigate the fundamental properties of numerical computation. The BSS model has been offered as a candidate for this purpose [4] (see [5, Chap. 1]). For error analysis, numerical analysts use the standard arithmetic model (see below) which falls short of a full-scale computational model. Perhaps numerical analysts see no need for a general model because most of their problems do not involve geometry or complex combinatorial structures. In the following exercise, I hope you will see some merit in taking up Smale's challenge.

¶13. *Duality in Numbers.* Numbers in $\mathbb{R}$ are dual citizens: they belong to an algebraic structure (a field), as well as to an analytic structure (a metric space). As in the particle-wave duality of light, numbers seem to vacillate between its particle-like (algebraic/discrete) and wave-like (analytic/continuous) natures. In many computation, we treat them exclusively as citizens of one or the other kingdom. But in order to address the central problems of continuous computation, we need a representation of numbers which expresses the dual nature of numbers.

Consider a concrete example: the number $\alpha = \sqrt{15 - \sqrt{224}}$ can be represented directly by the indicated radical expression. This is exact, but for the purposes of locating its proximity on the number line (e.g., is α in the range $[0.01, 0.02]$?), this representation alone is unsatisfactory. An approximation such as $\alpha = 0.0223$ would be useful for proximity queries. But no single approximation is universally adequate. If necessary, we should be able to improve the approximation to $\alpha = 0.02230498$, and so on. Thus, the analytic nature of α is captured by the *potential* to give arbitrarily good approximations for the locus of α. This is only a potential because we cannot reach its limit in finite time. In the analytic approach to real computation, this is the central concept [53, 27]. For us, this potential exists because we maintain an exact representation of α. Thus, we need a **dual representation** of α, comprising the exact expression plus a dynamic approximation process. Computationally, this is very interesting because iteration at run-time becomes necessary.

Of course, we have seen dual representations earlier in `Leda Real`. This representation can be generalized to geometric objects. For instance, to represent an algebraic curve C in $\mathbb{R}^3$ exactly, we store a pair of polynomials (f, g) in case the curve is defined by $f = g = 0$. To approximate the locus of C, we use any suitable explicitization. A simple solution is a polygonal line P that is ε-close to C (for any desired ε).

¶14. Some Virtues of Numerics. Let us next focus on numerics. I shall speak of "numerics" when I want to view numbers only as analytical objects, and ignoring their algebraic nature. In numerical analysis, they are fixed-precision floating point numbers, but for exact computation, we must transpose them to BigFloats. **BigFloats** (or dyadic numbers) form the set $\mathbb{F} := \{m2^n : m, n \in \mathbb{Z}\} = \mathbb{Z}\left[\frac{1}{2}\right]$. Practitioners instinctively know the virtues of numerics but it is easy for theoreticians to miss them. Implicit in our discussion of virtues is a comparison with numerical computing that is based on other number systems with more algebraic properties, for example $\mathbb{Q}$ or algebraic numbers. I do not claim to say anything new, but it is useful to collect these thoughts in one place. What is perhaps new is the audience, since I am talking numerics in the context of exact computation.

- Numerics is useful in exact computation. More precisely, approximations can often lead to the correct decisions, and when combined with zero bounds, such approximations will eventually lead to the right decisions. This is the *sine qua non* for exact computation.
- Numerics is relatively easy to implement. There is only one number type, the "real" numbers (which in computing is translated into floating point numbers). If you compute with algebraic numbers, the traditional approaches require data structures for manipulating polynomials and algorithms for manipulating polynomials. Most implementers avoid this if they could.

  Even the use of rational numbers $\mathbb{Q}$ for their analytical properties will introduce irrelevant algebraic properties that are expensive to maintain. Trefethen gave a striking example of this from Newton iteration [51]. There is a canonical reduced representation for elements of $\mathbb{F}$ and $\mathbb{Q}$: the numerator and denominator must be relatively prime. While performing a sequence of ring operations on a number, it is necessary to reduce its representation periodically in order to avoid exponential growth. This is computationally easy for $\mathbb{F}$, but not for $\mathbb{Q}$.
- Numerics are efficient. We know this for machine numerics, but even BigFloats are efficient. Essentially, BigFloats are as efficient as BigIntegers, and we regard the complexity of BigInteger arithmetic as the base line for exact computation.
- Numerics are easy to understand. This is an important consideration for implementations. For the most part, the analytical properties we need are the metric properties and total ordering of real numbers. In contrast, algebraic properties of numbers can be highly nontrivial (try simplifying nested radical expressions).
- Numerical computation has adaptive complexity. We saw this in CASE 4. There are applications in which the only possibility of obtaining any solution at all relies on adaptivity.

- Numerical approaches have wider applicability. Many problems of CS&E have no closed form solutions. In such situations, numerical solutions remain viable. But even when closed form solutions exist, the numerical solution might be preferred.
- In CS&E, the numerics may be an essential part of the solution. Such is the case with explicitization problems (¶6). An answer of the form "$\alpha \approx 0.022$" might be acceptable, but the form "$\alpha = \sqrt{15 - \sqrt{224}}$" is unacceptable (even though it is exact). This is a blind spot if you exclusively think in algebraic computational models.

¶15. Standard Numerical Model. Having established a place for numerics in exact computation, I will now discuss how we can incorporate it into our computational model. The numerical analysts are experts in this domain, so we first look at their treatment of numerics. The **standard arithmetic model** [26, p. 44] of numerical analysis is the following: if $\circ \in \{+, -, \times, \div\}$ is any arithmetic operation and x, y are floating point numbers, then the corresponding machine operation $\tilde{\circ}$ satisfies the following property:

$$x \tilde{\circ} y = (x \circ y)(1 \pm \mathbf{u})$$

where[10] $\mathbf{u}$ is the unit round-off error, provided $x \circ y \neq 0$. In the usual understanding, $\mathbf{u}$ is fixed. If we allow $\mathbf{u}$ to vary, we essentially obtain the multiprecision arithmetic model of Brent [9, p. 242-3].

More generally, I assume that for each operation $\tilde{\circ}$, an arbitrary non-zero relative error $\mathbf{u}$ can be explicitly given as an argument. Notice that the numerical analysts' model is only about arithmetic. It is agnostic about the nature of the base model. But to convert it into an abstract computational model, I choose pointer machines as the base model. We thereby obtain the **standard numerical model**. This can be classified as an analytic model.

¶16. Computational Ring. The standard numerical model is wonderful for developing the algorithms of numerical analysis, and especially for performing backwards error analysis. But this model is problematic for exact numerical computation (ENC) — it lacks the critical ability to decide zero. You can never be sure that any computed quantity is exactly zero. Zero as an algebraic object has been abolished. We have noted [56] that computing in the continuum puts a big "stress" on our computational models because we are trying to simulate an uncountable set $\mathbb{R}$ using only a countable domain ($\mathbb{N}$ or finite strings). The algebraic models [5] cope by making the zero problem trivial. The analytic models [53, 27] cope by making the zero problem undecidable. The standard numerical model represents a third solution, by making the zero problem meaningless.

To restore the place of zero, we must view BigFloats as an algebraic structure. In fact, it is useful to generalize BigFloats by an axiomatic treatment: let $D \subseteq \mathbb{R}$

---

[10] In our error notation, any appearance of "$\pm$" should be replaced by the sequence "$+\theta$" for some variable θ satisfying $-1 \leq \theta \leq 1$. E.g., $1 \pm \mathbf{u}$ translates to $1 + \theta \mathbf{u}$. Note that θ is an implicit variable.

be a countable set that is a ring extension of $\mathbb{Z}$, and which is closed under division by 2. Further, there is a representation[11] for D, viz., an onto partial function $\rho : \{0,1\}^* \dashrightarrow D$ relative to which there are algorithms to perform the ring operations, division by 2, and exact comparisons in D (see [56]). Call D a **computational ring**. We note that D is dense in $\mathbb{R}$, and we have mandated a minimal amount of algebraic properties in D. Computational rings provide our answer to the standard arithmetic model.

The smallest computational ring is the set of BigFloats $\mathbb{F}$. In practice, an important computational ring is $\mathbb{Z}[\frac{1}{2}, \frac{1}{5}]$ (see [35]). But $\mathbb{Q}$ or real algebraic numbers are also examples. We now construct an abstract pointer model in which elements of D are directly represented and the operations on D are available. The fundamental objects manipulated by our pointer machines are **numerical graphs**, i.e., tagged graphs in which each node stores an element of D. This constitutes our **basic numerical model**. Numerical graphs can directly represent $n \times n$ matrices $D^{n \times n}$, or polynomials with coefficients in D, etc. Under our classification scheme ¶8, this is both an analytic and an algebraic model.

Trefethen observed [52, Appendix] that numerical analysis has an undeserved reputation of being "the study of rounding errors", when its true subject matter is "the study of algorithms for continuous mathematics". I think this reputation is partly a function of the standard numerical model. What I found interesting [56] is that numerical analysts inevitably design algorithms in some exact algebraic model (check any numerical analysis text book). But they go on to address the implementation gap (¶9) between the exact model and the standard numerical model. This is the error analysis.

¶17. Exact Numerical Machines. We could design algorithms directly in the basic numerical model, but that would be programming in assembly language. So we explore some extensions of the basic numerical model. My goal is to introduce the capabilities needed to implement the algorithm of Plantinga-Vegter naturally. These capabilities will not affect computability, though they might affect complexity.

Functions will be the key abstraction for our model. Here, we see a major difference between algebraic and analytic thinking. In algebraic thinking, functions are seen as holistic objects within an algebraic structure, e.g., polynomials as elements of a ring. But in analytic thinking, functions are more versatile: they are objects which we can evaluate (query) at run-time, compute approximations of, compose with other functions, numerically differentiate, etc. In analytic complexity theory, functions viewed in this way are modeled by oracles [27].

---

[11] Here, $\{0,1\}^*$ is the set of binary strings. As ρ is a partial function, $\rho(w)$ may be undefined for some $w \in \{0,1\}^*$. If $\rho(w)$ is defined, then w is a "name" for the element $\rho(w) \in D$. Since ρ is onto, each element in D has at least one name. Our algorithms on D must directly operates names. See Weihrauch [53] for the theory of representations.

In the following definitions, let $f : \mathbb{R}^n \to \mathbb{R}$ be a real function.

- We say f is **sign computable** if the function $\mathtt{sign}(f) : D^n \to \{-1, 0, 1\}$ where $\mathtt{sign}(f)(\mathbf{x}) = \mathtt{sign}(f(\mathbf{x}))$ is computable by a basic numerical machine.
- Consider the approximations of functions. Any function of the form $\widetilde{f} : D^n \times \mathbb{N} \to D$ where $\widetilde{f}(\mathbf{x}, p) = f(\mathbf{x}) \pm 2^{-p}$ is called an **absolute approximation** of f. We say f is **absolutely approximable** if there is a basic numerical machine that computes such an $\widetilde{f}$.
- We need interval functions: let $\mathbb{I}(D)$ denote the set of intervals with endpoints in D. For $n \geq 1$, let $\mathbb{I}^n(D)$ denote the n-fold Cartesian product of $\mathbb{I}(D)$. Each $B \in \mathbb{I}^n(D)$ is called an n-box.
- We say $\Box f : \mathbb{I}^n(D) \to \mathbb{I}(D)$ is a **box function** for f if it is an inclusion function (i.e., $f(B) \subseteq \Box f(B)$) and whenever $\{B_i : i \in \mathbb{N}\}$ is a strictly monotone sequence of n-boxes with B_i properly containing B_{i+1}, and $\cap_i B_i$ is a point p, then $\cap_i \Box f(B_i) = f(p)$. We say that f is **box computable** if there is a basic numerical machine that computes such an $\Box f$. It is easy to see that box computable functions are (1) continuous and (2) absolutely approximable.
- We say f belongs to the class $\Box C^k$ ($k \geq 0$) if each partial derivative of f up to order k exists and is box computable. Thus $\Box C^0$ are just the box computable functions. Following [13], we say f is in the **class** PV if $f \in \Box C^1$ and f is sign computable.

The above notions of computability are all relative to the basic numerical model. This avoids issues of computability (cf. CASES 2 and 3). There are deeper issues which we do not take up, such as the dependence of these notions on D. Our goal is to incorporate such functions as $\mathtt{sign}(f)$, $\Box f$, and $\widetilde{f}$ as first class programming objects in our model. Recall that the basic numerical model operates on numerical graphs. A function whose input and output are numerical graphs is called a **semi-numerical function**.

Our **exact numerical model** (ENM) extends the basic numerical model by having extra-logical objects that are semi-numerical functions, and whose tagged graphs have nodes that can store either a semi-numerical function or an element of D. We have a built-in predicate to test nodes for the type of its stored value (the type is either D or a semi-numerical function). Suppose $u, v, w \in \Delta^*$ and G is a tagged graph (¶7). If a semi-numerical function F is stored in node $[u]_G$ and its argument is accessed through node $[v]_G$, then we can invoke an evaluation of F on this argument by executing the following instruction:

$$w \leftarrow EVALUATE(u, v)$$

See [56] for similar details. If we like, we could provide functors to construct semi-numerical functions from scratch, functors to compose two semi-numerical functions, etc. But for our simple needs here, we may assume the semi-numerical functions are simply available (passed as arguments to our numerical machines, like oracles).

¶18. *From Smooth Surfaces to Singular Ones* The preceding development was a build-up to state the following result:

THEOREM 1 (PLANTINGA-VEGTER) *There is an ENM algorithm which, given* $\varepsilon > 0$, *a box function and sign function for some function* $f : \mathbb{R}^3 \to \mathbb{R}$, *and* $B_0 \in \mathbb{I}^3(D)$, *will compute an isotopic ε-close mesh for the surface* $S : f = 0$ *provided S is non-singular and $S \subseteq B_0$.*

The statement of this theorem does not reveal the beauty and naturalness of the PV algorithm; to see this, refer to their original paper [36]. It suffices to say that their method uses standard subdivision of the box B_0 and cleverly exploits isotopy. This result can be extended in several ways: First, the surface S need not be confined within the box B_0, but we must slightly relax the correctness statement on the boundary of B_0. Box B_0 can be replaced by more complicated regions which need not be connected or simply-connected. The function f is allowed to have singularities outside B_0. See [13, 28] for these extensions in the plane. Extensions of the PV algorithm to higher than 3 dimensions are currently unknown, but recently Galehouse [24] introduced a new approach that is applicable in every dimension. All these extensions stayed within the ENM framework.

What if the surface S is singular? We can use the PV algorithm as a subroutine to locate and determine the singularities. This was done for the planar case in [13]. Let me sketch the basic idea: say $S : f = 0$ is a curve with only isolated singularities (if $f(X, Y)$ is a square free polynomial, this will be the case). Now apply the PV algorithm to $F = f^2 + f_x^2 + f_y^2 - \delta$ where $\delta > 0$. For sufficiently small δ, the curve $S_\delta : F = 0$ will be a nonsingular curve, i.e., a collection of ovals or infinite curves. Moreover, if the oval is sufficiently small, we know that it isolates a singularity. Once we have isolated singularities in sufficiently small boxes, we can run the PV algorithm on the original curve S but on a region that excludes these small boxes. We can determine the degree of each singularity (i.e., how many open arcs of S have endpoints in the singularity) by considering an annulus around its small box. All these can be carried out in the algebraic case, because we have computable zero bounds.

¶19. *Simple Real Root Isolation, or How to avoid zero.* The PV algorithm makes the fairly strong assumption that f is sign computable. For instance, we do not know whether this property holds for the class of hypergeometric functions (with rational parameters). But such hypergeometric functions can be shown to be box computable (cf. [19]). So it is desirable remove the sign computability condition all together.

I will now show this for the 1-D case (it will be the only technical result of this paper). In 1-D, meshing amounts to real root isolation and root refinement for a function $f : \mathbb{R} \to \mathbb{R}$. It is not hard to devise such a root isolation algorithm, which we call EVAL (see [14], but the original algorithm is from Mitchell [33]). EVAL depends on two interval predicates which we call C_0 and C_1:

$$C_0(I): \ 0 \notin \Box f(I),$$
$$C_1(I): \ 0 \notin \Box f'(I).$$

Clearly, these predicates are computable if $f \in PV$. Given a BigFloat interval I_0 in which f has only simple roots, EVAL will return a set of isolating intervals for each of the roots of f in I_0. We use a queue Q for processing the intervals:

EVAL(I_0):
    $Q \leftarrow \{I_0\}$
    While Q is non-empty
        Remove I from Q
        If $C_0(I)$ holds, discard I
        Else if $C_1(I)$ holds,
(*)   If f has different non-zero signs at end points of I, output I
(*)   Else, discard I
        Else
(**)   If $f(m) = 0$, output $[m, m]$ where m is the midpoint of I
(**)   Split I into I', I'' at m, and put both intervals into Q

Termination and correctness are easy to see. We now modify EVAL so that f does not have to be sign computable. There is a small price to pay, as there will be some indeterminacy at the boundary of the input interval.

LEMMA 2 *There is an ENM algorithm which, given $\varepsilon > 0$, a box function $\Box f$ for $f : \mathbb{R} \to \mathbb{R}$, and $[a, b] \in \mathbb{I}(D)$, will isolate all the real roots of f in some interval J where*

$$[a, b] \subseteq J \le [a - \varepsilon, b + \varepsilon]$$

provided f has only simple roots in $[a, b]$. Moreover, there is at most one output isolating interval that overlaps $[a - \varepsilon, a]$ and at most one that overlaps $[b, b + \varepsilon]$.

Proof. Observe that from the box function $\Box f$, we can easily construct an absolute approximation function $\widetilde{f}$ for f. Thus, for each $x \in D$ and $p \in \mathbb{N}$, we have $\widetilde{f}(x, p) = f(x) \pm 2^{-p}$. If $|\widetilde{f}(x, p)| > 2^{-p}$, then we know the sign of f at x.

    We modify the EVAL algorithm by omitting two lines marked (**) as we can no longer compute the sign of $f(m)$. We also replace the two lines marked (*) by the following subroutine: assume $I = [a, b]$ is the input to our subroutine. So $C_1(I)$ holds, and I has at most one root of f. The following subroutine will either decide that I has no root or some $J \subseteq [a - \varepsilon, b + \varepsilon]$ is an isolating interval:

1. We dovetail the absolute approximations of $f(a)$ and $f(b)$ with increasing precision until we see a non-zero sign of $f(a)$ or of $f(b)$.
   ◁ *This must halt because $C_1(I)$ holds.*
2. Wlog, say we see the sign of $f(a)$, and it is negative.
   If $f'(I) < 0$, then there are no roots in I. So assume $f'(I) > 0$.
3. For $i = 0, 1, 2, \ldots$, we check the predicate $C_1(J_i)$ where
   $J_i = [b, b + \varepsilon 2^{-i}]$.
   Halt at the first $k \geq 0$ where $C_1(J_k)$ holds.
   ◁ *This must halt because $f'(b) > 0$.*
4. This means $f'(b) > 0$ and $f'(b + \varepsilon 2^{-k}) > 0$.
   As before, do dovetailing to determine the sign of either $f(b)$ or $f(b + \varepsilon 2^{-k})$.
5. If we know the sign of $f(b)$, then I contains a root iff $f(a)f(b) < 0$.
   The other case of knowing the sign of $f(b + \varepsilon 2^{-k})$ is similar.

One final detail: the isolating intervals which this modified algorithm outputs might be overlapping. To clean up the intervals so that there is no ambiguity, observe that $C_1(I)$ holds at each output interval I. Therefore, if two outputs I and J overlap, we see that $I \cup J$ has a unique root which is found in $I \cap J$. So we may replace I, J by $I \cap J$.   **Q.E.D.**

We should be able to extend the PV algorithm in 2- and 3-D by a similar relaxation of the conditions on f. So what have we learned from this? It is (not surprisingly) that you can avoid the zero problem if there are no singularities. So you could have developed this algorithm in the standard numerical model. But should you have singularities (multiple roots in the 1-D case) this option is not available.

¶20. *Towards Numerical Computational Geometry.* Our exact numerical model satisfies the need for higher level abstractions in designing algorithms. Such algorithms will have adaptive complexity because of the use of numerics. Iterations is completely natural. As we saw in the PV Algorithm, domain subdivision will be useful in such algorithms. Another feature is that, unlike standard numerical algorithms, we can actively control the precision of individual operations. This can lead to a speedup [45, 44]. Another direction is in producing numerical algorithms that are "complete", i.e., do not have exceptional inputs for which the algorithm fails. E.g., see [55, 13, 16]. Currently, most geometric algorithms based on numerical primitives are "incomplete" because they are based on the standard numerical model.

Such algorithms represent a marked departure from the typical algorithms seen in computational geometry, and suggests the name "numerical computational geometry" for such activities. In fact, other researchers in interval computation are also producing similar kinds of algorithms. See particularly the work of Ratschek and Rokne [38, 39]. I think both lines of work may eventually converge, but the main gap between their view and ours is located in the difference between using the standard numerical model and our exact numerical model (cf. [40] and [28])

¶*21. What about Complexity Theory?* The most serious challenge for numerical computational geometry is the development of a complexity analysis of adaptive and iterative algorithms. Of course, the lack of analysis does not hamper the usefulness of such algorithms, but it discourages theoreticians from looking at this class of algorithms. Previous work on adaptive complexity analysis has stemmed from analysis of simplex algorithms in the 1980's [8]. All such analyses have depended on probabilistic assumptions. The acclaimed smoothed analysis of Spielman and Teng [48] tries to minimize such objections by "localizing" the probabilistic assumptions to each input instance.

Recently, we introduced the concept of **continuous amortization** [14]. This yields an analysis of adaptive complexity *without probabilistic assumptions*. The key idea is to bound the subdivision tree size in terms of an integral. If the input domain is a box B, the number of subdivisions can be bounded by an integral of the form $I = \int_{\mathbf{x} \in B} \phi(\mathbf{x}) d\mathbf{x}$. Amortization is a well-known computational paradigm and analysis technique in discrete algorithms [18]. We can view the integral approach as a "continuous" form of amortization. We applied this analysis to the EVAL algorithm ¶19, proving that the tree size is polynomial in the worst case depth. This is a mark of adaptivity since in the worst case, tree size is exponential in depth. We believe similar analyses are applicable to other subdivision algorithms.

6   Conclusion

This essay began with the accomplishments of Kurt in experimental computational geometry, and the significance of LEDA in the history of computing. I extracted from this work a unique mode of computation (exact numerical computation), and extrapolated it to general computing, and to computational geometry in particular. My motivation is to equip ourselves to address the host of interesting continuum problems in CS&E. But none of us plan to turn into applied mathematicians or numerical analysts to address these problems. Our strength is in exact/discrete thinking. We celebrate this, and rightly so. You probably agree with me that our discrete/exact views can bring something new to the problems of CS&E. But to do this, we need an analytic model of computation in which the exact views are captured. The clue lies in the zero problem, but more generally the "explicitization problems" of continuous-to-discrete computation. I described an exact numerical model that has many of the desired properties. This article (it turned out) spent much time discussing computational models because, as our case studies show, the wrong model can lead us astray. As a computer scientist, I have found extreme satisfaction in designing geometric algorithms in the exact numerical model. Some of these algorithms also seem quite practical. Perhaps you will find the same satisfaction.

Acknowledgments

I am grateful for the warm hospitality of Professor Subir Ghosh at Tata Institute of Fundamental Research, Mumbai and Professor Subhash Nandy at the Indian Statistical Institute (ISI), Kolkata. Discussions with them at WALCOM 2009 have provoked many of the thoughts expressed in this paper, and it was on the campus of ISI in which the term "numerical computational geometry" was first suggested. This paper has greatly benefited from insightful comments from Helmut Alt, David Bindel, Michael Burr, Richard Cole, Ker-I Ko, and Lloyd Trefethen.

References

1. Aho, A.V., Hopcroft, J.E., Ullman, J.D.: The Design and Analysis of Computer Algorithms. Addison-Wesley, Reading (1974)
2. Blömer, J.: Computing sums of radicals in polynomial time. IEEE Foundations of Computer Sci. 32, 670–677 (1991)
3. Blömer, J.: Simplifying Expressions Involving Radicals. PhD thesis, Free University Berlin, Department of Mathematics (October 1992)
4. Blum, L., Cucker, F., Shub, M., Smale, S.: Complexity and real computation: A manifesto. Int. J. of Bifurcation and Chaos 6(1), 3–26 (1996)
5. Blum, L., Cucker, F., Shub, M., Smale, S.: Complexity and Real Computation. Springer, New York (1998)
6. Boissonnat, J.-D., Cohen-Steiner, D., Mourrain, B., Rote, G., Vegter, G.: Meshing of surfaces. In: Boissonnat, J.-D., Teillaud, M. (eds.) Effective Computational Geometry for Curves and Surfaces, ch. 5. Springer, Heidelberg (2006)
7. Bokowski, J., Sturmfels, B.: Computational Synthetic Geometry. Lecture Notes in Mathematics, vol. 1355. Springer, Heidelberg (1989)
8. Borgwardt, K.H.: Probabilistic analysis of the simplex method. In: Lagarias, J., Todds, M. (eds.) Mathematical Developments Arising from Linear Programmng, vol. 114, pp. 21–34. AMS (1990); This volume also has papers by Karmarkar, Megiddo
9. Brent, R.P.: Fast multiple-precision evaluation of elementary functions. J. of the ACM 23, 242–251 (1976)
10. Buchberger, B., Collins, G.E., Loos, R. (eds.): Computer Algebra, 2nd edn. Springer, Berlin (1983)
11. Burnikel, C., Funke, S., Mehlhorn, K., Schirra, S., Schmitt, S.: A separation bound for real algebraic expressions. In: Meyer auf der Heide, F. (ed.) ESA 2001. LNCS, vol. 2161, pp. 254–265. Springer, Heidelberg (2001)
12. Burnikel, C., Funke, S., Seel, M.: Exact geometric computation using cascading. Int'l. J. Comput. Geometry and Appl. 11(3), 245–266 (2001); Special Issue
13. Burr, M., Choi, S., Galehouse, B., Yap, C.: Complete subdivision algorithms, II: Isotopic meshing of singular algebraic curves. In: Proc. Int'l. Symp. Symbolic and Algebraic Computation (ISSAC 2008), July 20-23, pp. 87–94. Hagenberg, Austria (2008)
14. Burr, M., Krahmer, F., Yap, C.: Integral analysis of evaluation-based real root isolation (submitted, February 2009)

15. Chang, E.-C., Choi, S.W., Kwon, D., Park, H., Yap, C.: Shortest paths for disc obstacles is computable. Int'l. J. Comput. Geometry and Appl. 16(5-6), 567–590 (2006); Special Issue of IJCGA on Geometric Constraints. (Gao, X.S., Michelucci, D. (eds.), Also: Proc. 21st SoCG, pp.116–125 (2005)

16. Cheng, J.-S., Gao, X.-S., Yap, C.K.: Complete numerical isolation of real zeros in general triangular systems. In: Proc. Int'l Symp. Symbolic and Algebraic Comp. (ISSAC 2007), Waterloo, Canada, July 29-August 1, 2007, pp. 92–99 (2007), http://doi.acm.org/10.1145/1277548.1277562 (in press); , Journal of Symbolic Computation

17. Cohen, H.: A Course in Computational Algebraic Number Theory. Springer, Heidelberg (1993)

18. Corman, T.H., Leiserson, C.E., Rivest, R.L., Stein, C.: Introduction to Algorithms, 2nd edn. The MIT Press and McGraw-Hill Book Company, Cambridge (2001)

19. Du, Z., Yap, C.: Uniform complexity of approximating hypergeometric functions with absolute error. In: Pae, S., Park, H. (eds.) Proc. 7th Asian Symp. on Computer Math. (ASCM 2005), pp. 246–249 (2006)

20. Edelsbrunner, H., Harer, J., Zomorodian, A.: Hierarchical Morse complexes for piecewise linear 2-manifolds. Discrete and Computational Geometry 30(1), 87–107 (2003)

21. Fabri, A., Giezeman, G.-J., Kettner, L., Schirra, S., Schoenherr, S.: The CGAL kernel: a basis for geometric computation. In: Lin, M.C., Manocha, D. (eds.) FCRC-WS 1996 and WACG 1996, vol. 1148, pp. 191–202. Springer, Heidelberg (1996); Proc. 1st ACM Workshop on Applied Computational Geometry (WACG), Federated Computing Research Conference 1996, Philadelphia, USA (1996)

22. Fortune, S.: Editorial: Special issue on implementation of geometric algorithms. Algorithmica 27(1), 1–4 (2000)

23. Funke, S., Mehlhorn, K., Näher, S.: Structural filtering: A paradigm for efficient and exact geometric programs. In: Proc. 11th Canadian Conference on Computational Geometry (1999)

24. Galehouse, B.: Topologically Accurate Meshing Using Spatial Subdivision Techniques. Ph.D. thesis, New York University, Department of Mathematics, Courant Institute (May 2009), http://cs.nyu.edu/exact/doc/

25. Hardy, G.: What is Geometry (presidential address to the mathematical association, 1925). The Mathematical Gazette 12(175), 309–316 (1925)

26. Higham, N.J.: Accuracy and stability of numerical algorithms. Society for Industrial and Applied Mathematics, Philadelphia (1996)

27. Ko, K.-I.: Complexity Theory of Real Functions. Progress in Theoretical Computer Science. Birkhäuser, Basel (1991)

28. Lin, L., Yap, C.: Adaptive isotopic approximation of nonsingular curves: the parametrizability and non-local isotopy approach. In: Proc. 25th ACM Symp. on Comp. Geometry, Aarhus, Denmark, June 8-10 (to appear, 2009)

29. Mehlhorn, K.: The reliable algorithmic software challenge RASC. In: Klein, R., Six, H.-W., Wegner, L. (eds.) Computer Science in Perspective. LNCS, vol. 2598, pp. 255–263. Springer, Heidelberg (2003)

30. Mehlhorn, K., Näher, S.: LEDA – a library of efficient data types and algorithms. In: Kreczmar, A., Mirkowska, G. (eds.) MFCS 1989. LNCS, vol. 379, pp. 88–106. Springer, Heidelberg (1989); Proc. 14th Symp. on MFCS, to appear in CACM

31. Mehlhorn, K., Näher, S.: LEDA: a platform for combinatorial and geometric computing. CACM 38, 96–102 (1995)

32. Mehlhorn, K., Schirra, S.: Exact computation with `leda_real` – theory and geometric applications. In: Alefeld, G., Rohn, J., Rump, S., Yamamoto, T. (eds.) Symbolic Algebraic Methods and Verification Methods, Vienna, pp. 163–172. Springer, Heidelberg (2001)

33. Mitchell, D.P.: Robust ray intersection with interval arithmetic. In: Graphics Interface 1990, pp. 68–74 (1990)

34. Nishida, T., Sugihara, K.: Voronoi diagram in the flow field. In: Ibaraki, T., Katoh, N., Ono, H. (eds.) ISAAC 2003. LNCS, vol. 2906, pp. 26–35. Springer, Heidelberg (2003)

35. Pion, S., Yap, C.: Constructive root bound method for k-ary rational input numbers. Theor. Computer Science 369(1-3), 361–376 (2006)

36. Plantinga, S., Vegter, G.: Isotopic approximation of implicit curves and surfaces. In: Proc. Eurographics Symposium on Geometry Processing, pp. 245–254. ACM Press, New York (2004)

37. Pohst, M., Zassenhaus, H.: Algorithmic Algebraic Number Theory. Cambridge University Press, Cambridge (1997)

38. Ratschek, H., J.G.R.G. Editors: Editorial: What can one learn from box-plane intersections? Reliable Computing 6(1), 1–8 (2000)

39. Ratschek, H., Rokne, J.: Geometric Computations with Interval and New Robust Methods: With Applications in Computer Graphics, GIS and Computational Geometry. Horwood Publishing Limited, UK (2003)

40. Ratschek, H., Rokne, J.G.: SCCI-hybrid methods for 2d curve tracing. Int'l J. Image Graphics 5(3), 447–480 (2005)

41. Research Triangle Park (RTI). Planning Report 02-3: The economic impacts of inadequate infrastructure for software testing. Technical report, National Institute of Standards and Technology (NIST), U.S. Department of Commerce (May 2002)

42. Richardson, D.: Zero tests for constants in simple scientific computation. Mathematics in Computer Science 1(1), 21–38 (2007); Inaugural issue on Complexity of Continous Computation

43. Schönhage, A.: Storage modification machines. SIAM J. Computing 9, 490–508 (1980)

44. Sharma, V.: Robust approximate zeros in banach space. Mathematics in Computer Science 1(1), 71–109 (2007); Based on his thesis

45. Sharma, V., Du, Z., Yap, C.: Robust approximate zeros. In: Brodal, G.S., Leonardi, S. (eds.) ESA 2005, vol. 3669, pp. 3–6. Springer, Heidelberg (2005)

46. Shoenfield, J.R.: Mathematical Logic. Addison-Wesley, Boston (1967)

47. Snyder, J.M.: Interval analysis for computer graphics. SIGGRAPH Comput. Graphics 26(2), 121–130 (1992)

48. Spielman, D.A., Teng, S.-H.: Smoothed analysis of algorithms: Why the simplex algorithm usually takes polynomial time. J. of the ACM 51(3), 385–463 (2004)

49. Stander, B.T., Hart, J.C.: Guaranteeing the topology of an implicit surface polygonalization for interactive meshing. In: Proc. 24th Computer Graphics and Interactive Techniques, pp. 279–286 (1997)

50. Traub, J., Wasilkowski, G., Woźniakowski, H.: Information-Based Complexity. Academic Press, Inc., London (1988)

51. Trefethen, L.N.: Computing with functions instead of numbers. Mathematics in Computer Science 1(1), 9–19 (2007); Inaugural issue on Complexity of Continous Computation. Based on talk presented at the Brent's 60th Birthday Symposium, Weierstrass Institute, Berlin (2006)

52. Trefethen, L.N., Bau, D.: Numerical linear algebra. Society for Industrial and Applied Mathematics, Philadelphia (1997)

53. Weihrauch, K.: Computable Analysis. Springer, Berlin (2000)
54. Yap, C.K.: Robust geometric computation. In: Goodman, J.E., O'Rourke, J. (eds.) Handbook of Discrete and Computational Geometry, ch. 41, 2nd edn., pp. 927–952. Chapman & Hall/CRC, Boca Raton (2004)
55. Yap, C.K.: Complete subdivision algorithms, I: Intersection of Bezier curves. In: 22nd ACM Symp. on Comp. Geometry, July 2006, pp. 217–226 (2006)
56. Yap, C.K.: Theory of real computation according to EGC. In: Hertling, P., Hoffmann, C.M., Luther, W., Revol, N. (eds.) Real Number Algorithms. LNCS, vol. 5045, pp. 193–237. Springer, Heidelberg (2008)
57. Yap, C.K., Dubé, T.: The exact computation paradigm. In: Du, D.-Z., Hwang, F.K. (eds.) Computing in Euclidean Geometry, 2nd edn., pp. 452–492. World Scientific Press, Singapore (1995)
58. Yap, C.K., Yu, J.: Foundations of exact rounding. In: Das, S., Uehara, R. (eds.) WALCOM 2009. LNCS, vol. 5431, Springer, Heidelberg (2009); Invited talk, 3rd Workshop on Algorithms and Computation, Kolkata, India
59. Yu, J.: Exact arithmetic solid modeling. Ph.D. dissertation, Department of Computer Science, Purdue University, West Lafayette, IN 47907, Technical Report No. CSD-TR-92-037 (June 1992)

Much Ado about Zero

Stefan Schirra

Otto von Guericke University Magdeburg,
Faculty of Computer Science,
Department of Simulation and Graphics,
39106 Magdeburg, Germany

Abstract. Zero separation bounds provide a lower bound on the absolute value of an arithmetic expression, unless the value is zero. Such separation bounds are used for verified identification of zero in sign computations with real algebraic numbers, especially with number types that record the computation history of a numerical value using expression dags. We summarize results on separation bounds and their use for adaptive sign computation with real algebraic numbers based on expression dags.

Keywords: zero separation bound, exact geometric computation paradigm, real algebraic number computation.

1 Introduction

Recognizing the signs and consequently taking the right decisions is crucial for success–particularly with regard to geometric algorithms and the exact geometric computation paradigm.

Naïvely implemented, theoretically correct geometric algorithms more or less frequently crash, loop, or compute garbage [12] due to rounding errors caused by inherently inexact floating-point arithmetic, which is still the standard substitution for exact real arithmetic in scientific computing. Implementations of geometric algorithms tend to crash if rounding errors lead to inconsistencies that their theoretical counterpart cannot handle, because they contradict real geometry, e.g., detection (according to floating-point arithmetic) of several intersection points of distinct straight lines. The basic idea of the exact geometric computation paradigm [31, 30] is to ensure that the control flow in the implementation is exactly as it would be in the underlying geometric algorithm executed with exact real arithmetic. Note that we do not require exact numerical values here. It suffices to ensure that all decisions in the branching steps of an algorithm are correctly taken by the implementation. The branching steps in geometric algorithms boil down to the comparison of numerical values and without loss of generality we may assume that one of the numerical values we compare is zero. Thus, once we know the correct sign of the other value, we know how to continue correctly. Thus, the exact computation of the sign of the numerical value of an arithmetic expression is crucial for the implementation of geometric algorithms in accordance with the exact geometric computation paradigm. Of course, numerical values far away from zero

S. Albers, H. Alt, and S. Näher (Eds.): Festschrift Mehlhorn, LNCS 5760, pp. 408–421, 2009.

do not cause much trouble for sign computation. There is much ado required only if the numerical values are about zero.

2   Arithmetic Filters

For most geometric algorithms exact sign computations for real algebraic numbers suffice for putting the exact geometric computation paradigm into action, provided the numerical input data are all rational. Since integers and floating-point values are rational numbers, the assumption usually holds.

If all computations are even rational, a straightforward but inefficient approach is to use a rational arithmetic based on arbitrary precision integer arithmetic. This is inefficient, because we always compute exact values no matter whether such ultimate accuracy is required or not. Since the beginning of the nineties, more efficient alternatives have been developed, especially so-called floating-point filters and related adaptive evaluation strategies [8, 11, 14, 28]. A floating-point filter evaluates an expression with fast floating-point arithmetic first and tries to verify the sign computation by an accompanying error analysis. If the verification fails, we switch to another method to compute the sign. Thereby, we filter those computations that are potentially delivering an incorrect sign. In case of failure, we can try a better error analysis, a floating-point computation with higher precision using a software floating-point number type that allows one to choose the mantissa length if needed, or we make use of techniques to extend the precision of floating-point computations [28], and so on. We can even cascade such approaches, at best, reusing previously computed values, resulting in adaptive sign computations, where the effort we make is directly related to how close the actual value is to zero. Fig. 1 illustrates a cascaded filter.

Even such cascaded filters have a small problem. Zero can be detected only if the computed floating-point approximation is zero and the error analysis or some other technique proves that the error involved is indeed zero, too. This is not impossible, but extremely rare. This is where zero separation bounds come into play.

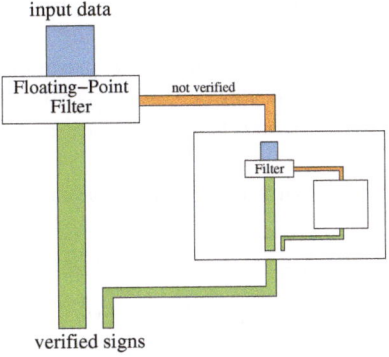

Fig. 1. Cascaded arithmetic filters

3   Zero Separation Bounds

Given an arithmetic expression E over a set of allowed operations, for example the basic arithmetical operations $+$, $-$, $\cdot$, $/$, and $\sqrt{}$, a constructive zero separation bound comprises an algorithm to derive a lower bound $\mathrm{sep}(E)$ on the absolute value of E. Such separation bounds are used for verified identification of zero in sign computations, in particular with number types supporting exact geometric computations with real algebraic numbers: Let ξ denote the value of E. Then, $\mathrm{sep}(E)$ is a separation bound if

$$\xi \neq 0 \quad \Rightarrow \quad \mathrm{sep}(E) \leq |\xi|.$$

Now let $\widetilde{\xi}$ be an approximation to ξ and Δ_{error} an upper bound on the error $|\xi - \widetilde{\xi}|$. If

$$|\widetilde{\xi}| + \Delta_{\mathrm{error}} < \mathrm{sep}(E)$$

we may conclude that $\xi = 0$, because otherwise the inequality above would rise a contradiction.

Note that, according to our definition, ξ itself does not necessarily provide separation bound, since the definition also asks for an algorithm to compute $\mathrm{sep}(E)$ and it might be impossible to compute ξ exactly.[1] However, even if an exact representation for ξ exists and is effectively computable, it might be too expensive to compute it. A separation bound must be easily and efficiently computable to be useful, such that, together with efficiently computed approximation and error bound, it provides a more efficient non-zero test, if the value is non-zero.

Let us look at a simple example. For an arithmetic expression E with integral operands, where E involves the operations $+$, $-$, $\cdot$, and $\sqrt[k]{}$ only, the so-called BFMS bound computes a quantity $U(E)$ which gives us an upper bound on the absolute value of ξ and its conjugates. We get $U(E)$ by replacing every $-$ in E by a $+$ and replacing every integer by its absolute value, see also Tab. 1. Then we have

$$\left(U(E)^{\deg(\xi)-1}\right)^{-1} \leq |\xi| \leq U(E)$$

where ξ denotes the value of E and $\deg(\alpha)$ denotes the algebraic degree of a real algebraic number α. Since we do not know the exact degree, we compute a straightforward upper bound on it, namely the product of all the radices of the radicals arising in the expression. The actual separation bound should be easily computable. Therefore, instead of $U(E)$ we compute an upper bound $u(E)$ on $\log U(E)$, see Tab. 1.

Let us use this separation bound to verify one of Ramanujan's equations on nested radicals.

$$3 \cdot \sqrt{\sqrt[3]{5} - \sqrt[3]{4}} = \sqrt[3]{20} - \sqrt[3]{25} + \sqrt[3]{2} \tag{1}$$

This is tantamount to confirming that the sign of

$$3 \cdot \sqrt{\sqrt[3]{5} - \sqrt[3]{4}} - \sqrt[3]{20} + \sqrt[3]{25} - \sqrt[3]{2}$$

---

[1] Of course, once you know, that there is no exact representation, you also know $\xi \neq 0$.

Table 1. Inductive computation of $U(E)$ and $u(E)$

	$U(E)$	$u(E)$
integer $N \neq 0$	$|N|$	$\lceil \log |N| \rceil$
0	0	0
$E_1 \pm E_2$	$U(E_1) + U(E_2)$	$\max(u(E_1), u(E_2)) + 1$
$E_1 \cdot E_2$	$U(E_1) \cdot U(E_2)$	$u(E_1) + u(E_2)$
$\sqrt[k]{E_1}$	$\sqrt[k]{U(E_1)}$	$\lceil u(E_1)/k \rceil$

is zero. The straightforward bound on the degree is $2 \cdot 3^5 = 1458$. Next,

$$u(\sqrt[3]{5} - \sqrt[3]{4}) = 2$$

and hence

$$u(3 \cdot \sqrt{\sqrt[3]{5} - \sqrt[3]{4}}) = 3.$$

Furthermore,

$$u((\sqrt[3]{20} - \sqrt[3]{25}) + \sqrt[3]{2}) = 4.$$

Finally, we get $u(E) = 5$, and thus $= 2^{-7285}$ is a separation bound. Therefore, it suffices to compute an approximation $\widetilde{\xi}$ and an error bound Δ_{error}, such that

$$|\widetilde{\xi}| + \Delta_{\text{error}} < 2^{-7285}$$

in order to verify (1).

4   Some Background on Algebraic Numbers

An element $\xi \in K \supseteq F$ is called *algebraic over* F, if there is a polynomial $P(X) \in F[X]$ such that $P(\xi) = 0$. A complex number $\xi \in \mathbb{C}$ is called *algebraic* if it is algebraic over $\mathbb{Q}$, i.e., if there is a $Q(X) \in \mathbb{Q}[X]$ such that $Q(\xi) = 0$. Such a polynomial exists if and only if there is a $P(X) \in \mathbb{Z}[X]$ such that $P(\xi) = 0$. The *minimal polynomial* of ξ is the unique, irreducible polynomial of smallest degree with this property. Note that this definition of minimal polynomial slightly deviates from other definitions where the monic polynomial of smallest degree with rational coefficients is called minimal polynomial. We prefer a polynomial with integer coefficients. The algebraic degree of an algebraic number is the degree of its minimal polynomial. An algebraic number ξ is an *algebraic integer* if its minimal polynomial is monic. Elements of $\mathbb{Z}$ are often called rational integers to better distinguish them from algebraic integers. Just as every rational number is an algebraic number, every rational integer is an algebraic integer as well.

Algebraic numbers form a subfield of $\mathbb{C}$, the real algebraic numbers a subfield of $\mathbb{R}$. Moreover, (real) algebraic integers form a ring. Furthermore, algebraic numbers and algebraic integers are closed under radical operations. We shall use the well-known fundamental theorem of symmetric functions to prove these facts. For the division operation, note already that if $\alpha \neq 0$ is a root of $A(X)$, then α^{-1} is a root of $X^m A(\frac{1}{X})$.

A polynomial is symmetric on variables $Y_1, \ldots, Y_n$ if it is unchanged by any permutation of these variables. The elementary symmetric functions on variables $Y_1, \ldots, Y_n$ are the polynomials

$$\sigma_1(Y_1, \ldots, Y_n) = \sum_{1 \leq i \leq n} Y_i$$

$$\sigma_2(Y_1, \ldots, Y_n) = \sum_{1 \leq i < j \leq n} Y_i Y_j$$

$$\sigma_3(Y_1, \ldots, Y_n) = \sum_{1 \leq i < j < k \leq n} Y_i Y_j Y_k$$

$$\vdots$$

$$\sigma_n(Y_1, \ldots, Y_n) = \prod_{1 \leq i \leq n} Y_i$$

The fundamental theorem of symmetric functions states that every polynomial symmetric on some variables can be written as a polynomial in the elementary symmetric functions on these variables.

Every monic polynomial $X^m + a_{m-1} X^{m-1} + \cdots + a_0 = \prod_{i=1}^m (X - \alpha_i)$ is symmetric in its roots α_i and hence a polynomial in the elementary symmetric functions on the α_i. Actually, we have

$$a_j = \sigma_{m-j}(\alpha_1, \ldots, \alpha_m) \tag{2}$$

Let $A(x) = \prod_{i=1}^m (X - \alpha_i)$ and $B(x) = \prod_{j=1}^n (X - \beta_j)$ be monic polynomials. The polynomials

$$\prod_{i=1}^m \prod_{j=1}^n (X - (\alpha_i + \beta_j))$$

$$\prod_{i=1}^m \prod_{j=1}^n (X - (\alpha_i - \beta_j))$$

$$\prod_{i=1}^m \prod_{j=1}^n (X - (\alpha_i \beta_j))$$

are all symmetric in both the α_i and the β_j and hence can be expressed as polynomials in the elementary symmetric functions of these two sets of conjugates. So if the coefficients of $A(X)$ and $B(X)$ are integers, it follows using (2) that the resulting polynomials have integer coordinates as well, which shows that algebraic integers are closed under the operations $+$, $-$, and $\cdot$.

Let $\alpha_{n-1}, \alpha_{n-2}, \ldots, \alpha_0$ be algebraic integers and let $\alpha_j^{(i_j)}$, $1 \le i_j \le \deg(\alpha_j)$, be the conjugates of α_j for $0 \le j \le n-1$. Consider

$$Q(X) = \prod_{i_0} \prod_{i_1} \cdots \prod_{i_{n-1}} \left(X^n + \alpha_{n-1}^{(i_{n-1})} X^{n-1} + \cdots + \alpha_0^{(i_0)} \right)$$

$Q(X)$ is symmetric in the $\alpha_j^{(i_j)}$ for all j. By applying the fundamental theorem on symmetric functions several times we see that $Q(X)$ can be expressed as a polynomial in the elementary symmetric functions of the conjugate sets. Hence the coefficients of $Q(X)$ are integers, since all the α_i are algebraic integers and hence by (2) all the elementary symmetric functions in each conjugate set deliver integral values. Hence we have

Lemma 1. *Let ϱ be the root of a monic polynomial*

$$P(X) = X^n + \alpha_{n-1}X^{n-1} + \alpha_{n-2}X^{n-2} + \cdots + \alpha_0$$

where the coefficients $\alpha_{n-1}, \alpha_{n-2}, \ldots, \alpha_0$ are algebraic integers. Then ϱ is an algebraic integer.

In particular, this shows that algebraic integers are closed under radical operations as well.

If α is a root of $P(x) = a_m X^m + \cdots a_0$ then $a_m \alpha$ is a root of

$$a_m^{m-1} P(\tfrac{X}{a_m})$$

and hence we have

Lemma 2. *Every algebraic number is the quotient of an algebraic integer and a (rational) integer.*

More precisely, it is the quotient of an algebraic integer and the leading coefficient of its minimal polynomial.

Next consider a root of a polynomial

$$P(x) = \gamma_m X^m + \gamma_{m-1} X^{m-1} + \cdots + \gamma_0$$

where the coefficients $\gamma_m, \gamma_{m-1}, \ldots, \gamma_0$ are algebraic numbers. Every coefficient γ_i is a quotient $\frac{\nu_i}{d_i}$ of an algebraic integer ν_i and a rational integer d_i. Now let $\Delta = \prod d_i$ and $\Delta_j = \Delta/d_j$. We multiply by Δ to get rid of the denominators of all coefficients and get a polynomial with algebraic integer coefficients $\Delta P(X)$ and leading coefficient $\nu_m \Delta_m$. As above, we construct a corresponding monic polynomial

$$Q(X) = (\nu_m \Delta_m)^{m-1} \Delta P(\tfrac{X}{\nu_m \Delta_m})$$

where $P(\xi) = 0$ implies $Q(\xi \nu_m \Delta_m) = 0$. By construction $Q(X)$ is monic and its coefficients are algebraic integers. Thus, by Lemma 1, its roots are algebraic integers and if ξ is a root of $P(X)$ then $\xi \nu_m \Delta_m$ is an algebraic integer. Summarizing, the roots of a polynomial with algebraic coefficients are algebraic again, or, in terms of field theory, if ξ is algebraic over an extension field $\mathbb{Q}(\gamma_0, \ldots, \gamma_m)$, then ξ is algebraic (over $\mathbb{Q}$).

5 Known Constructive Zero Separation Bounds

In this section we briefly review known constructive separation bounds for subsets of the real algebraic numbers. Let ξ be the value of an algebraic expression E. If the minimal polynomial $P \in \mathbb{Z}[X]$ for ξ would be known, one could use minimum root bounds for P to get a zero separation bound. The last statement is somewhat absurd because knowing the minimal polynomial requires to know whether ξ is zero. Nevertheless, this is a road taken by most separation bounds. Given an arithmetic expression E, most separation bounds derive bounds on some quantities that the minimal polynomial of ξ would fulfill, if ξ were nonzero. For example, the algebraic degree of ξ is (almost) always one of these quantities. The computed bounds on the quantities are then used to derive a minimum root bound for the roots of the minimal polynomial of ξ, if ξ were non-zero. Further quantities of (minimal) polynomials involved in separation bound computation are length, height, and Mahler measure. Length and height are L_1- and L_∞-norm of the coefficient vector of the polynomial, i.e., the sum of the absolute values of the coefficients and the maximum of the absolute values of the coefficients, respectively. For a polynomial $P(X) = a_n(X - \alpha_1) \cdots (X - \alpha_n)$, the Mahler measure is

$$|a_n| \prod_{i=1}^{n} \max(1, |\alpha_i|).$$

If ξ is an algebraic integer, an upper bound μ on the maximum absolute value of the conjugates of ξ, together with a bound d on the algebraic degree of ξ, gives us a zero separation bound:

$$\xi \neq 0 \quad \Rightarrow \quad |\xi| \geq (\mu^{d-1})^{-1}$$

Mignotte [18] discusses the *identification of algebraic numbers*, i.e., numerically verifying equality of real algebraic numbers given by different expressions. While, strictly speaking, he does not provide constructive zero separation bounds yet, his work is nevertheless a milestone regarding the development of such bounds. Below we list the constructive zero separation bounds we are currently aware of. We do not provide the technical details of these bounds, but refer the interested reader to the cited sources instead.

- Degree-height bound; as suggested by the name, bounds on degree and height of the minimal polynomial are derived from the given arithmetic expression [31].
- Degree-length bound; here, bounds on degree and length of the minimal polynomial are derived. This bound was initially used by Yap and Dubé [31] for adaptive sign computation with number types based on expression dags.
- Degree-measure bound as described in [3] and based on Mignotte's work [18, 19]; here, bounds on degree and Mahler measure of the minimal polynomial are derived.
- Canny's polynomial system bound [6] can be used to derive a separation bound automatically as described in [3].

- BFMS bound; Burnikel et al. [3] compute an upper bound on the magnitude of the conjugates for division-free expressions involving radicals. In the presence of division operations, the expression is virtually transformed to a quotient of two division-free expressions.
- Sekigawa's improved degree-measure bound [27]; here a refinement of the standard inequality on the Mahler measure for the sum of two algebraic numbers is used.
- Scheinerman's eigenvalue bound [22]; this bound for a subset of the algebraic integers is based on eigenvalues of integer matrices.
- Conjugate bound by Li and Yap [13]; Li and Yap compute both an upper bound and a lower bound on the absolute values of the conjugates as well as a bound on the leading coefficient of the minimal polynomial, where the latter involves the Mahler measure. Li and Yap allow for real algebraic operands that are defined as a certain root of a given polynomial with integer coefficients. Since most other bounds regard conjugates, too, we prefer to call this bound the LY bound, analogously to the BFMS bound.
- Improved degree-measure bound as described in [13]; here, the improvement addresses computation of the degree bound. It replaces the recursive rules presented in [3] by the method used in the BFMS bound.
- Improved BFMS bound described in [17]; this is basically the BFMS bound, besides that the range of permitted expressions is extended to allow for algebraic integers as operands, provided that an upper bound on the conjugates of the operands is available.
- Improved BFMS bound described in [16]; this is just a precursor of the BFMSS bound described next, without the improvement by Yap's symmetry rule for radicals [29].
- BFMSS bound; for division-free expressions, this bound is identical to the BFMS bound. Burnikel et al. [4] (a) modify the BFMS bound for expressions involving devisions, such that a much better degree bound can be achieved, and (b) extend the permitted operations to include the so-called diamond operator: For a polynomial $p \in \mathbb{R}[X]$ with real algebraic coefficients, $\diamond(p, j)$ computes the j-th largest real root of p, if p has that many real roots, and is undefined otherwise. This is similar to $\sqrt{}$-operations which are undefined for negative arguments. This way we can get all real algebraic numbers.
- BFMSS[k] bound; Pion and Yap [21] describe how to improve separation bounds if all rational operands are binary or decimal floating-point values. For such rational quotients, the denominators are powers of 2 and 10, respectively. The new method can improve known bounds drastically. Actually, Pion and Yap describe a more general method which applies to any number base k, and apply it to the BFMSS bound.
- Degree-measure[k] bound; similarly, Pion and Yap [21] apply their method for k-ary rational input numbers to the improved degree-measure bound.
- Schmitt's variant of the BFMSS[k] bound with improved rules for the diamond operator [25].

As mentioned above, the set of allowed operations and the supported subset of real algebraic numbers differs for the various bounds. In the presence of division operations, BFMS, BFMSS, and LY bound virtually rewrite ξ as a quotient $\frac{\nu}{\varrho}$ of two algebraic integers ν and ϱ. LY maintains estimates for the value ξ transformed to the quotient of an algebraic integer ν_{LY} and a rational integer. ξ and ν_{LY} have the same algebraic degree. BFMS maintains estimates for a quotient of two algebraic integers ν_{BFMS} and δ_{BFMS}, whose algebraic degree bound is quadratic in the degree bound of ξ. BFMSS improves on this. Here we have a quotient of two algebraic integers ν_{BFMSS} and δ_{BFMSS} such that the degree bound for each of them is the same as the degree bound for ξ.

A separation bound sep dominates another bound sep' for a class of expressions $\mathscr{E}$ if $\mathrm{sep}(E) \geq \mathrm{sep}'(E)$ for all $E \in \mathscr{E}$. The degree-measure bound dominates the degree-length bound. For division-free radical expressions (i.e., without diamond operator) Burnikel et al. show that BFMS dominates the degree-measure bound as well as a bound based on Canny's polynomial system bound [3]. Li and Yap [13] show that BFMS dominates Scheinerman's eigenvalue bound for such expressions as well. For division-free radical expressions BFMSS and LY bound are identical to BFMS. For expressions involving divisions, BFMSS, LY, and the improved degree-measure bound are incomparable [13].

6   Adaptive Sign Computation with Expression Dags

Constructive zero separation bounds are used in adaptive sign computation. In order to allow for re-evaluation with higher precisions, number types for exact decision computation with real algebraic numbers record the computation history in an expression dag, i.e., a directed acyclic graph. Internal nodes in such a dag correspond to subexpressions. An internal node is labeled with an operation and points to the subexpressions that represent the operands of the expression. Note that for non-commutative operations, the order of pointers to subexpressions matters.

The external nodes in such a dag represent operands from a base set, usually the integers. Whenever an operation op on operands $o_1, \ldots, o_k$ is executed, a new node is created and labeled by op. Pointers from the new node to the nodes representing the operands are created. Note that the same variable may show up more than once among the operands. Thus, several of the created pointers might point to the same node. Fig. 2 shows the dag built for the following piece of C++ code, where root(RealAlgebraic r, int k) computes the k-th root of r.

```
3 * sqrt(root(RealAlgebraic(5),3) - root(RealAlgebraic(4),3))
```

The actual work is done in the comparison operations. These are reduced to sign computations. There, a zero separation bound sep is calculated and, using the expression dag, iteratively more and more accurate approximations and error bounds are computed until the current approximation $\widetilde{\xi}$ for the exact value ξ and the current error bound Δ_{error} either fulfill

$$|\widetilde{\xi}| < \Delta_{\mathrm{error}} \quad \text{or} \quad |\widetilde{\xi}| + \Delta_{\mathrm{error}} < \mathrm{sep}.$$

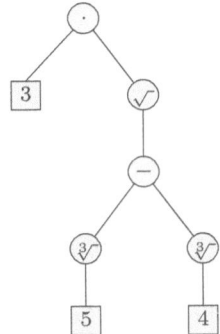

Fig. 2. An expression dag for the left hand side of (1)

The number types `CORE::Expr` [10] and `leda::real` [2, 5, 15] implement exact decisions via adaptive sign computation with expression dags in C++. The advantage of these number types is its ease of use. A user can use them like any built-in number type without knowing anything about zero separation bounds and required precision at all.

7   Some Ruminations and Observations

We conclude with some remarks regarding zero separation bounds and their use in number types with adaptive sign computations based on expression dags.

Combined Bounds. Since some of the bounds are incomparable, a good strategy is to compute all of them and take the best one. Besides this, as mentioned above, the LY bound computes lower bounds $\underline{\nu}(E)$ on the absolute value of the conjugates of the non-zero values of subexpressions E. Here, any available zero separation bound for E could be used instead of $\underline{\nu}(E)$. Note that this is different from simply computing two bounds and finally taking the better one. Here, the competition takes places for every subexpression.

Degree Bound Reduction by Restructuring. All the separation bound algorithms used for adaptive sign computation with expression dags in `CORE::Expr` and `leda::real` compute a crude bound on the degree, namely the product of all the radices of the radical expressions and the degrees of the polynomials in the diamond operations arising in the expression dag. So $\sqrt{a} \cdot \sqrt{b}$ would have degree four for integers a and b, while the actual degree is two. So restructuring this to $\sqrt{a \cdot b}$, or more generally, restructuring a dag where we have a product of two radical operations with same radix into a dag with a single radical operation on top is preferable, cf. Fig. 3.

Search for Common Subexpressions. Because of the crude degree bound computation, the separation bound for mathematically equivalent expressions

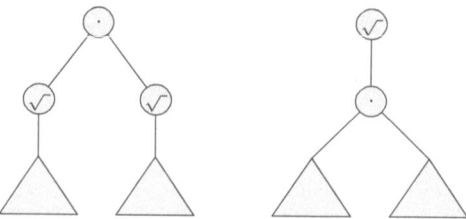

Fig. 3. Rearranging an expression dag to decrease the degree bound

given by code fragments can differ a lot. Observe, for example, that the degree bound computed for

```
RealAlgebraic  ra = sqrt( RealAlgebraic(2));
ra += sqrt( RealAlgebraic(2));
ra -= sqrt( RealAlgebraic(2));
ra -= sqrt( RealAlgebraic(2));
```

is 16, while the bound computed for

```
RealAlgebraic  rt2 = sqrt( RealAlgebraic(2));
RealAlgebraic  ra = rt2 + rt2 - rt2 -rt2;
```

is only 2, since the subexpression for $\sqrt{2}$ is reused in the latter code fragment and hence the corresponding radix is considered only once. Since number types CORE::Expr and leda::real are designed for ease of use, the observation above suggests to search for common subexpressions automatically. Such a search for common subexpressions has been implemented for leda::real [23, 24]. With this strategy enabled, leda::real compute degree bound 2 in both cases. The strategy also leads to a significant improvement for the shortest path example used by Burnikel et al. in [2]. However, since there is a lack of real-world examples where the strategy pays off, common subexpression search is switched off in the default version of leda::real nowadays. Note that we do not search for equivalent expressions, but for identical subexpressions only.

Zero Separation Bounds versus Root Separation Bounds. Whenever we compare α and β, we use estimates for $A(X)$ and $B(X)$ with $A(\alpha) = 0$ and $B(\beta) = 0$ to get estimates for $C(X)$ with $C(\alpha - \beta) = 0$. Then, we use these estimates to derive a bound on the gap between zero and the absolute values of the real roots of $C(X)$. As an alternative, we could also consider $A(X) \cdot B(X)$ and use a bound on the minimum separation of the distinct roots of this polynomial.

Let's have a closer look. For simplicity, assume we use the degree-measure bound and compare α and β with degree bounds d_α and d_β and measure bounds M_α and M_β, respectively. The standard approach bounds the measure of $\alpha - \beta$ by

$$2^{d_\alpha d_\beta} M_\alpha^{d_\beta} M_\beta^{d_\alpha}$$

and the inverse of this quantity is a zero separation bound. In the alternative approach, we consider a polynomial having both roots α and β. For a polynomial

$P(X)$ of degree d with integer coefficients, the minimum separation between distinct roots of P is at least

$$d^{-(d+2)/2} \cdot M(P)^{1-d} \tag{3}$$

where $M(P)$ denotes the Mahler measure of P, cf. [19]. In order to apply the root separation bound above, we need a bound on the degree of $A(X) \cdot B(X)$ and a bound on its measure. The degree is bounded by $d_\alpha + d_\beta$ and the measure by $M_\alpha M_\beta$. Applying (3), the inverse of

$$(d_\alpha + d_\beta)^{(d_\alpha+d_\beta+2)/2} M_\alpha^{d_\alpha+d_\beta-1} M_\beta^{d_\alpha+d_\beta-1}$$

is a root separation bound and hence a zero separation bound for $\alpha - \beta$. The alternative bound is better only if

$$d_\alpha d_\beta - (d_\alpha/2 + d_\beta/2 + 1) \log(d_\alpha + d_\beta) - (d_\alpha - 1) \log M_\alpha - (d_\beta - 1) \log M_\beta > 0.$$

This holds for $d_\alpha = d_\beta = 1$. In general, the alternative is better only if the (estimates for the) measures are very small.

Arithmetic versus Geometry. Exact decision number types with adaptive sign computation with expression dags are an easy-to-use general purpose tool. In many applications in geometric computing, however, fine-tuned solutions that exploit special restrictions, are more efficient. Moreover, recording computation history in terms of geometric operations instead of arithmetical operations often provides better performance [7, 9].

To Be or Not to Be Zero. If we are not interested in the sign of ξ, but only want to know whether ξ is zero or not, there are efficient probabilistic alternatives [1, 20, 26, 29] to the use of zero separation bounds. The best alternative, however, is to avoid zero testing at all. Sometimes, geometric considerations allow us to get around expensive zero testing. For example, assume that we want to compute the intersection points of two circles whose center points have rational coordinates. Using elimination, we can compute x- and y-coordinates of the intersection points, both of the form $\alpha \pm \sqrt{\beta}$. In order to find the correct combinations among the four possible pairs, one is tempted to plug pairs of coordinates

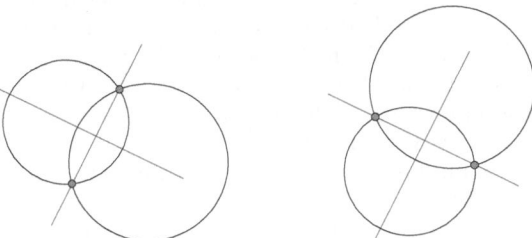

Fig. 4. Geometry helps to avoid zero testing

into the circle equations and to check whether this yields zero. However, this is
not necessary. Comparing the rational coordinates, which are usually distinct,
suffices to identify all valid combinations, cf. Fig. 4. Note that zero testing for a
probably non-zero rational value is much more efficient than zero testing for a
real algebraic number which is zero with probability $\frac{1}{2}$.

References

1. Blömer, J.: A Probabilistic Zero-Test for Expressions Involving Roots of Rational
   Numbers. In: Bilardi, G., Pietracaprina, A., Italiano, G.F., Pucci, G. (eds.) ESA
   1998. LNCS, vol. 1461, pp. 151–162. Springer, Heidelberg (1998)
2. Burnikel, C., Fleischer, R., Mehlhorn, K., Schirra, S.: Efficient Exact Geometric
   Computation Made Easy. In: 15th ACM Symposium on Computational Geometry,
   pp. 341–350. ACM, New York (1999)
3. Burnikel, C., Fleischer, R., Mehlhorn, K., Schirra, S.: A Strong and Easily Computable Separation Bound for Arithmetic Expressions Involving Radicals. Algorithmica 27(1), 87–99 (2000)
4. Burnikel, C., Funke, S., Mehlhorn, K., Schirra, S., Schmitt, S.: A Separation Bound
   for Real Algebraic Expressions. In: Meyer auf der Heide, F. (ed.) ESA 2001. LNCS,
   vol. 2161, pp. 254–265. Springer, Heidelberg (2001)
5. Burnikel, C., Könnemann, J., Mehlhorn, K., Näher, S., Schirra, S., Uhrig, C.: Exact
   Geometric Computation in LEDA. In: 11th ACM Symposium on Computational
   Geometry, pp. C18–C19. ACM, New York (1995)
6. Canny, J.F.: The Complexity of Robot Motion Planning. MIT Press, Cambridge
   (1988)
7. Fabri, A., Pion, S.: A Generic Lazy Evaluation Scheme for Exact Geometric Computations. In: 2nd Library-Centric Software Design, pp. 75–84. ACM, New York
   (2006)
8. Fortune, S., Van Wyk, C.J.: Efficient Exact Arithmetic for Computational Geometry. In: 9th ACM Symposium on Computational Geometry, pp. 163–172. ACM,
   New York (1993)
9. Funke, S., Mehlhorn, K.: Look: A Lazy Object-Oriented Kernel Design for Geometric Computation. Computational Geometry: Theory and Applications 22(1-3),
   99–118 (2002)
10. Karamcheti, V., Li, C., Pechtchanski, I., Yap, C.: A Core Library for Robust Nu-
    meric and Geometric Computation. In: 15th ACM Symposium on Computational
    Geometry, pp. 351–359. ACM, New York (1999)
11. Karasick, M., Lieber, D., Nackman, L.R.: Efficient Delaunay Triangulation Using
    Rational Arithmetic. ACM Trans. on Graphics 10, 71–91 (1991)
12. Kettner, L., Mehlhorn, K., Pion, S., Schirra, S., Yap, C.: Classroom Examples
    of Robustness Problems in Geometric Computations. Computational Geometry:
    Theory and Applications 40(1), 61–78 (2008)
13. Li, C., Yap, C.: A New Constructive Root Bound for Algebraic Expressions. In:
    12th ACM-SIAM Symposium on Discrete Algorithms (SODA), pp. 496–505 (2001)
14. Mehlhorn, K., Näher, S.: The Implementation of Geometric Algorithms. In:
    Pehrson, B., Simon, I. (eds.) IFIP 13th World Computer Congress, pp. 223–231.
    North-Holland, Amsterdam (1994)
15. Mehlhorn, K., Näher, S.: LEDA: A Platform for Combinatorial and Geometric
    Computing. Cambridge University Press, Cambridge (2000)

16. Mehlhorn, K., Schirra, S.: Generalized and Improved Constructive Separation Bound for Real Algebraic Expressions. Research Report MPI-I-2000-1-004, Max-Planck-Institut für Informatik (2000)
17. Mehlhorn, K., Schirra, S.: Exact Computation with leda_real - Theory and Geometric Applications. In: Alefeld, G., Rohn, J., Rump, S.M., Yamamoto, T. (eds.) Symbolic Algebraic Methods and Verification Methods, pp. 163–172. Springer, Wien (2001)
18. Mignotte, M.: Identification of Algebraic Numbers. J. Algorithms 3(3), 197–204 (1982)
19. Mignotte, M.: Mathematics for Computer Algebra. Springer, New York (1992)
20. Monagan, M.B., Gonnet, G.H.: Signature Functions for Algebraic Numbers. In: International Symposium on Symbolic and Algebraic Computation (ISSAC), pp. 291–296. ACM, New York (1994)
21. Pion, S., Yap, C.: Constructive Root Bound Method for k-ary Rational Input Numbers. In: 18th ACM Symposium on Computational Geometry, pp. 256–263. ACM, New York (2003)
22. Scheinerman, E.R.: When Close Enough is Close Enough. American Mathematical Monthly 107(6), 489–499 (2000)
23. Schmitt, S.: Common Subexpression Search in leda_reals. Report ECG-TR-243105-01, Effective Computational Geometry for Curves and Surfaces (2003)
24. Schmitt, S.: Common Subexpression Search in leda_reals – a Study of the Diamond Operator. Report ECG-TR-243107-01, Effective Computational Geometry for Curves and Surfaces (2004)
25. Schmitt, S.: Improved Separation Bounds for the Diamond Operator. Report ECG-TR-363108-01, Effective Computational Geometry for Curves and Surfaces (2004)
26. Schwartz, J.T.: Fast Probabilistic Algorithms for Verification of Polynomial Identities. J. ACM 27(4), 701–717 (1980)
27. Sekigawa, H.: Using Interval Computation with the Mahler Measure for Zero Determination of Algebraic Numbers. Josai University Information Sciences Research 9(1), 83–99 (1998)
28. Shewchuk, J.R.: Adaptive Precision Floating-Point Arithmetic and Fast Robust Geometric Predicates. Discrete and Computational Geometry 18, 305–363 (1997)
29. Tulone, D., Yap, C., Li, C.: Randomized Zero Testing of Radical Expressions and Elementary Geometry Theorem Proving. In: Richter-Gebert, J., Wang, D. (eds.) ADG 2000. LNCS, vol. 2061, pp. 58–82. Springer, Heidelberg (2001)
30. Yap, C.: Towards Exact Geometric Computation. Computational Geometry: Theory and Applications 7, 3–23 (1997)
31. Yap, C., Dubé, T.: The Exact Computation Paradigm. In: Du, D.Z., Hwang, F.K. (eds.) Computing in Euclidean Geometry. Lecture Notes Series on Computing, vol. 4, pp. 452–492. World Scientific, Singapore (1995)

Polynomial Precise Interval Analysis Revisited

Thomas Gawlitza[1], Jérôme Leroux[2], Jan Reineke[3], Helmut Seidl[1],
Grégoire Sutre[2], and Reinhard Wilhelm[3]

[1] TU München, Institut für Informatik, I2
80333 München, Germany
{gawlitza,seidl}@in.tum.de
[2] LaBRI, Université de Bordeaux, CNRS
33405 Talence Cedex, France
{leroux,sutre}@labri.fr
[3] Universität des Saarlandes, Germany
{reineke,wilhelm}@cs.uni-sb.de

Abstract. We consider a class of arithmetic equations over the complete
lattice of integers (extended with $-\infty$ and ∞) and provide a polynomial
time algorithm for computing least solutions. For systems of equations
with addition and least upper bounds, this algorithm is a smooth generalization of the Bellman-Ford algorithm for computing the single source
shortest path in presence of positive and negative edge weights. The
method then is extended to deal with more general forms of operations
as well as minima with constants. For the latter, a controlled widening is applied at loops where unbounded increase occurs. We apply this
algorithm to construct a cubic time algorithm for the class of interval
equations using least upper bounds, addition, intersection with constant
intervals as well as multiplication.

1 Introduction

Interval analysis tries to derive tight bounds for the run-time values of variables
[1]. This basic information may be used for important optimizations such as safe
removals of array bound checks or for proofs of absence of overflows [2]. Since the
very beginning of abstract interpretation, interval analysis has been considered
as an algorithmic challenge. The reason is that the lattice of intervals may have
infinite ascending chains. Hence, ordinary fixpoint iteration will not result in
terminating analysis algorithms. The only general technique applicable here is
the widening and narrowing approach of Cousot and Cousot [3]. If precision is
vital, also more expressive domains are considered [4, 5]. While often returning
amazingly good results, widening and narrowing typically does not compute the
least solution of a system of equations but only a safe over-approximation.

In [6], however, Su and Wagner identify a class of interval equations for which
the respective least solutions can be computed precisely and in polynomial time.
As operations on intervals, they consider least upper bound, addition, scaling
with positive and negative constants and intersection with constant intervals.
The exposition of their algorithms, though, is not very explicit. Due to the

S. Albers, H. Alt, and S. Näher (Eds.): Festschrift Mehlhorn, LNCS 5760, pp. 422–437, 2009.

importance of the problem, we present an alternative and, hopefully more transparent approach. In particular, our methods also show how to deal with arbitrary multiplications of intervals. Our algorithm demonstrates how well-known ideas need only to be slightly extended to provide a both simple and efficient solution.

We start by investigating equations over integers only (extended with $-\infty$ and ∞ as least and greatest elements of the lattice) using maximum, addition, scaling with positive constants and minimum with constants as operations. In absence of minima, computing the least solution of such a system of equations can be considered as a generalization of the single-source shortest path problem from graphs to grammars in presence of positive and negative edge weights. A corresponding generalization for positive weights has been considered by Knuth [7]. Negative edge weights, though, complicate the problem considerably. While Knuth's algorithm can be considered as a generalization of Dijkstra's algorithm, we propose a generalization of the Bellman-Ford algorithm.

More generally, we observe that the Bellman-Ford algorithm works for all systems of equations which use operators satisfying a particular semantic property which we call BF-property. Beyond addition and multiplication with positive constants, positive as well as negative multiplication satisfies this property. *Positive* multiplication returns the product only if both arguments are positive, while *negative* multiplication returns the negated product if both arguments are negative. In order to obtain a polynomial algorithm also in presence of minima with constants, we instrument the basic Bellman-Ford algorithm to identify loops along which values might increase unboundedly. Once we have short-circuited the possibly costly iteration of such a loop we restart the Bellman-Ford algorithm until no further increments are found.

In the next step, we consider systems of equations over intervals using least upper bound, addition, negation, multiplication with positive constants as well as intersections with constant intervals and arbitrary multiplication of intervals. We show that computing the least solution of such systems can be reduced to computing the least solution of corresponding systems of integer equations. This reduction is inspired by the methods from [8] for interval equations with unrestricted intersections and the ideas of Leroux and Sutre [9], who first proved that interval equations with intersections with constant intervals as well as full multiplication can be solved in cubic time.

The rest of the paper is organized as follows. In Section 2, we introduce basic notions and consider methods for general systems of equations over $\mathbb{Z}$. Then we consider two classes of systems of equations over $\mathbb{Z}$ where least solutions can be computed in polynomial time. In Section 3, we consider systems of integer equations without minimum. In Section 4, we extend these methods to systems of equations where right-hand sides are *Bellman-Ford* functions. These systems can be solved in quadratic time (if arithmetic operations are executed in constant time). In Section 5, we then present our cubic time procedure for computing least solutions of systems of integer equations which additionally use minima with constants. In Section 6, we apply these techniques to construct a cubic

algorithm for the class of interval equations considered by Su and Wagner [6] —
even if additionally arbitrary multiplication of interval expressions is allowed.

2   Notation and Basic Concepts

Assume we are given a finite set of variables $\mathbf{X}$. We are interested in solving
systems of constraints over the complete lattice $\mathbb{Z} = \mathbb{Z} \cup \{-\infty, \infty\}$ equipped
with the natural ordering:

$$-\infty < \ldots < -2 < -1 < 0 < 1 < 2 < \ldots < \infty$$

On $\mathbb{Z}$, we consider the operations addition, multiplication with nonnegative constants, minimum "$\wedge$" and maximum "$\vee$". All operators are commutative where
minimum, addition, and multiplication also preserve $-\infty$. Moreover for every
$x > -\infty$,

$$x + \infty = \infty \qquad\qquad 0 \cdot \infty = 0$$
$$x \cdot \infty \ = \infty \quad \text{whenever } x > 0 \qquad x \cdot \infty = -\infty \quad \text{whenever } x < 0$$

For a finite set $\mathbf{X}$ of variables, we consider systems of equations

$$\mathbf{x} = e, \quad \mathbf{x} \in \mathbf{X}$$

where the right-hand sides e are expressions built from constants and variables
from $\mathbf{X}$ by means of maximum, addition, multiplication with positive constants
and minimum with constants. Thus right-hand sides e are of the form

$$e \quad ::= \quad a \mid \mathbf{y} \mid e_1 \vee e_2 \mid e_1 + e_2 \mid b \cdot e \mid e_1 \wedge a$$

for variables $\mathbf{y} \in \mathbf{X}$ and $a, b \in \mathbb{Z}$ where $a > -\infty$ and $b > 0$. Note that we
excluded general multiplication since multiplication with negative numbers is no
longer monotonic. Similar systems of equations have been investigated in [10]
where polynomial algorithms for computing least upper bounds are presented
— but only when computing least solutions over nonnegative integers.

A function $\mu : \mathbf{X} \to \mathbb{Z}$ is called a *variable assignment*. Every expression e
defines a function $\llbracket e \rrbracket : (\mathbf{X} \to \mathbb{Z}) \to \mathbb{Z}$ that maps variable assignments to values,
i.e.:

$$
\begin{aligned}
\llbracket a \rrbracket \mu &= a & \llbracket \mathbf{x} \rrbracket \mu &= \mu(\mathbf{x}) \\
\llbracket e_1 \vee e_2 \rrbracket \mu &= \llbracket e_1 \rrbracket \mu \vee \llbracket e_2 \rrbracket \mu & \llbracket e_1 + e_2 \rrbracket \mu &= \llbracket e_1 \rrbracket \mu + \llbracket e_2 \rrbracket \mu \\
\llbracket b \cdot e \rrbracket \mu &= b \cdot \llbracket e \rrbracket \mu & \llbracket e \wedge a \rrbracket \mu &= \llbracket e \rrbracket \mu \wedge a
\end{aligned}
$$

for $a \in \mathbb{Z}$, $\mathbf{x} \in \mathbf{X}$, $b > 0$ and expressions e, e_1, e_2. For a system $\mathcal{E}$ of equations,
we also denote the function $\llbracket e \rrbracket$ by $f_\mathbf{x}$, if $\mathbf{x} = e$ is the equation in $\mathcal{E}$ for $\mathbf{x}$. A
variable assignment μ is called a *solution* of $\mathcal{E}$ iff it satisfies all equations in $\mathcal{E}$,
i.e. $\mu(\mathbf{x}) = f_\mathbf{x}\mu$ for all $\mathbf{x} \in \mathbf{X}$. Likewise, μ is a *pre-solution* iff $\mu(\mathbf{x}) \le f_\mathbf{x}\mu$ for
all $\mathbf{x} \in \mathbf{X}$. Since the mappings $f_\mathbf{x}$ are monotonic, every $\mathcal{E}$ has a unique least
solution. In the following, we denote by $|\mathcal{E}|$ the sum of expression sizes of right-
hand sides of the equations in $\mathcal{E}$. The following fact states bounds on the sizes
of occurring values of variables:

Proposition 1. *Assume that $\mathcal{E}$ is a system of integer equations with least solution μ^*. Then we have:*

1. *If $\mu^*(\mathbf{x}) \in \mathbb{Z}$ for a variable $\mathbf{x}$, then:*

$$-(B \vee 2)^{|\mathcal{E}|} \cdot A \leq \mu^*(\mathbf{x}) \leq (B \vee 2)^{|\mathcal{E}|} \cdot A$$

   where A and B bound the absolute values of constants $a \in \mathbb{Z}$ and constant multipliers $b \in \mathbb{N}$, respectively, which occur in $\mathcal{E}$.
2. *If $\mathcal{E}$ does not contain multiplication or addition of variables, the bounds under 1 can be improved to:*

$$\Sigma^- \leq \mu^*(\mathbf{x}) \leq \Sigma^+$$

   where Σ^- and Σ^+ are the sums of occurrences of negative and positive numbers, respectively, in $\mathcal{E}$. □

In order to prove these bounds, we observe that they hold for systems of constraints without minimum operators. Then we find that for every $\mathcal{E}$, we can construct a system of equations $\mathcal{E}'$ without minimum operators by appropriately replacing every minimum expression by one of its arguments in such a way that $\mathcal{E}'$ has the same least solution as $\mathcal{E}$.

Due to Proposition 1, the least solutions of systems of equations over $\mathcal{Z}$ are computable by performing ordinary fixpoint iteration over the finite lattice

$$\mathcal{Z}_{a,b} = \{-\infty < a < \ldots < b < \infty\}$$

for suitable bounds $a < b$. This results in practical algorithms if a reasonably small difference $b - a$ can be revealed. In the following, we consider algorithms whose runtime does not depend on the particular sizes of occurring numbers – given that operations and tests on integers take time $\mathcal{O}(1)$.

3   Integer Equations without Minimum

We first consider systems of integer equations without minimum. Let us call these systems *disjunctive*. Note that we obtain the equational formulation of the single-source *longest* path problem for positive and negative edge weights if we restrict systems of disjunctive equations further by excluding multiplication and addition of variables in right-hand sides. By replacing all weights a with $-a$, the latter problem is a reformulation of the single-source *shortest* path problem (see, e.g., [11]).

In [7], Knuth considers a generalization of the single-source shortest path problem with nonnegative edge weights to grammars. In a similar sense, computing least solutions of systems of disjunctive constraints can be considered as a generalization of the single-source shortest path problem with positive and negative edge weights. For the latter problem, only quadratic algorithms are known [11]. Here, we observe that quadratic time is also enough for systems of disjunctive constraints:

Theorem 1. *The least solution of a disjunctive system $\mathcal{E}$ of equations with n variables can be computed in time $\mathcal{O}(n \cdot |\mathcal{E}|)$.*

Proof. As a generalization of the Bellman-Ford algorithm [11] we propose alg. 1 for computing the least solution of the system $\mathcal{E}$. The algorithm consists of two

---

Algorithm 1

---
forall $(\mathbf{x} \in \mathbf{X})\ \mu(\mathbf{x}) = -\infty$;
for $(i = 0; i < n; i{+}{+})$
    forall $((\mathbf{x} = e) \in \mathcal{E})$
        $\mu(\mathbf{x}) = \mu(\mathbf{x}) \vee [\![e]\!]\mu$;
for $(i = 0; i < n; i{+}{+})$
    forall $((\mathbf{x} = e) \in \mathcal{E})$
        if $(\mu(\mathbf{x}) \not\geq [\![e]\!]\mu)\ \mu(\mathbf{x}) = \infty$;
return μ;

---

nested loops l_1, l_2 where the first one corresponds to n rounds of round robin fixpoint iteration, and the second one differs from the first in widening the value of a variable to ∞ whenever a further increase is observed. Let μ^* denote the least solution of $\mathcal{E}$. For a formal proof, let us define $F : (\mathbf{X} \to \mathcal{Z}) \to (\mathbf{X} \to \mathcal{Z})$ by

$$F(\mu)(\mathbf{x}) = [\![e]\!]\mu \qquad \text{if } (\mathbf{x} = e) \in \mathcal{E}$$

for $\mu : \mathbf{X} \to \mathcal{Z}$. Additionally we define the variable assignments μ_i for $i \in \mathbb{N}_0$ by

$$\begin{aligned}
\mu_0(\mathbf{x}) &= -\infty \qquad \text{for } \mathbf{x} \in \mathbf{X} \\
\mu_i &= F^i(\mu_0) \quad \text{for } i \in \mathbb{N}.
\end{aligned}$$

Thus $\bigvee_{i \in \mathbb{N}_0} \mu_i = \mu^*$ and in particular $\mu_i \leq \mu^*$ for all $i \in \mathbb{N}$. In order to prepare us for the proof, we introduce the following notion. Variable $\mathbf{x}$ μ-*depends on* $\mathbf{x}'$ iff

$$F(\mu \oplus \{\mathbf{x}' \mapsto \mu(\mathbf{x}') + \delta\})(\mathbf{x}) \geq F(\mu)(\mathbf{x}) + \delta$$

for all $\delta \geq 0$. Here, $\oplus$ denotes the update operator for variable assignments. We claim:

Claim 1: Let $k \geq 1$. Assume that $\mu_{k+1}(\mathbf{x}) > \mu_k(\mathbf{x})$. There exists a $\mathbf{y}$ s.t. $\mathbf{x}$ μ_k-depends on $\mathbf{y}$ with $\mu_k(\mathbf{y}) > \mu_{k-1}(\mathbf{y})$. $\qquad\qquad\Box$
The key observation is stated in the following Claim.

Claim 2: $\mu_n(\mathbf{x}) = \mu^*(\mathbf{x})$ whenever $\mu^*(\mathbf{x}) < \infty$.

Proof. Assume $\mu^*(\mathbf{x}) > \mu_n(\mathbf{x})$. Thus there exists an index $k \geq n$ s.t. $\mu_{k+1}(\mathbf{x}) > \mu_k(\mathbf{x})$. Claim 1 implies that there exist variables

$$\mathbf{x}_{k+1}, \mathbf{x}_k, \ldots, \mathbf{x}_1$$

where $\mathbf{x}_{k+1} = \mathbf{x}$ and $\mathbf{x}_{i+1}$ μ_i-depends on $\mathbf{x}_i$ for $i = 1, \ldots, k$. Since there are at least $n + 1$ elements in the sequence $\mathbf{x}_{k+1}, \ldots, \mathbf{x}_1$, the pigeon-hole principle implies that there must be a variable $\mathbf{x}'$ which occurs twice. W.l.o.g., let $j_1 < j_2$ s.t. $\mathbf{x}' = \mathbf{x}_{j_1} = \mathbf{x}_{j_2}$. Furthermore by assumption $\mu_{j_2}(\mathbf{x}') > \mu_{j_1}(\mathbf{x}')$.

By a straight forward induction it follows that

$$F^{j_2-j_1}(\mu_{j_1} \oplus \{\mathbf{x}' \mapsto \mu_{j_1}(\mathbf{x}') + \delta\})(\mathbf{x}') \geq \mu_{j_2}(\mathbf{x}') + \delta \qquad (1)$$

Let $\delta := \mu_{j_2}(\mathbf{x}') - \mu_{j_1}(\mathbf{x}') > 0$. Then

$$\begin{aligned}
\mu^*(\mathbf{x}') &\geq F^{i(j_2-j_1)}(\mu_{j_2})(\mathbf{x}') \\
&\geq F^{i(j_2-j_1)}(\mu_{j_1} \oplus \{\mathbf{x}' \mapsto \mu_{j_1}(\mathbf{x}') + \delta\})(\mathbf{x}') && \text{(monotonicity)} \\
&\geq F^{(i-1)(j_2-j_1)}(\mu_{j_1} \oplus \{\mathbf{x}' \mapsto \mu_{j_2}(\mathbf{x}') + \delta\})(\mathbf{x}') && (1) \\
&= F^{(i-1)(j_2-j_1)}(\mu_{j_1} \oplus \{\mathbf{x}' \mapsto \mu_{j_1}(\mathbf{x}') + 2\delta\})(\mathbf{x}') && \text{(def. } \delta) \\
&\geq \cdots \geq \mu_{j_2}(\mathbf{x}') + i\delta
\end{aligned}$$

for every $i \in \mathbb{N}$. Since $\mathbf{x}$ depends on $\mathbf{x}'$, we conclude that $\mu^*(\mathbf{x}) = \infty$. This proves claim 2. $\qquad \square$

Let $\hat{\mu}_i$ denote the value of the program variable μ after execution of the i-th nested loop. By construction $\mu_n \leq \hat{\mu}_1 \leq \mu^*$. Whenever a further increase in the second nested loop can be observed, we know that $\mu \leq \mu^*$ and by claim 2, that after the modification $\mu \leq \mu^*$ still holds. Thus, $\hat{\mu}_2 \leq \mu^*$.

To show that $\hat{\mu}_2 = \mu^*$ recall that there are n variables. Therefore, at most n variables can be set to ∞ — implying that the least fixpoint is reached after at most n rounds. $\qquad \square$

4   Extension with Positive and Negative Multiplications

Algorithm 1 can be generalized also to systems of equations which utilize a wider range of operators. We observe:

Proposition 2. *For any monotonic function* $f : (\mathbf{X} \to \mathcal{Z}) \to \mathcal{Z}$, *the two following conditions are equivalent:*

(i) *for any* $\mu : (\mathbf{X} \to \mathcal{Z})$ *and any* $\mathbf{Y} \subseteq \mathbf{X}$, *if* $f(\mu \oplus \{\mathbf{y} \mapsto -\infty \mid \mathbf{y} \in \mathbf{Y}\}) < f(\mu)$ *then there is* $\mathbf{y} \in \mathbf{Y}$ *such that* $f(\mu \oplus \{\mathbf{y} \mapsto \mu(\mathbf{y}) + i\}) \geq f(\mu) + i$ *for all* $i \geq 0$.
(ii) *for any* $\mu : (\mathbf{X} \to \mathcal{Z})$ *and any* $\mathbf{x} \in \mathbf{X}$, *if* $f(\mu \oplus \{\mathbf{x} \mapsto -\infty\}) < f(\mu)$ *then* $f(\mu \oplus \{\mathbf{x} \mapsto \mu(\mathbf{x}) + i\}) \geq f(\mu) + i$ *for all* $i \geq 0$.

Proof. (i) $\Rightarrow$ (ii) is trivial. For any $\mu : (\mathbf{X} \to \mathcal{Z})$ and any subset $\mathbf{Y} \subseteq \mathbf{X}$, we will write $\mu_{\mathbf{Y}}$ for $\mu_{\mathbf{Y}} = \mu \oplus \{\mathbf{y} \mapsto -\infty \mid \mathbf{y} \in \mathbf{Y}\}$. Assume that (ii) holds, and let us prove by induction on $|\mathbf{Y}|$ that (i) holds. The case of $\mathbf{Y} = \emptyset$ is trivial and the basis $|\mathbf{Y}| = 1$ follows from (ii). To prove the induction step, let $\mathbf{Y} \subseteq \mathbf{X}$ with $|\mathbf{Y}| > 1$ and assume that $f(\mu_{\mathbf{Y}}) < f(\mu)$. Pick some $\mathbf{y} \in \mathbf{Y}$ and let $\mathbf{Z} = \mathbf{Y} \setminus \{\mathbf{y}\}$. If $f(\mu_{\mathbf{Z}}) < f(\mu)$ then we derive from the induction hypothesis that there is $\mathbf{z} \in \mathbf{Z} \subseteq \mathbf{Y}$ such that $f(\mu \oplus \{\mathbf{z} \mapsto \mu(\mathbf{z}) + i\}) \geq f(\mu) + i$ for all $i \geq 0$. Otherwise, $f(\mu_{\mathbf{Z}} \oplus \{\mathbf{y} \mapsto -\infty\}) = f(\mu_{\mathbf{Y}}) < f(\mu) = f(\mu_{\mathbf{Z}})$, and we deduce from (ii) that $f(\mu_{\mathbf{Z}} \oplus \{\mathbf{y} \mapsto \mu(\mathbf{y}) + i\}) \geq f(\mu_{\mathbf{Z}}) + i$ for all $i \geq 0$. We come to $f(\mu \oplus \{\mathbf{y} \mapsto \mu(\mathbf{y}) + i\}) \geq f(\mu) + i$ for all $i \geq 0$ since $\mu \geq \mu_{\mathbf{Z}}$ and $f(\mu) = f(\mu_{\mathbf{Z}})$. We have thus shown that (i) holds for all $\mathbf{Y} \subseteq \mathbf{X}$. $\qquad \square$

We call a function $f : (\mathbf{X} \to \mathcal{Z}) \to \mathcal{Z}$ *Bellman-Ford* function (short: BF-function) when it is monotonic and it satisfies any (or equivalently all) of the above conditions.

We remark that the class of Bellman-Ford functions is incomparable to the class of *bounded-increasing* functions as considered in [9]. Bounded-increasing functions are monotonic functions $f : (\mathbf{X} \to \mathcal{Z}) \to \mathcal{Z}$ such that $f(\mu_1) < f(\mu_2)$ for all $\mu_1, \mu_2 : \mathbf{X} \to \mathcal{Z}$ with $\mu_1 < \mu_2$, $f(\lambda\mathbf{x}. -\infty) < f(\mu_1)$ and $f(\mu_2) < f(\lambda\mathbf{x}.\infty)$. However, for any bounded-increasing function $f : (\mathbf{X} \to \mathcal{Z}) \to \mathcal{Z}$, if (1) f is continuous (i.e. $f(\bigvee_k \mu_k) = \bigvee_k f(\mu_k)$ for every ascending chain $\mu_0 \leq \mu_1 \leq \cdots$) and (2) $f(\lambda\mathbf{x}.-\infty) = -\infty$ and $f(\lambda\mathbf{x}.\infty) = \infty$, then f is a Bellman-Ford function.

Let us call a k-ary operator $\square$ a BF-operator, if the function $f_\square(\mu) = \square(\mu(\mathbf{x}_1), \ldots, \mu(\mathbf{x}_k))$ (for distinct variables $\mathbf{x}_i$) is a BF-function.

Clearly, addition itself is a BF-operator as well as the least upper bound operation and the multiplication with constants. For simulating multiplication of intervals, we further rely on the following two approximative versions of multiplication:

$$x \cdot^+ y = \begin{cases} x \cdot y \text{ if } x, y > 0 \\ -\infty \text{ otherwise} \end{cases} \qquad x \cdot^- y = \begin{cases} -(x \cdot y) \text{ if } x, y < 0 \\ \infty \quad \text{ otherwise} \end{cases}$$

We call these *positive* and *negative* multiplication, respectively. Note that, in contrast to full multiplication over the integers, both versions of multiplication are monotonic. Additionally, they satisfy the conditions for BF-functions and therefore are BF-operators. By induction on the structure of expressions, we find:

Lemma 1. *Assume e is an expression built up from variables and constants by means of application of BF-operators. Then the evaluation function $[\![e]\!]$ for e is a BF-function.*

Let us call an equation $\mathbf{x} = e$ BF-equation, if $[\![e]\!]$ is a BF-function. Our key observation is that the Bellman-Ford algorithm can be applied not only to disjunctive systems of equations but even to systems of BF-equations. In order to adapt the proof of theorem 1, we in particular adapt the proof of claim 1. We use the same notations from that proof. Let $k \geq 1$ and assume that $\mu_{k+1}(\mathbf{x}) > \mu_k(\mathbf{x})$. Then $(\mathbf{x} = e) \in \mathcal{E}$ for some expression e. The monotonic function $f_{\mathbf{x}} = [\![e]\!]$ is a Bellman-Ford function where $\mu_{k+1}(\mathbf{x}) = f_{\mathbf{x}}(\mu_k)$ and $\mu_k(\mathbf{x}) = f_{\mathbf{x}}(\mu_{k-1})$, and recall that $\mu_k \geq \mu_{k-1}$.

Let $\mathbf{Y} = \{\mathbf{y} \in \mathbf{X} \mid \mu_k(\mathbf{y}) > \mu_{k-1}(\mathbf{y})\}$. Since $f_{\mathbf{x}}(\mu_k \oplus \{\mathbf{y} \mapsto -\infty \mid \mathbf{y} \in \mathbf{Y}\}) \leq f_{\mathbf{x}}(\mu_{k-1}) < f_{\mathbf{x}}(\mu_k)$, we get from Proposition 2 that there is some $\mathbf{y} \in \mathbf{Y}$ such that $f(\mu_k \oplus \{\mathbf{y} \mapsto \mu_k(\mathbf{y}) + i\}) \geq f(\mu_k) + i$ for all $i \geq 0$. Hence, $\mathbf{x}$ μ_k-depends on $\mathbf{y}$, and moreover, $\mu_k(\mathbf{y}) > \mu_{k-1}(\mathbf{y})$ as $\mathbf{y} \in \mathbf{Y}$. This completes the proof of this claim.

Altogether, we obtain:

Theorem 2. *The least solution of a system $\mathcal{E}$ of BF-equations with n variables can be computed in time $\mathcal{O}(n \cdot |\mathcal{E}|)$.*

It is important here to recall that we consider a uniform cost measure where each operator can be evaluated in time $\mathcal{O}(1)$. If besides addition, also positive and negative multiplication is allowed, then the sizes of occurring numbers may not only be single exponential, but even double in the occurring numbers. More precisely, assume that μ^* is the least solution of $\mathcal{E}$ $\mathbf{x}$ is a variable of $\mathcal{E}$ with $\mu^*(\mathbf{x}) \in \mathbb{Z}$. Then

$$(A \vee 2)^{|\mathcal{E}|^n} \leq \mu^*(\mathbf{x}) \leq (A \vee 2)^{|\mathcal{E}|^n}$$

where A is an upper bound to the absolute values of constants $c \in \mathbb{Z}$ occurring in $\mathcal{E}$, and n is the number of variables.

5   Integer Equations with Minimum

In this section, we extend the results in the previous section by additionally allowing minima with constants. For convenience, let us assume that all right-hand sides r in the system $\mathcal{E}$ of equations either are of the following simple forms:

$$r ::= a \mid \mathbf{y} \mid \square(\mathbf{y}_1, \ldots, \mathbf{y}_k) \mid \mathbf{y} \wedge a$$

for constants $a \in \mathbb{Z}$, variables $\mathbf{y}$ and BF-operators $\square$. Note that now the size $|\mathcal{E}|$ of $\mathcal{E}$ is proportional to the number of variables of $\mathcal{E}$. Our main result for systems of such equations is:

Theorem 3. *The least solution of a system $\mathcal{E}$ of integer equations using BF-operators and minima with constants can be computed in time $\mathcal{O}(|\mathcal{E}|^3)$.*

Proof. Let μ^* denote the least solution of $\mathcal{E}$. We introduce the following notions. We call a sequence $P = (\mathbf{y}_1, \ldots, \mathbf{y}_{k+1}) \in \mathbf{X}^*$ a path if for every $i = 1, \ldots, k$, variable $\mathbf{y}_{i+1}$ occurs in the right-hand side of the equation for $\mathbf{y}_i$ in $\mathcal{E}$. Thus, given a variable assignment μ, the path p represents the transformation $[\![p]\!]\mu : \mathcal{Z} \mapsto \mathcal{Z}$ defined by

$$[\![p]\!]\mu(z) = [\![e_1]\!](\mu \oplus \{\mathbf{y}_2 \mapsto [\![e_2]\!](\ldots [\![e_k]\!](\mu \oplus \{\mathbf{y}_{k+1} \mapsto z\}) \ldots)\})$$

where $\mathbf{y}_i = e_i$ is the equation for $\mathbf{y}_i$ in $\mathcal{E}$.

The path p is called a *cycle* iff $\mathbf{y}_{k+1} = \mathbf{y}_1$. The cycle p is called *simple* if the variables $\mathbf{y}_1, \ldots, \mathbf{y}_k$ are pairwise distinct.

In order to enhance alg. 1 for systems with minima, assume that an increase of the value of the variable $\mathbf{x}$ can be observed within the first iteration of the second nested loop. Then there exists a simple cycle $c = (\mathbf{y}_1, \ldots, \mathbf{y}_k, \mathbf{y}_1)$ that can be repeated until either all variables $\mathbf{y}_i$ receive values ∞ or the value of the argument e' in some minimum expression $\mathbf{y} \wedge a$ occurring along the cycle exceeds a. In order to deal with this, we provide the following modified version of the Bellman-Ford algorithm:

  i We initialize the variable assignment μ s.t. every variable is mapped to $-\infty$ and execute the first phase of alg. 1 which consists of n Round-Robin iterations.

ii Then we perform the second phase. If no increment in the second phase can be detected, we have reached the least solution and return μ as result.

iii Whenever an increment in the second phase under a current variable assignment μ is detected, we try to extract a simple cycle $c = (\mathbf{y}_1, \ldots, \mathbf{y}_k, \mathbf{y}_1)$ s.t. $f'_{c,\mu}(v) > v$ for some $v < \mu(\mathbf{y}_1)$. If this is possible, then we do an *accelerated* fixpoint computation on the cycle c to determine new values for the variables $\mathbf{y}_1, \ldots, \mathbf{y}_k$. We then update the variables with the new values and restart the procedure with step 2.

This gives us alg. 2.   Extra effort is necessary in order to extract cycles in the second phase which can be repeated. For that, the algorithm records in the variable *time*, the number of equations evaluated so far. Moreover for every variable $\mathbf{x}$, it records in *modified*($\mathbf{x}$) the last time when the variable $\mathbf{x}$ has received a new value, and in *evaluated*($\mathbf{x}$) the last time when the equation for $\mathbf{x}$ has been evaluated. Also, it records for every variable $\mathbf{x}$ in *pred*($\mathbf{x}$) a variable n the right-hand side of $\mathbf{x}$ which may have caused the increase and can give rise to an increase in the future. If no such occurrence exists, then *pred*($\mathbf{x}$) is set to $\bot$. This is implemented by the function **pred**($\mathbf{x}$). Let μ denote the current variable assignment, and assume that the right-hand side of $\mathbf{x}$ is r. Furthermore, let $\mathbf{Y}$ denote the set of variables $\mathbf{y}$ occurring in r which have been modified

---

Algorithm 2

forall $(\mathbf{x} \in \mathbf{X})$ $\mu(\mathbf{x}) = -\infty$;
do {
    done = **true** ; *time* = 0;
    forall $(\mathbf{x} \in \mathbf{X})$ {*modified*($\mathbf{x}$) = 0; *pred*($\mathbf{x}$) = $\bot$; *evaluated*($\mathbf{x}$) = 0;
    }
    for $(i = 0; i < n; i++)$
        forall $((\mathbf{x} = e) \in \mathcal{E})$ {
            time++;
            if $([\![e]\!]\mu > \mu(\mathbf{x}))$ {
                pred($\mathbf{x}$) = **pred**($\mathbf{x}$); $\mu(\mathbf{x}) = [\![e]\!]\mu$; *modified*($\mathbf{x}$) = *time*;
            }
            evaluated($\mathbf{x}$) = *time*;
        }
    forall $((\mathbf{x} = e) \in \mathcal{E})$
        if $([\![e]\!]\mu > \mu(\mathbf{x}))$ {
            $\mu(\mathbf{x}) = [\![e]\!]\mu$;
            if $(\mu(\mathbf{x}) < \infty)$ {
                widen($\mathbf{x}$); *done* = **false** ; **break**;
            } ;
        }
} **until** (*done*);
return μ;

after the last evaluation of $\mathbf{x}$, i.e., *modified*$(\mathbf{y}) \geq$ *evaluated*$(\mathbf{x})$. Since the value of $\mathbf{x}$ has increased, $\mathbf{Y}$ is non empty and, in particular, r cannot be equal to a constant. If $r = \mathbf{y}$, then $\mathbf{pred}(\mathbf{x}) = \mathbf{y}$. If $r = \mathbf{y} \wedge c$, then $\mathbf{pred}(\mathbf{x}) = \perp$ if $\mu(\mathbf{y}) \geq c$ and $\mathbf{pred}(\mathbf{x}) = \mathbf{y}$ otherwise. Finally, assume $r = \square(\mathbf{y}_1, \ldots, \mathbf{y}_k)$ and let $v_j = \mu(\mathbf{y}_j)$ for all j. Furthermore, let $v'_j = v_j$ if $\mathbf{y}_j \notin \mathbf{Y}$, i.e., has not been changed since the last evaluation of $\mathbf{x}$, and $v'_j = -\infty$ otherwise. Then $\square(v'_1, \ldots, v'_k) < \square(v_1, \ldots, v_k)$. Since $\square$ is a BF-operator, we thus can retrieve an index j such that $\square(v_1, \ldots, v_{j-1}, v_j + d, v_{j+1}, \ldots, v_k) \geq \square(v_1, \ldots, v_k) + d$ for all $d \geq 0$. Accordingly, we set $\mathbf{pred}(\mathbf{x}) = \mathbf{y}_j$. The following observation shows that **pred** can be computed in time $\mathcal{O}(1)$ if the maximal arity of the BF-operators is considered as a constant.

Lemma 2. *Consider a BF-function $f : (\mathbf{X} \to \mathcal{Z}) \to \mathcal{Z}$ and a variable assignment $\mu : \mathbf{X} \to \mathcal{Z}$ and a variable $\mathbf{x} \in \mathbf{X}$. We have $f(\mu \oplus \{\mathbf{x} \mapsto \mu(\mathbf{x})+1\}) \geq f(\mu)+1$ iff $f(\mu \oplus \{\mathbf{x} \mapsto \mu(\mathbf{x}) + i\}) \geq f(\mu) + i$ for all $i \geq 0$.*

Example 1. Let $\mu := \{\mathbf{x} \mapsto -10, \mathbf{y} \mapsto 0\}$ and consider the equation $\mathbf{z} = \mathbf{x} \vee \mathbf{y}$. Then for $\mathbf{Y} = \{\mathbf{x}, \mathbf{y}\}$, the function call $\mathbf{pred}(\mathbf{z})$ returns the variable $\mathbf{y}$.

Now we consider the second phase of alg. 2. Whenever a finite increase of the value of a variable $\mathbf{x}$ is detected, $\mathbf{widen}(\mathbf{x})$ is called (see alg. 3).

Algorithm 3. widen(x)

$c = (\mathbf{y}_1, \ldots, \mathbf{y}_k, \mathbf{y}_1) = \mathbf{extract\_cycle}(\mathbf{x});$

$\mu(\mathbf{y}_1) = \mu(\mathbf{y}_1) \vee \mathbf{eval\_cycle}(c);$
for $(i = k; \; i \geq 2; \; i--)$
$\quad \mu(\mathbf{y}_i) = \mu(\mathbf{y}_i) \vee f_{\mathbf{y}_i}(\mu);$

Within the procedure $\mathbf{widen}()$, the function $\mathbf{extract\_cycle}()$ is used to extract a cycle which has caused the increase and possibly causes further increases in the future. It works as follows. The call $\mathbf{extract\_cycle}(\mathbf{x})$ for a variable $\mathbf{x}$ looks up the value of *pred*$(\mathbf{x})$. If *pred*$(\mathbf{x}) \neq \perp$ a variable $\mathbf{x}_1$ in the right-hand side for $\mathbf{x}$ is returned. Then the procedure records $(\mathbf{x}_1)$ and proceeds with the value stored in *pred*$(\mathbf{x}_1)$ and so on. Thus, it successively visits a path according to the information stored in *pred* until it either reaches $\perp$ or visits a variable for the second time. In the latter case we obtain a simple cycle $(\mathbf{y}_1, \ldots, \mathbf{y}_k, \mathbf{y}_1)$. With a cyclic permutation *modified*$(\mathbf{y}_1)$ is assumed maximal. In the former case, the empty sequence will be returned.

The procedure $\mathbf{eval\_cycle}()$ does the accelerated fixpoint computation on a given cycle. The function $\mathbf{eval\_cycle}()$ takes a simple cycle $c = (\mathbf{y}_1, \ldots, \mathbf{y}_k, \mathbf{y}_1)$. Let $f := [\![c]\!]\mu$ and assume that $f(v) > v$ for some $v \leq \mu(\mathbf{y}_1)$. Then $\mathbf{eval\_cycle}()$ computes $\bigvee_{i \in \mathbb{N}} f^i(v)$. As monotonic functions over a linear order are distributive over $\wedge$, note that $f(z)$ can be written as

$$f(z) = f'(z) \wedge b'$$

for some unary BF-function f' and $b' \in \mathcal{Z}$. Since $f(v) > v$, $b' \geq f'(v) > v$. Therefore,

$$\bigvee_{i \in \mathbb{N}} f^i(v) = b' = f(\infty)$$

We conclude that $\bigvee_{i \in \mathbb{N}} f^i(v)$ can be computed in time linear to the size of the simple cycle c. Furthermore, $\bigvee_{i \in \mathbb{N}} f^i(v) \leq \mu^*(\mathbf{y}_1)$ by construction. Thus, we have shown the following claim:

Claim 1: Assume that c is a simple cycle which starts with the variable $\mathbf{y}_1$. Assume that $\mu \leq \mu^*$ and $v \leq \mu(\mathbf{y}_1)$ are s.t. $f(v) := [\![c]\!]\mu(v) > v$. Then $v' := \bigvee_{i \in \mathbb{N}} f^i(v) \leq \mu^*(\mathbf{y}_1)$ and v' can be computed in time linear to the size of c.   □
For a formal proof of correctness of the algorithm, let μ_i denote the variable assignment μ before the i-th extraction of a simple cycle and c_i the value of c after the i-th extraction of a simple cycle. Thereby c_i can be $\bot$. Let furthermore μ'_i denote the value of μ after the i-th call of the procedure **widen**().
    First, we show that the widening is correct, i.e., $\mu'_i \leq \mu^*$ for all i. For that, we only need to consider the case in which the i-th extraction leads to a simple cycle c_i and not to $\bot$. Thanks to Claim 1, we only need to show that the assertions of Claim 1 are fulfilled for every call of the procedure **widen**() in which **extract_cycle** extracts a simple cycle. Thus we must show:

Claim 2: Assume that $c_i \neq \bot$ is a simple cycle which starts with the variable $\mathbf{y}_1$. Then $([\![c_i]\!]\mu_i)(v) > v$ for some value $v < \mu_i(\mathbf{y}_1)$.

Proof. Assume that $c_i = (\mathbf{y}_1, \ldots, \mathbf{y}_k, \mathbf{y}_1)$ where $\mathbf{y}_j = e_j$ is the equation for $\mathbf{y}_j$. Observe that the algorithm always records an occurrence of a variable which possibly has caused the increase. Therefore, by monotonicity, $[\![e_j]\!]\mu_i$ is at least the current value $\mu_i(\mathbf{y}_j)$ of the left-hand side $\mathbf{y}_j$. This means for the cycle c_i that

$$\mu_i(\mathbf{y}_1) \leq [\![e_1[\mu_i(\mathbf{y}_2)/\mathbf{y}_2]]\!]\mu_i \wedge \ldots \wedge \mu_i(\mathbf{y}_{k-1}) \leq [\![e_{k-1}[\mu_i(\mathbf{y}_k)/\mathbf{y}_k]]\!]\mu_i$$

as well as

$$\mu_i(\mathbf{y}_k) \leq [\![e_k[v/\mathbf{y}_1]]\!]\mu_i$$

where v is the value of the variable $\mathbf{y}_1$ at the last point in time where the evaluation of the equation $\mathbf{y}_k = e_k$ lead to an increase. As *modified*($\mathbf{y}_1$) is maximal, we get $v < \mu_i(\mathbf{y}_1)$. Since by construction, $([\![c_i]\!]\mu_i)(v) \geq \mu_i(\mathbf{y}_1) > v$, the assertion follows.   □

Assume again that $c_i = (\mathbf{y}_1, \ldots, \mathbf{y}_k, \mathbf{y}_1)$ is a simple cycle and assume as induction hypothesis, that the variable assignment μ'_{i-1} after the $(i-1)$-th widening is less than or equal to the least solution μ^* of the system $\mathcal{E}$. Since the variable

assignment μ_i before the extraction of the cycle c_i is computed by fixpoint iteration, it follows that $\mu_i \leq \mu^*$. Let v' denote the values returned from the i-th call
of **eval_cycle**(). By Claim 1, $v' \leq \mu^*(\mathbf{y})$. Since the rest of procedure **widen**()
consists in ordinary fixpoint iteration, we obtain $\mu' \leq \mu^*$.

Thus by construction, alg. 2 returns μ^* — whenever it terminates. In order to
prove termination, let $M(\mathcal{E})$ denote the set of minimum expressions occurring
in $\mathcal{E}$. We show the following claims which imply that a progress occurs at each
increase of a variable's value in the second phase, i.e., either one further variable
receives the value ∞ or another minimum can (conceptually) be replaced by its
constant argument.

Claim 3: Assume that $c_i = \perp$. Then
- either there exists a variable $\mathbf{x}$ such that $\mu'_{i-1}(\mathbf{x}) < \infty$ and $\mu_i(\mathbf{x}) = \infty$;
- or there exists a subexpression $\mathbf{y} \wedge a$ from $M(\mathcal{E})$ s.t. $\mu'_{i-1}(\mathbf{y}) < a$ and
$\mu_i(\mathbf{y}) \geq a$.

Proof. $c_i = \perp$ implies that the procedure **pred**() returned $\perp$ for one of the
equations $\mathbf{x} = e$ in the *same* iteration of the main loop. This is because all
values of $pred()$ reachable within n steps by **extract_cycle**() have been modified
during this iteration. Longer paths would imply finding a simple cycle. However,
the procedure **pred**() only returns $\perp$ if some minimum a is reached which had
not been reached before.   □

From Claim 3 and the fact that the sequence (μ'_i) is increasing we conclude that
for every i,

$$\{\mathbf{x} \in \mathbf{X} \mid \mu'_i(\mathbf{x}) = \infty\} \supsetneq \{\mathbf{x} \in \mathbf{X} \mid \mu'_{i-1}(\mathbf{x}) = \infty\}$$

or

$$\{\mathbf{x} \wedge a \in M(\mathcal{E}) \mid [\![\mathbf{x}]\!]\mu'_i \geq a\} \supsetneq \{\mathbf{x} \wedge a \in M(\mathcal{E}) \mid [\![\mathbf{x}]\!]\mu'_{i-1} \geq a\}.$$

Accordingly, the algorithm can perform at most $\mathcal{O}(|\mathcal{E}|)$ iterations of the outer
while-loop. Since every iteration of the outer loop of the algorithm can be exe-
cuted in time $\mathcal{O}(n \cdot |\mathcal{E}|)$, the assertion follows.   □

Example 2. Consider the following system of equations:

$$\mathbf{x} = \mathbf{y} \wedge 5 \qquad \mathbf{y} = \mathbf{z} \wedge 3 \qquad \mathbf{z} = -17 \vee \mathbf{z} + 2$$

The first three rounds of Round-Robin iteration give us:

	0	1	2
x	$-\infty$	$-\infty$	-15
y	$-\infty$	-15	-13
z	-15	-13	-11

Since the value of **x** still increases during the next round of evaluation, we call the function **widen**() with the variable **x**. Within **widen**() the function **extract_cycle** is called which returns the simple cycle $(\mathbf{z}, \mathbf{z})$. — giving us the new value ∞ for **z**. Restarting the Round-Robin iteration for all variables, reveals the least solution:

$$\mu^*(\mathbf{x}) = 3 \qquad \mu^*(\mathbf{y}) = 3 \qquad \mu^*(\mathbf{z}) = \infty$$

6   Intervals

In this section, we consider systems of equations over the complete lattice of integer intervals. Let

$$\mathcal{I} = \{\emptyset\} \cup \{[z_1, z_2] \in \mathcal{Z}^2 \mid z_1 \leq z_2, z_1 < \infty, -\infty < z_2\}$$

denote the complete lattice of intervals partially ordered by the subset relation (here denoted by "$\sqsubseteq$"). The empty interval $\emptyset$ is also denoted by $[\infty, -\infty]$. It is the least element of the lattice while $[-\infty, \infty]$ is the greatest element, and the least upper bound "$\sqcup$" is defined by:

$$[a_1, a_2] \sqcup [b_1, b_2] = [a_1 \wedge b_1, a_2 \vee b_2]$$

Here, we consider systems of equations over $\mathcal{I}$ similar to the ones we have considered over $\mathcal{Z}$ with the restriction that at least one argument of every intersection is constant. Instead of multiplication with positive constants only, we now also support negation as well as *full* multiplication of interval expressions. For a fixed set **X** of variables, we consider expressions e of the form

$$e \quad ::= \quad a \mid \mathbf{y} \mid c \cdot e \mid -e \mid e_1 \sqcup e_2 \mid e_1 + e_2 \mid e \sqcap a \mid e_1 \cdot e_2$$

where $a \in \mathcal{I}$, $c > 0$ is a positive integer constant, and **y** is a variable from **X**.

As for expressions over $\mathcal{Z}$, we rely on an evaluation function $[\![e]\!]$ for interval expressions e built up from variables and constants by means of applications of operators. The function $[\![e]\!]$ then maps variable assignments $\mu : \mathbf{X} \to \mathcal{I}$ to interval values. Note that (in contrast to the integer case) full multiplication of intervals still is monotonic. Therefore, every system of interval equations has a unique least solution.

Our goal is to reduce solving of systems of equations over intervals, to solving of systems equations over integers. For that, we define the functions $(\cdot)^+, (\cdot)^- :$ $\mathcal{I} \to \mathcal{Z}$ which extract from an interval the upper and *negated* lower bound, respectively. These functions are defined by:

$$\emptyset^+ = \emptyset^- = -\infty \qquad [a, b]^+ = b \qquad [a, b]^- = -a$$

where $[a, b] \in \mathcal{I}$. Thus x^+ denotes the upper bound and x^- denotes the *negated* lower bound of $x \in \mathcal{I}$. In the following, we indicate how operations on intervals can be realized by means of integer operations on interval bounds.

Assume $x, y \in \mathcal{I}$ are intervals and $c > 0$. Then we have:

$$
\begin{aligned}
(c \cdot x)^- &= c \cdot x^- \\
(c \cdot x)^+ &= c \cdot x^+ \\[4pt]
(-x)^- &= x^+ \\
(-x)^+ &= x^- \\[4pt]
(x \sqcup y)^- &= x^- \vee y^- \\
(x \sqcup y)^+ &= x^+ \vee y^+ \\[4pt]
(x + y)^- &= x^- + y^- \\
(x + y)^+ &= x^+ + y^+ \\[4pt]
(x \sqcap y)^- &= (x^+ + y^-); (x^- + y^+); x^- \wedge y^- \\
(x \sqcap y)^+ &= (x^+ + y^-); (x^- + y^+); x^+ \wedge y^+ \\[4pt]
(x \cdot y)^- &= -(x^- \cdot y^-) \vee -(x^+ \cdot y^+) \vee x^- \cdot y^+ \vee x^+ \cdot y^- \\
&= (x^- \cdot^- y^- \wedge 0) \vee (x^+ \cdot^- y^+ \wedge 0) \vee x^- \cdot^+ y^+ \vee x^+ \cdot^+ y^- \\
(x \cdot y)^+ &= x^- \cdot y^- \vee x^+ \cdot y^+ \vee -(x^- \cdot y^+) \vee -(x^+ \cdot y^-) \\
&= x^- \cdot^+ y^- \vee x^+ \cdot^+ y^+ \vee (x^- \cdot^- y^+ \wedge 0) \vee (x^+ \cdot^- y^- \wedge 0)
\end{aligned}
$$

Here, the operator $x; y$ returns $-\infty$ if $x < 0$ and y otherwise. This operator can be expressed by means of positive multiplication together with a minimum with 0:

$$ x \,;\, y = (((x + 1) \cdot^+ 1) \wedge 0) + y $$

Additionally, we observe that w.r.t. the interval bounds, interval multiplication can be expressed through positive and negative multiplications together with minima with 0.

Every system $\mathcal{E}$ of interval equations gives rise to a system $\mathcal{E}^{\pm}$ of integer equations over $\mathcal{Z}$ for the upper and negated lower bounds for the interval values of the variables from $\mathcal{E}$. For every variable $\mathbf{x}$ of the interval system $\mathcal{E}$, we introduce the two integer variables $\mathbf{x}^-, \mathbf{x}^+$. The variable $\mathbf{x}^+$ is meant to receive the upper interval bound of $\mathbf{x}$ whereas the variable $\mathbf{x}^-$ is meant to receive the negated lower interval bound of $\mathbf{x}$.

Every equation $\mathbf{x} = e$ of $\mathcal{E}$ then gives rise to the equations $\mathbf{x}^- = [e]^-$ and $\mathbf{x}^+ = [e]^+$ of $\mathcal{E}^{\pm}$ for the new integer variables corresponding to the left-hand side $\mathbf{x}$ where the new right-hand sides $[e]^-$ and $[e]^+$ are obtained by the following transformations:

$$
\begin{aligned}
[[a_1, a_2]]^- &= -a_1 & [[a_1, a_2]]^+ &= a_2 \\
[\mathbf{x}]^- &= \mathbf{x}^- & [\mathbf{x}]^+ &= \mathbf{x}^+ \\
[c \cdot e]^- &= c \cdot [e]^- & [c \cdot e]^+ &= c \cdot [e]^+ \\
[-e]^- &= [e]^+ & [-e]^+ &= [e]^- \\
[e_1 \sqcup e_2]^- &= [e_1]^- \vee [e_2]^- & [e_1 \sqcup e_2]^+ &= [e_1]^+ \vee [e_2]^+ \\
[e_1 + e_2]^- &= [e_1]^- + [e_2]^- & [e_1 + e_2]^+ &= [e_1]^+ + [e_2]^+
\end{aligned}
$$

$$
\begin{aligned}
[e \sqcap a]^- &= ([e]^+ + a^-); ([e]^- + a^+); [e]^- \wedge a^- \\
[e \sqcap a]^+ &= ([e]^+ + a^-); ([e]^- + a^+); [e]^+ \wedge a^+ \\
[e_1 \cdot e_2]^- &= ([e_1]^- \cdot^- [e_2]^- \wedge 0) \vee ([e_1]^+ \cdot^- [e_2]^+ \wedge 0) \vee [e_1]^- \cdot^+ [e_2]^+ \vee [e_1]^+ \cdot^+ [e_2]^- \\
[e_1 \cdot e_2]^+ &= [e_1]^- \cdot^+ [e_2]^- \vee [e_1]^+ \cdot^+ [e_2]^+ \vee ([e_1]^- \cdot^- [e_2]^+ \wedge 0) \vee ([e_1]^+ \cdot^- [e_2]^- \wedge 0)
\end{aligned}
$$

We have:

Proposition 3. *Assume that $\mathcal{E}$ is a system of equations over the complete lattice of intervals, and $\mathcal{E}^{\pm}$ is the corresponding system for the negated lower and upper interval bounds of values for the variables of $\mathcal{E}$. Let μ and $\mu^{\pm}$ denote the least solutions of $\mathcal{E}$ and $\mathcal{E}^{\pm}$, respectively. Then for every variable $\mathbf{x}$ of $\mathcal{E}$, $(\mu(\mathbf{x}))^{-} = \mu^{\pm}(\mathbf{x}^{-})$ and $(\mu(\mathbf{x}))^{+} = \mu^{\pm}(\mathbf{x}^{+})$.* $\square$

Proposition 3 follows by standard fixpoint induction. By Proposition 3, computing least solutions of systems of interval equations reduces to computing least solutions of systems of equations over $\mathcal{Z}$ using the BF operators maximum, addition, multiplication with positive constants, positive and negative multiplications together with minima with constants. Thus, theorem 3 is applicable, and we obtain:

Theorem 4. *The least solution of a system $\mathcal{E}$ of interval equations can be computed in time $\mathcal{O}(|\mathcal{E}|^3)$.*

Note that before application of theorem 3, we must instroduce auxiliary variables for simplifying complex interval expressions in right-hand sides of $\mathcal{E}$. Furthermore, the transformations $[.]^{-}$ and $[.]^{+}$ may produce composite expressions which we again decompose by means of auxiliary variables. The number of these fresh variables, however, is linear in the number of occurring multiplications and thus altogether bounded by $\mathcal{O}(|\mathcal{E}|)$.

7   Conclusion

We presented a cubic time algorithm for solving systems of integer equations where minimum is restricted to always have at least one constant argument. The methods relied on a subtle generalization of the Bellman-Ford algorithm for computing shortest paths in presence of positive and negative edge weights. We also observed that this algorithm is still applicable when right-hand sides of equations not only contain maxima, addition and multiplication with constants, but additionally use positive and negative multiplications.

In the second step, we showed how solving systems of interval equations with addition, full multiplication and intersection with constant intervals can be reduced to solving systems of integer equations. In particular, the restricted variants of multiplication allowed us to simulate full interval multiplication as well as to construct tests whether or not the intersection of an interval with a constant interval is empty. The one hand, our methods thus clarifies the upper complexity bound for solving systems of interval equations with intersection with constant intervals as presented by Su and Wagner [6]; on the other hand the approach generalizes the system of equations considered in [6] by additionally allowing full multiplication of intervals.

Our algorithms were designed to be *uniform*, i.e., have run-times independent of occurring numbers — given that arithmetic operations are counted as $\mathcal{O}(1)$.

This is a reasonable assumption when multiplication is allowed with constants only. It is also reasonable in presence of full multiplication for intervals — given that numbers are from a fixed finite range only.

In [8], the ideas presented here have been extended to work also for systems of interval equations with full multiplication as well as with arbitrary intersections.

References

1. Cousot, P., Cousot, R.: Static Determination of Dynamic Properties of Programs. In: Second Int. Symp. on Programming, Dunod, Paris, France, pp. 106–130 (1976)
2. Cousot, P., Cousot, R., Feret, J., Mauborgne, L., Miné, A., Monniaux, D., Rival, X.: The ASTRÉE Analyser. In: Sagiv, M. (ed.) ESOP 2005. LNCS, vol. 3444, pp. 21–30. Springer, Heidelberg (2005)
3. Cousot, P., Cousot, R.: Comparison of the Galois Connection and Widening/Narrowing Approaches to Abstract Interpretation. In: JTASPEFL 1991, Bordeaux, vol. 74, pp. 107–110. BIGRE (1991)
4. Miné, A.: Relational Abstract Domains for the Detection of Floating-Point Run-Time Errors. In: Schmidt, D. (ed.) ESOP 2004. LNCS, vol. 2986, pp. 3–17. Springer, Heidelberg (2004)
5. Miné, A.: Symbolic Methods to Enhance the Precision of Numerical Abstract Domains. In: Emerson, E.A., Namjoshi, K.S. (eds.) VMCAI 2006. LNCS, vol. 3855, pp. 348–363. Springer, Heidelberg (2006)
6. Su, Z., Wagner, D.: A Class of Polynomially Solvable Range Constraints for Interval Analysis Without Widenings. Theor. Comput. Sci. (TCS) 345(1), 122–138 (2005)
7. Knuth, D.E.: A Generalization of Dijkstra's algorithm. Information Processing Letters (IPL) 6(1), 1–5 (1977)
8. Gawlitza, T., Seidl, H.: Precise Fixpoint Computation Through Strategy Iteration. In: De Nicola, R. (ed.) ESOP 2007. LNCS, vol. 4421, pp. 300–315. Springer, Heidelberg (2007)
9. Leroux, J., Sutre, G.: Accelerated Data-Flow Analysis. In: Riis Nielson, H., Filé, G. (eds.) SAS 2007. LNCS, vol. 4634, pp. 184–199. Springer, Heidelberg (2007)
10. Seidl, H.: Least and Greatest Solutions of Equations over $\mathcal{N}$. Nordic Journal of Computing (NJC) 3(1), 41–62 (1996)
11. Cormen, T.H., Leiserson, C.E., Rivest, R.L.: Introduction to Algorithms, 2nd edn. MIT Press, Cambridge (2001)

Author Index

Albers, Susanne   173
Alt, Helmut   235
Althaus, Ernst   199
Asano, Tetsuo   249

Bast, Hannah   355
Bereg, Sergey   249
Blum, Norbert   18

Constable, Robert L.   3

Degenbaev, Ulan   74
Doerr, Benjamin   99

Fleischer, Rudolf   368
Funke, Stefan   341

Garg, Naveen   187
Gawlitza, Thomas   422
Gebauer, Heidi   30

Hagerup, Torben   143
Hotz, Günter   55

Kaufmann, Michael   290
Kirkpatrick, David   249
Klau, Gunnar W.   199
Kohlbacher, Oliver   199

Lenhof, Hans-Peter   199
Leroux, Jérôme   422

Meyer, Ulrich   219
Moser, Robin A.   30
Munro, J. Ian   115
Mutzel, Petra   305

Näher, Stefan   261

Paul, Wolfgang J.   74
Preparata, Franco P.   158

Reineke, Jan   422
Reinert, Knut   199

Sanders, Peter   321
Scheder, Dominik   30
Schirmer, Norbert   74
Schirra, Stefan   408
Schmitt, Daniel   261
Seidel, Raimund   134
Seidl, Helmut   422
Smid, Michiel   275
Sutre, Grégoire   422

Tsakalidis, Athanasios K.   121

Welzl, Emo   30
Wilhelm, Reinhard   422

Yap, Chee K.   380

Author Index